INTRODUCTION TO
COMPUTING FOR ENGINEERS

INTRODUCTION TO COMPUTING FOR ENGINEERS

STEVEN C. CHAPRA, Ph. D.

Professor of Civil Engineering
Texas A&M University

RAYMOND P. CANALE, Ph. D.

Professor of Civil Engineering
The University of Michigan

McGraw-Hill Book Company

New York   St. Louis   San Francisco   Auckland   Bogotá
Hamburg   Johannesburg   London   Madrid   Mexico   Montreal   New Delhi
Panama   Paris   São Paulo   Singapore   Sydney   Tokyo   Toronto

INTRODUCTION TO COMPUTING FOR ENGINEERS
Copyright © 1986 by McGraw-Hill, Inc. All rights reserved.
Printed in the United States of America.
Except as permitted under the United States Copyright Act of 1976,
no part of this publication may be reproduced or distributed in any form or by any means,
or stored in a data base or retrieval system,
without the prior written permission of the publisher.

2 3 4 5 6 7 8 9 0   HALHAL   8 9 8 7 6

ISBN 0-07-010875-7

Part opening photograph credits:
Part 1 Courtesy of Lockheed Corporation; Part 2 Courtesy of Motorola, Inc.; Part 3 J&R Services, Inc.; Part 4 Courtesy of IBM Corporation; Part 5 Courtesy of Stephen Beekman; Part 6 Courtesy of Hewlett-Packard Company; Part 7 Courtesy Stephen Beekman; Part 8 Courtesy MAGI Synthavision® Solid Modeling System; Part 9 Courtesy Lotus Development Corporation.

This book was set in Times Roman by Waldman Graphics, Inc.
The editors were Kiran Verma, Cydney C. Martin, and David A. Damstra;
the designer was Charles A. Carson;
the production supervisor was Leroy A. Young.
The drawings were done by J&R Services, Inc.
Halliday Lithograph Corporation was printer and binder.

Library of Congress Cataloging-in-Publication Data

Chapra, Steven C.
    Introduction to computing for engineers.

    Bibliography: p.
    1. Engineering—Data processing.   I. Canale, Raymond P.   II. Title.
TA345.C46   1986   620'.00285   85-18160
ISBN 0-07-010875-7

ABOUT THE AUTHORS

Steven C. Chapra   teaches in the civil engineering department at Texas A & M University. His other books include two graduate texts on mathematical modeling and the undergraduate text *Numerical Methods for Engineers*.

Dr. Chapra received engineering degrees from Manhattan College and the University of Michigan. Before joining the faculty of Texas A & M, he worked as an engineer and computer programmer for the Environmental Protection Agency and the National Oceanic and Atmospheric Administration. In addition to his books, Dr. Chapra has published several computer programs and numerous journal articles. All deal with mathematical modeling and the application of computers for engineering problem solving.

Dr. Chapra is active in a number of professional societies and serves on several governmental and professional committees related to applied modeling and computer applications in engineering education. He is the recipient of the Department of Commerce Outstanding Achievement Award and the Texas A & M Outstanding Civil Engineering Professor Award.

Raymond P. Canale   received his Ph.D. in 1968 from Syracuse University. He is now a professor of civil and environmental engineering at the University of Michigan, where he teaches courses on computers and numerical methods. He also teaches and performs extensive research in the area of mathematical and computer modeling of environmental systems. He has authored books on mathematical modeling of aquatic ecosystems and is a coauthor of *Numerical Methods for Engineers*. He has published over 100 scientific papers and reports, and designed and developed personal computer software for engineers.

Professor Canale is also a practicing professional engineer. He is an active consultant for engineering firms as well as for industry and governmental agencies. He has served as an expert technical witness on numerous occasions.

Overall, Professor Canale has devoted over twenty years to his profession as a teacher, researcher, author, and practicing engineer.

To our friend

Gwen Guest

CONTENTS

PREFACE

Engineering practice has always been inextricably tied to the evolution of new tools and technologies. Whether developing the levers, ramps, and pulleys of ancient times or the machines of the industrial revolution, engineers have exploited new devices and concepts that have extended their capabilities. Today, microcomputer technology is providing a host of powerful new tools that will have a profound impact on the profession. Since the strength of the computer lies in its ability to manipulate and store data, engineering, with its dependence on math and its need for accurate information, is an ideal candidate for developing the potential of the machine.

Although its potential is just beginning to be tapped, modern computing technology can enhance the engineering student's education in a variety of ways. In the most immediate sense, the computer significantly increases the efficiency with which students can perform computations and process information, freeing them to devote more time to the conceptual, creative aspects of problem solving. Other benefits range from the use of computers for simulation and tutorial instruction to enhanced visualization through computer graphics.

We believe that the student's first computing course is critical to the effective exploitation of these opportunities. Admittedly, this is not an uncontroversial position. Some educators feel that a first course may be unnecessary because students attain computer literacy in high school. Although secondary education has ventured into this area, high school efforts often prove inadequate for a number of reasons, particularly where student engineers are concerned. Primary among the reasons is the critical shortage of competent teachers in all secondary technical areas, but particularly in computer science. Despite these difficulties, there are still students who, because of innate curiosity or exceptional school systems, acquire extensive computer skills during their high school years. However, this does not mean that these students have skills that are compatible with or attuned to engineering. For one thing, many of our incoming ''hackers'' are extremely game-oriented, and although games are excellent vehicles for developing creativity and programming cleverness, they do not necessarily foster the discipline required in engineering. For instance, we rarely encounter an incoming student of any skill level who can satisfy one of our most important criteria for computer literacy—being able to compose a well-structured program that can be readily used and modified by a colleague. The state of pre-college computer education today is similar to the state of high school sex education during the 1950s and 1960s. It was the rare and fortunate individual who received an accurate, comprehensive introduction to the subject. More often than not, information was gathered piecemeal in the street or through haphazard experimentation. Today's adolescents face a similar situation with respect to computer literacy—that is, many are acquiring their computer skills in the technological gutter!

Because so few freshmen are oriented toward engineering computing and because such a disparity of computer skills exists among incoming students, a freshman or sophomore course provides a degree of standardization. The roots of such a course are already in place in the form of the two courses traditionally available to provide orientation. The first type—introduction to computer programming—is intended to orient the student to the locally available hardware and to develop programming skills. Most courses of this type have, in the past, been configured around mainframe computers and FORTRAN. The second type—introduction to engineering—evolved from the old ''slide-rule course,'' which dealt with the rudiments of the slide rule, engineering drawing, and descriptive geometry. In time, this course incorporated engineering computations and principles, pocket calculators, modeling, design, and, more recently, material on digital computers and programming has been added to some versions. This book blends elements of both courses. Early chapters are aimed at developing sound programming skills and later parts deal with engineering applications and problem-solving software.

Although this book derives from the courses just described, our objective is to forge a new approach. The major thrust of this approach is to bring freshmen and sophomore engineers to a common level of computer literacy. In this spirit, the text is organized around four capabilities that we believe are the essential elements of engineering computer literacy. These capabilities are:

1. *Engineers must have a general understanding of what computers are and how they operate.* From their first day on the job until they retire, engineers will rely on computers as one of their primary tools. No area of engineering is immune. Just as a little mechanical knowledge can hold you in good stead when servicing or buying a car, knowing something about computers lets student engineers use them more effectively. Also, a familiarization with hardware and its operation goes a long way toward rendering computers less mysterious and intimidating.

The first two chapters of this book constitute a short orientation to computing. Chapter 1 provides illustrations of how computers are applied in engineering. Chapter 2 offers a brief history of computing, particularly with regard to the remarkable advances of the past 15 years. Chapters 3 and 4 are devoted to the hardware and systems software that make a computer work. This material is slanted toward personal computers because students are more likely to have hands-on experience with PCs than with mainframes. However, pertinent aspects of mainframe systems are also discussed because of their continuing relevance to engineering.

2. *Engineers must be able to develop, test, and document a structured computer program.* This is perhaps the most controversial aspect of our definition of computer literacy. Some educators contend that programming skills are unnecessary due to the proliferation of packaged or ''canned'' computer software. Although this contention may be true in a broad societal context or in other disciplines, we believe that engineers must have a working knowledge of programming. No matter how many good packaged programs are available, problems will invariably arise that will require tailor made

software. Additonally, a vast "literature" of algorithms has been developed over the past 20 years that has great utility in engineering. However, these algorithms are usually written in a high-level computer language such as BASIC or FORTRAN, and engineers who cannot read the codes will be unable to tap this valuable resource.

Moreover, one of the truly justifiable fears related to extensive application of canned software is that users will treat them as "black boxes." That is, they will apply them with little regard for their inner workings and attendant subtleties and limitations. We contend, however, that computer-literate engineers will be more discriminating about packaged programs and will be less naive about their capabilities and limitations. In the same spirit, although some of the innovative engineering software of the future may originate from professional programmers, a great deal more will undoubtedly arise from the profession itself. Computer literacy among engineers will stimulate and hasten the process.

Most importantly though, students learn to think critically and logically when they develop computer software. If a program is written incorrectly, it will not work. Students are forced to go back, retrace their steps in a logical fashion, and locate their mistakes. In the process, cognitive abilities and discipline are exercised that might otherwise lie dormant. In other words, students are not merely trained to program, they also learn how to think.

To be valuable, a program must be written so that it is easy to understand and maintain. Therefore, this text stresses the use of structured programming techniques. Years ago learning to program was more craft than art. Today structured programming has elevated computer programming to a higher plane. Structured programming is a set of rules that essentially prescribe good style habits for the programmer. Although structured programming is flexible enough to allow considerable creativity and personal expression, its rules impose enough constraints to render resulting codes far superior to unstructured versions. In particular, the resulting product is more elegant and easier to understand. The clarity of structured code is particularly important because a great deal of engineering work is performed by teams, where programs must be shared and modified by several persons.

Languages included in this book were selected in large part for their widespread availability, BASIC being the choice for personal computers and FORTRAN for mainframes. Because engineers must know how to function effectively in either of these environments, major sections are devoted to each. The other important factor in our choice of BASIC and FORTRAN was structure. Early BASIC dialects were deficient in this regard, but the MS BASIC used herein includes many of the essential features of a structured language: indention, long variable names, and adequate selection and repetition constructs. Although structure is not required, MS BASIC can be employed to develop structured, modular programs that are virtually as satisfactory as those developed with a language such as Pascal, which enforces structure. The same is true for the FORTRAN 77 employed herein.

Seventeen chapters are devoted to developing programming skills and language capabilities. Chapter 5 offers an overview of high-level languages. Chapters 6 and 7 concern the elements of programming style with strong emphasis placed on structured programming techniques. Chapters 8 to 14 cover BASIC, and Chapters 15 to 21 deal with FORTRAN. As stated above, the former focuses primarily on the MS BASIC employed on the IBM PC, whereas the latter emphasizes FORTRAN 77. However, material is also included on other important dialects such as Applesoft BASIC and FORTRAN IV. The sections on BASIC and FORTRAN are identical in organization; thus after mastering one language, students should be able to study the other in an efficient manner.

3. *Engineers must recognize the ways computers can be applied to solve problems.* Engineers are professional problem solvers, so this facet of computer literacy is essential to a freshman or sophomore computing course. Among other things, the applications provide motivation and make the subject come alive. Chapters 22 to 25 cover computer-oriented data analysis techniques, such as graphical methods, sorting, statistics, and simulation. Chapters 26 to 32 consider elementary computer mathematics, and cover the general areas of curve fitting (regression and interpolation), equation solving (root location and elimination), and computer calculus (differentiation, integration, and rate equations). The latter material is presented in simple, introductory fashion in order to be compatible with the mathematical sophistication of freshman engineers. All of these chapters include structured programs in BASIC and FORTRAN to implement the methods. Thus, students can begin to build software libraries that will be applicable in their subsequent coursework. In addition, all these chapters include examples and problems to illustrate the application of the techniques to engineering. These applications are reinforced by Chapters 33 to 37, which include problem-solving examples that focus on mathematical modeling in each of the major areas of engineering.

4. *Engineers should be able to use packaged software.* In addition to their own programs, engineers should be able to utilize software packages, a number of which, such as word processing programs and spreadsheets, have great value for the student engineer. We have, therefore, chosen to devote the last chapters of our book to packaged software. Although the rest of the text is compatible with both personal and mainframe environments, the section on canned programs focuses exclusively on the personal computer.

Chapter 38 deals with spreadsheet programs that can be employed for ''what if'' calculations which are valuable for building insight into computations and their sensitivity. Chapter 39 covers computer graphics that can be used to generate visual depictions of information. In Chapter 40, we turn to word processors and their application to prepare and edit texts and documents. Chapter 41 is devoted to three additional types of software: database management systems, teleprocessing, and integrated packages. In all cases, illustrations are furnished to demonstrate the utility of the programs in academic and professional engineering contexts. Chapter 42 provides guidance on how to purchase hardware and software, and a brief discussion of computing in the future, including fifth-generation machines and artificial intelligence.

A description of canned software can become a sterile exercise if the student is not provided with an opportunity to use actual packages. We have taken a number of steps to accommodate this need. First, we have developed software called ENGIN-COMP as a supplement to the text. This package includes three generic programs: descriptive statistics, linegraph generator, and spreadsheet. These programs provide a vehicle for introducing the student to this type of tool. The book has been designed to be compatible with ENGINCOMP but, because the programs are generic, they can be applied in other academic and professional contexts as well. In addition to ENGINCOMP, we have included descriptions and examples of three commercially available canned programs—NUMERICOMP, WordStar, and dBASE II—to illustrate numerical methods, word processing, and database management. Although the use of these and the ENGINCOMP software are optional, we feel that more rapid progress can be attained when the book and the software are employed together.

This book contains more material than can be covered in one semester. In fact, we usually spend two semesters on its contents; the first on programming and the second on applications and canned software. This mass of material was included intentionally because we wanted to give the instructor the flexibility to choose among a variety of topics when designing a specific version of the course. In addition, we hope that the book will serve not only as a text but also as a primary computer reference book.

We have included material on both personal and mainframe computing because we believe that today's engineer must be comfortable working in both environments. All the material in Chapters 22 through 37 can be applied on either personal or mainframe computers. However, we expect that students will first use this text with either one or the other of these environments, but not both. Thus, when students must inevitably operate in the other, the book can serve as a reference to facilitate their efforts. The same sort of strategy underlies our inclusion of both BASIC and FORTRAN.

The breadth of material in this text is also related to the fact that computing has experienced dramatic growth in the last decade. Twenty years ago, computing was monolithic; there were only mainframes, and these environments were dominated by large companies such as IBM and Control Data. Today's situation offers a great many more options and capabilities, but the rapid rate of change and the sheer number of choices can lead to confusion and frustration.

We hope our book helps young engineers successfully navigate these turbulent waters. We believe this generation of young engineers to be extremely important to our profession, to the nation, and to the world. Countries such as Japan and West Germany have made great strides and investments in engineering and technological development. Concurrently, the results of U.S. scholastic aptitude scores suggest a deterioration in the quality of American education. We believe microelectronic technology represents an opportunity to upgrade engineering education and perhaps mitigate the impact of this alarming trend.

This generation of young students is better suited to accomplish the needed upgrading than any previous generation of engineers. Today's freshmen and sophomores were TV babies who received the rudiments of their education on public television and spent their adolescence in video arcades. They are not afraid of computers and are, in fact, more comfortable sitting in front of a screen pushing buttons than almost anywhere else. If the enthusiasm and talent of this first generation of computer-oriented youth can be harnessed, the benefits will be reaped into the next century. We hope that this book is one positive contribution in this direction.

We would like to acknowledge reviews by Professors Barry Crittenden (Virginia Polytechnic Institute and State University), Lawrence J. Genalo (Iowa State University), Robert B. Jerard (Dartmouth College), Ginter Trybus (University of Southern California), and Thomas C. Young (Clarkson University).

A number of persons have been instrumental in creating the type of academic computing environments in which this project flourished. Dean James Duderstadt of the University of Michigan and Professor Donald McDonald of Texas A & M have demonstrated a commitment to computing and engineering that we found inspiring. In addition, the contributions of Professors Lee Lowery, Don Maxwell, Neilon Rowan, Brice Carnahan, and James Wilkes to computing at our institutions are appreciated. Finally, Professors E. B. Wylie, Roy W. Hann, Jr., Bill Batchelor, Chuck Giammona, Teddy Hirsch, Ken Strzepec, Harry Jones and Don Woods were very supportive of our efforts.

A number of others made explicit contributions. Paul Freedman, President of Limno-Tech Inc., made available some of the formidable computing and support services of his fine consulting firm. Doctor Harry Eckerson gave us the benefit of his comprehensive understanding of computer science. Tad Slawecki provided excellent assistance and creativity in the development of the supplementary software. Steve Verhoff read the manuscript several times, performed quality testing of the software, and developed many of the problems. Mike Glazner prepared most of the computer programs and Sallie Mullins developed the material on dBASE II. Susan Hulse, along with Vanessa Stipp and Linda Conte, did a masterful job of word processing the manuscript and MaryLou Jenkins assisted in a variety of support capacities.

Thanks are also due to our friends at McGraw-Hill. In particular, Bob Bowen, Dave Smith, Dave Serbun, Bill Willey, and B. J. Clark provided the executive leadership and vision that underlies McGraw-Hill's commitment to computing and engineering. On the editorial side, Cydney Martin, Kiran Verma, David Damstra, and Terri Zimmardi did much to expedite the project and make it a pleasurable experience. Ursula Smith did an outstanding job of copyediting the manuscript. The visual appearance of the book is due to the superb design work of Chuck Carson. Mel Haber and John Cordes provided guidance on our artwork, and Leroy Young supervised the book's production. Last but not least, Don Burden and the McGraw-Hill sales force continued to demonstrate to us why they are considered the best in the business.

Finally, we would like to thank our families, friends, and students for their enduring patience and support. In particular, Gwen Guest, Christian Chapra, and Jenne Castro were always there to provide the understanding, perspective, and love that helped us up and over this particular mountain.

Steven C. Chapra
Raymond P. Canale

INTRODUCTION TO COMPUTING FOR ENGINEERS

PART ONE

INTRODUCTION

From the beginning, computers have had a profound impact on engineering. Recent innovations in microelectronics have heightened this impact and augur far-reaching effects on our profession. Therefore, before we launch into the body of this book, it is useful to stop and take stock of the impact of computers and microelectronics on engineering. Chapter 1 serves this purpose by orienting you to how computers can be applied to engineering now and in the foreseeable future. Although this book also deals with engineering applications on large mainframe computers, the major focus of the chapter is on personal computers. Most engineers will possess or have easy access to these inexpensive and powerful machines in the near future. We believe that this "personal" aspect of the microelectronic revolution lies at the root of the present excitement over computing among students and professionals alike.

CHAPTER 1
COMPUTING AND ENGINEERING

An engineer carries a portable personal computer into a courtroom to present testimony regarding the pollution of a water-supply and recreational reservoir in Virginia. By projecting computer graphics onto a large screen, the engineer presents complex technical information to the court in a highly communicative fashion. In addition, the computer is used interactively to make predictions concerning the effect of anticipated land development on the reservoir's future water quality. The judge and both attorneys pose hypothetical scenarios, and the engineer uses the computer to generate immediate predictions of possible outcomes.

An engineer in Michigan uses special electronic pens and graphic display devices to prepare alternative design drawings for a new automobile. Each design can be tested by the computer to evaluate its structural strength and behavior under simulated road conditions. When the optimal version has been selected, the computer is used to prepare blueprints from the stored design. Finally, tapes are generated to control robots and automatic machine tools during the car's manufacturing phase.

A supertanker runs aground off the Texas coast. On shore, engineers from a spill-response team enter the latest water current and wind data into a microcomputer which makes hourly predictions of the point at which the spill is likely to hit the coastline. This information is relayed to cleanup crews who must be on the spot in order to minimize damage to the beaches.

A professional engineer from California who had formerly been employed by a large firm has started a private consulting practice. Because of a lack of capital, the engineer must initially work alone out of a small office. However, because of the acquisition of a microcomputer, the engineer can perform many of the normal functions that in the past required a staff of support personnel. Among other things, the com-

TABLE 1.1 Major Categories of Engineering Software and Applications for Computers

Software
   Homemade software
   Generic software
   Engineering-specific software
Application
   Computations
   Design, testing, and production
   Control and automation
   Business and management
   Communication
   Education

puter allows the engineer to perform computations, prepare reports and correspondence, and handle the accounting and billing needs of the firm. These capabilities provide the initial cash flow that will eventually allow the engineer to hire additional staff as the business expands.

Although the foregoing scenarios suggest that computers (and especially personal computers) are already having an effect on engineering, their potential is just beginning to be tapped. Interestingly, most of the major personal-computer applications to date have been oriented toward the business community. This might at first seem surprising because one would have assumed that the lion's share of applications would have originated in the seemingly more computer-oriented fields of engineering and science. But for a variety of reasons, business firms have been quicker to sense and to capitalize on the competitive edge provided by computers.

This situation is bound to change. Because of their speed, memory, and computational capabilities, computers have been allied with engineering for years. However, the enhanced accessibility offered by personal computers should stimulate major increases in the use of these tools by our profession.

The present chapter explores some of the ways that computers can serve the engineering community now and in the foreseeable future. Table 1.1 shows some major categories of computer software and applications. Notice that there is some overlap among the groupings. For example, one type of generic software—computer graphics—is used in nearly all the other categories in Table 1.1. Despite such overlaps, each of the groups is distinct enough to merit separate discussion.

1.1 SOFTWARE

The instructions, or programs, needed to direct the computer to perform tasks are called *software*. In order to effectively use computers, engineers require a wide variety of software, which can be generally divided into three major categories: homemade, generic, and engineering-specific software.

FIGURE 1.1
Twenty years ago, Professor John Kemeny of Dartmouth College stated that, in the future, "knowing how to use the computer will be as important as reading and writing." Today's microelectronic revolution is bringing Professor Kemeny's prophecy to reality.

1.1.1 Homemade Software

About 20 years ago, Prof. John Kemeny (Fig. 1.1), one of the developers of the BASIC computer language, stated that "knowing how to use a computer will be as important as reading and writing." At that time, Professor Kemeny's statement was greeted with some skepticism. This reaction was partially justified by the inaccessibility of computers. Today, in the light of what appears to be a microelectronic revolution, Kemeny's words sound prophetic. In particular, it is becoming increasingly evident that practicing engineers must attain computer literacy to maintain professional competence.

In a broad societal context, *computer literacy* can be defined as the ability to recognize applications where computers are appropriate and to be able to use computers for such applications. We believe that for engineers computer literacy should also include the capability of composing problem-solving programs in at least one computer language. Although many engineers will employ packaged or "canned" programs for major portions of their work, their ability to write their own homemade programs has several benefits.

First, no matter how many canned programs are available, there will inevitably be problems that will require tailor-made computer programs. In many of these cases, engineers must be capable of developing problem-specific software to fill their immediate needs.

Second, one of the truly justifiable fears related to extensive application of canned software is that users will treat the package as a "black box." That is, they will apply the software with little regard for its inner workings and attendant subtleties and limitations. Computer-literate engineers will be more likely to look critically at software designed by others and will be less naive regarding canned software and its possible limitations.

Finally, although some of the innovative engineering software of the future may originate from professional programmers, a great deal more will undoubtedly arise from the profession itself. Computer literacy among engineers will stimulate and hasten this process.

In their excellent little book, *The Beginner's Guide to Computers* (1982), Bradbeer, DeBono, and Laurie make the analogy that the present state of computer software is similar to that of literature when Gutenberg and Caxton developed the first printing presses in the fifteenth century. Prior to that time, books were the exclusive domain of a select group of clergymen and nobility. The vast majority of the populace had never seen a book, let alone learned how to read one. After the invention of the printing press, the situation changed and books began to pervade society as a whole. A century later, Shakespeare wrote Hamlet!

By analogy, the microelectronic revolution represents a quantum leap in the computer capabilities of the general public. It is inevitable that some time in the not-too-distant future, the Shakespeares of software will create their masterpieces. The premier creations in engineering will likely arise from the present pool of computer-literate students and professionals.

1.1.2 Generic Software

There are certain computer applications that are so general that they can be brought to bear on a wide variety of engineering problems. Rather than compose a new program every time one of these applications is required, generic software packages are available to implement these tasks.

Prime examples are the *word-processing packages* that are available for composing and editing texts. The utility of these packages does not necessarily depend on the particular problem context where they are applied. Thus we classify them as general tools.

Other examples are listed in Table 1.2. *Mathematical and graphics packages* are designed to perform calculations and to produce visual representations of information. *Spreadsheets* are a special type of mathematical software that allow the user to enter and perform calculations on rows and columns of data. As such, they are a computerized version of a large piece of paper or worksheet on which an involved calculation can be displayed or spread out. Because the entire calculation is automatically updated

TABLE 1.2   Several Examples of Generic Software
That Can Be Used by Engineers

Mathematical packages
   Statistics
   Numerical methods
   Optimization
Graphics packages
Spreadsheets
Word-processing packages
Database management systems
Teleprocessing
Integrated software

when any number on the sheet is changed, spreadsheets are ideal for ''what if?'' sorts of analyses. *Database management systems* are employed to efficiently store and manipulate large quantities of data. *Teleprocessing* provides users with the ability to communicate with other computers. *Integrated software* refers to packages that combine functions such as word processing, graphics, and spreadsheets into one easy-to-use package.

Although spreadsheets, database management systems, and teleprocessing were originally developed for business purposes, they can also be applied to great advantage in engineering. In the same sense, more engineering-oriented packages such as numerical methods and statistical software can also be applied to other fields. For example, business analysts make great use of statistics.

It should be noted that specific versions of all the packages can be tailored to the special needs of any particular profession. For instance, most word-processing software is not designed to handle mathematical symbols and formulas easily. Therefore an engineering and science word-processing package might be designed to fill this special need. Similarly, a graphics package could be developed for the specific drafting requirements of engineering. However, in most cases, generic software is flexible enough to be applied effectively in a variety of contexts.

1.1.3 Engineering-Specific Software

Although there are a great many areas that can be addressed with generic software, there are many other engineering applications that require tailor-made programs. Some of these programs will be so specific that they will be limited to a single problem setting. However, others will be similar in spirit to the generic software discussed in the previous section in the sense that they can be applied broadly. For example, engineers working in different locales could successfully apply the same structural design or electrical network analysis programs. Some other examples are listed in Table 1.3. Beyond these, there are even more specific programs that might be expressly tailored to one application.

TABLE 1.3 Some Examples of Engineering-Specific Software That are General Enough for Broad Application

Field	Software
Chemical engineering	Distillation tower design
	Integrated plant design
Civil engineering	Structural analysis
	Surveying
Electrical engineering	Electrical power system control
	Microprocessor design
Mechanical engineering	Pipe network design
Industrial engineering	Computer-aided manufacturing (CAM)
General engineering	Computer-aided design (CAD)
	Risk analysis
	Critical path method (CPM)
	Program evaluation and review technique (PERT)

1.2 APPLICATIONS

The microelectronic revolution has stimulated a variety of novel applications of computers in engineering. Some of the more important areas in which computers are playing an increased role are design, automation, communication, and education. However, before reviewing these somewhat new applications, we must first discuss the area where computers have traditionally shown their greatest utility, that is, in large-scale calculations, or ''number crunching.''

1.2.1 Computations, or ''Number Crunching''

From the earliest days, computers have been used to implement large-scale calculations. The earliest electronic computers—the ABC and the ENIAC—were expressly developed to perform otherwise arduous and tedious calculations. To this day, ''number crunching'' remains one of the primary applications areas for computers in engineering.

The advent and the widespread availability of personal computers is doing much to encourage and broaden these applications. In fact, as they increase in memory and speed, personal computers are not only enhancing our capabilities but also the manner in which calculations are performed. Their impact is directly attributed to their accessibility—that is, their low cost, their portability, and their graphic and interactive capabilities.

For example, in the past, engineers had to go to great expense to employ large mainframe systems. Today, a personal computer can perform many of the same tasks for a relatively small capital investment and minimal operating costs. Additionally, as seen in the oil-spill vignette at the beginning of this chapter, computational power can be transported out into the field.

Beyond these examples, there are more subtle changes linked to the graphic and interactive qualities of the machines. On mainframe systems, there is usually a lag between the time that a program is entered into the computer and the time that the results are obtained. On personal computers, the response is immediate for small- to moderate-size calculations. Thus the engineer has the capability of performing successive runs to gain immediate insight into the calculation and its implications. Such interactive or exploratory computations are extremely useful for developing intuition and holistic awareness regarding the particular engineering problem under study. This is especially effective when computer graphics are employed to express the results visually.

It is presently impossible to forecast where these changes will lead and how they will ultimately influence the profession. However, present trends suggest an overall upgrading and enhancement of our capabilities to perform and interpret computations.

1.2.2 Design, Testing, and Production

Today most of the basic arithmetic, logic, and storage elements of a computer can be contained on a single integrated circuit chip called a *microprocessor*. Because of their size, speed, memory, and portability, microprocessors are used increasingly to perform tasks connected with the effective design and operation of engineered works and products. One of the more prominent applications is in *computer-aided design and manufacturing* (CAD/CAM) of the type described in the second scenario at the beginning of this chapter. This application involves the use of microprocessors to formulate the design, to perform preliminary testing, and to control manufacturing and assembly operations. A significant aspect of CAD/CAM is that it is an integrated process. The computer serves as the focus for this integration.

Important tools of CAM are robots that can more efficiently perform manufacturing tasks formerly accomplished manually. The application of CAM and robotics is one reason for the explosive growth of industrial productivity in Japan and Europe. In the United States, engineers are applying these tools to a wide variety of manufacturing processes.

1.2.3 Control and Automation

Microprocessors have numerous potential applications throughout engineering and society at large. As is the case for robotics, many of these applications involve the control and automation of processes formerly handled by humans. For example, many engineers use computer-controlled sampling devices to collect and analyze test data. Integrated computer networks are increasingly employed for other control functions ranging from electric power to traffic to aerospace systems. Some structures are even being built that use computers to regulate their heating and cooling. Control and automation will be one of the major areas where microprocessors will have an immense, direct impact on society. Engineers will play a key role in the design and implementation of such applications.

1.2.4 Business and Management

Many engineers own firms or function professionally as private consultants. As such, they have a need for the many business-oriented software packages that are presently available. These programs perform functions—such as accounting, billing, and payroll management—that are essential for the effective operation of a company. Although they require an initial capital investment, their long-term payoff in increased speed and efficiency makes them an essential component of the modern engineering firm.

Beyond this direct use, one of the major tasks of the practicing engineer is to complete projects in a cost-effective manner. Engineers must therefore always be aware of the economic aspects of their work. Certain business-oriented software can serve these efforts. In addition, aids such as CPM (critical path method) and PERT (program evaluation and review technique) are available in computerized form to facilitate the scheduling and implementation of projects.

1.2.5 Communication

One definition of engineering labels it ''the art of executing a practical application of scientific knowledge.'' As such, engineers stand at the midpoint between science and the needs of society. Because of this, our work involves a substantial amount of societal interaction, and the ability to communicate effectively is a hallmark of most successful engineers. Whether marketing a project to a prospective client or presenting a report to a government agency, communication of technical information to non-technical audiences is a critical ingredient in a large amount of engineering work.

One of the most obvious examples is report and proposal generation. Many of the generic programs discussed in Sec. 1.1.2 can contribute greatly to quality publication of engineering documents. Integrated software is being developed that contains several capabilities such as word-processing, spreadsheet, and graphics programs in one package. Such integrated software greatly enhances the report-generation process. Together with electronic mail, such software provides a quantum leap in our capability to package and transmit technical information.

Another context in which personal computers enhance communication is in the area of presentations. Of course, they can serve the obvious function of preparing effective graphics. However, as seen in the courtroom vignette at the beginning of the chapter, they can also be directly utilized as communication tools. In particular, they can be used to establish a dialogue between the engineer and the client. In the courtroom example, the personal computer allowed the judge and the lawyers to pose questions that the computer could evaluate and answer. This sort of interactive involvement is a potent vehicle for making technical results less intimidating and more accessible to clients and the public. Such involvement can yield positive benefits as the client actively participates in the technical endeavor.

The ultimate extension of the interactive presentation is the computer game. Most of us are familiar with these from our experiences in arcades. Beyond these recreational applications, computer games are powerful communication devices. In a sense, the flight simulators used to train airplane pilots are sophisticated games. In the coming decades, engineers will employ games for training technicians and for communicating technical results to clients. As discussed in the next section, another area of application will be in the classroom.

1.2.6 Education

Communication is of critical importance to engineering. Nowhere is this more important than in the training of engineers. In Sec. 1.1.1 we made the case that computer literacy should be an absolute necessity for all practicing engineers. Additionally, related topics such as computer math (for example, numerical methods and statistics), CAD/CAM, and other computer-oriented subjects will undoubtedly gain more prominence in engineering curricula. As prices drop, more and more student engineers will possess their own personal computers. Courses are being modified to capitalize on this increased computational firepower.

The benefits will be great. For example, personal computers will remove some of the drudgery associated with large-scale calculations. Thus more time will be freed for studying problem formulation and solution interpretation. In this way, insight and creative intuition can be nurtured.

Beyond these impacts, computers may also be employed directly as learning tools. To date, two primary modes are used for this purpose: tutorials and simulations.

Tutorials set up an interactive dialogue between the student and the machine. In their most basic form, tutorials present material and ask questions. Depending on the answer to the question, the computer either reiterates the material or advances to another topic. In this way, concepts are clarified and students progress at their own pace.

Simulation is another name for computer games. In this case, the computer can be used to simulate a system for the student to study. Thus the student can explore the behavior of the system in response to a variety of conditions and in this way build up competence and intuition.

Both tutorials and simulations can be effective learning tools. This is particularly true because most students today have grown up in a video environment. Their first exposure was to the somewhat passive medium of television. Most students probably acquired the rudiments of reading, writing, and arithmetic from preschool educational television programs. In the late 1970s, video arcades came to provide a version of the technology that demanded active participation on the part of the player. Because of these experiences, most students today are very comfortable with video technology. Although there is a certain justification for those who lament the negative effects of television, computer education is one effort to glean something positive from this technology. By bridging the gap between video and engineering education and stressing the interactive nature of the medium, we can capitalize on the attraction and accessibility of television rather than continuing to lament some of its acknowledged faults.

We have now come full circle from voicing a plea for computer literacy at the beginning of the chapter to stressing the direct use of computers as educational tools. On this note, we move along to outline how this book can serve as the basis for a first course on computers for engineers.

1.3   AN INTRODUCTORY COURSE ON COMPUTING AND ENGINEERING

As stated above, the present book is designed to support a first computing course for engineers. The book has two major thrusts. First, it provides sufficient information for you to attain computer literacy in either one or both of the primary high-level programming languages used by engineers today, BASIC and FORTRAN. Second, it illustrates a wide variety of computer applications in engineering.

Part Two deals with fundamental information related to computer hardware and software. Just as automobile owners find it useful to learn a little about cars, we believe that it will be beneficial for you to know something about how computers work.

This introductory material is followed by *Parts Three, Four,* and *Five,* which are devoted to the prerequisites of programming and to programming in BASIC and FORTRAN, respectively. These parts include a sufficient level of detail to allow you to fulfill our definition of computer literacy. That is, they will provide you with enough information to write your own problem-solving programs.

After you have learned to program, the next step is to investigate how this skill can be applied to engineering. Thus the following parts of the book are devoted to a variety of problem-solving skills and engineering applications related to the computer. *Part Six* is devoted to data analysis. Here we discuss topics such as graphics, sorting, and statistics that contribute to the effective communication and analysis of information. In addition, we introduce the topic of simulation which, as the name implies, represents the use of the computer to simulate the behavior of physical systems.

In *Part Seven* we turn to numerical methods. These are general techniques for solving mathematical problems with computers. *Part Eight* illustrates how the numerical methods are used for engineering problem solving. This material concentrates on problems that would be intractable without the computer.

Part Nine deals with some of the major generic software that can be employed by engineers. These are spreadsheets, computer-graphics packages, word processors, database management, and telecommunications software systems. In addition, we describe integrated software that combines several generic programs into one package. Here we again focus on the engineering applications of these tools.

1.3.1 Computer Languages and Types of Machines

Today's engineer has a wide variety of computing tools at his or her disposal. During the course of your career, you will undoubtedly have occasion to make use of a spectrum of devices ranging from your own personal computer to large institutional machines. For this reason, and also because we anticipate that students will be using different types of machines in conjunction with their first computing course, we have designed this book so that it can be used with many of these systems. Aside from Part Nine, which is devoted to personal computer software, all other parts of the book are written so that the material can be implemented on either personal or large mainframe computers.

This approach extends to our choice of computer languages. Although a variety of languages are used in engineering today, the two dominant introductory languages are still BASIC for personal computers and FORTRAN for mainframes. We have, accordingly, devoted a major part of the book to each of these languages. In addition, most of the program listings throughout the book are presented in both languages.

Although we anticipate that your first use of this text will be limited to one or the other of these languages, this bilingual approach has some ancillary benefits. For years, most meaningful engineering software was written in FORTRAN. Recently, because of the influence of personal computers, an increasing number of programs are being composed in BASIC. Consequently, engineers routinely encounter software written in either of the languages. Therefore, although it is not a necessity, the ability to translate from one to the other can be a valuable asset. Thus we hope that once you attain fluency in one or the other of these languages, you will use this book as a reference or self-study guide to attain a working familiarity with the other. To facilitate such efforts, the parts on BASIC and FORTRAN have a similar structure and employ the same examples. Once one language has been mastered, it becomes a fairly simple proposition to study the other.

Although we have gone to great pains to ensure that this book is compatible with many types of computers, we have tried wherever possible to acknowledge and explore some of the exciting new applications that are directly associated with personal computers. This is why we have devoted Part Nine to personal computer software such as the spreadsheet and word-processing packages that represent powerful new tools for our profession. Even those who use a mainframe for this course can cover this material under the realistic assumption that most engineers will own or have ready access to a personal computer in the foreseeable future.

In the same spirit, we have designed a software package, called ENGINCOMP, that can be used in conjunction with the text. This package is available for the IBM PC. Although its use is optional, this software provides you with immediate capability to explore problem solving with graphics, statistics, and spreadsheets. As such, it can serve to illustrate how canned software can be applied in engineering practice.

With this as background, we can now begin to learn something about computing. We will start by investigating the components and working of the computer itself.

PROBLEMS

1.1   Distinguish between homemade, generic, and engineering-specific software.

1.2   Define computer literacy for engineers.

1.3   Give three reasons why engineers should learn to write computer programs.

1.4   What do we mean when we say that personal computers are ''accessible''?

1.5   What is CAD/CAM? What does it involve?

1.6   Distinguish between the two primary modes of educational software: tutorials and simulations.

1.7   Visit your local computer store.
(a) Make a list of the various computers available. Classify them according to type, cost, and capabilities.
(b) Make a list of the software available. Classify the engineering and business software according to application.

1.8   Determine what types of computers are used at your school or at a local business or engineering firm. Prepare a report that contains a list and description of the different types of hardware, software, and application areas.

PART TWO

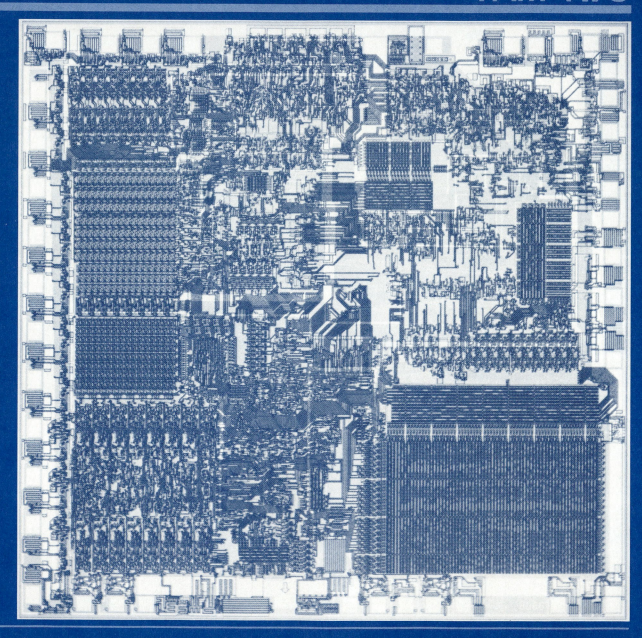

COMPUTERS

Before launching into a full-blown discussion of programming, it is essential to assimilate some background material on computers and how they work. This part of the book is devoted to such introductory material.

Chapters 2 and 3 deal with the actual equipment that constitutes a computer—that is, the hardware. *Chapter 2* presents the historical development of these machines, whereas *Chapter 3* describes the components and functions of the modern computer. *Chapter 4* begins our discussion of the instructions or programs (that is, the software) that must be developed to allow the computer to perform tasks such as data processing, numerical computations, and graphics. Chapter 4 deals with the systems programs that bring the machine to life and allow it to interact with users and to implement problem-solving programs.

CHAPTER 2
THE HISTORICAL DEVELOPMENT OF COMPUTERS

Today's microelectronic revolution is due in large part to the accelerating development of computer hardware over the past decade. The present section is intended to place these recent advances in a broad historical perspective. Beyond satisfying your innate curiosity, this material will help you to appreciate the magnitude of today's advances as well as to anticipate things to come. In addition, it should deepen your understanding of existing computers, since much of today's hardware evolved from the concepts and devices of the past.

2.1 PREELECTRONIC COMPUTING DEVICES

Humans have been interested in calculations and keeping track of information from the earliest days of recorded history. As will be clear from the following pages, the ability to calculate has always been closely related to the technology and energy

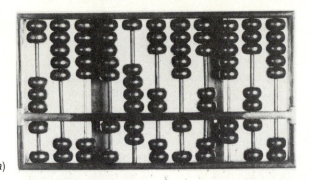

(a)

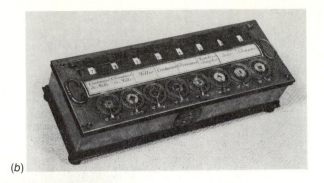

(b)

(c)

(d)

FIGURE 2.1
Preelectronic computing devices: (a) an abacus, (b) Pascal's adding machine, (c) Babbage's difference engine, and (d) the Mark I. (Courtesy of IBM Corporation.)

supplies of the day. For example, one of the earliest "computers" was the simple, muscle-powered device called the *abacus*. Developed in ancient China and Egypt, this device consists of rows of beads strung on wires in a rectangular frame (Fig. 2.1*a*). The beads are used to keep track of powers of 10 during the course of a computation. In the hands of a skilled operator, the abacus can rival a pocket calculator in speed.

Although manual devices such as the abacus certainly speed up computations, machines provide an even more powerful means for extending human calculating capability. Stimulated by the industrial revolution, seventeenth century scientists developed the first such mechanical computing devices. The French scientist and philosopher Blaise Pascal invented an adding machine in 1642. Called the *pascaline* (Fig. 2.1*b*), the device used gears with teeth to represent numbers. In the late 1600s, Gottfried Leibniz developed a similar mechanical calculator that could also multiply and divide.

The next major advances in computing technology came in the 1800s. In the early nineteenth century, Joseph Jacquard invented a loom that employed punched cardboard cards to automatically control the production of patterned cloth. To this day, punched cards are still used to transmit information to and from some computers.

In addition, Jacquard's idea of storing a sequence of instructions on the cards is conceptually similar to modern computer programs.

Although Jacquard's loom introduced some important new ideas, Charles Babbage's steam-driven calculators were more closely related to modern computers. Babbage was motivated by the large numbers of errors he found in mathematical tables. He developed a *difference engine* (Fig. 2.1*c*) in the early 1800s that could compute and print out tables automatically. He also conceived, but never built, an *analytical engine* that incorporated many features of modern computers, including punched card instructions, internal memory, and an arithmetic unit to perform calculations.

A final major contribution of the nineteenth century occurred when Herman Hollerith used punched cards to tabulate the U.S. census of 1890. This innovation reduced the time of tabulation from a decade to 3 years. Hollerith went on to found the Tabulating Machine Company, which in 1911 merged with several other firms to form IBM.

During the early part of the twentieth century, IBM and other manufacturers produced a variety of computing devices for business use. These were all *electromechanical;* that is, they were powered by electricity and had moving parts. The ultimate of these devices was the Mark I developed by Howard Aiken in collaboration with IBM. Although an awesome piece of technology (Fig. 2.1*d*), the slowness of the Mark I's electromechanical components made it quickly obsolete in the face of a more advanced technology—electronics.

2.2 HOW COMPUTERS "THINK" (HARDWARE)

In contrast to electromechanical devices, electronic machines have no moving parts. Their major logic elements operate on the basis of electricity. In order to appreciate the development of electronic computers, we must digress momentarily to describe the way in which these machines store and process information.

A mechanical device such as the pascaline kept track of quantities by a series of coupled gears and levers. But how can purely electrical instruments perform similar tasks? The answer lies in the two fundamental ways in which quantities can be handled: measuring and counting. Entities that exist on a continuous scale such as height, velocity, or mass are *measured*. In contrast, quantities that can be broken into indivisible units such as people, rivets, or beans are *counted*. The distinctions can be appreciated by realizing that if you measure someone's height as 71 in, he or she could just as well be 70.9999 or 71.0001 in. However, if you count 15 rivets in a package, there are 15 and only 15 rivets.

Electronic computers that measure are called *analog computers*. These devices use the level of an electrical quantity, such as voltage, to correspond to a numerical value that is to be measured. An example would be your car's gas gauge. As the float in your gas tank moves up or down, an electrical quantity is increased or decreased accordingly. The resulting changes are transmitted electrically to the gauge on your dashboard. Analog computers use a similar scheme to keep track of quantities and their interactions.

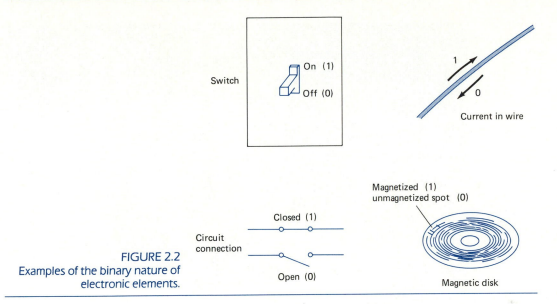

FIGURE 2.2
Examples of the binary nature of electronic elements.

Although analog computers have their place in engineering and science, they do not represent the primary way that electronic computers handle quantities. Most modern computers count. For this reason they are called *digital computers* after the primary way that humans count, with their fingers or "digits." However, because they have neither hands nor feet, computers count in their own unique way—that is, with electrical elements that are either "on" or "off." Several examples are shown in Fig. 2.2. Note that by convention, the "on" state corresponds to the number 1, whereas the "off" state corresponds to 0. Each element taking on a value of 1 or 0 is referred to as a *bit*, which is a contraction of *b*inary dig*it*. Electronic computers represent all their information in binary form. We will provide a description of how this is done when we discuss software in Chap. 4. For the time being, the important notion to understand is that all computer operations can be boiled down to a series of elements that are either "on" or "off." The importance of this notion to the present discussion is that the evolution of computers is closely linked to the type of electronic device or logic element used to keep track of these "ons" and "offs."

2.3 EARLY ELECTRONIC COMPUTERS

The first electronic computers employed vacuum tubes as their primary logic elements (Fig. 2.3). These machines were one-of-a-kind devices built for experimental purposes. For example, the ABC (Atanasoff-Berry computer) was developed at Iowa State University to solve simultaneous equations in the late 1930s.

A few years later, a more general purpose machine, the ENIAC (*e*lectronic *n*umerical *i*ntegrator and *c*alculator), was developed during World War II. Designed

FIGURE 2.3
Logic elements used in electronic computers: vacuum tubes (first generation), transistors (second generation), and integrated circuits (third generation). (Courtesy of IBM Corporation.)

by J. Presper Eckert, Jr., and John W. Mauchly, it was about 100 ft long, contained thousands of vacuum tubes, and consumed large quantities of electricity (Fig. 2.4). Although by present standards it was slow and unreliable, it represented the first successful general-purpose electronic calculator.

Although the ENIAC represented a major advance, it had to be rewired every time a new task was performed. A way to circumvent this problem was developed by the mathematician John Von Neumann. He suggested that instructions for processing information could be stored in the computer's memory. Thus, when the operator needed to perform a new task, a set of instructions could be fed into the machine along with the data. Two such *stored-program computers*, the EDSAC (*e*lectronic *d*elay *s*torage *a*utomatic *c*alculator) and the EDVAC (*e*lectronic *d*iscrete *v*ariable *au*tomatic *c*omputer), were developed at the end of the 1940s. With these machines, the stage was set for the modern computer age.

2.4   THE MODERN COMPUTER AGE

Early electronic computers were for the exclusive use of the military and of experimental scientists and engineers. In the early 1950s, computers began to be sold commercially. The development of a commercial computer industry truly represents the beginning of today's computer revolution.

FIGURE 2.4
The ENIAC. Built for military purposes, this early computer filled a room, generated large quantities of heat, and consumed great amounts of electricity. However, it represented the first successful general-purpose electronic computer. *(Courtesy of the Sperry Corporation.)*

The modern era can be conveniently broken down into four distinct generations distinguished by the primary electronic component or circuit element within the computer. As delineated in Fig. 2.5, each new logic unit has led to computers that are faster, smaller, and cheaper than previous devices. The magnitude of this improvement has been remarkable. It has been stated (Cougar and McFadden, 1981) that "if the automobile industry had been able to advance its technology as rapidly as the computer industry, a Rolls-Royce today would cost $2.50 and would get 2 million miles per gallon!"

The *first generation* began in 1951 when the U.S. Census Bureau purchased a UNIVAC I computer from the Remington-Rand Corporation. This computer, as did other early electronic devices, used *vacuum tubes* as its primary logic elements. Although the tubes were a major advance over electromechanical parts, they had a number of disadvantages, including excessive heat generation, size, and unreliability. Punched cards were the major medium for information transmission and storage. Most applications were limited to business-oriented data processing such as payrolls and accounting. The machine was usually run by a trained operator and by computer programs that were written in obscure languages and were extremely complicated, requiring expert programmers. All the above characteristics should make it obvious that, although the first-generation machines demonstrated utility for large-scale data processing, they were not yet appropriate for the broad, general use of professionals such as practicing engineers.

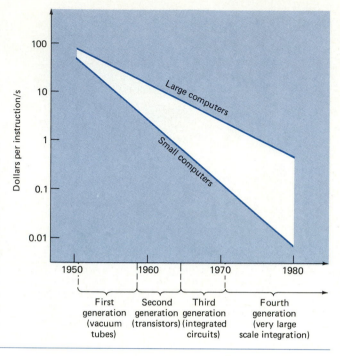

FIGURE 2.5
The progress in computer prices and performance. *(Adapted from Branscomb, 1982.)* The four generations of the modern computer age, along with their major circuit elements, are noted on the abscissa. The average rate of improvement for the large general-purpose computers has been less than that for comparable small machines because the newer large computers offer added functions at the cost of more computing power.

This situation began to change with the *second generation* of the early 1960s. The primary electronic elements of this generation were *transistors* (Fig. 2.5). These solid-state electronic devices performed the same function as vacuum tubes but were smaller, faster, and more reliable and had lower heat output and power requirements. Consequently, the computers of this generation were cheaper to operate than the earlier machines. Along with lower cost, the development of easier-to-use programming languages such as FORTRAN made computers accessible to large numbers of practicing engineers for the first time in history.

The next technological breakthrough, which ushered in the *third generation* of the late 1960s, was the integrated circuit (Fig. 2.5). An *integrated circuit*, or *IC*, consists of a tiny silicon chip on which thousands of transistors are fabricated. As with the previous advances, the IC resulted in large-scale computers that were faster, more efficient, and more reliable. However, other changes occurred that went beyond mere increases in computer power and that truly heralded the beginning of a microelectronic revolution.

Up to the mid-1960s only organizations such as corporations, universities, government agencies, and the military could afford to acquire and maintain a large-scale, or *mainframe*, computer (Fig. 2.6a). Because of a growing demand from smaller organizations and some individuals, *timesharing systems* were developed whereby several independent users could simultaneously utilize a large mainframe computer. Access was via telephone cables, and users were billed in a way similar to that by which a utility charges a customer.

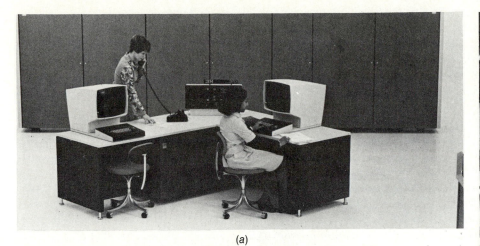

(a)

(c)

(b)

FIGURE 2.6
The three classes of computers used by today's engineer: *(a)* the large mainframe, *(b)* the medium-sized mini-computer, and *(c)* the microcomputer. *(Courtesy of IBM, Digital Equipment Corporation, and Texas Instruments Corporation, respectively.)*

Stimulated by the availability of these systems, programmers developed software to capitalize on the interactive nature of timesharing computers. This software allowed users to submit programs and data to the computer and receive a prompt reply. Thus an interactive dialogue between the user and the machine was established. Among the developers of this software were John Kemeny and Thomas Kurtz of Dartmouth College who at the same time developed the first major interactive programming language—BASIC.

Although timesharing boomed for awhile and certainly advanced the cause of the small user, the real breakthrough came in the late 1960s with the introduction of the minicomputer by the Digital Equipment Corporation (DEC). Made possible by the integrated circuit, the *minicomputer* (Fig. 2.6*b*) is essentially a scaled-down version of the large mainframe machines. Although less powerful, minicomputers are more than adequate for most engineering computations. Because of their low cost and convenience, they are widely used in small businesses, including many engineering firms.

The great increase in user control and access ushered in by timesharing and the minicomputer has continued and has actually accelerated in the *fourth generation* of the modern computer age. Although the IC is still the primary electronic element, *microminiaturization* has led to smaller and smaller chips holding more and more circuits. This process, which started in the early 1970s, is called *large-scale integration (LSI)* and *very large scale integration (VLSI)*. The most important product of these advances is the *microprocessor*. This is a single silicon chip that contains the complete circuitry of the computer's control unit. As such, the microprocessor is the heart of the *microcomputer* (Fig. 2.6*c*).

As might be expected, the evolution of the microprocessor and of the microcomputer is intertwined. The first microprocessor, the Intel 4004, was introduced in 1971. It was a four-bit microprocessor, so called in reference to the general manner in which it handled information—that is, in chunks of four bits at a time. We will delve more deeply into the power of microprocessors in Chap. 4. For the purposes of the present discussion, you can tentatively accept the notion that the power of a microcomputer is generally related to the number of bits handled by its microprocessor.

Although four-bit microprocessors such as the Intel 4004 could be used for devices such as calculators, they could not serve as the basis for a computer. It was only with the development of eight-bit processors that microcomputers became feasible. The first of these, the Intel 8008, was introduced in 1972.

The actual idea of the microcomputer originated in a talk by Alan Kay in 1972 (Kay, 1972). In mid-1974, an article was published in a magazine called *Radio Electronics* that described the technique for constructing a personal computer with a microprocessor at its heart. Soon after, in early 1975, the first microcomputer, the MITS Altair, appeared. This machine used an improved version of the Intel 8008, the Intel 8080, as its microprocessor.

The beginning of a commercial microcomputer industry can probably be dated to mid-1977 when Steve Jobs and Steve Wozniak put their first Apple II computer on the market. From that start, the company boomed and has made available updated models, the Apple IIe and IIc, as well as other products such as the Macintosh. The Apple II, as did the other early microcomputers, used an eight-bit microprocessor, the MOS Technology 6502, as its major component. Another popular eight-bit microprocessor is Zilog's Z-80 used in Radio Shack's TRS-80.

Another major watershed in the evolution of the microcomputer market was the entrance of IBM in 1981. Its major product, the IBM PC, has dominated the market to the extent that it can justifiably be considered the industry standard. The IBM PC uses a 16-bit version of the Intel 8080 microprocessor—the Intel 8088. Although the

Intel 8088 is generally more powerful than the earlier eight-bit chips, there is debate regarding its true power. This is because, although it stores data in 16-bit chunks, it transmits data eight bits at a time. True 16-bit microprocessors, such as the Intel 8086, are now available and are being used in more powerful microcomputers.

We will return to the topic of microprocessors and their power in Chap. 4. Here it is sufficient to know that more and more information and capability are being condensed onto silicon chips. The miracle of the microelectronic revolution is that for an affordable price, today's practicing engineer can have ready access to a micro- or minicomputer that is superior in performance and capability to most of the first- and second-generation mainframe machines.

2.5   MICROS, MINIS, AND MAINFRAMES

Before taking a closer look at the individual components that compose a typical computing system, we want to make sure that our terminology is clear. In most of the following material, specific discussions of personal computers will usually deal with microcomputers. However, in a broader sense, we must include minicomputers under the broad umbrella of personal computing. Although they are more costly and lack portability, they exhibit the accessibility, or hands-on operation, that is the hallmark of personal computing. Further, although it is unlikely that you will have a minicomputer in your study, it is possible that you will use one at your university or office. Finally, it must be stressed that the demarcation between minis and micros is not hard and fast. In all likelihood, the mini of today will become the micro of tomorrow.

Beyond personal computers, today's mainframe machines have also evolved into more powerful and efficient tools. Although micro- and minicomputers will fill a great portion of your needs, mainframe computers will continue to play an important role in engineering. Large machines are still the best tools for implementing the large-scale data processing and computations that engineers sometimes confront. The beauty of today's situation is that we have potential access to all the types of computers. For example, preliminary work on a large-scale computation might be performed on a microcomputer. Then a telecommunications link can be used to transfer the program to a mainframe for fast implementation. In this way, the strong points of each type of computer are exploited. Engineers with access to a range of machines will have enormous capability at their command. On this note, we can now turn to a discussion of the hardware that composes a modern computer system.

PROBLEMS

2.1   Explain the fundamental difference between electromechanical and electronic computing devices. Provide an example of each type.

2.2   Define hardware and software.

2.3 (a) Explain the difference between analog and digital computers.
(b) Characterize the following as either counted or measured:

Cities	Voltage
Pressure	Current
Temperature	Trees
Mass	Stars
Elevation	Words
Cars	Force

2.4 What does the word ''bit'' stand for? How does it relate to the way that a computer ''thinks''?

2.5 What are the primary electronic logic elements corresponding to the four generations of the modern computer age?

2.6 What is a microprocessor? Give examples of four-, eight-, and sixteen-bit microprocessors.

2.7 Define the terms ''mainframe computer,'' ''minicomputer,'' and ''microcomputer.'' Who would tend to own each type?

CHAPTER 3
COMPUTER HARDWARE

Computers are like automobiles in the sense that you need not know everything about them to operate them effectively. However, as with automobiles, a little knowledge about their workings can often help you to utilize them more intelligently and efficiently. The present chapter is an introduction to the physical equipment, or *hardware*, that constitutes a typical personal computing system. The emphasis of the following discussion will be on microcomputer hardware. We have chosen to emphasize these machines because they are the ones that you will likely own and operate in the coming years.

Although every type of computer will appear different at first glance, all include the four major components shown in Fig. 3.1. The present chapter is devoted to a brief description of each of these components.

3.1 THE CENTRAL PROCESSING UNIT (CPU)

When most people think of a computer they usually visualize input and output devices such as the keyboard and the monitor screen. This is understandable because these are the devices that are the primary communication interfaces between the user and the machine. However, the heart of the system, and by some definitions the actual computer, is the *central processing unit,* or the *CPU*.

The CPU is the control center or ''brain'' of the computer. Its functions can be divided into three categories:

1. *The control unit* This controls the electric signals passing through the computer. It performs the necessary function of coordination; that is, it directs and monitors the operation of the entire system.

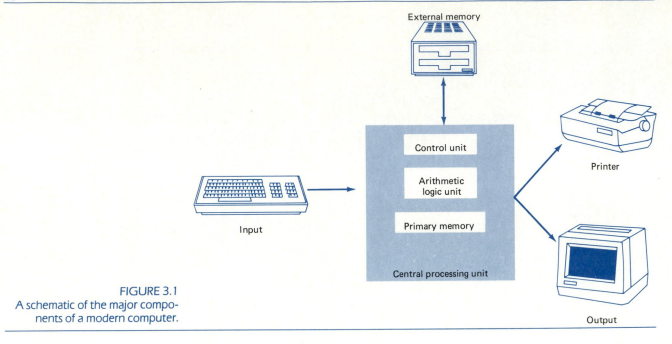

FIGURE 3.1
A schematic of the major components of a modern computer.

2. *The arithmetic/logic unit*   This is where the arithmetic and logical operations are performed.

3. *The primary memory*   This unit serves as a temporary retention device for information either being input to the system or resulting from internal calculations. In addition, it is used to store computer programs.

Primary memory is commonly referred to as *random-access memory,* or *RAM*. "Random access" means that the computer can go directly to the programs and data it wants, and the access time for any point in the memory is essentially the same. This is in contrast to *serial access,* where the computer must check each memory location in sequence to locate what it wants.

While some systems have a fixed amount of RAM, many allow expansion to meet the needs of the user. Along with other advances, memory chips have been developed to allow expanded RAM capacities in relatively inexpensive units. Many of the original personal computers had RAMs of 32K and 64K. The "K" stands for kilobytes, where a "byte" corresponds to eight bits. Although the prefix "kilo" conventionally designates 1000, in computer technology it actually represents 1024.* Therefore "64K" signifies that the computer has enough room to store 64 × 1024 × 8, or 524,288, bits of information. Interestingly, this is just about the capacity of many of the original first-generation computers that took up entire rooms. Personal

---

*As far as the computer is concerned, 1024 is a nice round number. This is because computers use bits to store information, as will be described subsequently in Sec. 4.1. Note that 1024 is equal to 2 raised to the power 10.

computers with capacities of 1M (1 megabyte, or 1 million bytes) are available. Although this is probably beyond the requirements of many users, the increased memory can be a definite advantage for certain engineering applications.

Aside from RAM, there is another type of memory that bears mention. *Read-only memory,* or *ROM,* holds computer programs that are so common that manufacturers wire them directly into the computer's hardware. In this sense, ROM is a combination of hardware and software and, as such, it is sometimes called *firmware*.

The actual innards of a particular microcomputer are displayed in Fig. 3.2. Although each model has its own unique layout, the components in this illustration are found in most machines.

One additional feature of the microcomputer that bears mention is that it contains an electronic clock. This has a number of functions, the primary of which is to synchronize the computer's operations. The computational power of a microprocessor is related to the frequency of this electronic clock. The trend in microcomputers is toward clocks with higher frequencies—that is, with more cycles per second.

FIGURE 3.2
The innards of a personal computer—the "motherboard."

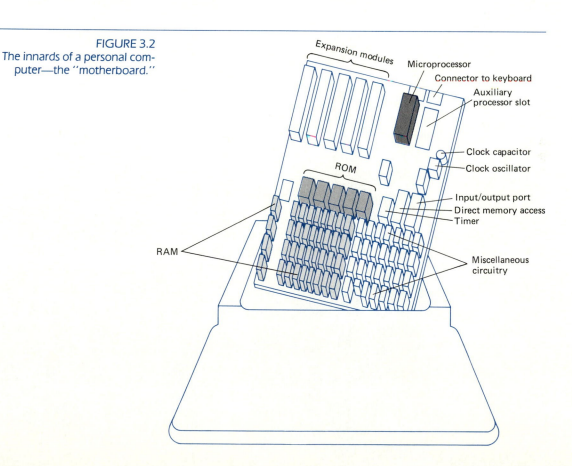

3.2   INPUT AND OUTPUT DEVICES

Although the CPU itself is sometimes referred to as "the computer," it is useless without some peripheral equipment to allow communication with the outside world. From the standpoint of human use, the vehicles for such communication are the input and output devices. The development of new devices each year precludes a complete description here of all available technology. However, three types are, and will continue to be, the primary equipment whereby engineers interact with personal computers. These include a device for input—the keyboard—and two for output—the monitor and the printer.

3.2.1   The Keyboard

The *keyboard* is the primary input device for the personal computer. Most keyboards are patterned after a conventional typewriter. However, other formats are available for some machines. For example, the keyboard of the conventional typewriter, the *QWERTY keyboard,* was deliberately designed to slow the typist up. This was because the original mechanical typewriters would jam if operated too quickly. Two alternatives, the *Dvorak simplified keyboard (DSK)* and the *American simplified keyboard (ASK),* have been developed which result in error reductions and speed increases of

FIGURE 3.3 *(a)* The conventional typewriter keyboard, which is sometimes referred to as the QWERTY after the sequence of letters at the upper left of the keyboard, and *(b)* the Dvorak keyboard.

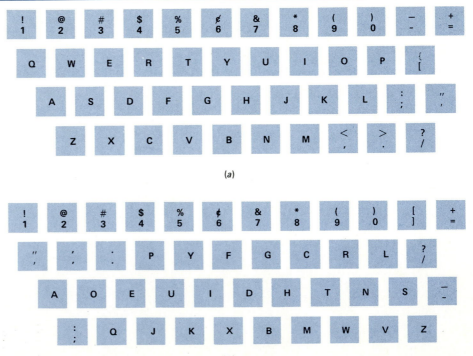

about 50 to 60 percent. Because a strong market demand for these innovations has yet to develop, they are difficult to obtain. However, their benefits could stimulate wider availability in the future. The QWERTY and the Dvorak keyboards are shown in Fig. 3.3.

In addition to the standard configuration, many personal computers include *function keys*. These enable you to input a set of instructions to the computer by depressing a single key. The number and placement of such keys vary among manufacturers.

Finally, there are a number of factors that influence the ease of use of keyboards. For example, keyboards vary with respect to touch and spacing. In addition, some keyboards are separate units from the rest of the computer and can be moved about at the user's convenience.

3.2.2   The Monitor

The *monitor* is a televisionlike unit that allows you to display your input as you enter it and your output as the computer responds. A variation is the *video display unit* that is a combination of a monitor and a keyboard. The most commonly available monitor, one related to the common television set, is the *cathode-ray tube (CRT)*. However, other technology such as the *liquid-crystal display (LCD)* is available. LCDs have the advantage that they are smaller and require lower power than CRTs. As a consequence they are preferable for small, portable personal computers. We will describe monitors in more detail when we discuss computer graphics in Chap. 39.

3.2.3   The Printer

Monitors have three primary limitations: The output is not permanent or portable, and only a small amount of information may be displayed at any one time. Therefore there are frequent occasions when you will require a *printer* to generate a permanent record, or "hard copy," of your output.

Printers can be categorized in a number of different ways. One method is according to the mechanisms whereby the characters are formed. *Dot-matrix printers* use a series of dots to form the individual characters (Fig. 3.4a). In contrast, *solid-font* mechanisms produce fully formed characters. A *daisy wheel* is one example of a solid font used with personal computers (Fig. 3.4b). Because the images they produce are sharper, solid-font mechanisms are usually considered superior to dot matrix and are preferred when "letter-quality" output is required.

Another way in which printers are categorized is by the manner in which the print is transferred to the page. *Impact printers* create characters by striking the type against an inked ribbon that presses the image onto the paper. The conventional typewriter is of this kind. *Nonimpact printers* do not involve physical contact but transfer the image to the paper using ink spray, heat, zerography, or laser.

Although there are some exceptions, nonimpact printers are usually quieter, faster, and cheaper than impact printers because they have fewer moving parts. However, because most nonimpact printers employ ink and dot-matrix mechanisms, they usually yield hard copies of lower quality than an impact-type, solid-font printer.

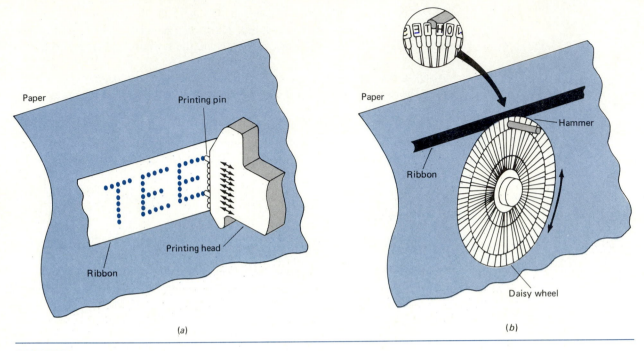

Paper

Printing pin

Printing head

Ribbon

Paper

Ribbon

Hammer

Daisy wheel

(a)   (b)

FIGURE 3.4
The two mechanisms for forming characters used on printers: *(a)* the dot-matrix and
(b) the solid-font mechanism called the "daisy wheel."

3.3   EXTERNAL MEMORY

Because the contents of the CPU's primary memory are lost when you turn off the
machine, you will require some means of storing programs or data for later use. This
is accomplished by storing the information on external memory media such as magnetic disks. Two of the most common media on microcomputers are floppy and hard
disks.

Floppy disks, which are also called *floppies* or *diskettes,* are made of a flexible
plastic that can be magnetized (Fig. 3.5). Data is stored on the disk as magnetized
spots on concentric tracks. The greater the number of tracks per inch or *disk density,*
the greater the amount of information that can be stored. Diskettes come in standard
sizes of $3\frac{1}{2}$-, $5\frac{1}{4}$-, or 8-in. However, some systems use nonstandard sizes. Floppy disks
usually require formatting for a particular system, and a disk formatted for one system
may not be compatible with another. Care must be exercised when handling floppies,
and they eventually need to be replaced. Back-up copies of all important disks should
always be maintained to prevent their loss in the event of damage. We will return to
floppy disks when we discuss program storage in later parts of the book.

For years, hard disks have been common secondary storage media on mainframe
computers. Recently, hard disks and drives are being marketed for some personal
computers. These have a number of advantages, including more reliability, larger

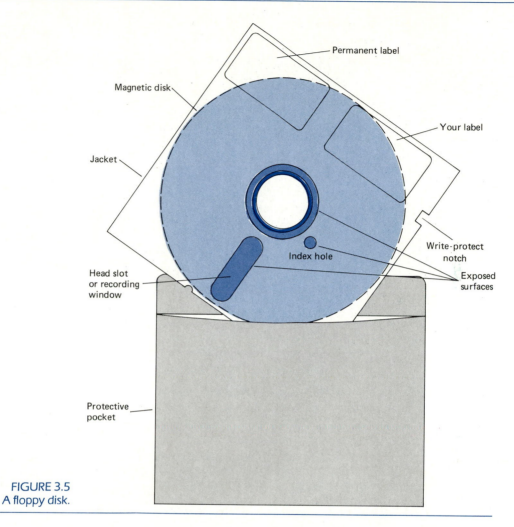

Permanent label

Magnetic disk

Your label

Jacket

Write-protect notch

Index hole

Exposed surfaces

Head slot or recording window

Protective pocket

FIGURE 3.5
A floppy disk.

storage, and faster access times. However, they are generally more expensive than floppy disks. In addition, the most common type, called a *Winchester disk* or ''Winnie,'' is permanently contained in its disk drive and does not have the portability of a floppy disk.

The significance of secondary storage lies in the fact that some microcomputers are limited by the speed and capacity of their disk storage devices. The computing speed and primary memory capacity of many micros are more than adequate for the needs of most individual users. However, with the addition of a Winchester disk, the microcomputer is elevated to a higher plane. With a hard disk, the machine is capable of holding a major database and implementing some of the larger applications software used by engineers. As such, it approaches a comprehensive computing tool for the professional engineer.

3.4   SUMMARY

This chapter has been designed to provide background on the physical devices that compose a typical personal computer. Because of the differences between models, we have had to limit our treatment to a general discussion of the most common devices. Thus we have not included some important devices and emerging technologies that could prove useful to you. Nevertheless, we hope that this introduction gets you started, and if you are employing a personal computer along with this book, we hope that you will take this opportunity to make an in-depth study of its configuration and workings. This will probably involve study of the machine's user's manual. We cannot stress strongly enough that the more you learn about your hardware, the more you will get out of your computer.

PROBLEMS

3.1   What are the components of the CPU and what are their primary functions?

3.2   What is RAM? How is it distinguished from serial-access memory? From which type do you think that information could be accessed faster? Why?

3.3   Distinguish between a bit and a byte. How many bits of information are contained in a 128K personal computer?

3.4   How does ROM differ from RAM?

3.5   Distinguish between dot-matrix and solid-font printing mechanisms and between impact and nonimpact printers.

3.6   The advertised storage capacity of a personal computer is 256K. Determine the storage capacity in the following units: (*a*) bits, (*b*) bytes, (*c*) kilobytes, and (*d*) megabytes.

3.7   Visit your local computer store. List the various types of printers that are sold. Characterize them as dot matrix, solid font, impact, or nonimpact. Which printers are considered "letter quality"?

3.8   Visit your local computer store. What kinds of magnetic disks are sold? How much do they cost and how is their capacity characterized? What type of disk is least expensive, measured in terms of K stored per dollar.

CHAPTER4
SYSTEMS SOFTWARE

When most people think of computers, they usually visualize the hardware described in the previous chapters. However, without human direction, hardware is a useless amalgam of metal and plastic. The human direction is provided by what are called computer programs. A *program* is merely a set of instructions that tells the computer what to do. All the programs that are necessary to run a particular computer are collectively called *software*.

The importance of software is reflected in the relative costs of hardware and programming (depicted in Fig. 4.1). Whereas early expenditures were primarily devoted to hardware, computer users have found that an increasing portion of computer costs is being devoted to programming. This includes both the programs needed to operate the computer itself, called *systems programs,* and the programs written by users to perform tasks, called *applications programs*.

The present chapter is devoted to systems programs. We start with these because they bridge the gap between the hardware you have just studied and the applications software with which most of the remainder of the book is concerned.

4.1   HOW COMPUTERS "THINK" (SOFTWARE)

To understand the distinction between systems and applications software, we have to briefly discuss the manner in which a computer is capable of holding and processing information. Recall that in Sec. 2.2 we introduced the notion that computer intelligence boils down to whether a bit is 0 or 1. Now we will see how such a simple concept can be used to represent more complicated information on a computer.

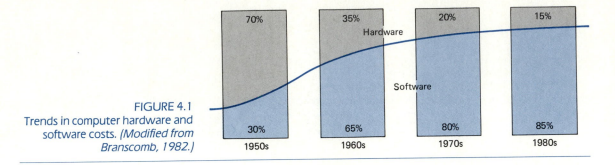

FIGURE 4.1
Trends in computer hardware and
software costs. *(Modified from
Branscomb, 1982.)*

The most elementary type of information that we might want to represent is a numerical quantity. To understand how a binary digit, or bit, is used for this purpose we have to understand how number systems work.

A *number system* is merely a way of representing numbers. Because we have ten fingers and ten toes, the number system that we are most familiar with is the *decimal,* or base 10, number system. A *base* is the number used as the reference for constructing the system. The base 10 system uses the ten digits—0, 1, 2, 3, 4, 5, 6, 7, 8, and 9—to represent numbers. By themselves, these numbers are satisfactory for counting from 0 to 9. For larger quantities, combinations of these basic digits are used, with the position or *place value* specifying the magnitude. The rightmost digit in a whole number represents a number from 0 to 9. The second digit from the right represents a multiple of 10. The third digit from the right represents a multiple of 100 and so on. For example, if we have the number 86,409 then we have eight groups of 10,000, six groups of 1000, four groups of 100, zero groups of 10, and nine more units, or

$$(8 \times 10^4) + (6 \times 10^3) + (4 \times 10^2) + (0 \times 10^1) + (9 \times 10^0) = 86,409$$

Figure 4.2*a* provides a visual representation of how a number is formulated in the base 10 system.

Now, because the decimal system is so familiar, it is not commonly realized that there are alternatives. For example, if human beings happened to have eight fingers and toes we would undoubtedly have developed an *octal,* or *base 8,* representation. In the same sense, our friend the computer is like a two-fingered animal who is limited to two states—either 0 or 1. Hence, numbers on the computer are represented with a *binary,* or *base 2,* system. Just as with the decimal system, each position represents higher powers of the base number. For example, the binary number 11 is equivalent to $(1 \times 2^1) + (1 \times 2^0) = 2 + 1 = 3$ in the decimal system. Figure 4.2*b* illustrates a more complicated example.

The use of bits to represent numbers is formally referred to as *true binary representation.* Computers often employ the true binary form to represent numbers when performing calculations. Although it works well for such applications, true binary representation is inadequate for specifying all the types of information stored and transmitted by the computer. Aside from numbers, other characters such as letters and

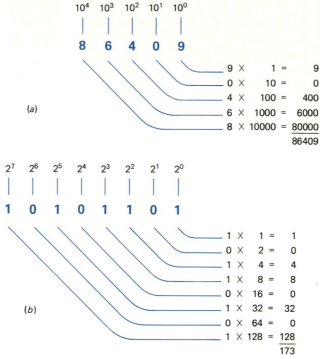

FIGURE 4.2
How (a) the decimal (base 10) system and (b) the binary (base 2) system work. In (b), the binary number 10101101 is equivalent to the decimal number 173.

symbols (periods, asterisks, etc.) are required. Therefore an alternative code that is capable of designating all three types of information—numbers, letters, and symbols— is needed.

In addition to not being able to directly designate letters and symbols, true binary representation also poses some problems related to data transmission. This is because numbers of different magnitude could have different-sized binary representations. For example, the decimal number 14 requires a binary number that is four bits long, 1110, whereas 173 requires eight bits, 10101101. Transmitting such numbers through a computer's circuitry is analogous to passing them through a pipe. If 14 and 173 were transmitted in binary form, the result at the other end would come out as 111010101101. The numbers are of different length, but there is no clue as to where one ends and the other begins. Thus the receiver would be unable to decipher their meaning.

For the above reasons, computers typically use a *fixed-length, binary-based code* to represent information. The fixed length is employed to facilitate transmission and storage of data. For example, if an eight-bit (that is, a byte) code is employed, the receiver knows that each group of eight bits that passes through the "pipe" represents one character.

The most popular codes are EBCDIC (*e*xtended-*b*inary-*c*oded *d*ecimal *i*nterchange *c*ode) and ASCII (*A*merican *s*tandard *c*ode for *i*nformation *i*nterchange). EBCDIC was developed by IBM and is employed in most of their computers. It uses eight bits

TABLE 4.1   EBCDIC and ASCII Bit Representation of Numbers and Letters

Character	EBCDIC	ASCII	Character	EBCDIC	ASCII
0	11110000	00110000	I	11001001	01001001
1	11110001	00110001	J	11010001	01001010
2	11110010	00110010	K	11010010	01001011
3	11110011	00110011	L	11010011	01001100
4	11110100	00110100	M	11010100	01001101
5	11110101	00110101	N	11010101	01001110
6	11110110	00110110	O	11010110	01001111
7	11110111	00110111	P	11010111	01010000
8	11111000	00111000	Q	11011000	01010001
9	11111001	00111001	R	11011001	01010010
A	11000001	01000001	S	11100001	01010011
B	11000010	01000010	T	11100010	01010100
C	11000011	01000011	U	11100011	01010101
D	11000100	01000100	V	11100100	01010110
E	11000101	01000101	W	11100101	01010111
F	11000110	01000110	X	11100110	01011000
G	11000111	01000111	Y	11100111	01011001
H	11001000	01001000	Z	11101000	01011010

per character. ASCII was developed under the guidance of the American National Standards Institute (ANSI) in an attempt to standardize the types of code used in computers. Both seven- and eight-bit versions are available to represent characters in ASCII. The seven-bit version was devised first. The eight-bit version was developed in order that the EBCDIC and ASCII codes would be of the same length. Although most of IBM's computers employ EBCDIC, its popular IBM PC, as well as other microcomputers, employs seven-bit ASCII.

Table 4.1 lists both the seven-bit ASCII and the EBCDIC codes for the digits (0 through 9) and letters (A through Z). Notice how the last four places of the number representations are in fact the true binary values. For the letters, notice that the bit representations are in ascending order. Thus the bit representation for A is less than that for B; that for B is less than that for C; and so on. We will find this useful in later sections when we want to place information in alphabetical order. In addition, although they are not shown in the table, lowercase letters can also be represented in ASCII. The lowercase letters come a fixed number of places after the uppercase. This fact can be exploited to easily convert text from uppercase to lowercase and vice versa.

Because there are $2^7 + 1$, or 256, possible seven-digit binary numbers (including zero), there are 256 possible characters that can be represented by the ASCII code. Aside from the 10 digits and 26 letters, there are a variety of other special symbols that can be represented. Table 4.2 shows a few that are available for the IBM PC. Appendix A includes a complete list.

Because one byte represents a single addressable storage location in many com-

TABLE 4.2
Some Symbols and Their ASCII Representations for the Extended Character Set of the IBM PC

Character	ASCII Value	
	Binary	Decimal
♥	00000011	3
→	00011010	26
⌂	01111111	127
β	11100001	225
π	11100011	227
≥	11110010	242

puters, manufacturers often use the number of bytes to signify their machine's storage capacity. Recall from Sec. 3.1 that a 64K computer can store about 524,288 bits of information. For computers where a byte is used to store a character, a more meaningful statement is that a 64K machine can store about 64 sets of 1024 bytes—that is, 65,536 characters.

Although many personal computers represent each character by a byte, other possibilities exist. The number of bits that constitute a common unit of information for a particular computer is called a *word*. As discussed previously in Chap. 2, most of the early personal computers used words that were one byte, or eight bits, long. These types, which are still prevalent today, were dubbed "eight-bit" machines. Today, 16- and even 32-bit personal computers are available. Most mainframe computers employ 32-bit words, and some supercomputers use 64-bit words.

Word size is important because when information is transmitted between locations in the computer, the transmission is performed a word at a time. Therefore a 16-bit machine operates faster than an 8-bit one, just as a large-diameter pipe transmits more water than a small-diameter one. In addition, larger word size yields a number of other benefits, including more primary memory, a larger set of available instructions, and greater precision.

Because of its importance for computations, we want to elaborate on the last advantage of large word size noted above—precision. Precision relates to the representation of numbers in the computer. Recall that although ASCII codes are employed to represent numbers, the true binary representations are used for calculations. For a 16-bit machine like the IBM PC, this means that there are 2^{16} different possible values. These can either be employed as unsigned numbers (ranging from 0 to 65,535) or, as is more common, as signed numbers (ranging from $-32,768$ to $32,767$). Numbers such as these, which are stored in their true binary form and which do not include decimals, are usually referred to as *integers*.

Although the integers from $-32,768$ to $32,767$ are certainly useful, larger numbers as well as those with decimal parts are also required for calculations. Accordingly, computers employ a variety of other schemes to store quantities. One common method is similar in spirit to scientific notation—that is, the number is stored in two parts, as a number and an exponent. Such numbers are referred to as *floating-point*, or *real, numbers*. An important practical aspect of storing numbers in this way is that the numbers are inexact or approximate. The degree of the approximation is referred to as the number's *precision*. The use of a larger word size allows more precision. In addition, there are often different degrees of precision available. The standard precision with which real numbers are represented is usually referred to as *single precision*, whereas the enhanced values are said to be in *double precision*.

The precision with which a computer handles numbers has great significance to engineers because of our heavy reliance on calculations. An example can illustrate why it is important. Suppose that you wanted to store the value one-third on the computer. As you know,

$$\tfrac{1}{3} = 0.3333333333333 \ldots$$

ad infinitum. A typical personal computer might be limited to storing only seven decimal places of this quantity in single precision as

$$\tfrac{1}{3} \cong 0.3333333$$

By omitting the 3s from the eighth decimal place to infinity, an error of 0.000000033333. . . has been introduced.

Approximations of this kind, which are attributable to the computer's ability to store only a finite number of significant figures, are referred to as *round-off errors*. For most cases, such round-off errors are negligible and will not have a significant effect on computations. However, for some engineering calculations, such single-precision numbers can be involved in millions and millions of arithmetic operations. In addition, many of these computations are sequential. That is, each calculation uses the result of previous ones. For such situations, it is possible for even small round-off errors to accumulate. In extreme cases, this can lead to erroneous results. For these situations, using double-precision real numbers can sometimes save the day. By allowing values to be approximated by more significant figures, round-off effects can be greatly mitigated. Thus a computer capable of higher precision can have real benefits in certain situations. In this respect, mainframe computers, which employ very high precision representations of numbers, have an advantage over personal computers.

4.2 MACHINE, ASSEMBLY, AND HIGH-LEVEL LANGUAGES

In the previous section, we showed how a series of 0s and 1s can be strung together to represent numbers, letters, and symbols within a computer. From these basic elements, all sorts of complex structures and commands can be developed. The process is akin to any language. For example, at first glance the English alphabet might seem too simple to be useful. However, from the 26 letters, a progression of words, sentences, and paragraphs can be developed to convey complex and subtle concepts. In the same spirit, a *machine language* composed of a limited number of "words" can be developed from computer bits. These words, which are called the computer's *instruction set,* allow us to instruct and manipulate the computer to perform tasks.

Although machine language allows us to instruct the computer to do our bidding, it is much too inefficient and cumbersome for human communication. A simple human language instruction such as "add A to B and print out the result" requires numerous machine-language words.

We therefore have a fundamental problem: Since computers and human beings speak different languages, how do we get them to communicate? At the beginning of modern computer development, specialists solved the problem by becoming fluent in machine language. This approach was impractical because the complexity of machine language made it indecipherable for all but the most dedicated computer experts. Consequently, intermediate languages were developed that were better suited for human comprehension but which were expressly designed so that they could be efficiently translated into machine language. The first type, called *assembly language,* still required that step-by-step instructions be written for the computer, but it used a more expressive vocabulary than binary digits for that purpose. The second type, called a *high-level language,* condensed groups of machine-language instructions into single statements. These statements were then given descriptive, human-oriented labels. Thus the human programmer could write a program using the high-level language, and another program could translate it into machine language for implementation by the computer.

The program that translates high-level programming languages into machine language and then translates machine language back into understandable human terms is one of the *systems programs* alluded to previously. In contrast, a program that an engineer might write or purchase to solve particular problems with the computer is an example of an *applications program*. It is this kind of program that would be written in a high-level language and translated into machine language by a systems program.

4.3   SYSTEMS PROGRAMS

Now that we have distinguished between systems and applications programs, we can discuss the former in more detail. *Systems software* consists of programs that perform background tasks that enable applications programs to run properly. For computers, the systems software can be broken down into three major categories: the operating system, utility programs, and language translators. First we will briefly introduce the function of each of these components. Then we will describe how they are marshalled to bring the computer to life and run applications software.

4.3.1   The Operating System

In the early years of electronic computing, almost every aspect of operating a system required some direct human supervision. Today many of the tasks formerly coordinated by a human operator are handled by the component of the systems software called the operating system.

The *operating system* is a group of programs that coordinates and controls the activities of the machine. In particular, it enables applications software to interface with the hardware of a particular computer. We will start by discussing the operating systems that are typical of mainframe and minicomputers. To this point, we have usually lumped micro- and minicomputers under the general heading of personal computers because of their common attribute of accessibility. For our discussion of operating systems, however, a different perspective will be adopted. This perspective is based on the fact that beyond accessibility, the three types of computers can be classified as to the number of individuals that can use them simultaneously. Employing this classification scheme places the single-user micros in a category distinct from the multiple-user minis and mainframes. Nowhere is this distinction more significant than as it relates to the capabilities of the operating system.

Mainframe and minicomputer operating systems.   The essential function of a mainframe or minicomputer operating system is coordination. A large mainframe computer may have upwards of 500 people using it simultaneously. Input, output, and storage of information is mediated by a whole host of diverse hardware. Mainframes can operate with card punches, card readers, magnetic tapes, and hard disks. They can transmit information via line printers, high-speed page printers, graphics plotters, and different kinds of telecommunications terminals. Once you realize that many of these devices will be needed concurrently by many users, the sheer magnitude of scheduling access to a particular device becomes mind-boggling. It is the job of the operating system to ensure that such coordination proceeds in a smooth and efficient manner.

Now that you have a general idea of the operating system's overall mission, we can take a closer look at its functions. Some of these, which are typical of mainframe computers, are listed here.

1. *Assignment of the system's resources*   The operating system must determine which resources are to be mobilized for any given job. A part of the operating system that contributes to this endeavor is the *job-control program*. This program reads instructions written by the user in job-control language (JCL) that specify the requirements for a particular job. These include the types of input/output devices, the type of language that is being employed, and other special requirements. In the absence of these instructions, the job-control program makes assumptions regarding some of these needs. Such *default options* will be implemented unless you specify otherwise with your JCL commands.

2. *Job and resource scheduling*   Not only must the operating system decide *what* resources must be used but also *when* they will be used. *Scheduling programs* determine the order in which jobs are implemented. This order is dictated by a number of factors, including the job's priority, the availability of required hardware, and the general level of activity.

There are a variety of sophisticated processing techniques that are employed to ensure that scheduling proceeds as smoothly and efficiently as possible. Timesharing and multiprogramming are two methods that allow the computer to work on several programs concurrently. In *timesharing,* the operating system moves from program to program, giving each a small slice of execution time on every cycle. Thus no single program is allowed to dominate the computer's attention. *Multiprogramming* also involves operating on one program at a time. However, rather than allocating each program a fixed slice of time, it will execute each until a logical stopping point is reached. Because computations and processing typically are much faster than input/output, a strategy might be to operate on one program until a slow input/output step is encountered. The operating system can then switch to another program and only return to the first when that program is ready for speedy activities again. Today many systems combine aspects of timesharing and multiprogramming in their scheduling schemes.

Because of the slowness of most input/output processes, another technique for expediting the implementation of programs is *spooling*. This involves the temporary holding of input and output in secondary storage. Jobs wait in a queue and are fed in and out of the machine in the order of their priority. In the meantime, the CPU concentrates on quick computations and processing without having to worry about slow input and output.

A final aspect of data processing that contributes to the efficient employment of a computer's resources is virtual storage. In the early days of computing, programs of even moderate length were often too large to fit in primary memory. Consequently, users had to split their programs into smaller pieces to be executed separately. Today many mainframe operating systems perform the same procedure automatically. *Virtual storage* refers to a special fast-access area of secondary storage which can be thought of as a large extension of primary memory. The operating system delivers large programs to virtual storage where they are divided into workable segments that can be executed one at a time in primary memory. In this way, programs can be implemented that are much larger than the actual extent of primary memory.

3. *Monitoring activities*   The final general function of a large-scale operating system is keeping track of activities. This involves three major tasks: security, bookkeeping, and detecting abnormalities. *Security* is needed to protect a user's data and programs as well as to ensure that individuals who do not belong on the system do not gain access. Among other things, the operating system screens identification information and passwords for this purpose. *Bookkeeping* involves keeping track of the appropriate costs for the services that users obtain. *Detecting abnormalities* entails terminating programs that contain errors or exceed time or storage allocations. For these cases, the operating system will send a message to the user describing the problem. In addition, the user will also be informed of other abnormalities, such as problems concerning input/output and other parts of the system.

Although minicomputers operate on a much smaller scale, the same types of coordination are required. This is in marked contrast to the single-user operating systems required on most microcomputers.

Microcomputer operating systems.   Most microcomputers and their operating systems have been designed to meet the needs of one user at a time. Usually the services and peripherals supported directly by the operating system are limited. Typically the operating system controls storage of information on a floppy disk or tape. It can both load the user's program and pass control of the machine to the program. The ability to edit input is often limited to a single line of data that specifies which applications program or system utility should be used next.

Usually the functions built into the operating system revolve around the handling of files on a floppy disk or tape. The functions usually included as a part of the operating system are the ability to display the data in a file on a monitor, name files, erase files, list the names of files, move the contents of RAM to a disk file, and run a program. Other tasks can be provided by utility programs that come with the operating system but are in fact distinct programs separate from the operating system. These utility programs reside on the disk or tape storage and are only brought into RAM when run by the user. In contrast, the core of the operating system resides in RAM at all times when the machine is in operation.

Operating systems for microcomputers were initially little more than disk-handling programs. This is why, as described in the next section, they are traditionally referred to as ''disk operating systems,'' or DOS. Today the functions supplied by mainframe or minicomputer operating systems are being incorporated into the microcomputer operating system. This has now become possible because the size of microcomputer RAM has increased as the cost of RAM has decreased. The additional features found on mainframes were always desirable, but the cost of these services prevented the microcomputer operating system from offering them.

Types of operating systems for personal computers.   A number of operating systems have been developed for personal computers. Because most operating systems conform to specific microprocessor chips, applications programs that work with one operating system do not usually work with another. Consequently, applications software for one machine often cannot be run on a different type of computer. Therefore the choice of an operating system is very important. Because so many operating systems are available and because new ones are continually developed a comprehensive review is inappropriate here. However, some are so ubiquitous or promising that they bear mention.

First, the most popular personal computers usually have their own operating systems. Thus most Apple II computers have *APPLE-DOS* and the TRS-80 computer has *TRS-DOS*. Because so many of these computers are in use, these operating systems have importance for no other reason than that they are prevalent. Of all these machine-specific operating systems, the most important is *Microsoft,* or *MS-DOS*. Because a version of this system is used on the highly popular IBM PC, a number of other computer companies have adopted MS-DOS as their operating system. Consequently, it is presently the most widely employed system on microcomputers.

Aside from computer-specific versions, a number of generic operating systems have been developed in an attempt to provide some sort of industry standard. *CP/M*—short for *c*ontrol *p*rogram/*m*icroprocessors—has been the most popular to date. Many computer manufacturers offer a version of CP/M that can be added to their machines as an alternative operating system. IBM, Texas Instruments, Radio Shack, and Apple all offer CP/M as an option.

The final operating system that will be mentioned was not originally designed for microcomputers. *UNIX* was developed by Bell Laboratories for minicomputers. It is especially well suited for timesharing systems where a number of users simultaneously employ the same minicomputer. Consequently, it is finding ready application in contexts such as small firms and university computing laboratories. Although originally developed for minis, it is being used in some microcomputers. In a sense, it is the next step in the natural evolution of disk-operating systems toward the true operating systems of the mainframe computer. Beyond its intrinsic value in this regard, its use will probably be further encouraged by the entrance of AT&T into the personal computer business.

As a final note, we must reiterate our original statement that the present rapid pace of change in the computer industry makes it next to impossible to definitively specify the most important operating systems. The key fact to remember is that the existence of so many systems means that software written for one system often will not work on another.

4.3.2   Utility Programs

Some operations are so common that it would be inefficient if every user had to write a program to perform them. Utility programs, which are supplied by the manufacturer or by software companies, are designed to perform these general housekeeping tasks. Some examples are programs for copying, merging data files, and sorting.

4.3.3 Language Translators

As mentioned at the beginning of this chapter, a number of high-level languages have been developed to facilitate computer programming by users. However, before programs written in these languages can be executed, they must be translated into the machine language understood by the computer. The systems software known as a language translator performs this task.

There are two common types of language translators on personal computers—interpreters and compilers. *Interpreters* perform the translation of the original high-level program, which is called the *source code,* one line a time. Thus the translation takes place while the program is being run. In contrast, *compilers* translate the program en masse and essentially generate a new machine-language version called the *object code*.

Each of the translators has advantages and disadvantages with respect to the other. Interpreters require much less storage space because they do not generate an object program that needs to be stored. Traditionally this has been a positive trait for personal computers, where memory has been at a premium. In addition, interpreters are much easier to use for program development and maintenance because errors can be detected and corrected as they occur.

Although compiled programs are not as conducive to debugging, they have the advantage that execution is much quicker and more efficient. This is because once a program is tested and compiled, the machine-language object code can be stored on a disk and run very quickly. This is in contrast to an interpreted language where each line must be reinterpreted on every run.

4.4   TURNING ON A PERSONAL COMPUTER

To this point, we have described all the major elements that compose a typical personal computer system. However, to gain a better appreciation of how the system actually operates, we will now describe what occurs when the machine is brought to life.*

The sequence of events in turning on a personal computer is depicted in Fig. 4.3. Before the process is initiated, the RAM component of the primary memory is empty. In this state, the computer can do nothing. The systems and applications programs that are needed to allow the computer to perform tasks initially reside on a diskette.

The first step in the process of starting up is to load the operating system from the disk to the primary memory (Fig. 4.3a). On personal computers, this is often accomplished with the help of ''bootstrap'' programs that reside on ROM chips. They are called bootstrap programs because they allow a computer, once it is turned on, to

*Appendixes C and E provide detailed description of how the procedures are implemented on the IBM PC and Apple II computers.

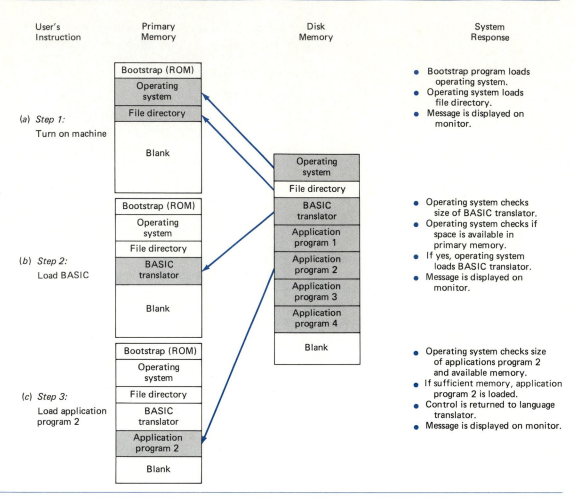

User's Instruction	Primary Memory	Disk Memory	System Response

(a) Step 1: Turn on machine

Primary Memory:
Bootstrap (ROM)
Operating system
File directory
Blank

Disk Memory:
Operating system
File directory
BASIC translator
Application program 1
Application program 2
Application program 3
Application program 4
Blank

System Response:
- Bootstrap program loads operating system.
- Operating system loads file directory.
- Message is displayed on monitor.

(b) Step 2: Load BASIC

Primary Memory:
Bootstrap (ROM)
Operating system
File directory
BASIC translator
Blank

System Response:
- Operating system checks size of BASIC translator.
- Operating system checks if space is available in primary memory.
- If yes, operating system loads BASIC translator.
- Message is displayed on monitor.

(c) Step 3: Load application program 2

Primary Memory:
Bootstrap (ROM)
Operating system
File directory
BASIC translator
Application program 2
Blank

System Response:
- Operating system checks size of applications program 2 and available memory.
- If sufficient memory, application program 2 is loaded.
- Control is returned to language translator.
- Message is displayed on monitor.

FIGURE 4.3 Visual depiction of how programs are loaded from disk memory to primary memory when a personal computer is in operation. *(Redrawn from Toong and Gupta, 1982.)*

get itself into a ready condition without any outside help—it is thus said to "pull itself up by its own bootstaps." For this same reason, the start-up process is sometimes referred to as "booting." After the disk containing the operating system is placed in the disk drive, the computer is switched on, and the bootstrap program automatically brings the operating system into primary memory. Then the operating system itself brings a file directory from the same disk and loads it into primary memory. This file directory contains information on all the programs and data files on the disk. After these operations are accomplished, the user is notified by a message on the monitor that the computer is ready to accept commands from the keyboard.

Now let's assume that you are interested in using the high-level language called BASIC to develop or update one of your computer programs. In the next step, you would type in the appropriate commands on the keyboard to direct the operating system to load the BASIC language translator (Fig. 4.3b). After checking to ensure that enough space is available, the translator would be moved to primary memory. Once this is accomplished, the user would again be notified by a message on the monitor that the operation was successful and that the machine is ready to accept commands. It should be mentioned that some personal computers perform the loading of the language translator automatically from ROM as part of the turn-on process.

Now, for the final step, let's assume that you are interested in loading an applications program (Fig. 4.3c). To do this, you would type in the appropriate commands on the keyboard. After checking to ensure that adequate space is available, the operating system would load the applications program into primary memory.

At this point you could work on the program or run it. The next chapter will begin to discuss such applications programs and how they can be developed to perform engineering tasks.

PROBLEMS

4.1 Distinguish between systems and applications programs.

4.2 Express the following base 10 numbers in binary: (*a*) 26, (*b*) 17, and (*c*) 47.

4.3 Express the following base 10 numbers in base 8: (*a*) 87, (*b*) 29, and (*c*) 53.

4.4 Express the following binary numbers in base 10: (*a*) 110011, (*b*) 101011101, and (*c*) 1000111.

4.5 What is an operating system and why is it important?

4.6 Distinguish between a mainframe and a microcomputer operating system.

4.7 Identify MS-DOS, CP/M, and UNIX.

4.8 Discuss the advantages and disadvantages of interpreters and compilers.

4.9 Discuss the advantages of 16-bit computers over 8-bit computers.

4.10 Distinguish between machine, assembly, and high-level computer languages.

4.11 Visit your local computer store. Name, describe, and briefly compare the operating systems of any three personal computers.

PART THREE

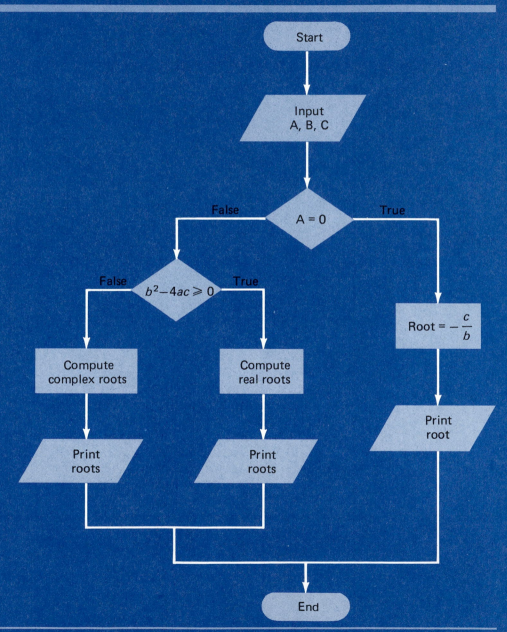

PROGRAMMING PREREQUISITES

The programs that you use to solve problems on a computer are called *applications software.* During your engineering career, there may be periods where your computer work will be limited to applications software developed by someone other than yourself. For instance, these programs might be developed by a software manufacturer or one of your assistants. On these occasions, knowledge of programming and high-level languages may not seem necessary. However, just as some knowledge of automobiles can be invaluable when buying a used car, so will an understanding of programming serve you well when you use someone else's software. We believe that every engineer should be capable of developing computer programs for small-scale problem solving. Once you have mastered sufficient skill to write such programs, you will have taken a large step toward more intelligently using software written by others.

The present part of the book is devoted to topics that should be assimilated before you begin to write computer programs. In *Chap. 5* we review programming languages that are useful for engineering applications. *Chapter 6* is devoted to an overview of the program-development process, with special emphasis on ensuring that the program has a strong logical foundation. Finally, *Chap. 7* deals with some general techniques that greatly facilitate the development of high-quality applications software.

CHAPTER 5

HIGH-LEVEL PROGRAMMING LANGUAGES

Hundreds of high-level languages have been developed since the modern computer age began. Out of these we have chosen to emphasize *BASIC* and *FORTRAN* as our major programming tools. Because of its relevance to engineering and its availability on personal computers, an additional language—*Pascal*—is also described briefly in the present chapter.

The following are thumbnail descriptions of these languages and some of their advantages and disadvantages. Notice that we have chosen to describe them in the order in which they were developed. Thus we can provide some sense of their historical evolution. In particular, you should recognize how each was developed to meet a specific need of users.

5.1 FORTRAN

Business and scientific programmers have related but primarily distinct computing needs. Business-oriented work usually involves sophisticated input/output, large amounts of data, and relatively simple computations. In contrast, scientific and engineering computing entails simple input/output, smaller quantities of data, and fairly intricate calculations. For this reason, there was a need in the 1950s for a high-level language expressly designed to accommodate technical computing.

FORTRAN, which stands for *for*mula *tran*slation, was introduced commercially by IBM in 1957 to meet this need. As the name implies, one of its distinctive features is that it uses a notation that facilitates the writing of mathematical formulas. Because of such features, FORTRAN is the native language of most engineers and scientists who attained computer fluency on large mainframe machines in the 1960s and 1970s.

Although most personal computers use BASIC, FORTRAN compilers are available for many machines. For this reason, and because of its continuing significance for engineering work, we have devoted a part of this book to the essentials of FORTRAN programming.

5.2 BASIC

Just as professional engineers and scientists have unique computing needs, so also do students. BASIC, which stands for *b*eginner's *a*ll-purpose *s*ymbolic *i*nstruction *c*ode, was developed expressly as an instructional language at Dartmouth College in the mid-1960s. Its originators, John Kemeny and Thomas Kurtz, recognized that, although FORTRAN was widely used, it had certain characteristics that posed problems for novice programmers. Kemeny and Kurtz came up with BASIC as an alternative that was easier to learn and use. However, because it retained many of the key features of FORTRAN and added some nice ones of its own, it is more than adequate for many engineering and scientific computations. As a consequence, it is presently among the most popular computer languages in the world. In addition, it is unquestionably the most commonly used language on personal computers.

5.3 PASCAL

As noted in Fig. 4.1, software development constitutes the lion's share of today's computing costs. Consequently, innovations that increase a programmer's productivity and efficiency can lead to significant savings. Pascal is one of the *structured* languages that has been developed to facilitate large-scale programming. Developed by Nicklaus Wirth and named after Blaise Pascal, the language uses a restricted set of design structures and other organizing principles. By essentially imposing a set of good style habits on the user, Pascal makes for programs that are easier to develop and maintain. These advantages become more pronounced for larger programs.

5.4 COMPARISON OF BASIC, FORTRAN, AND PASCAL

Table 5.1 shows BASIC, FORTRAN, and Pascal programs that perform the simple task of adding two numbers and printing the result. Although these programs hardly tap the capabilities of any of these languages, the example serves to highlight the major differences between them.

Of the three, the BASIC version is the simplest in structure and content. The lines are numbered in ascending order, and each instruction uses easy-to-understand commands such as INPUT and PRINT. About the only command that might not be immediately understood is the REM in line 10, which is merely an abbreviation for the word "remark." REM statements are used in BASIC programs to allow descriptive labels or remarks to be included within the program. Hence line 10 is intended to document the fact that this is a simple addition program.

TABLE 5.1   Comparison of BASIC, FORTRAN, and Pascal

	BASIC	FORTRAN	Pascal

Example Programs	10 REM SIMPLE ADDITION
	         PROGRAM
	20 PRINT ''ENTER TWO
	         NUMBERS''
	30 INPUT A,B
	40 LET TL = A + B
	50 PRINT ''SUM IS''; TL
	60 END
		C SIMPLE ADDITION PROGRAM
		  1 FORMAT(2F5.0)
		      READ(5,1)A,B
		      TOTAL = A + B
		  2 FORMAT(1X, ''SUM IS'', F5.0)
		      WRITE(6,2)TOTAL
		      STOP
		      END
			PROGRAM SUM (INPUT, OUTPUT);
			(*SIMPLE ADDITION PROGRAM*)
			VAR A,B,TOTAL : INTEGER;
			BEGIN
			    WRITELN('ENTER 2 NUMBERS');
			    READ(A,B);
			    TOTAL : = A + B;
			    WRITELN('SUM IS', TOTAL);
			END.

Pros	Simple and easy to use. ''Conversational'' or interactive orientation makes it ideal as a trial-and-error learning language. Well suited to and widely available on personal computers. Good for development purposes because errors are detected immediately Good graphics and use of sound.	Excellent for sophisticated computations, fast and precise. Standardized minimal and extended versions have been established; less variation among dialects. Relatively powerful input/output capabilities.	Structured orientation excellent for large-scale, complicated programs. Well suited for personal computers though not as ubiquitous as BASIC. Structured orientation is a plus from an educational perspective; instills good habits. Easily transferable from system to system.
Cons	Early versions not well suited for structured programming. Commonly available interpreted versions run slow. Although standardized minimal and extended versions have been established, much variation among dialects. Simplicity sometimes represents a limitation.	Not as good a learning language as BASIC. Noninteractive orientation and relative complexity are disadvantages in this regard. Early versions not well suited for structured programming. Graphics not a standard part of language.	Relative complexity makes it more difficult for the novice to master. Marginal input/output capabilities. Graphics not a standard part of language.

In contrast to the simple statements of the BASIC version, the FORTRAN and Pascal programs include labels, such as FORMAT, VAR, and WRITELN, that are less directly communicative. In addition to vocabulary, the FORTRAN and Pascal versions exhibit more structure. That is, the spacing and organization is somewhat more involved and apparent. In the case of FORTRAN, the spacing conventions are primarily due to the fact that programs in this language were originally typed on punched cards. Various columns of the cards were set aside for particular purposes. For example, the first column was set aside to designate that a particular card was devoted to documentation. Thus, just as the REM is used in BASIC, a C or an * in column 1 means that a descriptive remark or comment is included on a FORTRAN line.

While the structure of FORTRAN is to a certain extent an artifact of the punched card, the structure of Pascal programs is intentional and is, in fact, the language's greatest strength. When unstructured BASIC and FORTRAN programs are viewed from a distance, their organization is not apparent. In contrast, the design and configuration of Pascal programs are pronounced. This greatly facilitates their modification and use.

A number of other advantages and disadvantages of the three languages are delineated in Table 5.1. As time passes, new versions, or "dialects" of each language are being developed. Not surprisingly, the newer dialects of each language often incorporate some of the advantages of the others. For instance, recent dialects of BASIC have improved input/output capabilities that are closer in power to those of FORTRAN. Also, structured versions of both BASIC and FORTRAN are now available that incorporate some of the strengths of Pascal.

5.5 COMPUTER LANGUAGES IN THIS BOOK

Although the evolution of dialects means that the trade-offs in Table 5.1 are not hard and fast, the comparison should provide insight into our choice of BASIC and FORTRAN as the programming languages in this book. Both are adequate for the problem-solving programs that we will emphasize. In addition, both are widely available and will figure prominently in engineering for the foreseeable future. BASIC is the primary language of personal computers, and FORTRAN is the most fundamental engineering language associated with mainframe installations. Although FORTRAN is not quite as ubiquitous as BASIC on personal systems, FORTRAN compilers are available to implement the language on many machines.

Despite its excellence as both a programming and a learning language, we have omitted Pascal from this text. Because it has strong advantages for large-scale programs, Pascal is a very useful language for those engineering students who anticipate that computer programming will figure prominently in their careers. In addition, many of the strengths of Pascal are general enough to be transferred to another language. Wherever possible, we have integrated these strengths into our discussion of BASIC and FORTRAN. In particular, those features related to structured programming have been included.

We expect that either BASIC or FORTRAN (but not both) will be employed by most of those using this text. However, just as learning several conventional languages can represent a professional advantage, becoming bi- or multilingual in the computer sense can also serve you well. Therefore we encourage you to use this book as an opportunity to become a bilingual computer programmer. Once you learn either BASIC or FORTRAN, the other will be that much easier to master.

Finally, there are a variety of other languages beyond BASIC, FORTRAN, and Pascal that have potential utility in engineering. Table 5.2 provides brief descriptions of several of these languages.

PROBLEMS

5.1 Which high-level programming language would you choose for the following applications? Justify your choice.

(a) You are performing large-scale computations for NASA. The calculations

Table 5.2   *Other Computer Languages*

Ada (named after the world's first programmer, Augusta Ada Byron) A portable structured language sponsored by the Department of Defense. Developed to facilitate the composition and maintenance of large programs that would be created by teams, used repeatedly over long time periods, and subject to continual change.

ALGOL (*ALGOrithmic Language*) A structured language that is well-suited for scientific and mathematical projects but has limited file-processing capabilities. It is similar to FORTRAN and is popular in Europe. Considered to be a forerunner of Pascal and PL/I.

APL (*A Programming Language*) A highly compact, interactive language that is popular for short, mathematically oriented, problem-solving programs. Employs special symbols and can perform complex operations with a single command. However, these features also detract from the readability of APL programs.

C A general purpose, portable programming language featuring concise expressions and a design that permits well-structured programs. Originally developed at Bell Laboratories as a language for writing systems software. For example, the Unix operating system was written in C. However, it has many of the features of a high-level programming language and, thus, has a wide range of applications.

COBOL (*COmmon Business-Oriented Language*) A widely used machine-independent business language that allows programs to be written in an English-like style and to be self-documenting.

FORTH Similar to C in the sense that it is highly portable and has characteristics of both assembly and high-level languages. It is fast and extensible (you can add your own commands). Thus, it is a good language for controlling industrial processes and machines. However, the amount of freedom it allows to the programmer can be a disadvantage for large-scale programming efforts involving teamwork.

LISP (*LISt Processing Language*) The primary language of artificial intelligence research. Designed to process nonnumeric data such as characters or words.

Modula-2 A language written by Nicholas Wirth containing extensions and further improvements to Pascal.

PL/1 This language represents an effort to combine the strengths of FORTRAN and COBOL in order to meet the needs of both scientists and businessmen with a single language. Thus, it employs some of the computational concepts of FORTRAN and some of the file-handling capabilities of COBOL.

PROLOG A relatively recent artificial intelligence language. It has been adopted by the Japanese for their fifth-generation project (see Sec. 42.3).

must be performed rapidly and with high precision to provide guidance for reentry of the space shuttle into the earth's atmosphere.

(b) You are developing a complicated, large-scale program for transportation analyses of small cities. It will be utilized throughout the country on a variety of different personal computers.

(c) You are to teach computer programming to a group of freshman engineers. Although some have had some exposure to computers in high school, most have never written a program.

5.2   Visit your local computer store. Select any three personal computers. Determine which high-level languages each one supports.

CHAPTER 6
ALGORITHMS AND FLOWCHARTS

Although writing a program in a high-level language is obviously of critical importance, it is but one step in the overall development of high-quality software. Figure 6.1 delineates the five steps that constitute the entire process.

We should note that an important element is missing from Fig. 6.1. Before the process can even begin, there is the preliminary step of *problem definition and analysis*. This includes (1) identifying the problem and determining the solution technique and (2) specifying the objectives you want to accomplish with the software.

We have omitted this important step at this point for a number of reasons. First, problem solving is to a large extent an art that is acquired through a long-term process of education and experience. By definition, engineers are problem solvers. Thus, as you mature professionally, your problem-solving skills will be honed as a natural by-product of your development. Second, although problem definition is a critical prerequisite for developing effective software, this skill is not necessary in learning how to write a program in a high-level language. In fact, at this preliminary stage, it would probably complicate matters. Thus we will put off the task of problem solving until after you have mastered the rudiments of programming. Then in the later parts of the book we can begin to hone your problem-solving skills by using computer programming to approach a variety of engineering problems.

With this as background, we can begin the program-development process. As shown in Fig. 6.1, the first step is designing the underlying logic or algorithm of the program.

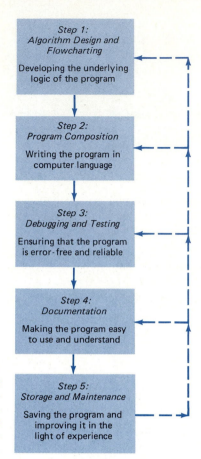

FIGURE 6.1
The five steps required to produce and maintain high-quality software. The feedback arrows indicate that the first four steps can be improved in the light of experience.

6.1 ALGORITHM DESIGN

The most common problems encountered by inexperienced programmers usually can be traced to the premature preparation of a program that does not encompass an overall strategy or plan. In the early days of computing, this was a particularly thorny problem because the design and programming components were not clearly separated. Programmers would often launch into a computing project without a clearly formulated design. This led to all sorts of problems and inefficiencies. For example, the programmer could get well into a job only to discover that some key preliminary factor had been overlooked. In addition, the logic underlying improvised programs was invariably obscure. This meant that it was difficult for someone to modify someone else's program. Today, because of the high costs of software development, much greater emphasis is placed on aspects such as preliminary planning that lead to a more coherent and efficient product. The focus of this planning is algorithm design.

An *algorithm* is the sequence of logical steps required to perform a specific task such as solving a problem. Aside from accomplishing its objectives, a good algorithm must have a number of specific properties:

1. Each step must be deterministic; that is, nothing can be left to chance. The final results cannot depend on who is following the algorithm. In this sense, an algorithm is analogous to a recipe. Two chefs working independently from a good recipe should end up with dishes that are essentially identical.

2. The process must always end after a finite number of steps. An algorithm cannot be open-ended.

3. The algorithm must be general enough to deal with any contingency.

Figure 6.2*a* shows an algorithm for the solution of the simple problem of adding a pair of numbers. Two independent programmers working from this algorithm might develop programs exhibiting somewhat different styles. However, given the same data, their programs should yield identical results.

Step-by-step English descriptions of the sort depicted in Fig. 6.2*a* are one way to express an algorithm. They are particularly useful for small problems or for specifying the broad major tasks of a large programming effort. However, for detailed representations of complicated programs, they rapidly become inadequate. For this reason, more versatile, visual alternatives, called flowcharts, have been developed.

6.2   FLOWCHARTING

A *flowchart* is a visual or graphical representation of an algorithm. The flowchart employs a series of blocks and arrows, each of which represents a particular operation

FIGURE 6.2
(*a*) Algorithm and (*b*) flowchart for the solution of a simple addition problem.

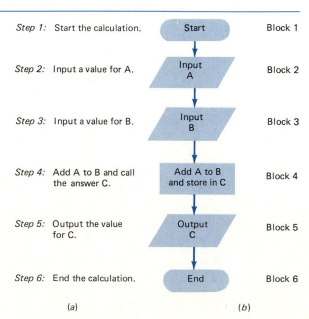

(*a*)   (*b*)

or step in the algorithm. The arrows represent the sequence in which the operations are implemented. Figure 6.2*b* shows a flowchart for the problem of adding two numbers.

Not everyone involved with computer programming agrees that flowcharting is a productive endeavor. In fact, some experienced programmers do not even use flowcharts. However, we feel that there are three good reasons for studying them. First, they are still used for expressing and communicating algorithms. Second, even if they are not employed routinely, there will be times when they will prove useful in planning, unraveling, or communicating the logic of your own or someone else's program. Finally, and most important for our purposes, they are excellent pedagogical tools. From a teaching perspective, they are ideal vehicles for visualizing some of the fundamental logical structures employed in computer programming. Additionally, developing flowcharts provides good basic training in the disciplined logical process that lies at the heart of successful computer programs.

6.2.1 Flowchart Symbols

Notice that in Fig. 6.2*b* the blocks have different shapes to distinguish different types of operations. In the early years of computing, flowchart symbols were nonstandardized. This led to confusion, since programmers used different symbols to represent the same operations. Today a number of standardized systems have been developed to facilitate communication. In this book, we use the set of symbols shown in Fig. 6.3. They are described below.

- *Terminal*   These signify the beginning and end of a program.
- *Flowlines*   These represent the flow of logic, with the arrows designating the

FIGURE 6.3 Common ANSI symbols used in flowcharts.

SYMBOL	NAME	FUNCTION
	Terminal	Represents the beginning or end of a program.
	Flowlines	Represents the flow of logic. The humps on the horizontal arrow indicates that it passes over and does not connect with the vertical flowlines.
	Process	Represents calculations or data manipulations.
	Input/output	Represents inputs or outputs of data and information.
	Decision	Represents a comparison, question, or decision that determines alternative paths to be followed.
	On-page connector	Represents a break in the path of a flowchart that is used to continue flow from one point on a page to another. The connector symbol marks where the flow ends at one spot on a page and where it continues at another spot.
	Off-page connector	Similar to the on-page connector but represents a break in the path of a flowchart that is used when it is too large to fit on a single page. The connector symbol marks where the algorithm ends on the first page and where it continues on the second.

direction. A flowchart is read by beginning at the start terminal and following the flowlines to trace the algorithm's logic. Thus, as shown in Fig. 6.2*b*, we would begin at block 1 and then follow the arrow sequentially until we arrive at the terminal block 6. If the flowchart is written properly there should never be any doubt regarding the correct path.

• *Process*   Arithmetic and data manipulations are generally placed in these boxes. This symbol can also be used to represent major program segments or modules.

• *Input/Output*   These represent either the input or the output of data and information. Thus at one point they might be used to designate the input of information from a keyboard, whereas at other points they might represent the output of data to a printer or monitor.

• *Decision*   This is the only svmbol that has more than one exit flowline. The proper exit path is determined by the correct answer to a question or conditional statement contained in the block.

• *On-page connector*   Although the flowlines in Fig. 6.3 show humps to depict lines that pass over each other without connecting, it is generally best to avoid crossing flowlines. Circular connectors allow the flow to be discontinued at one section of a flowchart and started again at another location. A number is placed within the exit- and entry-point connectors to specify each pair used on a given page (Fig. 6.4).

• *Off-page connector*   Some flowcharts are so large that they spread over several pages. The spade-shaped symbols in Figs. 6.3 and 6.4 are used to transfer flow between pages in a fashion similar to the circular on-page connectors.

Aside from these basic symbols, we will introduce a few more types in later chapters to represent more advanced programming operations. For the time being, this set is adequate.

FIGURE 6.4
Illustration of the use of on-page and off-page connectors. The dashed lines would not actually be a part of the charts but are used here to show how control is transferred between connectors.

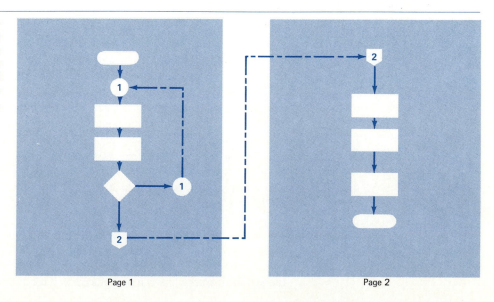

Page 1   Page 2

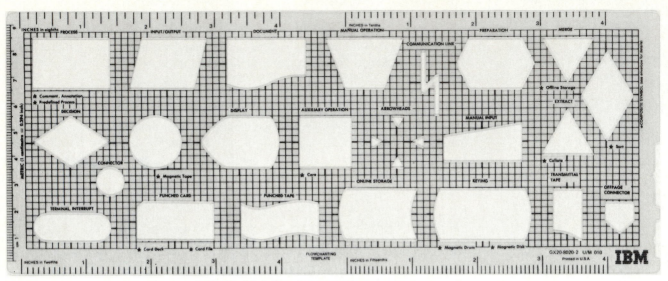

FIGURE 6.5 A flowchart template.

The construction of flowcharts is facilitated by use of a plastic template (Fig. 6.5). This stencil includes the standard symbols for constructing flowcharts. Templates are available at most office-supply shops and college bookstores.

6.2.2   Levels of Flowchart Complexity

There are no set standards for how detailed a flowchart should be. During the development of a major algorithm, you will often draft flowcharts of various levels of complexity. As an example, Fig. 6.6 depicts a hierarchy of three charts that might be used in the development of an algorithm to determine the grade-point average (GPA) or cumulative index of a student engineer. The system flowchart (Fig. 6.6*a*) delineates the big picture. It identifies the major tasks, or *modules,* and the sequence that is required to solve the entire problem. Such an overview flowchart is invaluable for ensuring that the total scheme is comprehensive enough to be successful.

Next the major modules can be broken off and charted in greater detail (Fig. 6.6*b*). Finally, it is sometimes advantageous to break the major modules down into even more manageable units, as in Fig. 6.6*c*. The process of subdividing an algorithm into major segments and then breaking these down into successively more refined modules is referred to as *top-down design*. We will return to this approach in the next chapter.

In the following sections we will study and improve the flowchart in Fig. 6.6*c*. This will be done to illustrate how the symbols in Fig. 6.3 can be orchestrated to develop an efficient and effective algorithm. In the process we will demonstrate how the power of computers can be reduced to three fundamental types of operations. The

FIGURE 6.6
A hierarchy of flowcharts dealing with the problem of determining a student's grade-point average. The system flowchart in (a) provides a comprehensive plan. A more detailed chart for a major module is shown in (b). An even more detailed chart for a segment of the major module is seen in (c). This way of approaching a problem is called *top-down design.*

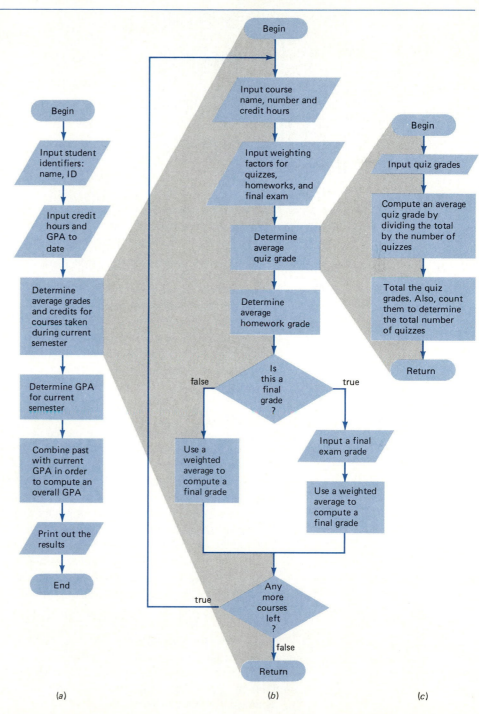

(a)

(b)

(c)

FIGURE 6.7
An example of an endless loop.
(*Kelly-Bootle, 1981.*) This loop will
persist forever because it
has no exit.

first, which should already be apparent, is that computer logic can proceed in a definite *sequence*. As depicted in Fig. 6.6*a*, this is represented by the flow of logic from box to box. The other two fundamental operations are *selection* and *repetition*.

6.2.3 Selection and Repetition in Flowcharts

Notice that the operations in Fig. 6.6*a* and *c* specify a simple sequential flow. That is, the operations follow one after the other. However, notice how the first diamond-shaped decision symbol in Fig. 6.6*b* permits the flow to diverge and follow one of two possible paths, or branches, depending on the answer to a question. For this case, if final grades are being determined, the flow follows the right branch, whereas if intermediate or midterm grades are being determined, the flow follows the left branch. This type of flowchart operation is called *selection*.

Aside from depicting a conditional branch, Fig. 6.6*b* demonstrates another flowchart operation that greatly increases the power of computer programs—*repetition*. Notice that the "true" branch of the second decision symbol allows the flow to return to the head end of the module and repeat the calculation process for another course. This ability to perform repetitive tasks is among the most important strengths of computers. The operation of "looping" back to repeat a set of operations is accordingly named a *loop* in computer jargon. The most elementary example is the endless or infinite loop in Fig. 6.7. Because it has no exit, this loop will persist interminably. In contrast, the presence of the diamond-shaped decision symbol permits the termination of the loop in Fig. 6.6*b* once all courses have been evaluated.

The operations of selection and repetition can be employed to great advantage when developing algorithms. This will be demonstrated in the next section where we will develop an improved version of Fig. 6.6*b*.

6.2.4 Accumulators and Counters in Flowcharts

In Fig. 6.6*c* we specified a box to total the quiz grades and keep count of the total number of quizzes. Now that you understand the notions of decisions and loops, we can develop a more detailed flowchart to accomplish these objectives (Fig. 6.8). In order to appreciate the more detailed version, we must introduce the concepts of accumulators and counters.

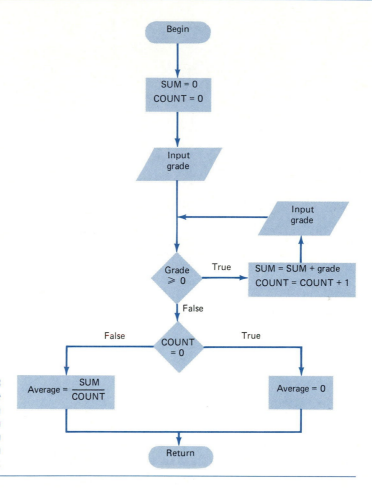

FIGURE 6.8
A detailed flowchart to compute the average quiz grade. Through the use of an accumulator and a counter, this example improves on the simple version shown in Fig. 6.6c.

An *accumulator* is a program-designated storage location that accepts and stores a running total of individual values as they become available during the operation of a program. In Fig. 6.8 the accumulator is given the symbolic name SUM. Notice that the contents of the storage location are initially set to zero. Then with each pass through the loop a new quiz grade is added to SUM in order to keep a running total.

A *counter* is merely a special type of accumulator that is employed to keep track of the number of passes through the loop. In Fig. 6.8 the symbolic name COUNT is used to represent the counter. As with the accumulator, the counter is initialized to zero. Then with each pass through the loop, 1 is added to COUNT in order to keep a running total of the number of passes. For Fig. 6.8 this corresponds to the number of quizzes that have been summed. Then, upon exit from the loop (which is triggered by entering a negative value for GRADE), SUM is divided by COUNT to determine the average quiz grade.

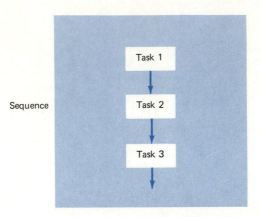

Sequence

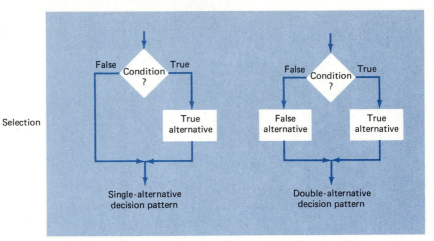

Selection

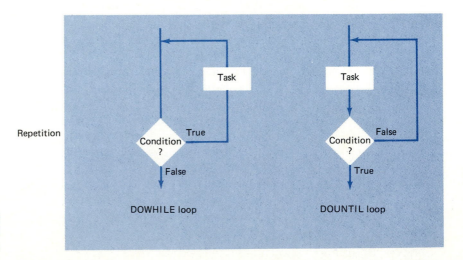

Repetition

FIGURE 6.9
The three fundamental control structures.

6.3   STRUCTURED FLOWCHARTS

Flowcharting symbols can be orchestrated to construct logic patterns of extreme complexity. However, programming experts have recognized that without rules flowcharts can become so complicated that they impede rather than facilitate clear thinking.

One of the main culprits that contributes to algorithm complexity is the unconditional transfer, also known as the GOTO. As the name implies, this control structure allows you to "go to" any other location in the algorithm. For flowcharts, it is simply represented by an arrow. Indiscriminate use of unconditional transfers can lead to algorithms that, at a distance, resemble a plate of spaghetti.

In an effort to avoid this dilemma, advocates of structured programming have demonstrated that any program can be constructed with the three fundamental control structures depicted in Fig. 6.9. As can be seen, they correspond to the three operations—sequence, selection, and repetition—that we have just been discussing.

All flowcharts should be composed of the three fundamental control structures shown in Fig. 6.9. If we limit ourselves to these structures and avoid GOTOs, the resulting algorithms will be much easier to understand.

6.4   PSEUDOCODE

The object of flowcharting is obviously to develop a quality computer program. As we will discuss in Parts Four and Five, a program is merely a set of step-by-step instructions to direct the computer to perform tasks. These instructions are also called *code*. Inspection of Figs. 6.6 and 6.8 illustrates how top-down flowcharting moves in the direction of a computer program. Recall that in top-down design, more detailed flowcharts are developed as we progress from the general overview (Fig. 6.6a) to specific modules (Fig. 6.6c). At the highest level of detail, the module flowcharts are almost in the form of a computer program. That is, as seen in Fig. 6.8, each flowchart compartment represents a single well-defined instruction to the computer.

An alternative way to express an algorithm that bridges the gap between flowcharts and computer code is called *pseudocode*. This technique uses Englishlike statements in place of the graphical symbols of the flowchart. Figure 6.10 shows pseudocode representations for the fundamental control structures.

Keywords such as BEGIN, IF, DOWHILE, etc., are capitalized, whereas the conditions, processing steps, and tasks are in lowercase. Additionally, notice that the processing steps are indented. Thus the keywords form a "sandwich" around the steps to visually define the extent of each control structure.

Figure 6.11 shows a pseudocode representation of the flowchart from Fig. 6.8. This version looks much more like a computer program than the flowchart. Thus one advantage of pseudocode is that it is easier to develop a program with it than with a flowchart. The pseudocode is also easier to modify. However, because of their visual form, flowcharts sometimes are better suited for designing algorithms that are complex. Pseudocode will not be used extensively in this book. However, because of its application for communicating algorithms, you should be familiar with it and capable of using it to develop a computer program.

Sequence

BEGIN *task*
  Task
END *task*

Selection

IF *condition*   IF *condition*
  True alternative   *True alternative*
ELSE ENDIF   ELSE
                                          False alternative
                                        ENDIF

Single-alternative   Double-alternative
decision pattern   decision pattern

Repetition

DOWHILE *condition*   DOUNTIL *condition*
  Task   *Task*
END DO   END DO

DOWHILE loop   DOUNTIL loop

FIGURE 6.10
Pseudocode representations of the
three fundamental control
structures.

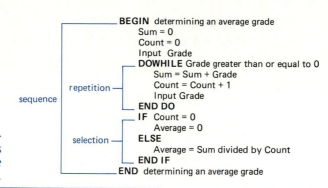

FIGURE 6.11
Pseudocode to compute the average quiz grade; this algorithm was represented previously by the flowchart in Fig. 6.8.

PROBLEMS

6.1 List and define six steps involved in the production of high-quality software in engineering.

6.2 What is an algorithm and why is it so critical to the effective development of high-quality software?

6.3 You are at the corner of 2nd Street and Avenue C (Fig. P6.3). Write an algorithm to direct a motorist to take the shortest route to 6th Street and Avenue D.

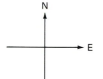

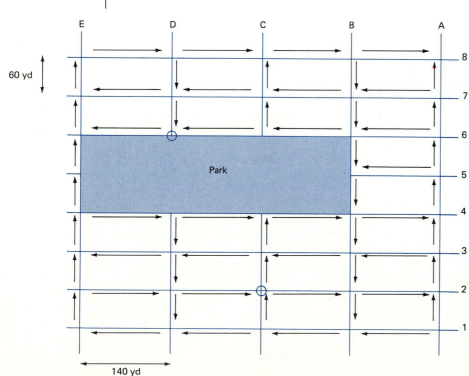

FIGURE P6.3
A street map for the business district of a city. The arrows designate one-way traffic flows.

6.4 A different number is written on each of ten index cards. Write an algorithm to sort through these cards in order to determine the lowest number.

6.5 Using the fundamental control structures from Fig. 6.9, write a flowchart for Prob. 6.4.

6.6 A value for the plasticity index of soil samples is recorded on each of a set of index cards. A card marked ''end of data'' is placed at the end of the set. Write an algorithm to determine the sum and the average of these values.

6.7 Using the fundamental control structures from Fig. 6.9, construct a flowchart for Prob. 6.6.

6.8 Write an algorithm to calculate and print out the real roots of a quadratic equation

$$ax^2 + bx + c = 0$$

where a, b, and c are real coefficients. The formula used to calculate roots is the quadratic formula

$$x = \frac{-b \pm \sqrt{b^2 - 4ac}}{2a}$$

Note that if the quantity within the square root sign (that is, the determinant) is negative, the root is complex. Also note that division by zero occurs if $a = 0$. Design your algorithm so that it deals with these contingencies by printing out an error message.

6.9 Using the fundamental control structures from Fig. 6.9, construct a flowchart for Prob. 6.8.

6.10 The exponential function e^x can be evaluated by the following infinite series:

$$e^x = 1 + x + \frac{x^2}{2} + \frac{x^3}{3!} + \frac{x^4}{4!} + \cdots$$

Write an algorithm to implement this formula so that it computes and prints out the values of e^x as each term in the series is added. In other words, compute and print in sequence the values for

$e^x \cong 1$
$e^x \cong 1 + x$
$e^x \cong 1 + x + x^2/2$
.
.
.

up to the order term of your choosing. For each of the above, compute and print out the percent relative error as

$$\% \text{ error} = \frac{\text{true} - \text{series approximation}}{\text{true}} \times 100\%$$

6.11 Using the fundamental control structures from Fig. 6.9, construct a flowchart for Prob. 6.10.

6.12 Figure P6.12 shows the reverse of a checking account statement. The bank has developed this sheet to help you balance your checkbook. If you look at it closely you will realize that it is an algorithm. Develop a step-by-step algorithm to accomplish the same task.

THE AREA BELOW IS PROVIDED TO HELP YOU BALANCE YOUR CHECKBOOK

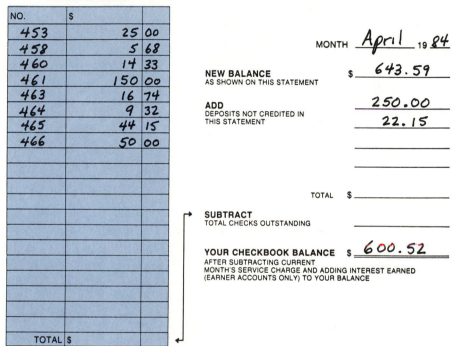

FIGURE P6.12
A form used to balance a checkbook.

6.13 Using the fundamental control structures from Fig. 6.9, construct a flowchart for Prob. 6.12.

6.14 Construct a pseudocode for Prob. 6.12.

CHAPTER 7
STRUCTURED DESIGN AND PROGRAMMING

Learning to program a computer is a little like learning to play the guitar. It is relatively easy to pick up the few chords that are required to plunk out a passable tune. However, disciplined training and practice are required to master the effortless technique and expressiveness of the accomplished musician. Before learning to program, it is important that you be aware of some general concepts and principles that will contribute to the quality of your efforts.

7.1 PROGRAMMING STYLE

In the early years of computing, programming was somewhat more of a craft than an art. Although all languages had very precise vocabularies and grammars, there were no standard definitions of good programming style. Consequently, individuals developed their own criteria for what constituted an excellent piece of software.

In the early years, these criteria were strongly influenced by the available hardware. Recall from Chap. 2 that the early computers were expensive and slow and had small memories. As a consequence, one mark of a good program was that it utilized as little memory as possible. Another might be that it executed rapidly. Still another was how quickly it could be written the first time.

The ultimate result was that programs were highly nonuniform. In particular, because they were strongly influenced by hardware limitations, programmers put little premium on how easy the programs might be to use and to maintain. Thus, if you wanted to effectively utilize and maintain a large program over long periods, you usually had to keep the original programmer close by. In other words, the program and the programmer formed a costly package.

Today many of the hardware limitations do not exist or are rapidly disappearing. In addition, as software costs constitute a proportionally larger fraction of the total computing budget (recall Fig. 4.1), there is a high premium on developing software that is easy to use and modify. In particular, there is a strong emphasis on clarity and readability rather than on short, terse, noncommunicative code.

Computer scientists have systematically studied the factors and procedures needed to develop high-quality software of this kind. Although the resulting methods are somewhat oriented toward large-scale programming efforts, most of the general techniques are also extremely useful for the types of programs that engineers develop routinely in the course of their work. Collectively, we will call these techniques *structured design and programming*. In the present chapter, we discuss three of these approaches—modular design, top-down design, and structured programming.

7.2 MODULAR DESIGN

Imagine how difficult it would be to study a textbook that had no chapters, sections, or paragraphs. Breaking complicated tasks or subjects into more manageable parts is one way to make them easier to handle. In the same spirit, computer programs can be divided into smaller subprograms, or *modules,* that can be developed and tested separately. This approach is called *modular design*.

The most important attribute of modules is that they be as independent and self-contained as possible. In addition, they are typically designed to perform a specific, well-defined function. As such, they are usually short (less than 50 lines) and highly focused.

One of the primary programming elements used to represent each module is the subroutine. A *subroutine* is a series of computer instructions that together perform a given task. A *calling,* or *main, program* invokes these modules as they are needed. Thus the main program orchestrates each of the parts in a logical fashion.

Modules have one input and one output. Decisions can be made within a module that impact the overall flow, but the branching itself should occur in the main program.

Modular design has a number of advantages. The use of small, self-contained units makes the underlying logic easier to devise and to understand for both the developer and the user. Development is facilitated because each module can be perfected in isolation. In fact, for large projects different programmers can work on individual parts. In the same spirit, you can maintain your own library of useful modules for later use in other programs. Modular design also increases the ease with which a program can be debugged and tested because errors can be more easily isolated. Finally, program maintenance and modification are facilitated. This is primarily due to the fact that new modules can be developed to perform additional tasks and then easily incorporated into the already coherent and organized scheme.

7.3   TOP-DOWN DESIGN

Although the previous section introduced the notion of module design, it did not specify how the actual modules are identified. The top-down design process is a particularly effective approach for accomplishing this objective.

Top-down design is a systematic development process that begins with the most general statement of a program's objectives and then successively divides it into more detailed segments. Thus the design proceeds from the general to the specific. We have already introduced this approach in our discussion of flowcharting (recall Fig. 6.6).

Most top-down designs start with an English description of the program in its most general terms. This description is then broken down into a few elements that define the program's major functions. Then each major element is divided into subelements. This procedure is continued until well-defined modules have been identified.

Besides providing a procedure for dividing the algorithm into well-defined units, top-down design has other benefits. In particular, it ensures that the finished product is comprehensive. By starting with a broad definition and progressively adding detail, the programmer is less likely to overlook important operations.

7.4   STRUCTURED PROGRAMMING

The previous two sections have dealt with effective ways to organize a program. In both cases, emphasis was placed on *what* the program is supposed to do. Although this usually guarantees that the program will be coherent and neatly organized, it does not ensure that the actual program instructions or code for each module will manifest an associated clarity and visual appeal. Structured programming, on the other hand, deals with *how* the actual program code is developed so that it is easy to understand, correct, and modify.

Not everyone agrees as to what constitutes structured programming. In the present context, we define it as a set of rules that prescribes good style habits for the programmer. The following are the most generally agreed upon rules:

1. Programs should consist solely of the three fundamental control structures of sequence, selection, and repetition (recall Fig. 6.9). Computer scientists have proved that any program can be constructed from these basic structures.
2. Each of the structures must have only one entrance and one exit.
3. Unconditional transfers (GOTOs) should be avoided.
4. The structures should be clearly identified with comments and visual devices such as indentation, blank lines, and blank spaces.

Although these rules may appear deceptively simple, they are extremely powerful tools for increasing program clarity. In particular, adherence to these rules avoids the spaghettilike code that is the hallmark of indiscriminate branching. Because of their benefits, these rules and other structured programming practices will be stressed continually in subsequent parts of this book.

```
10  S=0
20  I=0
30  INPUT G
40  IF G<0 THEN 80
50  S=S+G
60  I=I+1
70  GOTO 30
80  IF I=0 THEN 110
90  A=S/I
100 GOTO 120
110 A=0
120 PRINT A
130 END
```

(a)

```
100 REM  ***************************************************
110 REM  *              AVERAGE GRADE PROGRAM               *
120 REM  *                 (BASIC version)                  *
130 REM  *                 by  S. C. Chapra                 *
140 REM  *                                                  *
150 REM  *       This program determines the average        *
160 REM  *       value for a set of grades.                 *
170 REM  ***************************************************
180 REM
190 REM       .---------------------------------------.
200 REM       |          DEFINITION OF VARIABLES       |
210 REM       |                                        |
220 REM       |     GRADE = GRADE INPUT BY USER        |
230 REM       |     TOTAL = SUMMATION OF GRADES        |
240 REM       |     COUNT = NUMBER OF GRADES           |
250 REM       |     AVERAG = AVERAGE GRADE             |
260 REM       |                                        |
270 REM       '---------------------------------------'
280 REM  ****************** MAIN PROGRAM ******************
290 REM
300 LET GRADE = 0
310 LET TOTAL = 0
320 LET COUNT = 0
330 REM                     Accumulate Grade Data
340 REM
350 INPUT "GRADE = (ENTER NEGATIVE TO TERMINATE)";GRADE
360 WHILE GRADE >= 0
370     TOTAL = TOTAL + GRADE
380     COUNT = COUNT + 1
390     INPUT "GRADE = (ENTER NEGATIVE TO TERMINATE)";GRADE
400 WEND
410 REM                 Calculate Average Grade
420 REM
430 IF COUNT = 0 THEN 440 ELSE 460
440         AVERAG = 0
450         GOTO 480
460     REM ELSE
470         AVERAG = TOTAL/COUNT
480 REM ENDIF
490 PRINT "AVERAGE GRADE = ";AVERAG
500 END
```

(b)

FIGURE 7.1

(a) An unstructured and (b) a structured version of the same program. Both versions perform identical computations but differ markedly in terms of clarity. Note that the GOTO in line 450 of (b) had to be used in order to program a double-alternative decision in this dialect of BASIC. Otherwise, GOTOs should be avoided.

Figure 7.1 is a dramatic example of the contrast between unstructured and structured approaches. Both programs are designed to determine an average grade, the algorithm for which was discussed in Chap. 6. Although the programs compute identical values, their designs are in marked contrast. The purpose, logic, and organization of the unstructured program in Fig. 7.1a is not obvious. The user must delve into the code on a line-by-line basis to decipher the program. In contrast, the major parts of Fig. 7.1b are immediately apparent. Title and variable-identification modules are placed at the beginning to orient the user. Program execution progresses in an orderly fashion from the top to the bottom with no major jumps. Every segment has one entrance and one exit, and each is clearly delineated by comments, spaces, and indentation. In essence, the program is distinguished by a visual clarity that is reminiscent of a well-designed flowchart. Additionally, the program is a natural evolution of the pseudocode shown in Fig. 6.11.

On the debit side, there is no question that Fig. 7.1b takes longer to develop than Fig. 7.1a. Because many pragmatic engineers are accustomed to "getting the job done" in an efficient and timely manner, the extra effort might not always be justified. Such would be the case for a short program you might develop for a "one-time-only" computation. However, for larger more important programs that are intended for applications by yourself or others, there is no question that a structured approach will yield long-term benefits.

7.5   FINAL COMMENTS BEFORE YOU BEGIN TO PROGRAM

In his book *Megatrends,* Robert Naisbitt states that to be truly successful in the coming years Americans "will have to be trilingual: fluent in English, Spanish and computers." Aside from its obvious commentary on our shifting demography, Naisbitt's statement is interesting in its characterization of computer use as a linguistic process.

In fact, learning to program a computer is a lot like mastering a foreign language. One insight that can be gained from this analogy is that an efficient way to learn a foreign tongue is to use it to write and speak. Similarly, the best way to learn programming is to actually compose computer programs. The initial chapters in the next two parts of the book are designed to provide you with sufficient information to write problem-solving computer programs. With this knowledge as background, you can progress to the more advanced concepts covered in the subsequent chapters.

Although our primary intent is to get you on the machine as rapidly as possible, we must note one very important distinction between foreign and computer languages that will have a large bearing on your success. When speaking in a foreign language, your conversation will be with another human being. Therefore you can express yourself imperfectly and still be understood. The other person can interpret your tone of voice, your gestures, and the look in your eyes, and no matter how much you mangle the language's grammar and syntax, some communication is usually possible.

In contrast, you will be conducting your computer conversations with a machine that is incapable of gazing deeply into your eyes to search out what you are really trying to say. Although this might seem almost ridiculously obvious, it is one of the most important initial insights that must be grasped by the novice programmer. That is, in order to effectively communicate with the computer, you must express yourself precisely. If you don't, your programs will simply not work.

Fortunately, the rules of most high-level programming languages are easy to learn. This is due, in part, to the very fact that they are designed for communication with machines. Thus they do not involve the nuances and exceptions that represent the primary stumbling blocks to learning foreign languages. As described in the next two parts of the book, BASIC and FORTRAN are particularly well designed in this regard.

With this as background, we will now venture into the actual mechanics of composing a program in a high-level language. *Part Four* is devoted to the primary language of the personal computer—BASIC. *Part Five* deals with FORTRAN, which has historically been the major program language of engineers and scientists working on mainframe computers.

PROBLEMS

7.1   Define the following terms:
(a) Structured programming

(b) Modular design
(c) Subroutine
(d) Top-down design

7.2 Compare and contrast the learning of a computer language and a foreign language.

7.3 Name some changes in programming that have occurred because of the changing technology in hardware.

7.4 What has made structured programming techniques so helpful in the economical development of software?

PART FOUR

PROGRAMMING IN BASIC

After the development of a sound and comprehensive plan, the next step in the program-development process is to compose the program using a high-level language. This part of the book shows how BASIC is used for this purpose.

Chapter 8 is devoted to the most elementary information regarding BASIC programming. This material is designed to get you started by providing information sufficient to successfully write and execute simple programs. Chapters 9 through 14 cover additional material that will allow you to develop more complicated and powerful programs.

CHAPTER8
THE FUNDAMENTALS OF BASIC

As mentioned in Chap. 5, John Kemeny and Thomas Kurtz developed BASIC in the mid-1960s as a learning language. Although the ensuing years have seen the language evolve into a powerful problem-solving tool, it still retains most of the characteristics that originally made it easy to understand and use. The two most important of these characteristics are its interactive nature and its simplicity.

The fact that BASIC is an interactive, interpreted language has a number of benefits. From a learning perspective, it means that you receive instant feedback, since the computer interprets each program instruction immediately after it is entered. Thus you can progress rapidly in a trial-and-error fashion. This is in contrast to languages where the entire program must be completed, compiled, and executed before the user is provided feedback regarding success or failure.

The second strength of BASIC as a learning language is its simplicity. BASIC has a straightforward Englishlike vocabulary that is communicative and unintimidating. Consequently, you can compose working computer programs without having to wrestle with an alien vocabulary and syntax. The experience is analogous to learning to drive a car with a standard versus an automatic transmission. Over and above all the other skills required to operate an automobile, novices learning to drive with a standard transmission must simultaneously worry about shifting gears and mastering the clutch. In contrast, novices learning to operate a car with an automatic transmission can focus their attention on the more fundamental aspects of driving such as accelerating, stopping, and avoiding solid objects. BASIC is just like driving a car with an automatic transmission. Within a very short time, you will be ''on the road'' to successfully developing meaningful programs for engineering problem solving.

8.1 WHY BASIC?

Aside from the fact that it is easy to learn, BASIC is included in this text because of its widespread availability on personal computers and its utility for obtaining rapid answers to even moderately complex engineering questions. Although the syntax of BASIC may differ from that of other computer languages, it has most of their fundamental elements. Thus, once BASIC is mastered, learning a second or third language should present little difficulty.

The widespread availability of BASIC on personal computers can at least partially be ascribed to its "being in the right place at the right time." Most of the early microcomputers had limited amounts of memory and processing power. As a consequence, they could not accommodate a compiled language such as FORTRAN. In contrast, BASIC interpreters were more compatible with the limited capabilities of the early micros. Additionally, the "friendly" qualities of interactive BASIC made it philosophically consistent with a hands-on, personal computing environment. Consequently, the great majority of software developed for personal computers was composed in BASIC. Because the use of computer languages is to a certain extent self-perpetuating, a vast library of software is presently available in BASIC.

BASIC has been faulted for having deficiencies, but many of these can be overcome by the programmer. The primary complaint is that it does not enforce structure. As already mentioned in Chap. 7, lack of structure can lead to programs that are tangled webs of logic. However, most recent dialects of BASIC *can* be programmed in a structured, modular fashion. Although BASIC does not require that this be done, it is easily accomplished, and the result is virtually as satisfactory as with a language such as Pascal that does enforce structure. All the examples in this book will use structured programming practices. Since you are learning to program, you might as well learn to do it right. As you will see soon enough, the long-term benefits are considerable.

Darrell Fichtl expresses it nicely in the *Whole Earth Software Catalog:*

Let's set the record straight. I've worked with FORTRAN and own a C, a Pascal, and a BASIC compiler. All these work exceptionally well, but I like BASIC—it's the Chevy of the computer business. You'll also hear BASIC is sloppy. That depends on the person doing the programming. The impression that nothing "serious" can be written in BASIC is totally erroneous. If you do a cross-section of programs currently on the market, you'll find that a good percentage of them are written in BASIC. In BASIC, you can make an efficient program that is a joy to work with. It depends totally on you.

8.2 WHICH BASIC?

At present there are many versions of BASIC. Different computers often employ their own brand of the language. In addition, an evolution has occurred over time as new capabilities are incorporated into the various versions. Thus, as with any living language, BASIC has many dialects.

In an effort to enforce some uniformity on the language, the American National Standards Institute (ANSI) has proposed two standard versions: a minimal and an extended BASIC. Unfortunately, the minimal version is so simple that it has been extended by nearly every BASIC dialect. Thus, if we described ANSI minimal BASIC, you would be left far short of tapping the capabilities available on your machine. In contrast, the ANSI extended version is quite comprehensive and has many attractive attributes, not the least of which is that it provides for fully structured programs. Unfortunately, to date most of the widely available dialects have not evolved to the point that they are up to the ANSI extended standard. Thus if we adhered strictly to the extended version, many of you would be frustrated by the limitations of your dialect.

For all the above reasons, we have chosen to focus the following description on the Microsoft BASICA employed on the IBM PC. We chose this dialect because of its widespread availability and because it is a fairly "typical" BASIC. From here on, we will refer to this IBM PC BASICA as Microsoft BASIC.

Because of IBM's predominant position in the marketplace, its Microsoft BASIC is as near to an industry standard as is presently available. Machines such as the Macintosh and the TRS-80 all use BASICs that are similar to the IBM version. Thus our choice of this dialect encompasses a great many more machines than just those with the IBM label.

As might be expected, the other major BASIC presently in use is the one employed on the popular Apple II computers. This dialect, which is called Applesoft, differs in many ways from IBM BASIC. Other popular computers, such as the Commodore, utilize a BASIC that is quite similar to Applesoft.

Significant differences between Microsoft BASIC and other popular dialects are tabulated at appropriate places in the text.* Thus users of these dialects should be able to employ this book to learn BASIC. In addition, because engineers have occasion to use several systems in the course of their work, this material serves as a reference for programming in BASIC on these machines.

8.3   "TALKING" TO THE COMPUTER IN BASIC

Before you begin to program, you must perform two tasks. You must format a diskette,† and you must make sure that BASIC is available for your use in the computer's primary memory. Procedures for implementing both tasks can be found in your user's manual. The former task is required so that you can retain your finished programs. The latter is necessary because, as mentioned in Secs. 4.3.3 and 4.4, a BASIC language translator or interpreter must be loaded into primary memory so that the translator can convert your BASIC commands into machine language. For some computers, BASIC is in ROM and therefore is automatically ready for use when the

---

*Although our description focuses on BASICA for the IBM PC, material on the following personal computer systems is also included: Apple II, Commodore, Macintosh, and TRS-80.

†Some computers use tape to save programs and data. Although the IBM PC can use tape, it is a slow storage medium in contrast to a disk because the tape has to proceed sequentially to locate information. In any event, the steps required to record a program are similar, even though the medium is different.

machine is turned on. For the IBM PC, BASIC or BASICA can be loaded into primary memory by the disk operating system. The procedure is described in Appendix C.

Once the BASIC interpreter is in primary memory, you can begin to develop a computer program. As mentioned previously, a program is a set of instructions to tell a computer what to do. In BASIC, each instruction is usually written as a separate line. To input an instruction to the computer, you type it as a line on the keyboard. On most personal computers, the instructions are displayed simultaneously on the monitor. If a hard copy is required, the instructions can also be output to a printer.

After typing an individual line, the ENTER key (which on some systems is called RETURN or is marked with a special symbol such as ◄┘) is struck in order to advance to the next line. This act causes the line to be "entered" into the computer. In the remainder of the book, we use the symbol ® to represent the act of striking the ENTER key.

Instructions can be typed in two fundamental ways in BASIC: programming (or indirect) mode and immediate (or direct) mode. *Programming mode*, as the name implies, is used for entering the instructions that compose a computer program. Each instruction is identified by a line number followed by the body of the instruction in the BASIC language. Figure 8.1 shows six lines that constitute a very simple BASIC program to add two numbers. Each of these lines is referred to as a *statement*. The line numbers are used to identify the statements and to help determine the order in which the computer executes them. Within one program, every line number must be unique and must be a whole number. No decimal numbers can be used. It is good practice to start numbering BASIC programs at 100 and to number the lines in increments of 10 or 20 when the program is initially written, as is done in Fig. 8.1. This allows additional lines to be easily added between existing lines and at the beginning of the program. BASIC progresses from one line to the one with the next-higher line number, unless programmed to do otherwise. Thus the order of typing the lines does not determine their order of execution by the computer.

In contrast to the program mode, *immediate mode* instructions, which are called *commands*, are unnumbered. Their primary characteristic is that they are carried out or executed immediately after they are entered. This allows you to see the effect of the command at once. For example, if we type the word PRINT followed by a message in quotes such as

```
PRINT "HOWDY" ®
```

the computer will immediately print the message

```
HOWDY
OK
```

FIGURE 8.1
A simple six-line BASIC program to add two numbers.

```
100 REM      SIMPLE ADDITION PROGRAM
110 LET A = 28
120 LET B = 15
130 LET C = A + B
140 PRINT C
150 END
```

The same sort of result occurs for a number:

```
PRINT 55 ®
55
Ok
```

Note that for the number the quotes are not employed. In both examples, the ''Ok'' signifies that the instruction has been implemented without encountering a BASIC syntax error and that the computer is ready for another instruction.

The immediate mode is sometimes referred to as the ''calculator mode'' because, when it is used, the computer behaves like a high-powered calculator. For example, we could add two numbers, as in

```
PRINT 28 + 15 ®
43
Ok
```

The disadvantage of using a BASIC statement without a line number is that it is not retained by the computer for later use. In contrast, the numbered BASIC statements of the program mode are retained and can be reused, modified, and made to work in concert with other lines. That is, they can be made into a computer program.

Although unnumbered commands will not be used in the actual body of a computer program, they serve an important role in BASIC. As described in the next section, they are used to direct the computer to perform a number of tasks related to the proper execution of the program.

8.4   HOW TO PREPARE AND EXECUTE BASIC PROGRAMS

System commands are unnumbered instructions that direct the computer to take immediate action to perform specific tasks pertaining to the program. Some of the commands available in Microsoft BASIC are LIST, SAVE, LOAD, FILES, KILL, NEW, RUN, DELETE, EDIT, RENUM, AUTO, and MERGE. As outlined in Table 8.1, these commands provide you with tools you will need to create, store, test, and run programs. The next sections provide detailed descriptions of each command.

8.4.1   Viewing Programs

The LIST command.   After typing in a number of instructions, you might want to list the program to check it for mistakes. The LIST command is used for this purpose, as in

```
LIST
LIST -130
LIST 110-
LIST 110-130
```

TABLE 8.1 BASIC System Commands

Purpose	Command		
	IBM PC	Applesoft	Your Computer
Allows all or part of a program to be listed on the monitor	LIST LIST 10-40	LIST LIST 10,40	
Saves a program on diskette	SAVE "SIMPLADD"	SAVE SIMPLADD	
Loads a program from the diskette back to primary memory	LOAD "SIMPLADD"	LOAD SIMPLADD	
Displays the names of files on your diskette	FILES	CATALOG	
Deletes a program from the diskette	KILL "SIMPLADD.BAS"	DELETE SIMPLADD	
Deletes a program currently in primary memory	NEW	NEW	
Executes the program currently in primary memory	RUN	RUN	
Deletes a program line	DELETE 10	DEL 10,10	
Deletes program lines	DELETE 10-40	DEL 10,40	
Allows editing of a portion of a program line without retyping the entire line	EDIT 10	Not standard*	
Allows renumbering of the lines of the program in primary memory	RENUM	Not standard*	
Generates a line number automatically after every carriage return	AUTO	Not standard*	
Allows two files to be merged; provides a means to combine programs	MERGE "SIMPLMLT"	Not standard*	

*This capability may usually be acquired as a utility program.

These are four ways to use the LIST command. In the first case, the whole program will be displayed on the screen. LIST -130 causes all of the lines up to and including line 130 to be displayed, whereas LIST 110- causes the lines from 110 to the end to be shown. Lastly, LIST 110-130 displays lines 110 through 130. You can test these commands by trying them out on the program in Fig. 8.1. The main idea is that the LIST command permits you to view your program in convenient portions.

Note that depressing the CTRL and the BREAK keys simultaneously will interrupt the listing and return you to the command level. This action can also be employed to interrupt some of the other commands. The CTRL and C keys are sometimes used for the same purpose on other personal computers.

8.4.2 Retaining Programs

The SAVE command.   Once an initial version of a program has been typed, it is important to record it permanently on a diskette. Saving the program prior to executing it is vital! Although it does not happen frequently, there are instances when your computer could stop responding to the keyboard (and you) during program execution. If this occurs, your only recourse may be to turn off the machine and start over again, in which case you would have lost the program you were working on. To guard against this possibility, it is critical to record major revisions of your program prior to testing. This can be done with the SAVE command:

```
SAVE  "program name"
```

where *program name* is a descriptive name that you assign to the program. For example, you could name the program from Fig. 8.1 SIMPLADD, as a contraction for ''the simple addition program.'' Thus to save this program on diskette you would type

```
SAVE  "SIMPLADD"
```

When this command is implemented, the file will be stored on the disk in a compressed binary format. This file does not contain a readable version of the program unless you are working in BASIC. The program is stored in a compact form that is more rapidly transferred between RAM and the diskette and in a form that is ready to be executed by BASIC. An alternative option, available on the IBM PC, is to store it in ASCII format by adding an A, as in

```
SAVE  "SIMPLADD",A
```

Although the ASCII format takes up more room on the diskette, it is sometimes advantageous to store programs in this way because some disk-access operations require it. For example, as we will describe in Sec. 8.4.7, the MERGE command requires that a program be stored in ASCII. In addition, if you want to edit your program with a word processing package such as WordStar, you must store it in ASCII format.

Program names.   Program names on the IBM PC can have a maximum of eight characters prior to a period and three after. The three characters after the period are called the *extension.* In our SIMPLADD example, BASIC supplies the extension BAS, which indicates that the program is written in BASIC. If you consult the file directory on your diskette (this can be done with the FILES command as described below), you will find the program listed as

```
SIMPLADD.BAS
```

The LOAD command.   Once the program is successfully saved on diskette, it will be retained after the computer is turned off. The LOAD command is used to bring the program back from the disk into primary memory. LOAD clears any program currently in memory before bringing a new program into RAM memory. For the example program from Fig. 8.1, this is done by typing

```
LOAD  "SIMPLADD"
```

Once RETURN is struck, the simple addition program is loaded into primary memory. You could type LIST ® at this point to verify that this has occurred.

The FILES command. The command FILES is used to list the names of the files stored on your diskette. If you merely type in FILES, all the files will be listed. In order to limit the listing to your BASIC programs you would type

```
FILES "*.BAS" ®
```

The asterisk is sometimes referred to as a ''wild card'' in that it is taken to represent any and every file name on your disk. Thus every BASIC file—that is, every file with an extension of BAS—would be displayed.

8.4.3   Deleting Programs

The KILL command. A final operation dealing with disk storage of programs is to remove a program from the disk. To do this, the KILL command is used, as in

```
KILL "program name.BAS"
```

or, for the simple addition problem from Fig. 8.1,

```
KILL "SIMPLADD.BAS"
```

Notice how in contrast to the SAVE and LOAD commands, the KILL command requires that you include the period and the three-letter extension BAS after the program name.

The NEW command. A related command, NEW, is used to delete the program that is currently in primary memory. This command is useful when you want to wipe the primary memory clean so that you can start typing a new program. NEW does not affect copies of programs already stored on the diskette. Obviously, any command that erases programs or files such as NEW or KILL must be used carefully lest you inadvertently delete a valuable file.

8.4.4   Executing Programs

The RUN Command. Now comes the crucial test: Does the program work? This can be verified by executing the program with the RUN command. With the simple addition program in primary memory, implementing the RUN command causes the computer to start at the first numbered line and to execute each line in sequence until the END statement is encountered. The result for the program in Fig. 8.1 looks like this:

```
RUN ®
43
Ok
```

Thus the program adds the two numbers and prints out the result, 43.

If the simple addition program is saved on the diskette but not in primary memory, a RUN command can be given to both move it into RAM and begin its execution, as in

```
RUN "SIMPLADD" ®
43
OK
```

The program to be loaded and run is given in quotations following the RUN command. This form of the command clears primary memory prior to bringing in the specified program. RUN with a file name is an abbreviated way of saying

```
LOAD "program name" ®
RUN ®
```

If the program in primary memory is valuable, be sure and save it before using this form of the RUN command lest you lose it when the new program is loaded.

8.4.5   Modifying Programs

The DELETE command.   Once the RUN command has been implemented, the computer will try to carry out the statements in your program. If it encounters an erroneous statement, such as a mistyped line, a syntax error message will be printed along with the line number where the error occurs. For certain errors, Microsoft BASIC will show the exact location in the line where the error occurs and will actually print out the line in a form that is ready for correction or editing.

Whether you are correcting an error or merely making improvements in the program, there will be numerous occasions when you will want to delete or modify the lines of a program. Deletion is accomplished in one of two ways—that is, by removing one line at a time or a group of lines. An individual line can be deleted by merely typing its line number and striking RETURN. A single line or a group of consecutive lines can be purged from the program with the DELETE command. The DELETE command uses the same syntax rules for line numbers as the LIST command. For example, an individual line can be deleted with

```
DELETE 150 ®
```

whereas a group can be deleted with

```
DELETE 120-200 ®
```

Because the DELETE command is irreversible it should be applied with caution. Before purging a large section of a program, it is good practice to save the program on the diskette in the event that you mistakenly delete a valuable section.

Modifying lines. Aside from deleting lines, you will also have frequent occasion to replace or modify existing lines. The simplest approach is merely to retype the line. When the new version is entered, it automatically replaces the old line. On the other hand, if you notice the error in the course of typing a line, the computer has a key by which you can backspace to retype the line correctly. Such a key is very useful when you are typing a long line and realize that you have made an error at the beginning of the line. The backspace key allows you to move from right to left in order to return to the point of the error. The mistake can then be corrected by retyping the section that was erroneous.

The IBM PC, as is the case for most personal computers, has two types of backspace keys. The key marked BACKSPACE actually erases characters as you move back along a line. In contrast, the arrows on the cursor pad at the right of the keyboard allow you to move in any direction around the screen without erasing characters.

Two additional keys that are extremely useful for modifying lines on the IBM PC bear mentioning. The delete key (marked DEL) deletes the character where the cursor is positioned. The insert key (marked INS) is used to insert characters within an already-typed line. It is called a ''toggle'' key because every time it is pressed it switches between one of two modes. When the computer is turned on, it is in its normal state, which is called the *overstrike mode*. In this state, any keystrokes typed replace existing text on the screen. When the insert key is struck once, the computer shifts to *insert mode*. Then, any characters typed are inserted at the location of the cursor, and all text to the right of the new character is shifted to the right. If the insert key is struck again, the computer shifts back to the overstrike mode.

The EDIT command. Often times it will be necessary to change just a few characters in a particular line of your program. While you could retype the whole line and replace it with the new one, it is much more convenient to simply modify the part of the line that requires correction. The EDIT command lets you do this. Editing a BASIC statement involves specifying what line of the program you want to modify and then making the changes. In order to specify the line, you type

```
EDIT 290 ⏎
```

This command tells BASIC that you want to edit line 290. BASIC will display the current contents of the line and be ready for you to make your changes. On the IBM PC, the cursor is moved to the desired location on the line using the cursor arrow keys. Characters are replaced by typing over the old ones, deleted with the delete key, and inserted with the insert key. Pressing the RETURN key is required to record the changes you have made. BASIC will then be ready to accept other commands. The IBM PC also allows for editing a whole screen of information. This is done using the LIST command and the cursor movement keys. Again, pressing the RETURN key is required to record the changes, but now this must be done for each line that is changed. Not all dialects of BASIC can be edited this conveniently. Some require that you obtain separate utility programs for the purpose.

8.4.6 Numbering Programs

The RENUM command.   What happens when you have to add a new line between lines 140 and 150? It is easy to merely add a new line numbered, for example, 145. However, because BASIC only allows positive integer line numbers, it would not be possible to add a line between lines 110 and 111. To take care of this situation, Microsoft BASIC has the command RENUM, which renumbers the lines of a program and changes all references to numbers within the program to the new numbering scheme. The general form of the command is

> RENUM *new first line, old first line, new increment* ®

Some examples of the RENUM command are

```
RENUM
RENUM 100
RENUM 100,85
RENUM 100,85,20
```

These commands renumber the program currently in RAM. RENUM with no modifiers renumbers the program from the beginning; the first line of the renumbered program would be 10 and each line would be incremented by 10. The first modifier after RENUM, the 100, is the first line number for the renumbered program. Thus the second example would produce a new numbering scheme starting at 100, and each succeeding line would be incremented by 10. The second modifier, 85, is the point in the old numbering scheme to start renumbering. Thus if there was a line 85 in the current program it would become line 100, the next 110, and so forth. When the three modifiers are used with RENUM, the third is the increment between lines for the new numbering scheme. Line 85 would again become line 100, followed by 120, 140, and so forth to the end of the program.

The AUTO command.   The AUTO command causes the computer to automatically generate a line number after each carriage return. Its general form is

> AUTO *first line number, increment*

where *first line number* is the line at which numbering is to begin and *increment* is the gap desired between successive line numbers. The default values (that is, the values the computer assumes if you do not enter them) for both *first line number* and *increment* are 10. For example, entering the AUTO command by itself yields

```
AUTO ®
10 _
```

Thus the computer prints a 10 and waits for you to type in the rest of the line. For example, if you use Fig. 8.1 as a model, the first line could be typed in and entered with the result

```
AUTO ®
10 REM SIMPLE ADDITION PROGRAM ®
20_
```

The computer prints the line number for the next line, 20, and waits for you to enter the next line. This process is repeated until you type in the entire program. In order to terminate automatic line numbering you merely hold down the CTRL key and press the C or the BREAK key.

8.4.7   Combining Programs

It is very useful to be able to combine programs. For example, you may have developed a plotting program to produce graphs. This software would be of great use in a variety of other programs. Therefore you would like to be able to take this program off your disk and merge it with other programs.

The MERGE command is used for this purpose. For example, suppose that the simple addition program were in primary memory and that a simple multiplication program were stored on a diskette with the name SIMPLMLT. The two would be merged by typing

```
MERGE "SIMPLMLT" ®
```

Several important rules should be noted. First, the file on diskette must be stored in ASCII format (recall Sec. 8.4.2). Second, the line numbers of the file in diskette should not overlap those of the file in primary memory. This is because each line loaded from the diskette will be inserted in its proper position according to its line number in the program in primary memory. If a line number from the diskette is the same as one in primary memory, the line from the diskette replaces and, hence, destroys the one in primary memory.

8.4.8   Obtaining Hard Copy of a LIST or RUN

To this point, we have only mentioned displaying your program on the monitor. You may also desire a list and/or a run on a printer. This sort of output is called a *hard copy*. On the IBM PC several options are available for this purpose. To obtain a list, simply type the system command

```
LLIST ®
```

In addition, a hard copy of either a list or a run may be obtained by holding down the CTRL key and then pressing the PRTSC key. Then, if LIST ® or RUN ® is typed, a listing or a run will be output on the printer. To terminate this printer output, the CTRL and the PRTSC keys would again be pressed simultaneously. Comparable commands for other popular personal computer systems are summarized in Table 8.2.

TABLE 8.2   Operations to Obtain Hard Copies of Listings and Runs for a Number of Popular Personal Computers

Computer System	Listing	Listing and Run
Apple II	PR#1	PR#1
	LIST	LIST
	PR#0	RUN
		PR#0
Commodore	OPEN 4,4	OPEN 4,4
	CMD 4	CMD 4
	LIST	LIST
	PRINT#4	RUN
	CLOSE 4	PRINT#4
		CLOSE 4
IBM PC	LLIST	CTRL/PRT SC
		LIST
		RUN
		CTRL/PRT SC
Macintosh	LLIST	SHIFT/special feature/key 4
TRS-80	LLIST	Differs from system to system
Your machine		

8.5   A SIMPLE BASIC PROGRAM

Now that you are familiar with some useful BASIC commands, we can begin to discuss the actual mechanics of developing a computer program. The remainder of this chapter will be devoted to expanding Fig. 8.1 into a quality BASIC program. Because of the simple nature of the algorithm underlying Fig. 8.1, the following material does not represent a detailed description of the capabilities of BASIC. A more in-depth discussion of the language will be deferred to subsequent chapters. For the time being, we will focus on the fundamentals of BASIC programming. In addition, we will introduce some practices that will add to the quality of your finished product.

8.5.1   Adding Descriptive Comments: The REM Statement

As we will reiterate time and time again, an important attribute of a high-quality computer program is how easy it is to read and to understand. One deficiency of the program in Fig. 8.1 is that it is almost devoid of documentation. That is, it consists merely of a series of instructions to get a job done. Although this is not a major problem for such a simple program, inadequate documentation can detract immensely from the utility of more complicated programs. If the user has to expend a great deal of time and energy to decipher your program's operation and logic, it will diminish the impact and usefulness of your work.

One of our primary tools for the internal documentation of a program is the REM (short for *rem*ark) statement. The general form of this statement is

ln REM *comment*

```
100 REM      ************************************************
110 REM      *               SIMPLE ADDITION PROGRAM        *
120 REM      *                  (BASIC version)             *
130 REM      *                  by  S. C. Chapra            *
140 REM      *                                              *
150 REM      *       This program adds two variables and    *
160 REM      *       and prints out the results.            *
170 REM      ************************************************
180 REM
190 REM ****************** MAIN PROGRAM ******************
200 REM
210 REM            Assign Values to Variables
220 REM
230 LET A = 28
240 LET B = 15
250 REM
260 REM
270 REM        Perform Computation and Print Result
280 REM
290 LET C = A + B
300 PRINT C
310 END
```

FIGURE 8.2
A revised version of the simple addition program from Fig 8.1. Here REM statements serve to document the program.

where *ln* is the line number and *comment* is any descriptive information you would like to include. For example, the following REM statement is included at the beginning of Fig. 8.1 in order to provide a title for the program

```
100  REM  SIMPLE ADDITION PROGRAM
```

Thus the user can immediately identify the general subject of the program.

Beyond setting forth titles, REM statements can be employed to incorporate additional documentation. Figure 8.2 is an example of how this can be done for the simple addition program. Notice that a group of REM statements has been used to form a title module at the program's beginning. Along with the title, this module includes the author's name and a concise description of what the program is to accomplish.

After the title module, REM statements can be used to interject descriptive comments as well as to make the program more readable. As seen in Fig. 8.2, this latter objective can be accomplished by using REMs to clearly delineate the program's major modules. Notice how the comment for each REM is centered. This is one style that makes it easier to pick out the comments. Also notice how several blank REM statements and a REM followed by a line of asterisks are included to set off and separate some of the comments and modules. The judicious use of lines and blank spaces is one effective tool for making the program more readable.

In some dialects of BASIC, an apostrophe (') or an exclamation point (!) are used to designate a remark. This is just an abbreviated way of specifying a REM statement. For example,

```
' SIMPLE ADDITION PROGRAM
```

and

```
! SIMPLE ADDITION PROGRAM
```

are acceptable alternatives to the REM in some dialects. We will limit ourselves to REMs because many dialects do not allow the use of apostrophes or exclamation points for comments, whereas all allow REMs.

It should be noted that several statements can be written on a single line by separating, or ''delimiting,'' them with colons. For example, lines 230 and 240 in Fig. 8.2 could be represented by the single line

```
230 LET A = 28 : LET B = 15
```

The indiscriminate use of colons to write several statements on a single line should be avoided because such a tactic can obscure the sequential structure of the statements. However, it is acceptable in certain cases. For example, the colon is handy for adding a REM statement to a line to elaborate on its meaning:

```
290 LET C = A + B : REM THE ADDITION EQUATION
```

Additionally, when the apostrophe or exclamation point is employed to designate a remark, a colon is unnecessary and the remark can be included on the same line as the BASIC statement; for example,

```
290 LET C = A + B   'THE ADDITION EQUATION
```

Thus, when someone reads the code, he or she will recognize the intent of the equation. Admittedly, this is not necessary for such a simple equation. However, there are situations where such labeling can prove useful in clarifying a program line.

While the use of REM statements is optional in BASIC, their use is strongly encouraged. But do not overuse remark statements. If the action of a line or lines is obvious, do not add excess verbiage in the form of remarks. A program can quickly become cluttered with unneeded comments that actually make the program harder to read and follow. As described in the next section, careful selection of variable names helps to minimize the need for remarks.

8.6   NUMERIC CONSTANTS, VARIABLES, AND ASSIGNMENT

As an engineer, one of the primary uses you will have for a computer is to manipulate numbers. Consequently, it is important to understand how numeric quantities are represented on the computer.

There are two fundamental ways in which numeric data is expressed in BASIC: directly as constants and indirectly as variables. *Constants* are values that remain fixed during program execution, whereas *variables* are symbolic names to which a value is assigned that can be changed during execution.

Numeric constants.   Numeric constants are numbers that remain fixed throughout program execution. They may be either positive or negative and may be expressed in integer or real form. *Integers* have no decimal point, as in

508   67   −1507   883

where, in Microsoft BASIC, the values can range from −32,768 to 32,767. (Recall our discussion in Sec. 4.1.)

The *real constants* are those with decimal points:

896.   −4.78   0.0056   −100.4

or those that are greater than 32,767 or less than −32,768—for example, 50,000 or −600,000.

In addition, real constants can be expressed in BASIC in a form similar to mathematical scientific notation. *Exponential constants* are expressed in such a form and are handy to use for very large or very small numbers. To review, a number is expressed in scientific notation by converting it to a number between 1 and 10 multiplied by an appropriate integral power of 10. This can be done by moving the decimal point to the right or left until there remains a number between 1 and 10. This remaining number must then be multiplied by 10 raised to a power equal to the number of places the decimal point was moved. If the decimal has been moved right, the power is negative, and if it has been moved left, the power is positive. The general form is

$$b \times 10^n$$

where b is the decimal or integer value that remains after the number has been converted to a value between 1 and 10 and n is the integer number of places the decimal has been moved. For example, the number 8,056,000 would be converted to 8.056 by moving the decimal six places to the left. Therefore in scientific notation it would be expressed as 8.056×10^6.

BASIC employs the same concept but uses a different notation:

bEn

Thus in BASIC the number 8,056,000 is expressed as 8.056E6.

Note that letters (other than those used to specify exponential constants) and symbols such as dollar signs and commas are not allowed in numeric constants. Therefore $56.23 and 1,760,453 are illegal numeric constants and would constitute a syntax error in BASIC. In addition, only one minus sign may be used at the front of a constant to designate that it is negative. Minus signs cannot be used as delimiters (that is, separators), as is commonly done in social security numbers. For example, 120-44-5605 is an illegal constant in BASIC.

It should be noted that real numbers in Microsoft BASIC can be expressed in either single or double precision. *Single-precision numbers* are stored with seven significant digits, whereas *double-precision numbers* are stored with seventeen. If a number is written with more than seven digits, it will automatically be stored as a double-precision variable. Examples are

3.1415926536   2.56001753E-6   −2050000768

In addition, if the # sign is added at its end, the number will be forced to be stored in double precision:

49.7#   2000#   −8.86#

Finally, the double-precision version of the exponential form uses a D rather than an E, as in

1.6D5   −4.83702D−5   15D10

Table 8.3 summarizes the levels of precision that are available on a number of popular computers. Note that the precision of some machines differs according to whether the constant is in storage or is being displayed on an output device such as a monitor or printer.

Just as the # sign forces the computer to make a real variable into the double-precision form, an integer or double-precision constant can be forced to be a single-precision constant by appending an exclamation point, as in

30!   −1050!   3.1415926536!   2.71828183!

You may be wondering at this point why so many different forms are needed and why we would want to force the computer to store a constant in a particular form. There are a number of reasons why we might want to go to the trouble. For example, an integer takes up less space in memory and can be processed faster during computations than a real constant. Similarly, single-precision numbers have the same advantages over double-precision constants. Thus, if we use integers rather than single precision, and single precision rather than double precision, the resulting programs will run faster and use up less memory. On the other hand, double-precision constants are advantageous when a computation must be performed with a high degree of accuracy. As discussed in Chap. 4, such situations often occur in engineering.

Although the above is true, it should be noted that in most cases you will not be concerned with how numbers are stored on the computer. You will just write the constants naturally and let the computer determine how they should be stored. With

TABLE 8.3   Number of Significant Digits of Constants for Some Popular Personal Computer Systems

	Single Precision		Double Precision	
	Stored	**Displayed**	**Stored**	**Displayed**
Apple	10	9	Not available	
Commodore	10	9	Not available	
IBM PC	7	6	17	16
Macintosh	6	6	14	14
TRS-80	7	6	16	16
Your machine				

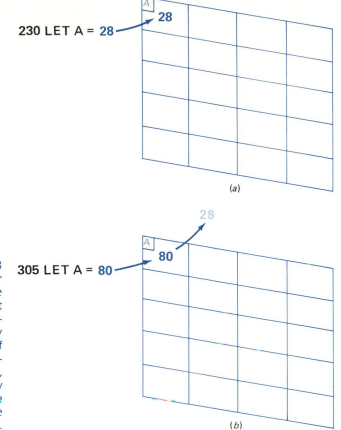

230 LET A = 28

(a)

305 LET A = 80

(b)

FIGURE 8.3
Analogy between computer
memory locations and post-office
boxes. (a) The BASIC statement
230 LET A = 28 assigns the variable name A to one of the memory
locations and stores a value of
28 in that location. (b) A subsequent assignment statement,
305 LET A = 80, places a new
value of 80 in the location, and the
process destroys the prior value
of 28.

this as background, we can now turn to the way in which the constants are actually stored.

Assignment. In addition to constant quantities, it is useful to have variable quantities that can change during the execution of a program. These variable quantities are designated by symbolic names. It is common practice to refer to the variable name as the variable itself. For example, in the simple addition programs from Figs. 8.1 and 8.2, we used three variables—A, B, and C.

To understand how variables work, it is convenient to think of the computer memory as consisting of a number of cells that are similar to post-office boxes (Fig. 8.3). The variable is the name that the computer assigns to each of these cells. For example, in the simple addition problem each of the variables—A, B, and C, would correspond to one of the cells. When we used line 230 from Fig. 8.2 to assign a value of 28 to A, the computer stored the value in the memory location for A (Fig. 8.3a). Then later in the program when we performed the addition in line 290 the computer "remembered" that A was equal to 28 by referring to the memory cell where it was stored. Now suppose that after the addition we redefined A by another assignment statement such as

```
305 LET A = 80
```

When this line was executed, the computer would replace the old value by this new one. Using our analogy of the postal boxes, this act would be akin to placing the new number in the box and pushing the old one out the back and into oblivion (Fig. 8.3*b*). This ability to change value is the reason symbolic names are called *variables*.

Because lines 230 and 305 assign a constant value to a variable, they are called *assignment statements*. They can be written in general form as

ln LET *variable = expression*

where *ln* stands for the line number, *variable* stands for the variable name, and *expression* can be a constant, another variable, or an arithmetic expression. Line 290 is an example of the use of an arithmetic expression in an assignment statement:

```
290 LET C = A + B
```

When such a statement is executed, the numeric value of the expression is automatically stored in the storage location for C. Note that variables can also be subtracted, multiplied, and divided, as in

```
292 LET C = A - B
294 LET C = A * B
296 LET C = A/B
```

The LET in an assignment statement is optional. For example, line 290 could also be written correctly as

```
590 C = A + B
```

Although the LET is optional, we will usually employ it throughout this text. One practical reason for doing this is to clearly communicate the significant difference between computer and algebraic equations.

Algebraic vs. computer equations.   The equal sign in the assignment statement can be thought of as meaning ''be replaced by'' or, if the LET is omitted, ''is replaced by.'' This is very significant because it signifies that computer equations differ in a very important way from standard algebraic equations. That is, computer equations must always have a single variable on the left-hand side of the equal sign. Thus, because it violates this rule, the following legitimate algebraic equation is illegal in BASIC:

```
10 A + B = C
```

In the same sense, although

```
20 X = X + 1
```

is not a proper algebraic equation, it is a perfectly acceptable BASIC assignment statement. This is evident when the words ''is replaced by'' are substituted for the equal sign. For example, if X had a prior value of 5, the new value after execution of line 20 would be 6.

Variable names. The symbolic name for the variable should reflect, as simply as possible, what information is stored by the variable. Mnemonic names that are easy to remember are best. Names that are short enough to type quickly and accurately are helpful, as long as the name is of sufficient length to convey the meaning or purpose of the variable.

The actual symbolic names used to designate variables must follow rules (Table 8.4). In the simplest forms of BASIC, these rules are somewhat restrictive. For systems using Microsoft BASIC, variables must begin with a letter, not a number. After the first letter, they may have either letters or numbers. BASIC reserved words or keywords may not be used as variable names. For example, REM always identifies a comment and is never a variable name. However, the name may include imbedded keywords—for example, TREMOR would be acceptable even though it includes two keyboards, REM and OR. Such is not the case for simpler forms of BASIC such as those commonly used on the Apple or Commodore (Table 8.4).

For Microsoft BASIC, variable names can be of any length, but only the first 40 are employed to identify the variable. Any characters after the first 40 are ignored. Because most BASIC dialects allow less than 40 characters in a name, variable names with six or less characters will routinely be used here.

TABLE 8.4 Rules for Naming Variables on Some Popular Personal Computing Systems

System	Naming Rules
Apple II	A variable name begins with a letter followed by letters and numbers. Although the name may be of any length, only the first two characters are actually used for identification. Keywords may not be used as names, nor may they be imbedded in other names.
Commodore	Same as for Apple II.
IBM PC	A variable name begins with a letter followed by letters, numbers, and decimal points. Although the name may be of any length, only the first 40 characters are actually used for identification. Keywords may not be used as names, but they may be imbedded in other names.
Macintosh	Same as for IBM PC.
TRS-80	Same as for Apple II.
Your machine	

Variable names should not contain characters other than letters, numbers, and decimal points except for their last character, if the last character is one of the special type identifiers—!, #, %, or $ (Table 8.5). Variable names end when a character that is not a letter, a number, or decimal point is encountered. Names allowed in Microsoft BASIC include

ONE	VARI	X1938	Z2001V
PRODUCT	I%	SUM#	SUM!
STRING$	VAR!	LETTER$	I

All of these variable names conform to the above rules. Not all may be ideal variable names because their use may not be obvious. Examples of variable names that are not acceptable for BASIC are

8THVAR	4567	!TEST	THE&1ST
LET	INT%1	A*VAR	

8THVAR, 4567, and !TEST do not begin with a letter. THE&1ST has in it a character, the ampersand, that is neither a letter nor a number. LET is a BASIC reserved word. INT%1 has one of the special characters that identify type, but it is not the last character of the name. A*VAR is actually two variable names and the multiplication operator *. Thus, although both A and VAR are legitimate names, A*VAR cannot be used to name a single variable because the computer would "think" that it signifies A multiplied by VAR.

BASIC has four types of variables (Table 8.5). A variable can contain a real number, a double-precision real number, only whole (integer) numbers, or characters. The variable type most commonly used in BASIC programs is the real type, which is identified by having either an exclamation point as the last character of the name or no special character at the end of the name. Real variables of this type are also called *single-precision real variables* because they store constants of up to seven digits, the sign of the number, and an exponent. A double-precision variable is similar to the single-precision version but ends in a pound sign and can store 17 digits of information. Variables ending with the percent character contain only integer numbers. Integers can range from −32,768 to 32,767 and never have decimal portions.

TABLE 8.5   Special Types of Variables Used in Microsoft BASIC

Type	Identifier	Description
Real	!	Contains a single-precision real number, which can be either a decimal, integer, or exponential; variable names that do not end in a special type identifier are also used to contain single-precision real numbers.
Double precision	#	Contains a double-precision real number.
Integer	%	Contains an integer.
Character	$	Contains character or alphanumeric information.

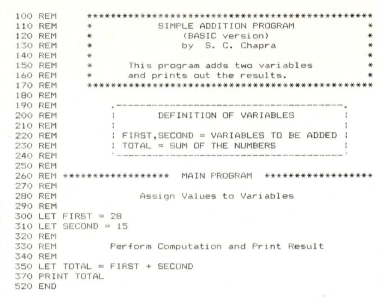

```
100 REM       ***********************************************
110 REM       *              SIMPLE ADDITION PROGRAM        *
120 REM       *                  (BASIC version)            *
130 REM       *                  by  S. C. Chapra           *
140 REM       *                                             *
150 REM       *        This program adds two variables      *
160 REM       *        and prints out the results.          *
170 REM       ***********************************************
180 REM
190 REM       .-------------------------------------------.
200 REM       |          DEFINITION OF VARIABLES          |
210 REM       |                                           |
220 REM       | FIRST,SECOND = VARIABLES TO BE ADDED      |
230 REM       | TOTAL = SUM OF THE NUMBERS                |
240 REM       '-------------------------------------------'
250 REM
260 REM ******************  MAIN PROGRAM  ******************
270 REM
280 REM                Assign Values to Variables
290 REM
300 LET FIRST = 28
310 LET SECOND = 15
320 REM
330 REM             Perform Computation and Print Result
340 REM
350 LET TOTAL = FIRST + SECOND
370 PRINT TOTAL
520 END
```

FIGURE 8.4
A revision of the simple addition program from Fig. 8.2. A module has been included to define the variables used in the program. Notice that more-descriptive variable names are employed.

Variables ending in a dollar sign store character information. We will discuss this type of information in the next section. For the time being, it is sufficient to note that these variables are used to store nonnumerical information such as labels, names, dates, etc., that consist of letters, numbers, and symbols. These sorts of nonnumerical information are formally referred to as *string constants*. (Note that string constants are enclosed in quotes.) The variables in which they are stored are called *string variables*.

Although single-precision and double-precision real numbers and integers can be used together in mathematical statements in BASIC, variables ending in the $ cannot. The latter have special operators designed for character manipulation and cannot be directly used in a mathematical computation with variables holding numeric data. This restriction is not illogical because the sum of a variable containing the letters "Hello" and one containing the numbers "123.07" makes no sense. Actually there are times when it is useful to be able to convert a single character into a number and back, and there are operators in BASIC to allow this to be done. Some of these operators are described in Appendix B.

A few notes of caution are in order when naming variables. The variables I, I#, and I% are three distinct variables that could hold different information. The variable names VAR1 and VAR1! both refer to the same single-precision real variable, and they would always contain the same data. Microsoft BASIC ignores the distinction between upper and lower case letters in variables so that if you typed 'var1' and 'VAR1', both would refer to the same variable. This point is driven home by the BASIC editor, which converts all of your BASIC program to uppercase except string constants and the comments of REM statements.

Figure 8.4 is a revised version of the simple addition program that reflects a more communicative use of variable names. Notice how a module (lines 190 to 240) has been incorporated to define the meaning of the variables at a single location at the program's beginning. Also, more descriptive names have been used to specify the variables. FIRST, SECOND, and TOTAL are much more suggestive of the function of the variables than the original names A, B, and C.

8.7 INTRODUCTION TO INPUT/OUTPUT

''Input/output'' refers to the means whereby information is transmitted to and from a program. You have already been introduced to output with the PRINT command in the simple addition program. Now we will elaborate on the topic so that you can perform more descriptive input and output.

The INPUT statement.   One disadvantage of the LET statement is that it assigns a specific value to a variable. Thus in the simple addition program we used LET statements to set FIRST equal to 28 and SECOND equal to 15. If we want to add two different numbers, say, FIRST = 8 and SECOND = 33, we have to retype lines 300 and 310 in order to assign new values to the variables.

The INPUT statement is an alternative that allows new data to be entered every time the program is run. As an example of its use, suppose that we want to assign a different value to FIRST every time we run the program. We could rewrite line 300 as

```
300 INPUT FIRST
```

Then when we execute the program by typing RUN ®, the computer will respond by displaying a ''prompt'' on the monitor. A *prompt* is a signal from the computer that you must input information. For Microsoft BASIC, the prompt is represented by a question mark. For the present example, the prompt appears because the computer has reached line 300, which specifies that a value must be input for FIRST. Suppose that you want to set FIRST equal to 8. Then by merely typing in an 8 and striking RETURN you would have instructed the computer that FIRST = 8. It would then continue executing the program and print out the correct answer of TOTAL = 8 + 15 = 23.

More than one value may be entered with an INPUT statement. For example, to input values for both FIRST and SECOND, you could type

```
300 INPUT FIRST, SECOND ®
```

and delete line 310. Then after you type RUN ®, a prompt will result on the screen and two values separated by a comma can be typed to enter values for FIRST and SECOND. If FIRST = 8 and SECOND = 33, the result will look like

```
RUN ®
? 8, 33 ®
41
```

In all the above examples, until the RETURN key is pressed, the BACKSPACE and other editing keys can be used to alter the input data. Once the RETURN key is pressed, BASIC examines what was typed to determine if it matches the variables in the INPUT statement. If the data entered does not match the number and type of variables in the INPUT statement, then BASIC will give the error message, ?REDO FROM START, discard what had been entered, and redisplay the ? until the data is entered correctly. Here are INPUT statements with acceptable entries:

INPUT Statement	Acceptable Entry
INPUT VAR1	429.07
INPUT VAR1, VAR2	429.07,943
INPUT VAR1, VAR2$	429.07, "Hello there."
INPUT VAR1, VAR2#, VAR3	− 9.067E-4,1094,15968.8

In contrast, the following entries would be rejected by BASIC and cause the error message and the ? to be displayed:

INPUT Statement	Unacceptable Entry	Problem
INPUT VAR1	"Hello"	Entry type does not match type of VAR1.
INPUT VAR1, VAR2	429.07 943	There is no comma between data entries.
INPUT VAR1, VAR2$	4.9,"Hi",5	There are too many data entries.
INPUT VAR1, VAR2#, VAR3	− 9.067E-4,1094	There are insufficient data entries for the number of variables in the INPUT statement.

The PRINT statement. The PRINT statement is the major vehicle for outputting data. It can be used to print out the values of a series of variables, as in

```
50 PRINT A, B, C
```

When the variables are delimited by commas, as in this example, the computer automatically spreads the values out evenly across the page. For example, if A = 28, B = 15, and C = 43, the output will look something like

```
28              15              43
```

However, if a semicolon is used as delimiter, as in

```
50 PRINT A; B; C
```

the computer will pack the numbers closely together, as in

```
28  15  43
```

You might wonder why it would ever be preferable to jam numbers together in this fashion. Actually, it is rarely desirable to do this in its own right. However, when used in conjunction with the TAB function, semicolon delimiters provide a means to control the spacing of the output.

The TAB function.   Although delimiting data with commas automatically results in evenly spaced output, there may be cases where you will desire to control the output yourself. This can be done using the TAB function that can be included in the PRINT statement to specify the column at which printing starts. For example,

```
50 PRINT TAB(10);C
```

will cause the value of C to be printed starting at column 10. Several TABs can be used in a single PRINT statement to space out a series of values, as in

```
50 PRINT A; TAB(8); B; TAB(16); C
```

Thus if A = 28, B = 15, and C = 43, the following output results:

```
28      15      43
                 └col. 16
        ↑
        └────────col. 8
```

Hence the TAB function works just like a typewriter tab setting.

String constants and variables.   Now that we have introduced the INPUT and PRINT statements, we will show how they can be used to input and output data in a more descriptive fashion. To do this, we have to elaborate more on the concept of string constants and variables.

Aside from numbers, computers are also designed to handle alphanumeric information (that is, groups of letters, numbers, and symbols) such as names, addresses, and other miscellaneous labels. *String constants* are specific groupings of alphanumeric information. In BASIC they are enclosed in quotes, as in

```
"JOHN DOE"    "8/5/48"    "THE FIRST RESULT IS"
```

String variables work in the same way as numeric variables in the sense that they are names that designate a specific location in the computer's memory. As with numeric variables, in the simplest forms of BASIC the naming conventions are somewhat restrictive. For these cases, a string variable is usually designated by a single letter (or one letter followed by a single digit) followed by a dollar sign, as in

```
N$    D$    R1$
```

Microsoft BASIC allows longer variable names. However, it still requires that a dollar sign be placed at their ends, for example,

```
NAM$    DATE$    SUM1$
```

String variables can be given constant values in the same manner as numeric variables. One way is via an assignment statement, as in

```
100 LET NAM$ = "JOHN DOE"
200 LET DATE$ = "JULY 4, 1976"
300 LET SUM1$ = "THE FIRST RESULT IS"
```

In addition, an INPUT statement can be used for the same purpose. For example,

```
55 INPUT NAM$
```

would result in a prompt appearing on the monitor during execution. In response to the prompt you could type in the appropriate value, as in

```
? "JOHN DOE" ®
```

After you have struck the RETURN, the storage location corresponding to NAME$ would hold the string constant JOHN DOE.

On most systems character strings can be input without using quotes, for example,

```
? JOHN DOE ®
```

However, if the character string includes commas, the quotes must be used. This is because the computer "perceives" commas that are not enclosed by quotes as delimiters. For example,

```
? "JULY 4, 1976"
```

is correct because the date includes a comma. If the quotes were not used, as in

```
? JULY 4, 1976
```

the computer would "perceive" that the character string constant was JULY 4 and that 1976 was a separate value.

Descriptive input.   One problem with the INPUT statements described to this point is that the user has been provided no clue as to what information is supposed to be input. For example,

```
300 INPUT FIRST
```

would cause a ? to be printed when the program is run. Unless the user knew beforehand that a value of FIRST was to be entered, the question mark provides no clue as to what variables are to be input. One way to remedy this situation is to use a

PRINT statement and string constants to display a message that alerts the user to what is required, for example,

```
300 PRINT "FIRST ="
305 INPUT FIRST
```

When these lines are executed, the computer will print

```
FIRST =
?
```

Thus the message lets the user know what is required. The descriptive input can be improved further by adding a semicolon at the end of line 300, as in

```
300 PRINT "FIRST = ";
305 INPUT FIRST
```

The semicolon acts to suppress the carriage return on the computer output. Thus when the program is run, rather than being printed on the next line, the prompt appears at the end of the message, as in

```
FIRST = ?
```

There is one final improvement that may be made. On most systems it is possible to include one character string message directly in the body of an INPUT statement. For example, rather than use lines 300 and 305 as above, we could simply type

```
300 INPUT "FIRST = "; FIRST
```

When the program is run the result would be the same as if both lines 300 and 305 were used, that is,

```
FIRST = ?
```

The question mark prompt can be eliminated by using a comma in place of a semicolon.

Descriptive output.   Just as input can be made more user-friendly, output can also be made more descriptive using character strings, semicolon delimiters, TAB functions, and blank lines. For example, in the simple addition program (Fig. 8.4) the statement

```
360 PRINT TOTAL
```

merely results in the answer, 43, being printed out. A message can easily be incorporated into the statement, as in

```
100 REM     *********************************************
110 REM     *           SIMPLE ADDITION PROGRAM         *
120 REM     *              (BASIC version)              *
130 REM     *              by  S. C. Chapra             *
140 REM     *                                           *
150 REM     *     This program adds two variables       *
160 REM     *     and prints out the results.           *
170 REM     *********************************************
180 REM
190 REM     .------------------------------------------.
200 REM     |          DEFINITION OF VARIABLES         |
210 REM     |                                          |
220 REM     | FIRST,SECOND = VARIABLES TO BE ADDED     |
230 REM     | TOTAL = SUM OF THE NUMBERS               |
240 REM     '------------------------------------------'
250 REM
260 REM ****************** MAIN PROGRAM  ******************
270 REM
280 REM            Clear Screen and Identify Program
290 CLS
300 PRINT:PRINT
310 PRINT TAB(25);"PROGRAM TO ADD TWO NUMBERS"
320 REM
330 REM                Obtain Numbers from User
340 REM
350 PRINT:PRINT
360 PRINT TAB(30);"INPUT THE NUMBERS"
370 PRINT:PRINT TAB(30);
380 INPUT "FIRST = ";FIRST
390 PRINT TAB(30);
400 INPUT "SECOND = ";SECOND
410 REM
420 REM          Perform Computation and Print Result
430 REM
440 LET TOTAL = FIRST + SECOND
450 PRINT
460 PRINT TAB(30);"TOTAL = ";TOTAL
470 REM
480 REM          Signal that Program was Successful
490 REM
500 PRINT
510 PRINT TAB(25);"COMPUTATION COMPLETED"
520 END
```

FIGURE 8.5
A final revision of the simple addition program. Descriptive input and output are incorporated into this version.

```
360 PRINT "TOTAL =";  TOTAL
```

When the program is run the result is

```
TOTAL = 43
```

In addition, several constants, variables, or messages can be output in a single PRINT statement. For example, line 360 could be reformulated as

```
360 PRINT "THE SUM OF";
361 PRINT FIRST;"PLUS";SECOND;"=";TOTAL
```

which, when executed, results in the output

```
THE SUM OF 28 PLUS 15 = 43
```

Figure 8.5 is a final revised version of the simple addition program. We have added a variety of descriptive input and print statements to make the program more

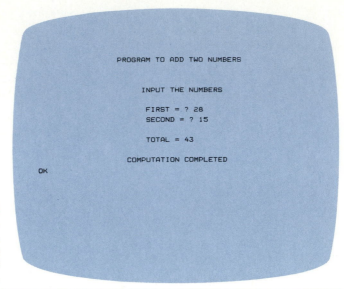

```
          PROGRAM TO ADD TWO NUMBERS


             INPUT THE NUMBERS

           FIRST = ? 28
           SECOND = ? 15

           TOTAL = 43

          COMPUTATION COMPLETED

 OK
```

FIGURE 8.6
A run of the program from Fig. 8.5.
Note that the command, KEY OFF,
was entered prior to this run to
erase the function key prompts
from the bottom of the screen.

readable and easier to use. In this spirit, a new type of statement, the CLS, is included on line 290. This statement clears the screen except for the function-key prompts on the bottom line. Additionally, the KEY OFF statement can be used to clear the function-key prompts (which can be restored with KEY ON). Together with the TAB, these statements allow an attractive output when the program is run (Fig. 8.6). Table 8.6 summarizes the various statements that are required to clear the screen on a number of popular personal computing systems.

8.8   QUALITY CONTROL

To this point, you have acquired sufficient information to compose a simple BASIC computer program. Using the information from the previous section, you should now be capable of writing a program that will generate some output. However, the fact that a program prints out information is no guarantee that these answers are correct.

TABLE 8.6   Statements Required to Clear the Screen
on a Number of Popular Personal Computer Systems

Computer System	Statements to Clear the Screen
Apple	HOME
Commodore	PRINT "Press Shift and CLR HOME keys"
IBM PC	CLS
Macintosh	CLS
TRS-80	CLS
Your machine	

Consequently, we will now turn to some issues related to the reliability of your program. We are emphasizing this subject at this point to stress its importance in the program-development process. As problem-solving engineers, most of our work is used directly or indirectly by clients. For this reason, it is essential that our programs be reliable—that is, that they do exactly what they are supposed to do. At best, lack of reliability can be embarrassing; at worst, for engineering problems involving public safety, it can be tragic.

8.8.1 Errors or "Bugs"

During the course of preparing and executing a program of any size, it is likely that errors will occur. These errors are sometimes called *bugs* in computer jargon. This label was coined by Capt. Grace Hopper, one of the pioneers of computer languages. In 1945 she was working with one of the earliest electromechanical computers when it went dead. She and her colleagues opened the machine and discovered that a moth had become lodged in one of the relays. Thus the label "bug" came to be synonymous with problems and errors associated with computer operations, and the term *debugging* came to be associated with their solution.

When developing and running a program, three types of bugs can occur:

1. *Syntax errors* violate the rules of the language such as spelling, number formation, line numbering, and other conventions. These errors are often the result of mistakes such as typing IMPUT rather than INPUT. They usually prompt the computer to print out an error message. Such messages are called *diagnostics* because the computer is helping you to "diagnose" the problem.

2. *Run-time errors* are those that occur during program execution, for example, the case where there is an insufficient number of data entries for the number of variables in an INPUT statement. For such situations a diagnostic such as ?REDO FROM START will be printed, and you must then reenter the data properly. For other run-time errors the computer may simply terminate the run or print a message informing you at what line the error occurred. In any event, you will be cognizant of the fact that an error has occurred.

3. *Logic errors*, as the name implies, are due to faulty program logic. These are the worst types of errors because no diagnostics will flag them. Thus your program will appear to be working properly in the sense that it executes and generates output. However, the output will be incorrect.

All the above types of errors must be removed before a program can be employed for engineering applications. We divide the process of quality control into two categories: debugging and testing. As mentioned above, *debugging* involves correcting known errors. In contrast, *testing* is a broader process intended to detect errors that are not known. Additionally, testing should also determine whether the program fills the needs for which is was designed.

As will be clear from the following discussion, debugging and testing are interrelated and are often carried out in tandem. For example, we might debug a module to remove the obvious errors. Then, after a test, we might uncover some additional bugs that require correction. This two-pronged process is repeated until we are convinced that the program is absolutely reliable.

8.8.2 Debugging

As stated above, debugging deals with correcting known errors. There are three ways in which you will become aware of errors. First, an explicit diagnostic can provide you with the exact location and the nature of the error. Second, a diagnostic may be printed, but the exact location of the error is unclear. Third, no diagnostics are printed, but the program does not operate properly. This is usually due to logic errors such as infinite loops or mathematical mistakes that do not necessarily yield syntax errors. Corrections are easy for the first type of error. However, for the second and third types some detective work is required in order to identify the errors and their exact locations.

Unfortunately, the analogy with detective work is all too apt. That is, ferreting out errors is often difficult and involves a lot of "legwork," collecting clues and running up some blind alleys. In particular, as with detective work, there is no clear-cut methodology involved. However, there are some ways of going about it that are more efficient than others.

First, it should be noted that interpreted BASIC on a personal computer is an ideal tool for debugging. Because of its interactive, or hands on, mode of operation, the user can efficiently perform the repeated runs that are sometimes needed to zero in on the location of errors. One key to finding bugs is to print out intermediate results. With this approach, it is often possible to determine at what point the computation went bad. Similarly, the use of a modular approach will obviously help to localize

FIGURE 8.7
An example of a TRON-activated trace.

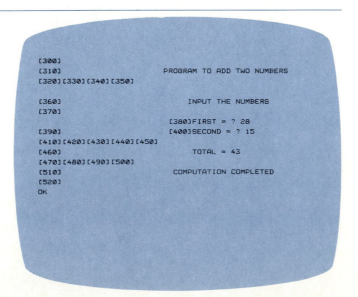

```
[300]
[310]                          PROGRAM TO ADD TWO NUMBERS
[320][330][340][350]

[360]                              INPUT THE NUMBERS
[370]
                               [380]FIRST = ? 28
[390]                          [400]SECOND = ? 15
[410][420][430][440][450]
[460]                              TOTAL = 43
[470][480][490][500]
[510]                          COMPUTATION COMPLETED
[520]
OK
```

errors in this way. You could put a PRINT statement at the start of each module and in this way focus on the module in which the error occurs. In addition, it is always good practice to debug (and test) each module separately before integrating it into the total package.

A more formal way to determine the location of an error is with the TRON (or on some systems the TRACE) command. Typing in this command before a run causes the computer to print out every line number as it is executed, along with any other input or output for the program (Fig. 8.7). Although for long programs this can result in large quantities of output it is sometimes the most efficient way to zero in on an error. The trace is discontinued by typing in the TROFF (or in some systems the NO TRACE) command.

Although the above procedures can help, there are often times when your only option will be to sit down and read the code line by line, following the flow of logic and performing all the computations with a pencil, paper, and pocket calculator. Just as a detective sometimes has to think like the criminal in order to solve a case, there will be times you will have to ''think'' like a computer to debug a program.

8.8.3 Testing

One of the greatest misconceptions of the novice programmer is the belief that if a program runs and prints out results, it is correct. Of particular danger are those cases where the results appear to be reasonable but are in fact wrong. To ensure that such cases do not occur, the program must be subjected to a battery of tests for which the correct answer is known beforehand.

As mentioned in the previous section, it is good practice to debug and test modules prior to integrating them into the total program. Thus testing usually proceeds in phases. These phases consist of module tests, development tests, whole-system tests, and operational tests.

Module tests, as the name implies, deal with the reliability of individual modules. Because each is designed to perform a specific, well-defined function, the modules can be run in isolation to determine whether they are executing properly. Sample input data can be developed for which the proper output is known. This data can be used to run the module, and the outcome can be compared with the known result to verify successful performance.

Development tests are implemented as the modules are integrated into the total program package. That is, a test is performed after each module is integrated. An effective way to do this is with a top-down approach that starts with the first module and progresses down through the program in execution sequence. If a test is performed after each module is incorporated, problems can be isolated more easily, since new errors can usually be attributed to the latest module added.

After the total program has been assembled, it can be subjected to a series of *whole-system tests*. These should be designed to subject the program to (1) typical data, (2) unusual but valid data, and (3) incorrect data to check the program's error-handling capabilities.

Finally, *operational tests* are designed to check how the program performs in a realistic setting. A standard way to do this is to have a number of independent subjects (including the client) implement the program. Operational tests sometimes turn up bugs. In addition, they provide feedback to the developer regarding ways to improve the program so that it more closely meets the user's needs.

Although the ultimate objective of the testing process is to ensure that programs are error-free, there will be complicated cases where it is impossible to subject a program to every possible contingency. In fact, for programs of even moderate complexity, it is often practically impossible to prove that an error does not exist. At best, all that can be said is that, under rigorous testing conditions, no error could be found. It should also be recognized that the level of testing depends on the importance and magnitude of the problem context for which the program is being developed. There will obviously be different levels of rigor applied to testing a program to determine the batting and fielding averages of your intramural softball team as compared with a program to regulate the operation of a nuclear reactor. In each case, however, adequate testing must be performed, testing consistent with the liabilities connected with a particular problem context.

8.9 DOCUMENTATION

Documentation is the addition of English-language descriptions to allow you or another user to implement the program more easily. (Remember, along with other people who might employ your software, you are also a user.) Although a program may seem clear and simple when you first compose it, after the passing of time the same code may seem inscrutable. Therefore sufficient documentation must be included to allow you and other users to immediately understand your program and how it can be implemented. This task has both internal and external aspects.

8.9.1 Internal Documentation

Picture for a moment a text with just words and no paragraphs, subheadings, or chapters. Such a book would be more than a little intimidating to study. The way the pages of a text are structured—that is, all the devices that serve to break up and organize the material—serve to make the text more effective and enjoyable. In the same sense, internal documentation of computer code can greatly enhance the user's understanding of a program and how it works.

We have already broached the subject of internal documentation when we introduced the REM statement in Sec. 8.5.1. The following are some general suggestions for documenting programs internally:

1. Include a module at the head end of the program, giving the program's title and your name and address. This is your signature and marks the program as your work.
2. Include a second module to define each of the key variables.
3. Select variable names that are reflective of the type of information the variables are to store.

4. Insert spaces within statements to make them easier to read.

5. Use REM statements liberally throughout to program for purposes of labeling and clarifying.

6. In particular, use REM statements to clearly label and set off all the modules.

7. Use indentation to clarify the program's structures. In particular, indentation can be used to set off loops and decisions (to be discussed in Chaps. 11 and 12).

8.9.2 External Documentation

External documentation refers to instructions in the form of output messages and supplementary printed matter. The messages occur when the program runs and are intended to make the output attractive and user-friendly. This involves the effective use of spaces, blank lines, and special characters to illuminate the logical sequence and structure of the program output. Attractive output simplifies the detection of errors and enhances the communication of program results.

The supplementary printed matter can range from a single sheet to a comprehensive user's manual. Figure 8.8 is an example of a simple documentation form that we recommend you prepare for every program you develop. These forms can be maintained in a notebook to provide a quick reference for your program library. The user's manual for your computer is an example of comprehensive documentation. This manual tells you how to run your computer system and disk operating programs. Note that for well-documented programs, the title information in Fig. 8.8 is unnecessary because the title module is included in the program.

8.10 STORAGE AND MAINTENANCE

Recall that when the computer is shut off, the contents of RAM are destroyed. In order to retain a program for later use, you must transfer it to a secondary storage device such as a diskette before turning the machine off.

TABLE 8.7   Suggestions and Precautions for Effective Maintenance of Software on Diskettes

- Maintain paper printouts of all finished programs in a file or binder.
- Always maintain at least one backup copy of all diskettes.
- Handle diskettes gently. Do not bend them or touch their exposed surfaces.
- Affix a temporary label describing the contents to all diskettes (see Fig. 3.5). Once the label is attached, use only soft felt-tipped pens to add more information.
- Once a diskette is complete, cover the write-protect notch (Fig. 3.5) with foil tape.
- Keep the diskette in its protective envelope when not in use.
- Store diskettes upright in a place that is neither too hot nor too cold.
- Keep diskettes away from water and contaminants such as smoke and dust.
- Store diskettes away from magnetic devices such as televisions, tape recorders and electric motors.
- Remove diskettes from the drive before turning the system off.
- Never remove a diskette when the red light on the drive is lit.

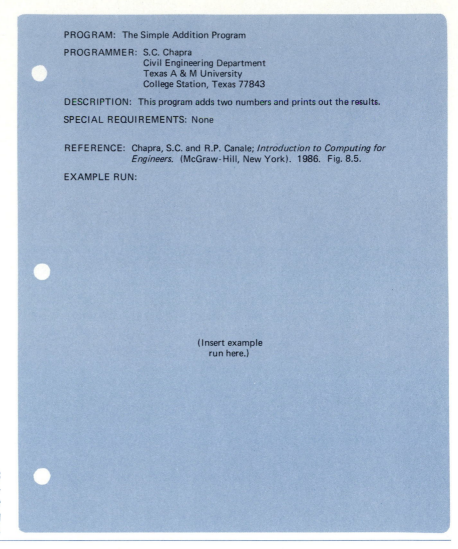

PROGRAM: The Simple Addition Program

PROGRAMMER: S.C. Chapra
                          Civil Engineering Department
                          Texas A & M University
                          College Station, Texas 77843

DESCRIPTION: This program adds two numbers and prints out the results.

SPECIAL REQUIREMENTS: None

REFERENCE: Chapra, S.C. and R.P. Canale; *Introduction to Computing for Engineers.* (McGraw-Hill, New York). 1986. Fig. 8.5.

EXAMPLE RUN:

(Insert example
run here.)

FIGURE 8.8
A simple one-page format for program documentation. This page can be stored in a binder along with a printout of the program.

We have already discussed how to SAVE a program on a diskette in Sec. 8.4.2. Aside from this physical act of retaining the program, storage and maintenance consist of two major tasks: (1) upgrading the program in light of experience and (2) ensuring that the program is safely stored. The former is a matter of communicating with users to gather feedback regarding suggested improvements to your design.

The question of safe storage of programs is especially critical when using diskettes. Although these devices offer a handy means for retaining your work and transferring it to others, they can be easily damaged. Table 8.7 outlines a number of tips for effectively maintaining your programs on diskettes. Follow the suggestions on this list carefully in order to avoid losing your valuable programs through carelessness.

PROBLEMS

8.1 Describe some of the strengths of BASIC.

8.2 Learn how to turn on your computer, format a disk, and load BASIC (if it is not loaded automatically on your system).

8.3 The arithmetic operator for BASIC multiplication is the asterisk. Therefore a simple multiplication program can be developed by reformulating line 130 of Fig. 8.1 as

```
130 LET C = A * B
```

Develop a version of this program patterned after Fig. 8.5. Save the program on your diskette and generate a paper printout or hardcopy of the output. Document this program using the format specified in Fig. 8.8. Employ values of FIRST = 20 and SECOND = 16 for your example run.

8.4 Develop, test, and document a program to input the length, width, and height of a box. Have the program compute the volume and the surface area of the box. Also, allow the user to input the units (for example, feet) for the dimensions of the box as string constants and print out the correct units with the answers. Document this program using the format specified in Fig. 8.8. Employ values of length, width, and height equal to 5, 1.5, and 1.25 m, respectively, for your example run.

8.5 Examine Table 8.3. Determine the significant figures available on your computer.

8.6 Examine Table 8.1. List equivalent system commands (if available) for your personal computer.

8.7 Use the BASIC program given in Fig. 8.1 to practice using each of the system commands given in Table 8.1.

8.8 Examine the list of computers in Table 8.2.
  (a) Complete the table entries for obtaining a hard copy of a list and a run on your personal computer if it is not given in the table.
  (b) Obtain a hard copy of a list and a run of the program in Fig. 8.1 if you have a printer.

8.9 Examine Table 8.4.
  (a) Determine the rules for naming variables in BASIC on your computer if it is not listed in the Table.
  (b) Make a list of five acceptable and five unacceptable variable names in BASIC for your personal computer.

8.10 Write a BASIC program comparable to the one given in Fig. 8.5. The program should prompt the user to input a value for the radius of a circle and should respond with values for the diameter, circumference, and area, printed in a neat format.

8.11 Write a BASIC program comparable to the one given in Fig. 8.5. The program should prompt the user for input values for *a, b,* and *c* and should respond with calculated values of *y*, where

$$y = ac + ba + bc$$

8.12 Write a BASIC program comparable to the one given in Fig. 8.5. The program should prompt the user for input values for the velocity of a car and the duration of a trip and should respond with the calculated distance traveled.

8.13 Repeat Prob. 8.4, except calculate the volume and surface area of a cylinder, given its radius and height. Use a radius of 1 m and a height of 10 m to test your program.

CHAPTER 9

BASIC: COMPUTATIONS

As stated on several occasions prior to this point, one of the primary engineering applications of computers is to manipulate numbers. In the previous chapter, you were introduced to some fundamental concepts that lie at the heart of this endeavor: constants, variables, and assignment. Now we will expand on these concepts and introduce some additional material that will allow you to exploit the full potential of BASIC to perform computations.

9.1 ARITHMETIC EXPRESSIONS

Obviously, in the course of engineering problem solving, you will be required to perform more complicated mathematical operations than the simple process of addition described in the previous chapter. A key to understanding how these operations are accomplished is to realize that a primary distinction between computer and conventional equations is that the former are constrained to a single line. In other words, a multilevel, or "built-up," equation such as

$$X = \frac{A + B}{Y - Z^2} + \frac{33.2}{X^3}$$

must be reexpressed so that it can be written as a single-line expression in BASIC.

The operators that are used to accomplish this are listed in Table 9.1. Notice how the two operations that are intrinsically multilevel—that is, division and exponentiation—can be expressed in single-line form using these operators. For example, A divided by B, which in algebra is

TABLE 9.1   *Arithmetic Symbols and Their Associated Priorities*

Symbol	Meaning	Priority
( )	Parentheses	1
$\wedge$ or $\uparrow$ or **	Exponentiation	2
−	Negation	3
*	Multiplication	4
/	Division	4
+	Addition	5
−	Subtraction	5

$$\frac{A}{B}$$

is expressed in BASIC as A/B. Similarly, the cube of A, which in algebra demands that the 3 be raised as a superscript, as in A^3, is expressed in BASIC as A $\wedge$ 3. Note that the power to which the number is raised can be a positive or negative integer or decimal. However, a negative number can only be raised to an integer power.

In order to construct more complicated expressions, a number of rules must be followed. These involve the priorities that are given to the operations. As shown in Table 9.1, parentheses are given top priority in the sense that the calculations within parentheses are implemented first. For example, in

```
C * (A + B)
```

the addition is performed first and the result is then multiplied by C. If several sets are nested within each other the operations in the innermost parentheses are performed first.

Note that parentheses cannot be used to imply multiplication as is done in conventional arithmetic. For example, the above multiplication could not be expressed as C(A + B) in BASIC. The asterisk must always be included to indicate multiplication.

Parentheses must always come in pairs. A good practice when writing expressions is to check that there is always an equivalent number of open and closed parentheses. For example, the expression

$$((X + 2 * (3 - B - (4 * A/B) - 7))$$

is incorrect because there is one too many open parentheses. Checking the number of open and closed parentheses does not ensure that the expression is correct, but if the number of parentheses are not equal, the expression is definitely erroneous.

After operations within parentheses, the next highest priority goes to exponentiation. If multiple exponentiations are required, as in $X^{\wedge}Y^{\wedge}Z$, they are implemented from left to right. This is followed by negation—that is, the placement of a minus sign before an expression to connote that it is negative.* Next, all multiplications and divisions are performed from left to right. Finally, the additions and subtractions are performed from left to right. The use of these priorities to construct mathematical expressions is best shown by example.

EXAMPLE 9.1   Arithmetic Expressions

PROBLEM STATEMENT: Determine the proper values of the following arithmetic expressions in BASIC:

(a) $5 + 6 * 7 + 8$
(b) $(5 + 6) * 7 + 8$ (f) $5 \wedge 2 \wedge 3$
(c) $(5 + 6) * (7 + 8)$ (g) $5 \wedge (2 \wedge 3)$
(d) $-2 \wedge 4$ (h) $8/2 * 4 - 3$
(e) $(-2) \wedge 4$ (i) $(10/(2 - 3) * 5 \wedge 2) \wedge 0.5$

SOLUTION:
(a) Because multiplication has a higher priority than addition, it is performed first to yield

$$5 + 42 + 8$$

Then the additions can be performed from left to right to yield the correct answer, 55.
(b) Although multiplication has a higher priority than addition, the parentheses have even higher priority, and so the parenthetical addition is performed first:

$$11 * 7 + 8$$

followed by multiplication

$$77 + 8$$

and finally addition to give the correct answer, 85.
(c) First evaluate the additions within the parentheses:

$$11 * 15$$

and then multiply to give the correct answer, 165.
(d) Because exponentiation has a higher priority than negation, the 2 is first raised to the power 4 to give 16. Then negation is applied to give the correct answer, -16. (Note that for Applesoft the answer would be 16.)

---

*Note that, in Applesoft, negation has higher priority than exponentiation.

(*e*) In contrast to the expression in (*d*), the parentheses here force us to first transform 2 into -2. Then the -2 is raised to the power 4 to give the correct answer 16.

(*f*) Because the order of exponentiation is from left to right, 5 is first squared to give 25, which in turn is raised to the power 3 to yield the correct answer, 15625.

(*g*) In contrast to the expression in (*f*), the normal order of exponentiation is overruled, and 2 is first cubed to give $5\,^{\wedge}\,8$, which in turn is evaluated to yield the correct answer, 390625.

(*h*) Because multiplication and division have equal priority, they must be performed from left to right. Therefore, 8 is first divided by 2 to give

$$4 * 4 - 3$$

Then 4 is multiplied by 4 to yield

$$16 - 3$$

which can be evaluated for the correct answer, 13. Note that if the left-to-right rule had been disobeyed and the multiplication performed first, an erroneous result of -2 would have been obtained.

(*i*) Because the parentheses in this expression are nested (that is, one pair is within the other), the innermost is evaluated first:

$$(10/5 * 5\,^{\wedge}\,2)\,^{\wedge}\,0.5$$

Next the expression within the parentheses must be evaluated. Because it has highest priority, the exponentiation is performed first:

$$(10/5 * 25)\,^{\wedge}\,0.5$$

Then the division and multiplication are performed from left to right to yield

$$(2 * 25)\,^{\wedge}\,0.5$$

and

$$50\,^{\wedge}\,0.5$$

which can be evaluated to give the correct answer, 7.071.

When dealing with very complicated equations, it is often good practice to break them down into component parts. Then the components can be combined to come up with the end result. This practice is especially helpful in locating and correcting typing errors that frequently occur when entering a long equation.

EXAMPLE 9.2   *Complicated Arithmetic Expressions*

PROBLEM STATEMENT: Write the comparable BASIC version for the algebraic expression

$$y = \frac{\dfrac{x^2 + 2x + 3}{\sqrt{5 + z}} - \sqrt{\dfrac{15 - 77x^3}{z^{3/2}}}}{x^2 - 4xz - 5x^{-4/5}}$$

Treat each major term of the equation as a separate component in order to facilitate programming the equation in BASIC.

SOLUTION: The two terms in the numerator and the denominator can be programmed separately as

```
10 PART1 = (X^2+2*X+3)/((5+Z)^0.5)
20 PART2 = ((15-77*X^3)/(Z^(3/2)))^0.5
30 DENOM = X^2-4*X*Z-5*X^(-4/5)
```

Then the total equation could be formed from the component parts, as in

```
40 Y = (PART1 - PART2)/DENOM
```

Notice that we included unnecessary pairs of parentheses in lines 10 and 20. We could just as correctly have written these expressions as

```
10 PART1 = (X^2+2*X+3)/(5+Z)^0.5
20 PART2 = ((15-77*X^3)/Z^(3/2))^0.5
```

We included the extra parentheses because, in our judgment, they make the equation easier to read. As long as they are matched, extra parentheses are allowable. As in this example, it is a good stylistic practice to include them in order to make the expression more readable. In addition, spaces can be employed to make the expressions clearer. For example, line 30 could be written as

```
30 DENOM = X^2 - 4*X*Z - 5*X^(-4/5)
```

9.2   INTRINSIC, OR BUILT-IN, FUNCTIONS

In engineering practice many mathematical functions appear repeatedly. Rather than require the user to develop special programs for each of these functions, manufacturers include them directly as part of the computer's interpreter. These *built-in* (or *intrinsic* or *library*) *functions* are of the general form

TABLE 9.2   Commonly Available Built-in Function in BASIC (The argument of the function X can be a constant, variable, another function, or an algebraic expression.)

Function	Description
SQR(X)	Calculates the square root of X where X must be positive
ABS(X)	Determines the absolute value of X
LOG(X)	Calculates the natural logarithm of X where X must be positive
EXP(X)	Calculates the exponential of X
SIN(X)	Calculates the sine of X where X is in radians
COS(X)	Calculates the cosine of X where X is in radians
TAN(X)	Calculates the tangent of X where X is in radians
ATN(X)	Calculates the arctan of X where X is dimensionless and returns a value in radians
INT(X)	Determines the greatest integer value that is less than or equal to X
FIX(X)	Determines the truncated value of X, ignoring the decimal part
SGN(X)	Determines a value dependent on the sign of X (If X is positive, $+1$ is returned; if X is zero, 0 is returned; if X is negative, -1 is returned.)
RND	Generates a random number

name (*argument*)

where the *name* is the three-letter word that designates the particular intrinsic function (Table 9.2) and the *argument* is the value that is to be evaluated. The argument can be a constant, variable, another function, or an algebraic expression.

The available types of built-in functions vary from system to system. You should consult your computer's user's manual to determine the types that can be implemented on your machine. In the present chapter, we will discuss the more commonly available functions listed in Table 9.2.

9.2.1   General Mathematical Functions: SQR, ABS, LOG, and EXP

The first group of functions deals with some standard mathematical operations that have broad application in engineering.

The square root function (SQR).   To this point we have learned one way to determine the square root of a number by raising the number to the $1/2$ power, as in

```
50 Y = X^(1/2)
```

After this statement is executed, Y will have the value of the square root of X.

Because this is such a common operation, the built-in function named SQR is included on most BASIC interpreters to perform the same task. An example is

```
50 Y = SQR(X)
```

The absolute value function (ABS).   There are many situations where you will be interested in the value of an expression or variable but not its sign. The ABS function

```
50 Y = ABS(X)
```

provides the value of X without any sign. If X is greater than or equal to 0, ABS(X) will equal X. If it is less than 0, ABS(X) will equal X times -1.

EXAMPLE 9.3   Using SQR and ABS Functions to Determine the Complex Roots of a Quadratic Equation

PROBLEM STATEMENT: Given the quadratic equation

$$f(x) = ax^2 + bx + c$$

the roots may be determined by the formula

$$x = \frac{-b \pm \sqrt{b^2 - 4ac}}{2a}$$

If $b^2 - 4ac \geq 0$, this equation can be used directly. However, if $b^2 - 4ac < 0$, then the roots are complex and a revised solution is

$$x = \text{Re} \pm \text{Im } i$$

where Re is the real part of the root, which can be calculated as

$$\text{Re} = \frac{-b}{2a}$$

Im is the imaginary part, which is calculated as

$$\text{Im} = \frac{\sqrt{|b^2 - 4ac|}}{2a}$$

and $i = \sqrt{-1}$. Use the ABS and SQR functions to determine the complex roots for the case where $b^2 - 4ac \leq 0$.

SOLUTION: The code to determine the complex roots of a quadratic equation is listed in Fig. 9.1. Notice how the ABS function is located within the argument of the SQR function.

FIGURE 9.1
A BASIC program to determine the complex roots of a quadratic equation.

```
10 INPUT "A,B AND C = ";A,B,C
20 REAL = -B/(2*A)
30 IMAGIN = SQR(ABS(B^2 - 4*A*C))/(2*A)
40 PRINT REAL, IMAGIN
50 END
```

The natural logarithm function (LOG). Logarithms have numerous applications in engineering. Recall that the general form of the logarithm is

$$y = \log_a x$$

where a is called the *base* of the logarithm. According to this equation, y is the power to which a must be raised to give x. The most familiar and easy-to-understand form is the *base 10,* or *common, logarithm,* which corresponds to the power to which 10 must be raised to give x; for example,

$$\log_{10} 10{,}000 = 4$$

because 10^4 is equivalent to 10,000.

Common logarithms define powers of 10. For example, common logs from 1 to 2 correspond to values from 10 to 100. Common logs from 2 to 3 correspond to values from 100 to 1000 and so on. Each of these intervals is called an *order of magnitude.*

Besides the base 10 logarithm, the other important form in engineering is the *base e logarithm* where $e = 2.71828. \ldots$. The base e logarithm, which is also called the *natural logarithm,* is extremely important in calculus and other areas of engineering mathematics. For this reason, most BASIC interpreters compute the natural logarithm, as in

```
50 Y = LOG(X)
```

Unfortunately, many BASIC interpreters do not include a similar function to compute the base 10 logarithm. However, it is simple to convert the base e to the base 10 version using the following formula:

$$\log_{10} x = \frac{\log_e x}{\log_e 10} = \frac{\log_e x}{2.3025}$$

Therefore in BASIC the base 10 logarithm can be computed by dividing the natural logarithm by 2.3025, as in

```
50 Y = LOG(X)/2.3025
```

The exponential function (EXP). An additional mathematical function that has widespread engineering applications is the exponential function, or e^x, where e is the base of the natural logarithm. In BASIC it is computed as

```
50 Y = EXP(X)
```

Aside from its direct use, the exponential function provides a means to compute other values such as the hyperbolic trigonometric functions as described in the next section.

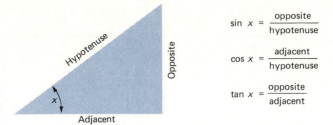

FIGURE 9.2
Definitions of the trigonometric functions.

$$\sin x = \frac{\text{opposite}}{\text{hypotenuse}} \qquad \csc x = \frac{1}{\sin x}$$

$$\cos x = \frac{\text{adjacent}}{\text{hypotenuse}} \qquad \sec x = \frac{1}{\cos x}$$

$$\tan x = \frac{\text{opposite}}{\text{adjacent}} \qquad \cot x = \frac{1}{\tan x}$$

FIGURE 9.3
Relationship of radians to degrees: (a) the angle x of right triangle placed in a standard position at the center of a circle; (b) the definition of a radian; (c) π = 180°.

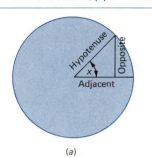

(a)

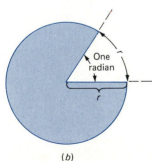

(b)

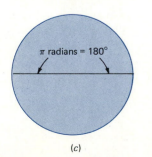

(c)

9.2.2　Trigonometric Functions: SIN, COS, TAN, and ATN

Another set of mathematical functions that have great relevance to engineering is the trigonometric functions. In BASIC the sine (SIN), cosine (COS), and tangent (TAN) functions are commonly available. As outlined in Fig. 9.2, other trigonometric functions can be derived by inverting these three. In addition, the inverse or arctangent (ATN) function is available on many systems. If the value of the tangent is known, the arctangent can be used to determine the corresponding angle.

For all of these functions, the angle is expressed in radians (Fig. 9.3). A *radian* is the measure of an angle at the center of a circle-that corresponds to an arc that is equal to the circle's radius (Fig. 9.3b). The conversion from degrees to radians is (Fig. 9.3c)

$$\pi \text{ radians } = 180°$$

Therefore, if X is expressed originally in degrees, the sine of X can be computed in BASIC as

```
50 Y = SIN (X * 3.14159/180)
```

Aside from the standard trigonometric functions, hyperbolic trigonometric functions can be computed with the EXP function using the following relationships:

$$\sinh x = \tfrac{1}{2}(e^x - e^{-x})$$

$$\cosh x = \tfrac{1}{2}(e^x + e^{-x})$$

$$\tanh x = \frac{\sinh x}{\cosh x}$$

A group of other trigonometric functions that can be evaluated on the basis of intrinsic functions is listed in Table 9.3.

9.2.3　Miscellaneous Functions: INT, FIX, SGN, and RND

Aside from general mathematical and trigonometric functions, there are other built-in functions that are usually available and that have utility to engineers.

TABLE 9.3   Derived Trigonometric Functions that Can Be Developed from the Intrinsic
Functions of BASIC

Function	Equivalent
Secant	1/COS(X)
Cosecant	1/SIN(X)
Cotangent	1/TAN(X)
Inverse sine	ATN(X/SQR(−X * X + 1))
Inverse cosine	ATN(X/SQR(−X * X + 1)) + 1.5708
Inverse secant	ATN(SQR(X * X − 1)) + (SGN(X) − 1) * 1.5708
Inverse cosecant	ATN(1/SQR(X * X − 1)) + (SGN(X) − 1) * 1.5708
Inverse cotangent	−ATN(X) + 1.5708
Hyperbolic sine	(EXP(X) − EXP(−X))/2
Hyperbolic cosine	(EXP(X) + EXP(−X))/2
Hyperbolic tangent	(EXP(X) − EXP(−X))/(EXP(X) + EXP(−X))
Hyperbolic secant	2/(EXP(X) + EXP(−X))
Hyperbolic cosecant	2/(EXP(X) − EXP(−X))
Hyperbolic cotangent	(EXP(X) + EXP(−X))/(EXP(X) − EXP(−X))
Inverse hyperbolic sine	LOG(X + SQR(X * X + 1))
Inverse hyperbolic cosine	LOG(X + SQR(X * X − 1))
Inverse hyperbolic tangent	LOG((1 + X)/(1 − X))/2
Inverse hyperbolic secant	LOG((SQR(−X * X + 1) + 1)/X)
Inverse hyperbolic cosecant	LOG((SGN(X) * SQR(X * X + 1) + 1)/X)
Inverse hyperbolic cotangent	LOG((X + 1)/(X − 1))/2

Functions to determine integer values (INT and FIX).

The *integer function* gives
the greatest integer (that is, whole number) that is less than or equal to the argument.
Because of this definition, it is sometimes referred to as the *greatest integer function.*
For example,

```
30 X = 3.9
40 Y = INT(X)
```

will give Y equal to 3. Also, it is important to realize that

```
30 X = -2.7
40 Y = INT(X)
```

will give Y equal to −3, because it is the greatest integer that is less than −2.7.

The *FIX function* discards the decimal part of a number and returns the integer
part. For positive numbers, this gives the same result as does the INT function. For
example,

```
30 X = 3.9
40 Y = FIX(X)
```

will give Y equal to 3. However, in contrast to the INT function,

```
30 X = -2.7
40 Y = FIX(X)
```

will give Y equal to -2, because the FIX simply truncates, or chops off, the 0.7 to attain its result. It must be noted that the Apple II and the Commodore do not carry the FIX function.

EXAMPLE 9.4   *Using the INT Function to Truncate or Round Positive Numbers*

PROBLEM STATEMENT: It is often necessary to round numbers to a specified number of decimal places. One very practical situation occurs when performing calculations pertaining to money. Whereas an economic formula might yield a result of $467.9587, in reality we know that such a result must be rounded to the nearest cent. According to rounding rules, because the first discarded digit is greater than 5 (that is, it is 8), the last retained digit would be rounded up, giving a result of $467.96. The INT function provides a way to accomplish this same manipulation with the computer.

SOLUTION: Before developing the rounding formula, we will first derive a formula to truncate, or "chop off," decimal places. First, we can multiply the value by 10 raised to the number of decimal places we desire to retain. For example, in the present situation we desire to retain two decimal places, and therefore we multiply the value by 10^2, or 100, as in

$$467.9587 \times 100 = 46795.87 \tag{E9.4.1}$$

Next we can apply the INT function to this result to chop off the decimal portion:

```
INT(46795.87) = 46795
```

Finally we can divide this result by 100 to give

$$46795/100 = 467.95$$

Thus we have retained two decimal places and truncated the rest. The entire process can be represented by the BASIC equation

```
X = INT(X * 10^D)/10^D
```
$$\tag{E9.4.2}$$

where X is the number to be truncated and D is the number of decimal places to be retained.

Now it is a very simple matter to modify this relationship slightly so that it will round. This can be done by looking more closely at the effect of the multiplication in Eq. (E9.4.2) on the last retained and the first discarded digit in the number being rounded (Fig. 9.4). Multiplying the number by 10^D essentially places the last re-

FIGURE 9.4
Illustration of the retained and discarded digits for a number
rounded to two decimal places.

tained digit in the unit position and the first discarded digit in the first decimal place. For example, as shown in Eq. (E9.4.1), the multiplication moves the decimal point between the last retained digit (5) and the first discarded digit (8). Now it is simple to see that if the first discarded digit is above 5, adding 0.5 to the quantity will increase the last retained digit by 1. For example, for Eq. (E9.4.1)

$$46795.87 + 0.5 = 46796.37 \tag{E9.4.3}$$

If the first discarded digit were less than 5, adding 0.5 would have no effect on the last retained digit. For instance, suppose the number were 46795.49. Adding 0.5 yields 46795.99, and therefore the last retained digit (5) is not changed.

To complete the process, the INT function is applied to Eq. (E9.4.3):

```
INT(46796.37) = 46796
```

and the result divided by 100 to give the correct answer, 467.96. The entire process is represented by

```
X = INT(X * 10^D + 0.5)/10^D
```
(E9.4.4)

It should be noted that numbers can be rounded to an integer value by setting $D = 0$, as in

```
X = INT(467.9587 * 10^0 + 0.5)/10^0 = 468
```

or to the tens, hundreds, or higher places by setting D to negative numbers. For example, to round to the 10's place, $D = -1$ and Eq. (E9.4.4) becomes

```
X = INT(467.9587 * 10^(-1) + 0.5)/10^(-1) = 470
```

The sign function (SGN).   There are occasions where you will need to know the sign of a quantity without regard for its magnitude. The sign function fills this need. If the quantity being evaluated is positive, the SGN function will be $+1$. If it is zero, SGN will be 0. If it is negative, SGN will be -1. For example,

```
30 X = -3.14159
40 Y = SGN(X)
```

will yield a value of -1 for Y.

EXAMPLE 9.5   *Using the FIX and SGN Functions to Truncate and Round Both Positive and Negative Numbers*

PROBLEM STATEMENT: Notice that we did not truncate or round negative numbers in Example 9.4. Develop relationships for truncating and rounding that hold for both positive and negative numbers.

SOLUTION: Because of the way that the integer function operates on negative values, Eq. (E9.4.2) is inadequate for truncating negative numbers. For example, -467.9587 truncated to two decimal places would be [according to Eq. (E9.4.2)]

```
X = INT(-467.9587 * 10^2)/10^2 = -467.96
```

which is erroneous. A more general truncation formula that works for either positive or negative values can be developed by substituting the FIX for the INT function in Eq. (E9.4.2) to give

```
X = FIX(X * 10^D)/10^D
```
      (E9.5.1)

For example, using the same values as before,

```
X = FIX(-467.9587 * 10^2)/10^2 = -467.95
```

which is correct. Because FIX and INT yield identical results for positive values, Eq. (E9.5.1) holds for both positive and negative values.

    For rounding, Eq. (E9.4.4) works properly for most negative numbers. For example, -467.9587 is rounded to two decimal places, as in

```
X = INT(-467.9587 * 10^2 + 0.5)/10^2 = -467.96
```

which is correct. However, there is one special case where it yields an erroneous result. If the first discarded digit is a 5 and all other discarded digits are zeros, Eq. (E9.4.4) rounds negative numbers improperly. For example, rounding -467.95 to one decimal place gives

```
X = INT(-467.95 * 10^1 + 0.5)/10^1 = -467.9
```

Thus, instead of rounding to the correct answer of -468.0, it rounds to -467.9. Therefore a slight adjustment of the equation using the FIX and the SGN functions can be made to correct this problem:

```
X = FIX(X * 10^D + SGN(X) * 0.5)/10^D
```
      (E9.5.2)

For the example,

```
X = FIX(-467.95 * 10^1 + SGN(-467.95) * 0.5)/10^1
```

which when evaluated yields the correct result of -468.0. Thus, Eq. (E9.5.2) is superior to Eq. (E9.4.4) in that it adequately rounds both negative and positive numbers.

The random-number function (RND).   In a later chapter, we will introduce you to the subject of statistics. One important aspect of this subject is the random number. To understand what this means, suppose that you toss an honest coin over and over again and record whether it comes up heads (H) or tails (T). Your result for 20 tosses might look like

HTTHHTHHHHTHTTTHTHTT

This is called a *randomly generated sequence* because each letter represents the result of an independent toss of the coin. If we repeated the experiment, we would undoubtedly come up with a different sequence because for an honest coin the outcome of each toss is random. That is, there is an equal likelihood that each toss will yield either a head or a tail, and there is no way to predict the outcome of an individual toss beforehand.

The same sort of experiment could be performed with a die (that is, one of a pair of dice). For this case, every toss would result in either 1, 2, 3, 4, 5, or 6, and the sequence of 20 tosses might look like

53122341626444153566

Random-number sequences are very important in engineering because many of the systems that you will study have aspects that vary in a random fashion. We will investigate some of these processes in depth in later chapters. For the time being, it is useful to know that most BASIC interpreters have a random-number function. For Microsoft BASIC, it is represented as

```
40 Y = RND
```

Every time this function is implemented, a random decimal fraction somewhere between 0 and 1 is returned.

EXAMPLE 9.6   Simulating a Random Coin Toss with the Computer

PROBLEM STATEMENT: Use the RND function to simulate the random toss of a coin.

SOLUTION: The following four-line program can be used to simulate the random toss of a coin:

```
10 X = RND
20 Y = INT(X * 10^0 + 0.5)/10^0
30 PRINT X, Y
40 END
```

Line 10 generates a random number between 0 and 1. Then line 20 rounds the number to an integer. If X is less than 0.5, the rounded value will be 0 and we will consider the toss a tail. If X is greater than or equal to 0.5, the rounded value will be 1 and we will consider the toss a head. A run yields

```
RUN
   .1213501          0
```

Therefore a tail has been tossed. To obtain additional values, the statements can be repeated. For example, for three tosses a simple program would be

```
10 X = RND
20 Y = INT(X * 10^0 + 0.5)/10^0
30 PRINT X, Y
40 X = RND
50 Y = INT(X * 10^0 + 0.5)/10^0
60 PRINT X, Y
70 X = RND
80 Y = INT(X * 10^0 + 0.5)/10^0
90 PRINT X, Y
100 END
RUN
   .1213501          0
   .651861           1
   .8688611          1
```

Thus for this case the second and third tosses come out heads.

Every time the program at the end of Example 9.6 is run on a particular computer, the same series of random numbers will be generated. This is because the computer generates the random number on the basis of a starting value called a *seed*. Unless the seed is changed, a particular computer will generate the same sequence of random numbers every time it is run. The seed can be changed with a RANDOMIZE statement. This statement can be written in two ways. One way is to include any integer number between $-32,766$ and $32,767$, as in

```
5 RANDOMIZE 26
```

TABLE 9.4   Statements to Generate Random Numbers

System	Statements
Apple	10 Y = RND(X) If X > 0, a unique sequence will be generated on each run. If X < 0, the same sequence will be generated for each negative value of X.
Commodore	Same as for Apple.
IBM PC	10 RANDOMIZE 20 Y = RND The RANDOMIZE statement can be followed by a number which acts as the seed for the random number generator.
Macintosh	Same as for IBM PC.
TRS-80	Same as for IBM PC, except RANDOM instead of RANDOMIZE.
Your machine	

This changes the seed to 26, and if this line were included in the program at the end of Example 9.6, a different sequence of random numbers would result. A second way to change the seed is to merely write

```
5 RANDOMIZE
```

When the statement is executed the computer will query you, as in

```
RANDOM NUMBER SEED (-32768 to 32767)?
```

whereupon you could enter 26 or any other integer number within the limits.

The operation and syntax of the RND statement differ on various personal computers. Table 9.4 summarizes these differences for some popular systems.

9.2.4   Other Library Functions

Aside from the standard library functions described in the previous sections, each computer has its own special functions. For example, the IBM PC has several built-in functions that are used to perform operations on string data. These and other specialized functions are described in Appendix B.

9.3   USER-DEFINED FUNCTIONS

Aside from built-in functions, you might be interested in defining functions of your own. The DEF, short for *def*inition, statement is used for this purpose and has the general form

ln DEF FN*variable* (*argument*) = *expression*

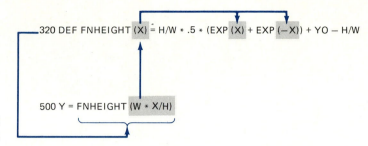

FIGURE 9.5
An example of a DEF statement.

where the FN is followed by any legal *variable* name, the *argument* is the value at which the function is to be evaluated, and the *expression* is the actual arithmetic expression. For example, in the simple addition program given in Chap. 8 (Fig. 8.5), we could have defined a function for the addition of FIRST and SECOND as

```
265 DEF FNADD(X,Y) = X + Y
```

Then line 440 could have been rewritten as

```
440 TOTAL = FNADD(FIRST,SECOND)
```

For Microsoft BASIC, several arguments may be passed into a DEF statement.* Other variables in the statement have the same value as they had in the main program at the point the user-defined function was invoked (Fig. 9.5).

A user-defined function must always be placed before the point in the program at which the function is to be used. A good stylistic practice is to place all DEF statements together at the beginning of the program.

EXAMPLE 9.7   A User-Defined Function

PROBLEM STATEMENT: The height of a hanging cable is given by the formula (Fig. 9.6)

$$y = \frac{H}{w} \cosh\left(\frac{w}{H}x\right) + y_0 - \frac{H}{w} \tag{E9.7.1}$$

where H is the horizontal force at $x = a$, y_0 is the height at $x = a$, and w is the cable's weight per unit length. In Sec. 9.2.2, we noted that the hyperbolic cosine can be computed by

$$\cosh x = \tfrac{1}{2}(e^x + e^{-x}) \tag{E9.7.2}$$

Compose a computer program to calculate the height of a cable at equally spaced

*Note that Applesoft permits only one argument to be passed.

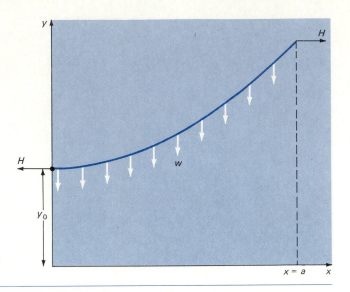

FIGURE 9.6
A section of a hanging cable.

FIGURE 9.7
A program to determine the height
of a hanging cable.

```
100 REM      *************************************************
110 REM      *        PROGRAM TO COMPUTE THE HEIGHT OF A      *
120 REM      *             HANGING CABLE                      *
130 REM      *             (BASIC version)                    *
140 REM      *             by  S. C. Chapra                   *
150 REM      *                                                *
160 REM      *     This program computes the height of a      *
170 REM      *     hanging cable at four equally-spaced       *
180 REM      *     points along the X-dimension.              *
190 REM      *************************************************
200 REM
210 REM      .-----------------------------------------------.
220 REM      |     DEFINITION OF VARIABLES                   |
230 REM      |                                               |
240 REM      |   W    = WEIGHT PER LENGTH                    |
250 REM      |   H    = HORIZONTAL FORCE                     |
260 REM      |   Y0   = INITIAL HEIGHT                       |
270 REM      |   XMAX = MAXIMUM LENGTH                       |
280 REM      |   XINC = INCREMENT OF X WHERE Y COMPUTED      |
290 REM      |   FNHEIGHT = FUNCTION TO COMPUTE CABLE HEIGHT |
300 REM      '-----------------------------------------------'
310 REM
320 DEF FNHEIGHT(X) = H/W*.5*(EXP(X) + EXP(-X)) + Y0 - H/W
330 REM
340 REM ******************  MAIN PROGRAM  ******************
350 REM
360 INPUT "WEIGHT PER LENGTH = ";W
370 INPUT "HORIZONTAL FORCE,H = ";H
380 INPUT "INITIAL HEIGHT,Y0 = ";Y0
390 INPUT "MAXIMUM X = ";XMAX
400 XINC = XMAX/4
410 PRINT:PRINT "DISTANCE        HEIGHT":PRINT
420 REM
430 REM                Print Initial Values
440 REM
450 PRINT X,Y0
460 REM
470 REM    Calculate and Print Remaining Values
480 REM
490 X = X + XINC
500 Y = FNHEIGHT(W*X/H)
510 PRINT X,Y
520 X = X + XINC
530 Y = FNHEIGHT(W*X/H)
540 PRINT X,Y
550 X = X + XINC
560 Y = FNHEIGHT(W*X/H)
570 PRINT X,Y
580 X = X + XINC
590 Y = FNHEIGHT(W*X/H)
600 PRINT X,Y
610 END


RUN
WEIGHT PER LENGTH = ? 10
HORIZONTAL FORCE,H = ? 2000
INITIAL HEIGHT,Y0 = ? 10
MAXIMUM X = ? 40

DISTANCE          HEIGHT

0                 10
10                10.25006
20                11.00084
30                12.25423
40                14.01335
Ok█
```

values of x from $x = 0$ to a maximum x, given w and H. Employ a user-defined function to compute the hyperbolic cosine. Test the program for the following case: $w = 10$, $H = 2000$, and $y_0 = 10$ for $x = 0$ to 40. Print out values of y at four equally spaced increments from $x = 0$ to 40.

SOLUTION: Figure 9.7 shows a program to compute the height of a hanging cable. Notice how line 320 has been set up to calculate the height as a user-defined function. Lines 500, 530, 560, and 590 access this function to compute the cable height. Notice how the argument of FNHEIGHT in these lines is an arithmetic expression, $W * X/H$, that is passed to line 320. There its value is ascribed to the dummy variable X which is used to evaluate the height. This result is then passed back and substituted for FNHEIGHT ($W * X/H$) in lines 500, 530, 560, and 590. X is called a *dummy variable* because its use as the argument in line 320 does not affect its value in the rest of the program. In fact, the value of X in the user-defined function is actually equal to $W * X/H$.

Although the foregoing example shows how a user-defined function is employed, it does not really illustrate all of its advantages. These are seen when the function is extremely long and complicated and is used numerous times throughout the program. In these instances, the concise representation of the user-defined function is a potent asset.

PROBLEMS

Which of the following are valid assignment statements in BASIC? State the errors for those that are incorrect.

9.1 $R1 = -4 ^\wedge 1.5$

9.2 $C = C + 22$

9.3 $A + B = C + D$

9.4 $X - Y = 7$

9.5 $4 = X + Y$

If $R = 3$, $S = 5$, and $T = -2$, evaluate the following. Perform the evaluation step by step, following the priority order of the individual operations.

9.6 $-S * R + T$

9.7 $S + R/S * T$

9.8 $S + T ^\wedge T * T ^\wedge R$

9.9 $R * (R + (R + 1) + 1$

9.10 $T + 10 * S - S \wedge T \wedge R$

Write BASIC statements for the following algebraic equations.

9.11

$$x = a \div c + \frac{c - d^2}{g} \frac{r^5}{2f^2 - 4}$$

9.12

$$r = \frac{cb^3}{r^2}(3d + 8x) - \frac{4}{9}x$$

9.13 In civil engineering the vertical stress under the center of a loaded circular area is given by

$$s = q\left[1 - \frac{1}{(r^2/z^2 + 1)^{3/2}}\right] ,$$

Write this equation in BASIC.

9.14 In engineering economics the annual payment, A, on a debt that increases linearly with time at a rate G for n years, is given by the equation

$$A = G\left[\frac{1}{n} - \frac{n}{(1 + i)^n - 1}\right]$$

Write this equation in BASIC.

9.15 In chemical engineering the residence time of five equal-sized reactors in series is given by

$$t = \frac{5}{k}\left[\left(\frac{c_0}{c_5}\right)^{1/5} - 1\right]$$

Write this equation in BASIC.

9.16 In fluid mechanics the pressure of a gas at the stagnation point is given by

$$p = p_0\left(1 + m^2\frac{k - 1}{2}\right)^{k/(k-1)}$$

Write this equation in BASIC.

9.17 In electrical engineering the electric field at a distance x from the center of a ring-shaped conductor is given by

$$E = \frac{1}{4\pi\epsilon_0}\frac{Qx}{(x^2 + a^2)^{3/2}}$$

Write this equation in BASIC.

9.18 In chemical engineering van der Waals equation gives pressure of a gas as

$$p = \frac{RT}{(V/n) - b} - a\left(\frac{n}{V}\right)^2$$

Write this equation in BASIC.

9.19 The saturation value of dissolved oxygen in fresh water is calculated by the equation

$$c_s = 14.652 - (4.1022 \times 10^{-1})T_C + (7.9910 \times 10^{-3})\,T_C^2 \\ - (7.7774 \times 10^{-5})T_C^3$$

where T_C is temperature in degrees Celsius. Saltwater values can be approximated by multiplying the freshwater result by $1 - (9 \times 10^{-6})n$, where n is salinity in milligrams per liter. Finally, temperature in degrees Celsius is related to temperature in degrees Fahrenheit, T_F, by

$$T_C = \tfrac{5}{9}(T_F - 32)$$

Develop a computer program to compute c_s, given n and T_F. Test the program for $n = 18{,}000$ and $T_F = 70$.

9.20 A projectile is fired with a velocity v_0, at an angle of θ with the horizon, off the top of a cliff of height y_0 (Fig. P9.20). The following formulas can be used to compute how far horizontally (x) the projectile will carry prior to impacting on the plane below.

$$(v_y)_0 = v_0 \sin \theta$$

$$(v_x)_0 = v_0 \cos \theta$$

$$t = \frac{-(v_y)_0 - \sqrt{(v_y)_0^2 + 2gy_0}}{g}$$

$$x = (v_x)_0 t$$

where $(v_y)_0$ and $(v_x)_0$ are the vertical and horizontal components of the initial velocity, t is the time of flight, and g is the gravitational constant (or -9.81

FIGURE P9.20
The path of a projectile fired off the top of a cliff.

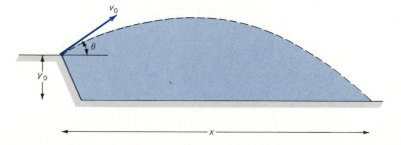

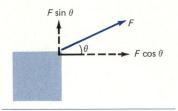

FIGURE P9.23

m/s^{-2}, where the negative sign designates that we are assigning a negative value to downward acceleration). Prepare a program to compute x, given v_0, θ, and y_0. If $v_0 = 180$ m/s, $\theta = 30°$, and $y_0 = -150$ m, x will be equal to 3100 m. Use this result to test your program. Then employ the program to determine what value of θ results in the greatest value of x. Use trial and error to determine this optimal value of x.

9.21 Examine Table 9.2 and compare the functions given there with the built-in functions available for your personal computer.

9.22 Examine Table 9.4. Determine how your personal computer can be used to generate a random number.

9.23 Figure P9.23 shows a force F acting on a block at an angle θ. The horizontal and vertical components of the force are given by $F \cos \theta$ and $F \sin \theta$, respectively. Write a program similar to the one shown in Fig. 9.7 to compute the horizontal and vertical components of F for given values of F and θ. Use the following values for F and θ:

Force, lb	Angle, degrees
1000	30
50	65
1200	130

Your program should have at least one application of each of the following BASIC statements: REM, INPUT, PRINT, and DEF.

9.24 Write a computer program that requests numerical values to be rounded and the number of decimal points to be retained. Employ a DEF statement to compute a rounded number. Use the following data to test your program:

Number to Be Rounded	Number of Decimal Points Retained
-3.14159265	3
86,400.761	1
0.666666	4

9.25 Write a computer program that requests numerical values for the following constants and computes and displays a value of x using a DEF statement. Values of x are computed according to the equation from Prob. 9.11.

$$a = 3 \qquad f = 7$$
$$c = 4 \qquad g = 5$$
$$d = 1 \qquad r = 2$$

9.26 Repeat Prob. 9.25, except compute A using a DEF statement and the formula in Prob. 9.14. ($G = 1000$, $n = 5$, and $i = 0.3$.)

9.27 Repeat Prob. 9.25, except compute E using a DEF statement and the formula in Prob. 9.17. ($a = -2$; $\epsilon_0 = 2.7$; $\pi = 3.2$; $Q = 100$; and $x = 2$.)

9.28 Repeat Prob. 9.25, except compute p using a DEF statement and the formula in Prob. 9.18. ($a = 3.592$; $b = 0.04267$; $n = 1$; $R = 0.082054$; $T = 300$; and $V = 2.4616$.)

9.29 Population growth in an industrial area is given by the following formula:

$$p(t) = P(1 - e^{-at})$$

where $p(t)$ = population at any time
P = maximum population
a = constant, yr^{-1}
t = time, years

Write a BASIC computer program to compute the population at $t = 5$ years with $P = 100,000$ for $a = 0.05$, 0.2, and 0.8. Use REM, INPUT, PRINT, and DEF statements in your program.

9.30 The maximum velocity of a rocket $\bar{v}$ is given by

$$\bar{v} = a \ln\left(\frac{m_0}{m_0 - m_1}\right)$$

where a = relative velocity of exhaust gases
m_0 = total mass of rocket plus fuel at start of flight
m_1 = mass of fuel

Write a BASIC computer program to compute $\bar{v}$ for $m_1 = 25,000$, $30,000$, and $35,000$, given $m_0 = 40,000$ and $a = 25,000$. Use REM, INPUT, PRINT, and DEF statements in your program.

9.31 A triangle is defined by sides ADJ, OPP, and HYP (Fig. P9.31). Write a BASIC computer program to calculate HYP and θ using two DEF statements. Use the following data to test your program.

ADJ	OPP
10	10
5	0
2	6

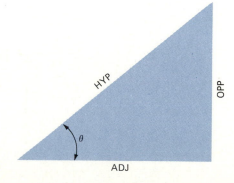

FIGURE P9.31

CHAPTER 10

BASIC:
INPUT/OUTPUT

Because it represents the interface between the user and the program, input/output is a critical aspect of software design. You have been introduced to two statements that are employed for this purpose: the INPUT and the PRINT statements. In Chap. 8 you were shown how these statements can be used in conjunction with string constants to devise descriptive input and output. Now we turn to some additional statements and concepts that will extend your capabilities in this area.

10.1 FORMATTED OUTPUT

There are many cases in engineering where you would want to control the format whereby information is displayed on the screen or on printed output. This is particularly important when the program is to be used for report generation. You have already been introduced to several ways to do this using the PRINT, TAB, and CLS commands. Now we will delve more deeply into the topic within the context of a very important programming task—outputting data in tabular form.

10.1.1 Column Output with Commas

The simplest way to produce tabular output of data is to use the PRINT statement, with the variables separated by commas. The output device, such as the monitor screen, can then be thought of as consisting of vertical columns or print zones that are an equal number of characters wide (Fig. 10.1). The first variable in a PRINT statement is placed in the first zone, and subsequent values, delimited by commas, are placed sequentially in the zones to the right. Negative numbers and string data are printed starting in the first column of each 14-character-wide zone. In contrast,

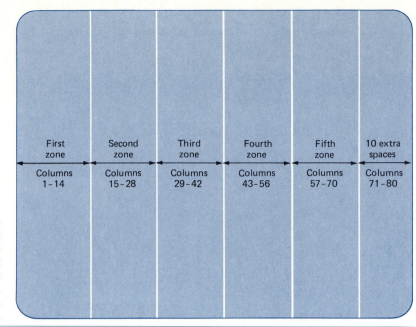

First zone	Second zone	Third zone	Fourth zone	Fifth zone	10 extra spaces
Columns 1–14	Columns 15–28	Columns 29–42	Columns 43–56	Columns 57–70	Columns 71–80

FIGURE 10.1
When values are delimited with commas in a PRINT statement, the output device can be thought of as consisting of equal-width print zones. For Microsoft BASIC, the zones are 14-characters wide. Other systems, such as the Apple, use somewhat different widths.

positive numbers start in the second position because Microsoft BASIC leaves a space for the + sign but does not actually print it. Computers, such as the Apple, do not leave a space for the + sign but print all data flush against the left side of the zone. Some other rules and aspects of comma-delimited PRINT statements are explored in the following example.

EXAMPLE 10.1   Using the Comma-Delimited PRINT Statement

PROBLEM STATEMENT: Investigate the behavior of the series of PRINT statements shown in Fig. 10.2a. Figure 10.2b shows the resulting output.

SOLUTION: Notice how statements 100 to 140 produce output that fits within the five print zones. In contrast, line number 160 produces results that extend beyond this limit. Because each line consists of only five complete 14-character print zones, the last three numbers, 16, 17, and 18, do not fit on the first line. Consequently, the output continues, or is "wrapped around," to the next line. On some computers, this continuation can lead to nonalignment of the columns. For Fig. 10.2, however, the computer maintains correct alignment.

   The PRINT statement with commas always goes to the next unused area to the right to print the next piece of information. In lines 180 and 200 the width of the item to be printed extends over more than one 14-character-wide area, forcing the subsequent data to be printed in the next zone to the right. Actually +123456789.123 in line 200 is exactly 14 characters long, and so it does not quite extend into the second zone. Nonetheless, −1.57711E-9 is not printed in the second zone, but it is printed

```
100 PRINT 1,2,3,4,5
110 PRINT
120 PRINT,10,20,30,40
130 PRINT
140 PRINT,,,99
150 PRINT
160 PRINT-11,12,13,14,15,16,17,18
170 PRINT
180 PRINT"This is a long data label for a little data.",1.23E+11
190 PRINT
200 PRINT 123456789.123#,-1.55771E-09,-45
210 LET A# = 123456789.123#
220 LET B#=-1.57711E-09
230 PRINT
240 PRINT A#,B#,-45
250 END
```
(a)

```
RUN
 1                 2                 3                 4                 5

                  10                20                30                40

                                                      99

-11               12                13                14                15
 16               17                18

This is a long data label for a little data.                          1.23E+11

 123456789.123                      -1.55771E-09   -45

 123456789.123                      -1.577109998152082D-09             -45
Ok▪
```
(b)

FIGURE 10.2
(a) A group of PRINT statements
using commas as delimiters; (b) the
resulting output.

in the third. If BASIC had put -1.57711E-9 in the second zone, then the display might have looked like this:

```
123456789.123-1.57711E-9
```

which makes little sense because no spaces are present. The PRINT statement of Microsoft BASIC always leaves at least one blank following a number unless you use one of the advanced formatting features (explained in a later section) to explicitly override this requirement.

Although the PRINT statement of BASIC tries to display numbers in as simple and consistent a form as possible, that is, using 4 instead of 4.00 and $1.23\text{E}+11$ instead of $12.3\text{E}+10$, it always displays the values without rounding. Thus, when the constant $-.0157711\text{E-7}$ is displayed, it is printed in proper scientific notation, -1.57711E-9. When this constant is assigned to the double-precision real variable B#, its form is changed, and the PRINT statement outputs the value exactly as stored in the variable B# without correcting or rounding the value. These extra, insignificant digits can cause the results which follow in the PRINT statement to be printed one area more to the right than they would have otherwise. This can create a problem if you are trying to print results in a tabular form. In line 200, -45 is printed in the fourth column, whereas in line 240 it is in the fifth column of the table.

If care is taken to use the comma-delimited PRINT properly, data can be output simply and attractively. Such is the case in the following example

EXAMPLE 10.2   Outputting a Table with Commas

PROBLEM STATEMENT: The students in a class have obtained the following grades:

Student Name	Average Quiz	Average Homework	Final Exam
Smith, Jane	82.00	83.90	100
Johnson, Fred	96.55	92.00	77

Their final grade (FG) is computed with the following weighted average:

$$FG = AQ/3 + AH/3 + FE/3$$

where AQ, AH, and FE are the average quiz, average homework, and final exam grades, respectively. Develop a computer program to determine the final grade for each student. Print out the students' grades in a tabular format as shown above, with the final grade added as the last column.

SOLUTION: The computer code to determine the final grade and to output the table is shown in Fig. 10.3. Notice that by delimiting the output with commas, the table is printed nicely.

FIGURE 10.3
The computer code and output
from Example 10.2.

```
10 INPUT NAME1$,AQ1,AH1,FE1
20 INPUT NAME2$,AQ2,AH2,FE2
50 FG1 = AQ1/3 + AH1/3 + FE1/3
60 FG2 = AQ2/3 + AH2/3 + FE2/3
80 PRINT "STUDENT","AVERAGE","AVERAGE","FINAL","FINAL"
90 PRINT "NAME","QUIZ","HOMEWORK","EXAM","GRADE"
95 PRINT
100 PRINT NAME1$,AQ1,AH1,FE1,FG1
110 PRINT NAME2$,AQ2,AH2,FE2,FG2
300 END


RUN
? "SMITH,JANE",82.00,83.90,100.00
? "JOHNSON,FRED",96.55,92.00,77.00
STUDENT       AVERAGE      AVERAGE       FINAL        FINAL
NAME          QUIZ         HOMEWORK      EXAM         GRADE

SMITH,JANE    82.00        83.90         100.00       88.63333
JOHNSON,FRED  96.55        92.00         77.00        88.51666
Ok
```

10.1.2 Column Output with Semicolons, TAB, and SPC Commands

As described in Chap. 8, an alternative way to output data in tabular form is provided by the TAB command. This command is used in a PRINT statement when the data is delimited by semicolons. When semicolons are used as delimiters, the automatic spacing associated with commas does not occur. The TAB command can then be used like a typewriter tab setting to specify the starting position of each value. For example, the following would be used to align headings for a table:

```
80 PRINT TAB(2);"STUDENT";TAB(25);"AVERAGE"
85 PRINT TAB(3); "NAME", TAB(25); "QUIZ"
```

When these statements are executed, the result is

```
STUDENT                 AVERAGE
   NAME                 QUIZ
```

A command that is similar in spirit to the TAB is the SPC command, which is of the general form

$$ln \quad \text{PRINT} \quad value_1 ; \quad SPC(n_1) ; value_2$$

where n_1 is the number of spaces to be inserted between the end of $value_1$ and the beginning of $value_2$. Thus this command acts just like a typewriter space bar. Codes that give output equivalent to that obtained with the TAB commands above are

```
80 PRINT SPC(1);"STUDENT";SPC(16);"AVERAGE"
85 PRINT SPC(2);"NAME";SPC(18);"QUIZ"
```

EXAMPLE 10.3   Outputting a Table with Semicolons and Tabs

PROBLEM STATEMENT: Suppose that a third student is to be added to the table. The student's name is John Brysmialkiewicz and his grades are AQ = 97.3, AH = 92, and FE = 99. Because this student's last name exceeds the 14 spaces in a comma-delimited print zone, adding him to the table will pose problems. The results will look like

```
run
STUDENT             AVERAGE       AVERAGE       FINAL     FINAL
NAME                QUIZ          HOMEWORK      EXAM      GRADE
SMITH,JANE          82            83.9          100       88.63333
JOHNSON,FRED        96.55         92            77        88.51666
BRYSMIALKIEWICZ,JOHN              97.3          92        99
 96.1
Ok
```

Because the name extends into the second print zone, all subsequent output is shifted one print zone to the right. Because there are only five comma-delimited print zones per line, the final grade will be pushed to the next line. We will use TAB commands and semicolons to print the table correctly.

SOLUTION: Lines 80–110 in Fig. 10.3 can be deleted and the following lines added:

```
80  PRINT "STUDENT";TAB(25);"AVERAGE";TAB(36);
81  PRINT "AVERAGE";TAB(47);"FINAL";TAB(58);"FINAL"
85  PRINT "NAME";TAB(25);"QUIZ";TAB(36);"HOMEWORK";
86  PRINT TAB(47);"EXAM";TAB(58);"GRADE"
90  PRINT
100 PRINT NAME1$;TAB(25);AQ1;TAB(36);AH1;TAB(47);FE1;TAB(58);FG1
110 PRINT NAME2$:TAB(25);AQ2;TAB(36);AH2;TAB(47);FE2;TAB(58);FG2
120 PRINT NAME3$;TAB(25);AQ3;TAB(36);AH3;TAB(47);FE3;TAB(58);FG3
```

The results when the program is run are

```
STUDENT                 AVERAGE AVERAGE  FINAL FINAL
NAME                    QUIZ    HOMEWORK EXAM  GRADE
SMITH,JANE              82      83.9     100   88.63333
JOHNSON,FRED            96.55   92       77    88.51666
BRYSMIALKIEWICZ,JOHN    97.3    92       99    96.1
Ok
```

10.1.3 Rounding

Although the table developed in Example 10.3 is certainly acceptable, it has some deficiencies. Most of the values have a different number of significant figures. Some have decimal points and others do not. Because the values are left-justified (that is, they start at the left side of the 14-character print zones), the final-exam column is not aligned perfectly. That is, the units, tens, and hundreds columns of the final exam grades are not lined up.

One option for improving the appearance of the output would be to round the numbers to a set number of decimal places before printing them. Equation (E9.5.2) can be employed for this purpose. Although the format of the output can be partially controlled in this way, there are two shortcomings to this approach. First, it will do nothing about the improper alignment of columns with numbers of different orders of magnitudes, such as the final-exam column from Example 10.3. Second, once you round the numbers using a relationship such as Eq. (E9.5.2), they have been permanently changed. If they are to be used for subsequent computations, you will have introduced a round-off error. Thus care should be exercised to ensure that numbers are rounded only after they are not needed for further calculations. If they are required for further calculations, you must save them in their original form prior to rounding.

Although the rounding option certainly has disadvantages, it is sometimes the only feasible way to print out tabular information on computers such as the Apple and the Commodore. For systems with Microsoft BASIC, such as the IBM PC, the PRINT USING statement is available to exert more direct control over the format of output.

10.1.4   THE PRINT USING Command

For machines such as the IBM PC and the TRS-80, the PRINT USING command allows the direct formatting of output. For Microsoft BASIC, the general form of the PRINT USING statement is

$$ln \; \text{PRINT USING } \textit{format string} \, ; \textit{item}_1, \textit{item}_2, etc.$$

where the *format string* is either a string constant or a string variable that specifies the format of the items that are being printed. The PRINT USING statement can be employed to format either numeric or string data. Separate rules apply to each case.

Numeric data.   An example of an actual PRINT USING statement to output numbers is

```
20 PRINT USING "#####";8,99
```

The string constant "#####" defines the *field*—that is, the number of spaces on the printed line reserved for an item. The symbol # is employed to reserve a space for a digit. Thus, for this example, a five-digit-length field is set aside for each number that is printed. Note that the statement could also be written using a string variable:

```
10 FORMAT$ = "#####"
20 PRINT USING FORMAT$;8,99
```

If the items do not fill the field, each is printed right-justified (that is, placed at the far-right end of the field) and the unused portion filled with blanks. For our example, the result would be

```
    8   99
```

Thus the 8 is printed in column 5 and the 99 in columns 9 and 10. Spaces between numbers can be interjected by merely leaving blanks in the *format string*, as in

```
20 PRINT USING "##### ";8,99
    8    99
```

In addition to controlling the spacing, the PRINT USING also allows decimal points to be placed in the output. The decimal is merely located in the desired position in the format string, for example,

```
20 PRINT USING "###.##";3.56,.85,20.697
   3.56  0.85 20.70
```

Notice that if the format string specifies that a digit is to precede the decimal point, it is always printed (even as a 0 as in the case of 0.85). Also notice that because 20.697 has one too many decimals, it is automatically rounded to 20.70 when it is printed. Some additional symbols that are employed in PRINT USING statements and that have utility for engineering are described in Table 10.1.

TABLE 10.1   Symbols Employed in Numeric Fields of the PRINT USING Statement That Are Useful in Engineering

Symbol	Description	Examples
+	Positioned at the beginning or end of a format string, it causes the sign of the number to be printed before or after the number.	PRINT USING"+##.##";−29,66.3,−.97 −29.00+66.30 −0.97 PRINT USING"##.##+";−29,66.3,−.97 29.00− 66.30+ 0.97−
−	Positioned at the end of a format string, it causes negative numbers to be printed with a trailing negative sign.	PRINT USING"##.##−";−29,66.3,−.97 29.00− 66.30 0.97 −
,	Positioned anywhere to the left of the decimal point of a format string, it causes a comma to be printed to the left of every third digit to the left of the decimal.	PRINT USING"#########,."; −1.5E6 −1,500,000. PRINT USING"#####,";1760,−5830 #1,760−5,830
	Positioned at the end of a format string after the decimal, the comma is printed as part of the string.	PRINT USING "##.#,";55.86,−7,.49 55.9,−7.0, 0.5,
∧∧∧∧	Exponential format is specified by four exponentiation signs at the end of the field.	PRINT USING"##.##∧∧∧∧";201678,−0.11745 #2.02E+05−1.17E−01

EXAMPLE 10.4   Outputting a Table with PRINT USING Statements

PROBLEM STATEMENT: Employ PRINT USING statements to design a table to output the information from Example 10.3.

SOLUTION: For line 100, we would substitute

```
100 PRINT NAME1$;TAB(25);
101 PRINT USING "###.##      ";AQ1,AH1,FE1,FG1
```

Similar substitution would be made for lines 110 and 120. The resulting printout would look like

```
STUDENT                 AVERAGE AVERAGE   FINAL   FINAL
NAME                    QUIZ    HOMEWORK  EXAM    GRADE
SMITH,JANE              82.00     83.90   100.00  88.63
JOHNSON,FRED            96.55     92.00    77.00  88.52
BRYSMIALKIEWICZ,JOHN    97.30     92.00    99.00  96.10
Ok
```

Notice that now all columns are aligned properly and all numbers have the same number of decimal places. It is this kind of control that makes the PRINT USING so valuable when preparing tables for reports and other engineering documents.

String data.   When the PRINT USING statement is employed to print strings, three characters are used to control formatting: !, \, and &. The exclamation point is utilized if only the first character of the string is to be output, as in

```
PRINT USING "!";"MOE","LARRY","CURLY"
MLC
```

Spaces can be inserted by placing spaces within the quotes specifying the format.
    The reversed slash is used to specify that a set number of characters is to be output. For example, \\ specifies that two characters are to be printed and \ \ (one blank between the slashes) specifies three. If the string values to be printed are greater than the field specified by the format string, then the extra characters are ignored. If the string values to be printed are less than the field, then the string is left-justified and blanks are used to fill out the field, for example,

```
PRINT USING "\     \";"MOE","LARRY","CURLY"
MOE LARRCURL
```

The final symbol, &, is employed when the entire string value is to be output exactly as it is stored:

```
PRINT USING "&";"MOE";"LARRY";"CURLY"
MOELARRYCURLY
```

Spaces can be added within the quotes to leave spaces between the string values when they are printed:

```
PRINT USING " &";"MOE";"LARRY";"CURLY"
 MOE LARRY CURLY
```

10.1.5 Screen Design

An exciting aspect of personal computing is the interaction that can be developed between the user and the machine. By effectively designing input and output, the programmer can cause the computer to become a more efficient and enjoyable tool for engineering problem solving.

To this point, you have been introduced to a number of statements that can be used for designing an effective user-software interface. These are screen clearing (CLS), line and space skipping (PRINT, TAB, and SPC), data formatting (delimiters, rounding, and PRINT USING), and string messages. With these tools at your disposal, the screen can be conceived of as an artist's canvas where you can position your input and output in an effective and aesthetically pleasing fashion.

An additional tool is the screen layout form shown in Fig. 10.4. This particular form is 80 columns wide and 24 rows deep.* These are common dimensions for many personal computers. If your computer has other dimensions, you can construct your own layout form.

The layout form provides a vehicle for visualizing how information will appear on a monitor screen. As shown in Fig. 10.4, you can sketch your desired output on the form and then use it as a guide for developing the corresponding BASIC code.

The LOCATE statement.   One additional type of statement can facilitate screen design. To this point, you have the capability to move from left to right (TAB) and from top to bottom (PRINT) to position data on the screen. This is a somewhat limited state of affairs. The situation would be similar to that of an artist who had to paint a canvas starting at the upper left corner and then proceed from left to right and top to bottom.

For Microsoft BASIC, the LOCATE statement allows you to position the cursor at any of the 25 × 80 cells on the screen. Its general form is

ln LOCATE *row number, column number*

where *row number* specifies the cell's row (1 to 25) and *column number* specifies its column (1 to 80). For example, the following statements would move the cursor to the middle of the screen and print an asterisk:

```
10 LOCATE 12,40
20 PRINT "*"
```

*Actually, for the IBM PC the screen is 25 rows deep, but the twenty-fifth row is taken by the function-key prompts. These can be suppressed by the KEY OFF command.

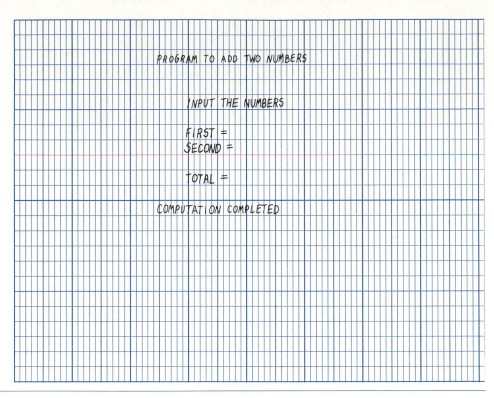

FIGURE 10.4
Layout form.

For computers such as the Apple II, the comparable operation is accomplished with the commands HTAB and VTAB, which have the general form

ln HTAB *column number*

and

ln VTAB *row number*

where *column number* is the column (1 to 40) and *row number* is the row (1 to 24). For the Apple II, the commands to print an asterisk at the middle of the screen are

```
10 VTAB 12
15 HTAB 20
20 PRINT "*"
```

The concept of screen design is closely related to the topic of computer graphics. In Chaps. 22 and 39, we will discuss this topic and elaborate further on some of the ideas we have introduced in this section.

10.2   READ and DATA Statements

Data can be entered to a BASIC program in a variety of ways. For example, the INPUT statement allows the user to enter different values of a variable on each run without modifying the program itself. Although this capability has its advantages, there will be times when it will be inconvenient to enter all or a part of the data from the keyboard. This is particularly true when entering large amounts of data.

One remedy might be to assign values to the variables within the body of the program using assignment or LET statements. Although this circumvents the problem of retyping the data for each run, it can add quite a few lines to your program. The combination of DATA and READ statements provides a more concise and efficient alternative for incorporating large amounts of information into the program code. It allows the data to be kept together, apart from the body of the program so that it is not cluttered. Having data in a discrete place may also facilitate program maintenance, especially if the data is to be changed. The general forms of these statements are

ln READ var_1, var_2, . . .

and

ln DATA $const_1$, $const_2$, . . .

where $const_i$ is the ith numeric or string constant that is being entered and var_i is the ith numeric or string variable name to which $const_i$ is being assigned.

The DATA statement contains the actual information, whereas the READ statement assigns the information to a particular variable. Functionally, the READ and DATA statements

```
100 READ NUMBER
110 DATA 7
```

are equivalent to the LET statement

```
100 LET NUMBER = 7
```

The READ statement must be placed before the lines where the data is used. The DATA statement can appear anywhere in the program. A good stylistic practice is to place it just before the END statement.

You can include as many DATA lines as necessary, and as many values as are permissible may appear on each line. When a READ statement is executed, the values are read in order from the DATA statement. Figure 10.5 presents a visual depiction of this sequential process. Conceptually it is akin to having an imaginary pointer that moves along the DATA statement as each READ statement is implemented.

It is not necessary to read all the DATA values; for example, the value of 800 at the end of line 510 in Fig. 10.5 is not used. However, if the READs exceed the

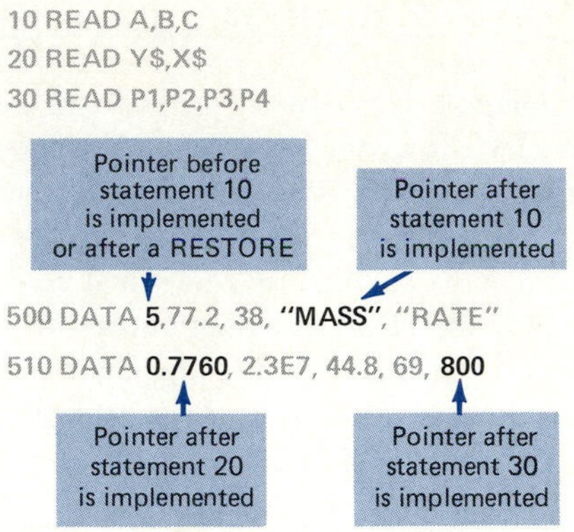

```
10 READ A,B,C
20 READ Y$,X$
30 READ P1,P2,P3,P4
```

Pointer before statement 10 is implemented or after a RESTORE

Pointer after statement 10 is implemented

FIGURE 10.5
Illustration of how information is input with READ and DATA statements. An imaginary pointer indicates how the DATA is entered in sequence as each READ is implemented.

```
500 DATA 5,77.2, 38, "MASS", "RATE"

510 DATA 0.7760, 2.3E7, 44.8, 69, 800
```

Pointer after statement 20 is implemented

Pointer after statement 30 is implemented

DATAs, an error message such as OUT OF DATA will appear on your monitor, and the run will be terminated.

EXAMPLE 10.5   Using READ and DATA Statements

PROBLEM STATEMENT: Employ READ and DATA statements to input the values to the program developed in Examples 10.2 through 10.4.

SOLUTION: The revised program is shown in Fig. 10.6. The inclusion of these statements means that the names and grades do not have to be input on each run as was the case with the programs in Examples 10.2 through 10.4. This feature would become particularly useful for large amounts of data.

FIGURE 10.6
A revised program to input student grades using the READ and DATA statements.

```
10  READ  NAME1$,AQ1,AH1,FE1
20  READ  NAME2$,AQ2,AH2,FE2
30  READ  NAME3$,AQ3,AH3,FE3
50  FG1 = AQ1/3 + AH1/3 + FE1/3
60  FG2 = AQ2/3 + AH2/3 + FE2/3
70  FG3 = AQ3/3 + AH3/3 + FE3/3
80  PRINT  "STUDENT";TAB(25);"AVERAGE";TAB(36);
81  PRINT  "AVERAGE";TAB(47);"FINAL";TAB(58);"FINAL"
85  PRINT  "NAME";TAB(25);"QUIZ";TAB(36);"HOMEWORK";
86  PRINT  TAB(47);"EXAM";TAB(58);"GRADE"
90  PRINT
100 PRINT  NAME1$;TAB(25);
101 PRINT  USING "###.##         ";AQ1,AH1,FE1,FG1
110 PRINT  NAME2$;TAB(25);
111 PRINT  USING "###.##         ";AQ2,AH2,FE2,FG2
120 PRINT  NAME3$;TAB(25);
121 PRINT  USING "###.##         ";AQ3,AH3,FE3,FG3
200 DATA  "SMITH, JANE", 82,83.9,100
210 DATA  "JOHNSON, FRED",96.55,92,77
220 DATA  "BRYSMIALKIEWICZ, JOHN",97.3,92,99
300 END
```

Note that once you read the data, a RESTORE statement of the form

ln `RESTORE`

can be added to your code to restore the pointer to the first data item (Fig. 10.5). This statement can prove useful for programs where you want to use the same set of data more than once. In Microsoft BASIC, a line number may be typed after RESTORE to bring the pointer back to the beginning of a particular line. This option is not available on the Apple and Commodore systems.

PROBLEMS

10.1 Members of a sales force receive commissions of 5 percent for the first $1000, 7 percent for the next $1000, and 10 percent for anything over $2000. Compute and print out the commissions and social security numbers for the following salespeople: John Smith, who has $5325.76 in sales and a social security number of 080-55-6642; Priscilla Holottashakingoinon, who has $8050.63 in sales and whose number is 306-77-5520; and Bubba Schumacker, who has $6643.96 in sales and a social security number of 707-00-1122. Output the results in a neat, tabular format. Use DATA and READ statements to input the information.

10.2 Write, debug, and document a computer program to determine statistics for your favorite sport. Pick anything from softball to bowling to basketball. Design the program so that it is user-friendly and provides valuable and interesting information to anyone (for example, a coach or player) who might use it to evaluate athletic performance. In particular, print out the results in a neat tabular format. Use DATA and READ statements to input the information.

10.3 Expand the program in Examples 10.2 through 10.5 so that it also includes a class participation grade (PG). The final grade would be recalculated as

$$FG = 0.9 * (AQ + AH + FE) + 0.1 * PG$$

Give John and Jane PGs equal to 95 and give Fred a 75. Include this new information on the table along with their identification numbers (Jane: 739936E; Fred: 440169S; and John: 139936E).

10.4 In Prob. 9.19, you were given equations to predict the saturation value of oxygen for various values of temperature and salinity. Use those formulas and the capabilities you have gathered in the present chapter to compute and print out the following table of saturation concentrations:

	$n = 0$	$n = 5000$
$T = 40°F$		
$T = 50°F$		

10.5 Employ the PRINT, CLS, KEY OFF, and LOCATE commands to form a design on the screen. The design can be your initials, your school's block letter, your fraternity's or sorority's Greek letters, or any other symbol, image, or logo. Use the screen layout form given in Fig. 10.4.

10.6 Select 10 of your favorite issues from the New York Stock Exchange. Write, debug, and document a BASIC program that reads the name of the stock, its selling price today, and its 52-week low. The program should then print out the name of the stock, its selling price, and the percentage gain over the 52-week low. Output the results in a neat, tabular format.

10.7 Repeat Prob. 9.23. Print all inputs and outputs using techniques given in this chapter.

10.8 Repeat Prob. 9.24. Print all inputs and outputs using techniques given in this chapter.

10.9 Repeat Prob. 9.25. Print all inputs and outputs using techniques given in this chapter.

10.10 Repeat Prob. 9.26. Print all inputs and outputs using techniques given in this chapter.

10.11 Repeat Prob. 9.27. Print all inputs and outputs using techniques given in this chapter.

10.12 Repeat Prob. 9.28. Print all inputs and outputs using techniques given in this chapter.

10.13 Repeat Prob. 9.29. Print all inputs and outputs using techniques given in this chapter.

10.14 Repeat Prob. 9.30. Print all inputs and outputs using techniques given in this chapter.

10.15 Repeat Prob. 9.31. Print all inputs and outputs using techniques given in this chapter.

CHAPTER 11
BASIC:
SELECTION

To this point we have written programs that perform instructions sequentially. That is, the program statements are executed line by line starting with the lowest-numbered line and ending with the highest. However, a strict sequence is only one of the ways in which a program can be executed. As discussed in Chap. 6, the flow of logic can branch (selection) or loop (repetition). The present chapter deals with the statements required to branch, or to reroute the flow of logic to a line number other than the next highest one.

Although selection can greatly enhance your power and flexibility, its indiscriminate and undisciplined application can introduce enormous complexity to a program. Because this greatly detracts from program maintenance, computer scientists have developed a number of conventions to impose order on selection. Some of these structured-programming conventions involve the way in which the statements are applied—that is, they are voluntary and involve the ''style'' with which the programmer chooses to construct the code. Others actually involve new types of BASIC statements that provide very clear and coherent ways to perform selection. Unfortunately, some of these new types of statements are unavailable in certain BASIC dialects. Because of their great utility, these innovations will undoubtedly be incorporated into all major dialects in the future. For the time being, however, their absence complicates our description of selection. On the one hand, we want to introduce you to the best and most highly advanced capabilities of BASIC. On the other hand, we do not want you to be frustrated by your inability to implement advanced constructs on your machine. To resolve this dilemma, we have chosen to include the advanced constructs in the following sections. However, in order that all readers will be capable of using them, we have included ways to reformulate the constructs employing simpler statements available in all forms of BASIC. Although this may sometimes involve a little extra effort, the long-term positive impact on the quality of your programs will more than justify the investment.

11.1   NONSEQUENTIAL PROGRAM FLOW: THE GOTO STATEMENT

The GOTO is the simplest of all transfer statements. It allows you to directly override the numerical sequence in which the program is executed. The general form is

$$ln_1 \text{ GOTO } ln_2$$

where ln_1 is the line number of the GOTO statement and ln_2 is the line number that is executed next, for example,

```
40 GOTO 10
```

Under normal circumstances, the program would advance to the next-highest line number after executing this statement. However, because of the GOTO statement, rather than transferring to line 41 or higher, the program here would jump back or ''go to'' line 10.

By allowing transfer to any other line, the GOTO greatly increases the program's flexibility. However, a word of caution is in order. All the jumping around that results from the indiscriminate use of GOTO statements can lead to programs that are extremely difficult to understand. For this reason, structured programming considerations mandate that they be employed only when absolutely necessary.

Negative feelings about GOTO statements are so strong that in some programming languages (Pascal for one) the GOTO construct is discouraged and is rarely seen. Languages that shun the GOTO statement usually have other constructs that control the flow of program execution in a more coherent and structured manner. As mentioned at the beginning of this chapter, many dialects of BASIC do not presently include a wealth of constructs for controlling program flow in this fashion. For this reason, we will use the GOTO to compensate for the lack of certain types of structured control statements in most BASICs. Our use of GOTOs will be limited according to the following guidelines:

1. The GOTO statement will be used only when it functions as a part of a structured control construct.

2. The line to which the GOTO transfers control will always be a REM statement to clarify the reason for the transfer.

3. The GOTO statement will not cause a program loop to have more than one exit point.

The impact and importance of these limitations will be clearer once we begin to show how selection can be expressed in a structured manner. However, we must first review the most fundamental statements for implementing selection in BASIC.

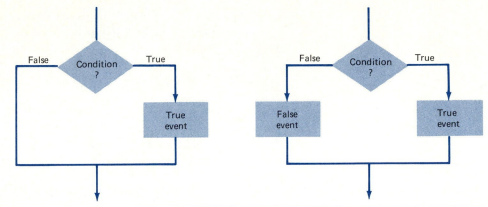

FIGURE 11.1
Two kinds of constructs, the single- and double-alternative decision patterns, are the fundamental building blocks for all types of selection. The combination of these two patterns allows any possible selection process to be implemented.

11.2 CONDITIONAL TRANSFER: THE IF/THEN STATEMENT

The IF/THEN statement represents the diamond-shaped decisions in the flowchart representations shown in Fig. 11.1. It is called a *conditional transfer* because it only transfers control if a particular condition is true. The fundamental formats for the IF/THEN statement are

$$ln_1 \ \text{IF} \ condition \ \text{THEN} \ ln_2$$

and

$$ln_1 \ \text{IF} \ condition \ \text{THEN} \ statement$$

where ln_1 is the line number of the IF/THEN statement, the *condition* represents a relation between expressions that is either true or false, ln_2 is a line number to which control is transferred if the relation is true, and *statement* is a BASIC statement that is to be implemented if the relation is true. If the relation is false, neither transfer to ln_2 nor implementation of *statement* occurs, and control passes automatically to the line following ln_1.

The IF/THEN statement controls which path a program takes at a branch point. The chosen path could theoretically lead anywhere in the program. However, this could lead to the same type of confusion that arises from the indiscriminate application of GOTOs. Consequently, the paths should always lead to statements immediately following the IF/THEN statement. This concise branching is exhibited by the flowchart representations for the two types of selection—the single- and double-alternative decision patterns—illustrated in Fig. 11.1.

An example of the first form of the IF/THEN is

```
100 IF VAR = 1 THEN 120
```

For this case, if the value of VAR is equal to 1, control will transfer to line 120. If not, control passes to the next line after 100 (say, 110). Line 100 could also be written as

```
100 IF VAR = 1 THEN GOTO 120
```

Because BASIC assumes the number following the THEN is a line number to branch to, the GOTO is implied and does not have to be included. If the number following the THEN is not an existing line number in your program then an error message will result. Likewise, a THEN statement cannot appear anywhere in a program other than after an IF—thus our use of the term IF/THEN to describe the whole statement.

An example of the second form of the IF/THEN statement is

```
200 IF NUMBER > 0 THEN LET AVERG = SUM/NUMBER
```

For this example, if the value of the variable NUMBER is greater than zero, then the variable AVERG will be computed as the SUM divided by the NUMBER. After the computation is implemented, control will transfer to the line number following 200. If NUMBER is equal to or less than zero, the statement AVERG = SUM/NUMBER will be ignored, and control will transfer immediately to the next line.

In both examples, the condition to be tested is located between the IF and the THEN. Now we will take a closer look at how these conditions can be formulated.

11.3 FORMULATING THE CONDITION OR DECISION OF AN IF/THEN STATEMENT

The condition in an IF/THEN statement represents the relation between two expressions. Therefore a more detailed representation of the condition (which occurs between the IF and the THEN) is

expression₁ relation expression₂

where the *expressions* can be variables, constants, or formulas. The *relation* is any one of the relational operators in Table 11.1. These include equality, inequality, and testing whether one expression is greater or less than another.

The distinction between the = operator and the LET = statement now becomes important. The IF VAR = 0 portion of statement 100 in Table 11.1 does not cause a value of zero to be assigned to VAR. Rather, it merely tests whether VAR is currently equal to zero. In contrast, the LET = statement following the THEN causes VAR to take on the value of 99 if it has previously been equal to zero. If VAR has not been equal to zero when line 100 is reached, it will still maintain its previous value after line 100 is executed. The difference between the equality operator = and the assignment statement LET = is obvious and logical if the LET portion of the assignment statement is present. Remember that the LET of the assignment statement

TABLE 11.1   Relational Operators

Operator	Relationship	Example
=	Equal	100 IF VAR = 0 THEN LET VAR = 99
<>	Not equal	110 IF C$ <> "9" THEN PRINT C$
<	Less than	120 IF VAR < 0 THEN LET VAR = −VAR
>	Greater than	130 IF X% > Y THEN LET Z = Y/X%
<=	Less than or equal to	140 IF 3.9 <= X/3 THEN 200
>=	Greater than or equal to	150 IF VAR >= 0 THEN 160

is not required by BASIC. However, because it helps clarify situations such as the present one, its use is strongly encouraged.

The inequality tests whether the expression on the right side of the $<>$ is different from that on the left. Because of line 110 the value of C$ will be printed only if C$ is not equivalent to the character string "9".

Testing whether one value is less than another is accomplished with the $<$ operator. Likewise, the $>$ determines if the value on the left side of the $>$ is greater than that on the right side. The operators $>=$ allow the left-hand value to be either exactly equal to or greater than the right-hand value. Similarly, the operators $<=$ will be true if the two are equal or if the left-hand value is less than the right-hand value. Remember the rules for working with inequality operators, which require the operator to change if the values on the left and right side of inequality are exchanged. $X > Y$ is the same test as $Y < X$.

The order of evaluation of the IF/THEN statement is important to note. First the condition is tested. If the condition is true, then the statements following the THEN are carried out. In line 100, VAR is tested to determine whether it is equal to zero prior to performing the assignment following THEN. Otherwise VAR can never equal zero. It will always equal 99.

Arithmetic and relational operators can be combined in a condition being tested. The arithmetic operations are always implemented before testing the relationship. In line 140 of Table 11.1, X is first divided by 3, and then 3.9 is tested to see if it is equal to or less than the quotient.

Strings and character data can also be tested for their relationship to other string and character data. Most frequently the test for this type of data will be for equality or inequality, as in line 110. Character data cannot be directly compared to numeric data. That is why the 9 in line 110 has to be enclosed in quotation marks. Enclosing the 9 in quotation marks causes it to be treated as a string character, like an A or a B, and overrides its numeric value. If the 9 is not set in quotations, BASIC will stop processing the line and report an error. Just as string variables cannot have numbers assigned to them or vice versa, relational operations require that both sides of the relation have either numeric data or string data, but not both in one operation.

Be cautious when testing for the equality or inequality of two real expressions. As mentioned in Sec. 4.1, real numbers are stored in the computer as series of binary digits and an exponent. If the value to be tested cannot be exactly represented by the seven digits of a single-precision real variable then the test may fail, even though it is, in fact, "almost" true. For example, the following condition would test as false:

$$\frac{1}{3} = 0.333$$

Even though 0.333 is approximately equal to $\frac{1}{3}$, the computer must sense an exact equality for the condition to be true. More subtle differences can occur, as illustrated in the following example.

EXAMPLE 11.1   *Problems with Testing Equalities*

PROBLEM STATEMENT: Divide 1 by 82 and then test whether the result multiplied by 82 is equal to 1.

SOLUTION: The following code was written to investigate the testing of equalities between real numbers:

```
10 A = 1/82
20 B = A * 82
30 IF B = 1 THEN 60
40 PRINT "EQUALITY DOES NOT HOLD"
50 GOTO 70
60 PRINT "EQUALITY HOLDS"
70 PRINT B
80 END
RUN
EQUALITY DOES NOT HOLD
 .9999999
OK
```

Although we know that 1/82 multiplied by 82 should be exactly equal to 1, such is not the case for this code. Because of the approximate way that the computer stores real numbers, the test in line 30 is false.

One way to circumvent this problem is to reformulate it. Rather than testing whether the two numbers are equal, you can test whether their difference is small. For example, line 30 could be reformulated as

```
30 IF ABS(B - 1) < .00001 THEN 60
```

This statement tests whether the absolute value of the difference between B and 1 is less than 0.00001—that is, less than 0.001 percent. Formulated in this way, line 30 would test true, as desired. Although you should be aware of this problem, testing whether one value is less than ($<$ or $<=$) or greater than ($>$ or $>=$) another can almost always be performed without running into this difficulty.

TABLE 11.2 Logical Operators

Operator	Meaning	Example
NOT	This negates a condition. In the example, the condition is true if X <= 0 is false, and vice versa.	100 IF NOT (X <= 0) THEN PRINT "X IS POSITIVE"
AND	Both conditions must be true; otherwise the whole operation is considered false.	110 IF X > 0 AND X < 5 THEN PRINT "X BETWEEN 0 and 5"
OR	If either or both of the conditions are true, the whole operation is true.	120 IF (L$ = "n") OR (L$ = "N") THEN PRINT "NO"
XOR	If either (but not both) condition is true, the whole operation is true.	130 IF TEMP < 200 XOR PRESS < 150 THEN 170

11.4 COMPOUND IF/THEN STATEMENTS AND LOGICAL OPERATORS

Many dialects of BASIC allow the testing of more than one logical condition in the same IF/THEN statement. These *compound IF/THEN statements* employ the logical operators AND, OR, XOR, and NOT to simultaneously test conditions (Table 11.2). For example, the AND is employed when both conditions must be true, as in the general form

$$ln_1 \quad \text{IF} \quad condition_1 \quad \text{AND} \quad condition_2 \quad \text{THEN} \quad \begin{matrix} ln_2 \\ or \\ statement \end{matrix}$$

If both conditions are true then this statement will transfer to ln_2 or the *statement* will be performed. If either or both conditions are false, neither the transfer nor *statement* will be carried out, and the program will advance to the line directly following ln_1.

In contrast, the OR is employed when either one or the other or both of the conditions are true, as in

$$ln_1 \quad \text{IF} \quad condition_1 \quad \text{OR} \quad condition_2 \quad \text{THEN} \quad \begin{matrix} ln_2 \\ or \\ statement \end{matrix}$$

If either or both conditions are true, the transfer to ln_2 will be executed or the *statement* will be performed.

Microsoft BASIC also employs the XOR operator that tests whether one or the other but not both of the conditions are true. It is called the *exclusive or*. As shown in line 130 of Table 11.2, if the temperature is less than 200 or the pressure is less than 150, the test is true and the IF/THEN transfers to line 170. However, if both the temperature and the pressure are less than 200 and 150, respectively, the test is false and the transfer is not made.

TABLE 11.3   Complements for the Relational Operators

Operator	Complement	Identities
=	<>	(A = B) is equivalent to NOT (A <> B).
<	>=	(A < B) is equivalent to NOT (A >= B).
>	<=	(A > B) is equivalent to NOT (A <= B).
<=	>	(A <= B) is equivalent to NOT (A > B).
>=	<	(A >= B) is equivalent to NOT (A < B).
<>	=	(A <> B) is equivalent to NOT (A = B).

Finally, the NOT is used for the case where the transfer or the statement is completed if a condition is false, as in

$$ln_1 \quad \texttt{IF NOT } condition \texttt{ THEN} \quad \begin{matrix} ln_2 \\ or \\ statement \end{matrix}$$

If a condition is false, the complement of this condition is true, and vice versa. For example, the complement of $X <= 0$ is $X > 0$. Thus line 100 from Table 11.2 is equivalent to

```
100 IF X > 0 THEN PRINT "X IS POSITIVE"
```

The complements for the relational operators are summarized in Table 11.3. Note that in most cases, it is clearer and more convenient to use the complement of a relational operator than to employ the NOT.

Logical expressions can contain more than one of the above operators, for example,

```
80 IF A>0 OR B<0 AND C=0 THEN 50
```

However, just as with arithmetic expressions, the order in which the operations are evaluated must follow rules. The priorities for the operators (along with all the other priorities established to this point) are specified in Table 11.4. Thus, all other things

TABLE 11.4   The Priorities of Operators Used in BASIC

Type	Operator	Priority
Arithmetic operators	Parentheses, ( )	Highest
	Exponentiation, $\wedge$	↑
	Negation, $-$	
	Multiplication, $*$, and division, /	
	Addition, $+$, and subtraction, $-$	
Relational operators	$=$, $<$, $>$, $<=$, $>=$, $<>$	
Logical operators	NOT	
	AND	
	OR, XOR	Lowest

being equal, the NOT would be implemented prior to the AND and the AND prior to the OR.

For complicated logical expressions, parentheses and position also dictate the order. Thus operations within parentheses are implemented first. Then the priorities in Table 11.4 are used to determine the next step. Finally, within the same priority class, operations are executed from left to right.

EXAMPLE 11.2   *Priorities in Compound Logical Expressions*

PROBLEM STATEMENT: Determine whether the following compound logical expressions are true or false:

```
(a) A < 0 OR B = 2 AND X > 7 XOR Y > 9
(b) A < 0 OR B = 2 AND (X > 7 XOR Y > 9)
```

Note that $A = -1$, $B = 2$, $X = 1$, and $Y = 12$.

SOLUTION:

(*a*) Using the priorities from Table 11.4, we can determine that the expression is false by the following sequence of evaluations:

```
A < 0 OR B = 2 AND X > 7 XOR Y > 9
```

```
                      False
```

```
         True
```

```
                    False
```

First all the relational operators are evaluated because they have higher priority than the logical operators. Of the logical operators, the AND comparison is implemented first because it has a higher priority than OR and XOR. It is false because X is less than 7. The OR is implemented next because of the left-to-right rule. It is true because, even though the second condition (B = 2 AND X > 7) is false, A is less than zero, and therefore the whole operation is considered true. Finally, the XOR comparison is false because both conditions being compared are true.

(*b*) Because of the introduction of the parentheses, we are forced to evaluate the XOR comparison first. As a result, the entire operation yields a true result, as in

```
A < 0 OR B = 2 AND (X > 7 XOR Y > 9)
```

```
                         True
```

```
                 True
```

```
               True
```

11.5   THE IF/THEN/ELSE STATEMENT

An alternative form of the IF/THEN statement is the IF/THEN/ELSE statement. In this statement, an alternative branch or statement is executed when the conditional expression of the ELSE is false. The general form of the IF/THEN/ELSE is

$$ln_1 \text{ IF } condition \text{ THEN } \begin{matrix} ln_2 \\ or \\ statement \end{matrix} \text{ ELSE } \begin{matrix} ln_3 \\ or \\ statement \end{matrix}$$

Thus both the true and the false options can either implement a statement or transfer to a line number. Examples of all four possibilities are

```
100 IF VAR = 1 THEN 200 ELSE LET VAR = 2
110 IF VAR <> 0 THEN 200 ELSE 300
120 IF VAR1 >= VAR2 THEN PRINT VAR1 ELSE PRINT VAR2
130 IF STRVAR$ <> "DONE" THEN INPUT VAR1$ ELSE 200
```

Unfortunately, some of the BASICs employed on some popular personal computers (for example, the Apple and the Commodore) do not include the IF/THEN/ELSE. For these dialects the standard IF/THEN with a GOTO must be employed, as described in Sec. 11.6.1.

11.6   STRUCTURED PROGRAMMING OF SELECTION

The statements introduced in the previous sections can be orchestrated to introduce all sorts of complicated logic into your computer programs. However, as cautioned at the beginning of this chapter, their indiscriminate application can lead to code that is so complex that it is practically indecipherable. In order to avoid this situation, computer scientists have developed a series of rules and suggested practices that when followed can lead to easy-to-understand code. In the present section, we will show how the two fundamental decision constructs of Fig. 11.1 can be programmed according to these rules. We will first use the simple IF/THEN statement to express these constructs. Then we will show how the IF/THEN/ELSE statement provides a more efficient vehicle for the same purpose. Finally, we will briefly discuss the block-structured IF/THEN/ELSE construct that is available in the most advanced forms of BASIC and which is expressly designed to program the selection constructs in a concise and clear fashion.

11.6.1   Selection with the IF/THEN Statement

There are several ways that the IF/THEN construct can be employed to program a double-alternative decision pattern. Figure 11.2 shows a flowchart that will serve to illustrate this point. Figure 11.2*a* through *d* presents four sets of BASIC code that all implement the algorithm in the flowchart. Figure 11.2*a* and *b* both employ the form

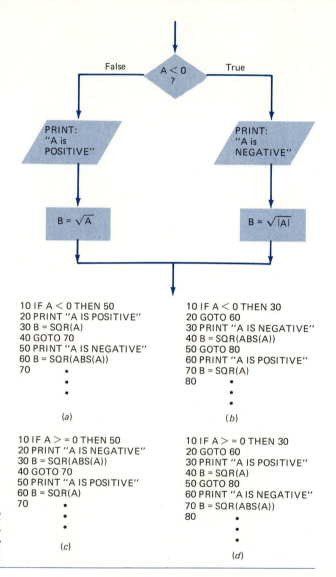

FIGURE 11.2
Four sets of BASIC code that implement the algorithm specified by the flowchart.

of the condition shown in the flowchart—that is, they test whether A is less than zero. They differ in that Fig. 11.2*a* has the false alternative first, whereas Fig. 11.2*b* has the true alternative first. Figure 11.2*c* and *d* are alternatives that result from utilizing the complement of the original condition—that is, they test whether A is greater than or equal to zero. They also differ in that Fig. 11.2*c* has the false alternative first, whereas Fig. 11.2*d* has the true alternative first. Notice how in all cases, GOTO statements are required to transfer control properly. As mentioned in Sec. 11.1, this is the only way in which GOTO statements will be employed in this book.

The fact that four separate versions can be developed to perform the same double-alternative decision is one example of how selection can introduce flexibility and an associated complexity into a program. The result is that the logic of such statements is often very difficult to decipher. In order to reduce this potential confusion, computer scientists have suggested that use of the IF/THEN construct for selection be limited to the forms shown in Fig. 11.2b and d. That is, they suggest that the true alternative always be placed first. Notice that this means that two GOTOs are required. However, this is more than compensated for by the clarity introduced in having the true alternative directly following the IF/THEN. Thus every time you see an IF/THEN you will automatically expect that the true alternative will be first. The construct can be made even clearer by adding REM statements to document its logic. For example, for Fig. 11.2b,

```
10 IF A < 0 THEN 30
20 GOTO 60
30 PRINT "A IS NEGATIVE"
40 B = SQR (ABS(A))
50 GOTO 90
60 REM ELSE
70 PRINT "A IS POSITIVE"
80 B = SQR (A)
90 REM ENDIF
```

Line 60 has been added to mark the beginning of the false alternative. In addition, line 90 is included to designate the end of the IF/THEN construct. The documentation can now be taken a final step by using indentation to visually express the structure.

```
10 IF A < 0 THEN 30
20       GOTO 60
30       PRINT "A IS NEGATIVE"
40       B = SQR(ABS(A))
50       GOTO 90
60    REM ELSE
70       PRINT "A IS POSITIVE"
80       B = SQR(A)
90 REM ENDIF
```

Note how the indentation clearly sets off the true and the false alternatives. The same style can also be applied to the single-alternative pattern, as will be described in the next section. Unfortunately, some dialects of BASIC (notably those used on the Apple II and Commodore) do not allow the user to control indentation. The lack of structure due to this deficiency makes the selection pattern much more difficult to decipher.

11.6.2 The Block-Structured IF/THEN/ELSE Construct

The IF/THEN construct in the previous section can be simplified further if the IF/THEN/ELSE statement is available for your computer. The result is shown in Fig.

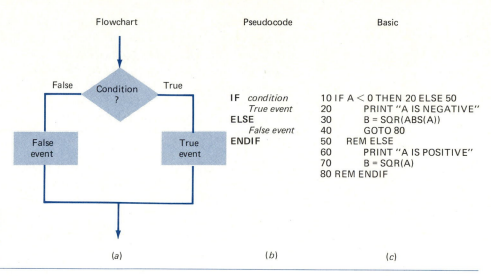

Flowchart Pseudocode Basic

```
IF   condition        10 IF A < 0 THEN 20 ELSE 50
       True event     20      PRINT "A IS NEGATIVE"
ELSE                  30      B = SQR(ABS(A))
       False event    40      GOTO 80
ENDIF                 50    REM ELSE
                      60      PRINT "A IS POSITIVE"
                      70      B = SQR(A)
                      80 REM ENDIF
```

FIGURE 11.3
A block-structured IF/THEN/ELSE construct for a double-alternative decision pattern.

(a) (b) (c)

11.3. The same style can also be developed for the single-alternative pattern. For example, the IF/THEN/ELSE construct in Fig. 11.4c can be developed to characterize the single-alternative pattern in Fig. 11.4a. Note that because there is no false alternative, the line branched to by the ELSE is labeled with the remark ENDIF and the THEN branch is directed to the line immediately following the IF/THEN/ELSE statement.

The formats shown in Fig. 11.3 and 11.4 are referred to as *block-structured IF/THEN/ELSE constructs*. The rules for building these constructs are summarized in Table 11.5. The benefits of using these constructs are great. Each IF/THEN/ELSE construct has only one entry point and one exit. The BASIC statements that are a part of the IF/THEN/ELSE construct are immediately apparent, and the true and false alternatives are clearly delineated. Although this organization uses the GOTO state-

FIGURE 11.4
(a) A flowchart for a single-alternative decision pattern; (b) the pseudocode and (c) the corresponding block-structured IF/THEN/ELSE construct in BASIC.

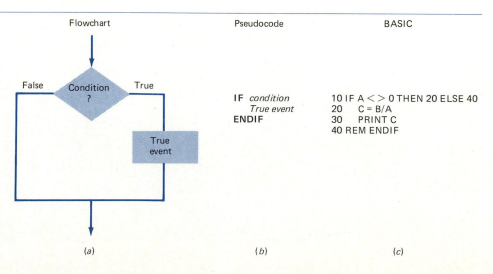

Flowchart Pseudocode BASIC

```
IF   condition        10 IF A < > 0 THEN 20 ELSE 40
       True event     20      C = B/A
ENDIF                 30      PRINT C
                      40 REM ENDIF
```

(a) (b) (c)

ment, it does so in a very regular, predictable manner so that the resulting program does not become a tangled web. The primary objective here is to make the resulting code coherent and easily understandable to someone other than the original author of the program. These conventions help standardize the writing of programs so that others may more readily grasp the organization of an unfamiliar program.

TABLE 11.5 Rules for Writing Block-Structured IF/THEN/ELSE Constructs

Rule 1 The last statement is always the remark ENDIF.
Rule 2 The true alternative always comes first.
Rule 3 The line branched to by the ELSE clause (that is, the false alternative) is always a remark labeled with either REM ELSE for a double-alternative pattern or REM ENDIF for a single-alternative pattern.
Rule 4 For a double-alternative pattern, the last statement of the true alternative will be a GOTO that branches to the REM ENDIF at the end of the construct.
Rule 5 The REM ENDIF is indented to the same degree as the start of its IF/THEN/ELSE statement.
Rule 6 The REM ELSE statement is indented three spaces more than the IF/THEN/ELSE statement.
Rule 7 The BASIC statements comprising the true and false alternatives are indented three spaces from the level of the REM ELSE or the IF/THEN/ELSE statement to which they refer, whichever is greater.

All but the simplest of IF/THEN/ELSE statements should be constructed as block structures for easier readability. If there is only one BASIC statement associated with the THEN condition, a one-line IF/THEN statement is reasonable; however, if there is more than one statement in the THEN condition or if there is a statement or statements under the ELSE condition, then a block-structured IF/THEN/ELSE construct with proper indentation is much easier to understand. Figure 11.5, which reprints Figs. 11.2a and 11.3c, should make this evident. Even though this is a very simple example, the superiority of the structured approach is obvious. For more complicated code, the benefits become even more pronounced.

FIGURE 11.5 A side-by-side comparison of Figs. 11.2a and 11.3c. Both sets of code accomplish the same objective, but the structured version in (b) is much easier to read and decipher.

```
10 IF A < 0 THEN 50
20 PRINT "A IS POSITIVE"
30 B = SQR(A)
40 GOTO 70
50 PRINT "A IS NEGATIVE"
60 B = SQR(ABS(A))
70 REM CONTINUE
                (a)
```

```
10 IF A < 0 THEN 20 ELSE 50
20       PRINT "A IS NEGATIVE"
30       B = SQR(ABS(A))
40       GOTO 80
50    REM ELSE
60       PRINT "A IS POSITIVE"
70       B = SQR(A)
80 REM ENDIF
                (b)
```

11.6.3   Nesting of Block IF/THEN/ELSE Constructs

An IF/THEN/ELSE construct can have another IF/THEN/ELSE statement as part of its THEN or ELSE statements. The inclusion of a construct within a construct is called *nesting*. For the case of the block IF/THEN/ELSE construct, nesting provides an alternative to having involved conditional expressions with many AND, OR, or XOR clauses. By using simpler conditional expressions it may be easier to follow the logic of the decisions.

The more complex nested IF/THEN/ELSE structures are much easier to understand when proper rules for indentation are followed. The rules for forming and indenting the nested IF/THEN/ELSE statements are essentially the same as those for a single IF/THEN/ELSE structure. Just remember that indentation of the conditional statement REM ELSE and the REM ENDIF is always determined by the IF/THEN/ELSE statement to which they refer. An example using nested IF/THEN/ELSE statements will demonstrate the use of nesting and proper indentation.

EXAMPLE 11.3   *Nesting of IF/THEN/ELSE Constructs*

PROBLEM STATEMENT: Develop a program to determine the real and complex roots of the quadratic equation.

SOLUTION: A structured flowchart for this problem is shown in Fig. 11.6. Note that there are two nested selections. The first selection is designed to avoid division by zero in the quadratic equation for the case where $a = 0$. The second selection branches to compute the real or complex roots depending on whether the term within the square root (that is, the determinant) is positive or negative.

A structured code to implement the flowchart is listed in Fig. 11.7. Indentation is used to clearly show the true and false options of each IF/THEN/ELSE.

At this point you may wonder why all BASIC programs have not been written in a structured form. What are the disadvantages of using this type of program organization? For one thing, the amount of typing is increased in the structured code. When typing structured code, the line number referenced by the GOTO that branches to the REM ENDIF is not known until the lines that follow have been determined. This tends to slow the typing process. In addition, the speed of execution of the structured code may be slightly slower than that of the nonstructured version because the structured code has more lines of BASIC to interpret. Thus there is a trade-off here. The structured program is easier to understand and change later, but more time is consumed developing it initially. However, in the overall scheme of program development and use, the time and effort are almost always a worthwhile investment. As dialects of BASIC that have block-structured IF/THEN/ELSE constructs formally built into the language become widely available some of these disadvantages will be mitigated. We will now briefly discuss these constructs.

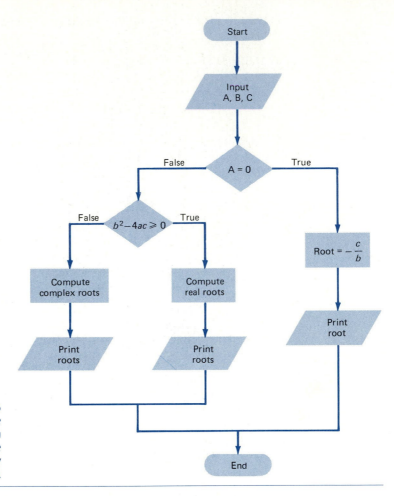

FIGURE 11.6
Structured flowchart to determine the real and complex roots of a quadratic equation. Note that the first selection prevents division by zero in the equation if $a = 0$.

11.6.4   The Formal Block IF/THEN/ELSE Construct

To this point we have tried to illustrate the progression from the unconditional GOTO to the unstructured IF/THEN to the block-structured IF/THEN/ELSE. In some of the advanced dialects of BASIC, the progression is taken one final step to what we will call a *formal*, or *single-line, block IF/THEN/ELSE*.

The major difference is that the whole construct has a single line number. An example of the single-line version for the same structure seen in Fig. 11.5*b* is

```
10 IF A < 0 THEN                 &:
       PRINT "A IS NEGATIVE"     &:
       \LET B = SQR(ABS(A))      &:
   ELSE                          &:
       PRINT "A IS POSITIVE"     &:
       \LET B = SQR(A)
```

```
100 REM       ************************************************
110 REM       *         ROOTS OF A QUADRATIC EQUATION        *
120 REM       *            (BASIC version)                   *
130 REM       *              by  S. C. Chapra                *
140 REM       *                                              *
150 REM       *      This program determines the real        *
160 REM       *      and complex roots of a quadratic.       *
170 REM       ************************************************
180 REM
190 REM             .---------------------------------.
200 REM             |    DEFINITION OF VARIABLES      |
210 REM             |                                 |
220 REM             | A = COEFFICIENT OF X SQUARED    |
230 REM             | B = COEFFICIENT OF X            |
240 REM             | C = CONSTANT                    |
250 REM             '---------------------------------'
260 REM
270 REM ****************** MAIN PROGRAM  ******************
280 REM
290 INPUT "A,B AND C = ";A,B,C
300 IF A = 0 THEN 310 ELSE 340
310         ROOT = -C/B
320         PRINT "SINGLE ROOT = ";ROOT
330         GOTO 490
340    REM ELSE
350       IF (B^2 - 4*A*C ) >= 0 THEN 360 ELSE 420
360            REM COMPUTE REAL ROOTS
370            ROOT1 = (-B + SQR(B^2 - 4*A*C))/(2*A)
380            ROOT2 = (-B - SQR(B^2 - 4*A*C))/(2*A)
390            PRINT "REAL ROOT = ";ROOT1
400            PRINT "REAL ROOT = ";ROOT2
410            GOTO 480
420         REM ELSE
430            REM COMPUTE COMPLEX ROOTS
440            REAL = -B/(2*A)
450            IMAGIN = SQR(ABS(B^2 - 4*A*C))/(2*A)
460            PRINT "COMPLEX ROOT =";REAL;"+(";IMAGIN;"i)"
470            PRINT "COMPLEX ROOT =";REAL;"-(";IMAGIN;"i)"
480       REM ENDIF
490 REM ENDIF
500 END
```

FIGURE 11.7
Structured BASIC program to implement the algorithm from Fig. 11.6. Notice how indentation clearly sets off the nested loops.

The & must be used at the end of each line that is continued. The \ must be used to separate executable statements of both the true and the false statement groups. Notice that aside from using a single line number, the other major contrast with Fig. 11.5*b* is that the above construct is devoid of GOTOs.

The above example of a block IF/THEN/ELSE construct is employed in extended dialects such as VAX-11 BASIC and BASIC-PLUS-2 by the Digital Equipment Corporation. Other extended dialects (for example, Dartmouth's True BASIC), allow programming of the block IF/THEN/ELSE construct without the special continuation and separation symbols. True BASIC can now be obtained for the IBM PC (Kemeny, 1985). Because of their many advantages, such extended dialects will soon become the standard for BASIC programming.

11.7 THE CASE CONSTRUCT

The CASE construct is a special type of IF/THEN/ELSE block. The CASE construct selects one path from a group of alternative paths. This is a very useful construction in a situation where, for example, the user is presented with a menu and makes a

```
CASE variable
    value-1:
    statements
    value-2:
    statements
        ⋮
    value-N:
    statements
        ELSE
    statements
ENDCASE
```

FIGURE 11.8
Generalized form of the CASE
construct.

selection, after which the program takes an action that depends upon the choice of the user.

As shown in Fig. 11.8, the CASE statement tests to find which of the values, 1 through *n*, matches the value of the variable. Only the statements associated with the matched value are then executed. The statements following ELSE would be carried out if none of the values matched the contents of the variable.

Although it is a part of many advanced languages such as Pascal, the CASE statement is not formally a part of Microsoft BASIC. However, it can be simulated with a series of IF/THEN/ELSE statements. To make this construction more obvious to the reader, slightly different indenting and branching in the IF/THEN/ELSE structure are helpful. The approach is depicted in Fig. 11.9. Again, statements following the condition where the *variable* equals the *value* (*value-1*. . .*value-N*) are executed. The *endcase* refers to the line number of the REM ENDCASE.

The rules governing the forming of a CASE construction using the IF/THEN/ELSE statements of BASIC, which are different from the rules for the usual IF/THEN/ELSE block structure, are summarized in Table 11.6.

FIGURE 11.9
Generalized format for the CASE
construct using IF/THEN/ELSE
statements.

```
100  REM CASE
110      IF variable = value-1 THEN 120 ELSE 140
120          BASIC statement
130          GOTO 220
140      IF variable = value-2 THEN 150 ELSE 170
150          BASIC statement
160          GOTO 220
170      IF variable = value-N THEN 180 ELSE 200
180          BASIC statement
190          GOTO 220
200      REM ELSECASE
210          BASIC statement
220  REM ENDCASE
```

TABLE 11.6   Rules for Writing the CASE Construct Using IF/THEN/ELSE Statements

Rule 1   The CASE begins with REM CASE and ends with REM ENDCASE.
Rule 2   The GOTO at the end of the THEN statements branches to the REM ENDCASE.
Rule 3   The ELSE statement branches directly to the next IF/THEN/ELSE statement unless it is the last ELSE clause; then the branch is to the REM ELSECASE.
Rule 4   REM ELSE and REM ENDIF are not used.
Rule 5   The IF/THEN/ELSE statements are all at the same level of indentation, three spaces in from the level of the REM CASE.

EXAMPLE 11.4   The CASE Construct

PROBLEM STATEMENT: Suppose that each class at a university is to go to a different room to register for courses. The classroom assignments are

Class	Identification Code	Classroom
Freshmen	1	108
Sophomores	2	306
Juniors	3	214
Seniors	4	215
Graduates	5	107

The identification code is a number that signifies the student's class. Design a program that uses the CASE construct to print out the proper classroom number in response to the student's identification code.

SOLUTION: The BASIC program to accomplish the objective is listed in Fig. 11.10. Notice that the CASE construct is used to print out the correct message on the basis of the student's code.

Note that another way to formulate the CASE construct is with the ON GOTO statement which is of the general form

ln ON *expression* GOTO *ln₁, ln₂, . . . , ln_n*

where the *expression* is a variable or formula that may take on values from *1* to *n*. If the value is *1*, the program goes to *ln₁*; if it is *2*, the program goes to *ln₂*; and so on. If the *expression* in the ON GOTO is a noninteger, the fractional part is rounded. After it is rounded, if the *expression* is zero or greater than *n,* control transfers automatically to the next line. If the *expression* is less than zero or greater than 255, an error message results.

The use of the ON GOTO statement is discouraged by some advocates of structured programming because its undisciplined use can lead to code that is very confusing. However, if employed for a structured construct such as the CASE, it is perfectly acceptable. In fact, because it allows direct transfer to particular lines, it is

```
100 REM      Program fragment to illustrate        330       GOTO 430
110 REM      CASE statement.                        340     IF CLASS = 4 THEN 350 ELSE 370
120 CLS                                             350       PRINT "PROCEED TO ROOM 215"
130 REM      Create a Menu                          360       GOTO 430
140 REM                                             370     IF CLASS = 5 THEN 380 ELSE 400
150 PRINT:PRINT:PRINT                               380       PRINT "PROCEED TO ROOM 107"
160 PRINT "SELECT ONE OF THE FOLLOWING OPTIONS"     390       GOTO 430
170 PRINT TAB(10);"1. FRESHMAN"                     400     REM ELSECASE
180 PRINT TAB(10);"2. SOPHOMORE"                    410       PRINT "NO SUCH CLASS"
190 PRINT TAB(10);"3. JUNIOR"                       420       PRINT "PLEASE RUN PROGRAM AGAIN"
200 PRINT TAB(10);"4. SENIOR"                       430 REM ENDCASE
210 PRINT TAB(10);"5. GRADUATE"                     440 END
220 INPUT "ENTER YOUR CLASS";CLASS                  RUN
230 CLS                                             SELECT ONE OF THE FOLLOWING OPTIONS
240 REM CASE                                                    1. FRESHMAN
250     IF CLASS = 1 THEN 260 ELSE 280                          2. SOPHOMORE
260       PRINT "PROCEED TO ROOM 108"                           3. JUNIOR
270       GOTO 430                                              4. SENIOR
280     IF CLASS = 2 THEN 290 ELSE 310                          5. GRADUATE
290       PRINT "PROCEED TO ROOM 306"              ENTER YOUR CLASS? 4
300       GOTO 430                                 PROCEED TO ROOM 215
310     IF CLASS = 3 THEN 320 ELSE 340            Ok
320       PRINT "PROCEED TO ROOM 214"
```

FIGURE 11.10 A program using the CASE construct.

actually somewhat more efficient than the IF/THEN/ELSE CASE construct of Fig. 11.9. Problem 11.9 at the end of this chapter involves reprogramming Fig. 11.10 so that it employs an ON GOTO version of the CASE.

PROBLEMS

11.1 The following code is designed to print out a message to a student according to his or her final exam grade (FINAL) and total points accrued in a course (POINTS).

```
400 IF FINAL <= 60 THEN PRINT "FAILING GRADE"
410 IF FINAL > 60 THEN PRINT "PASSING GRADE"
420 IF FINAL > 60 AND POINTS < 200 THEN PRINT
    "POOR"
430 IF FINAL > 60 AND POINTS >= 200 AND POINTS <=
    400 THEN PRINT "GOOD"
440 IF FINAL > 60 AND POINTS > 400 THEN PRINT
    "EXCELLENT"
```

Reformulate this code using block-structured IF/THEN/ELSE constructs. Test the program for the following cases:
(*a*) FINAL = 50, POINTS = 250
(*b*) FINAL = 70, POINTS = 321
(*c*) FINAL = 90, POINTS = 561

11.2 As an industrial engineer you must develop a program to accept or reject a rod-shaped machine part according to the following criteria: its length must be not less than 7.75 cm nor greater than 7.85 cm; its diameter must be not less than 0.335 cm nor greater than 0.346 cm. In addition, under no circumstances may its mass exceed 5.6 g. Note that the mass is equal to the volume (cross-sectional area times length) multiplied by the rod's density (7.8 g/cm^3). Write a structured computer program to input the length and diameter of a rod and print out whether it is accepted or rejected. Test it for the following cases. If it is rejected, print out the reason (or reasons) for rejection.

(a) Length = 7.71, diameter = 0.338
(b) Length = 7.81, diameter = 0.341
(c) Length = 7.83, diameter = 0.343
(d) Length = 7.86, diameter = 0.344
(e) Length = 7.63, diameter = 0.351

11.3 An air conditioner both cools and removes humidity from a factory. The system is designed to turn on (1) between 7 A.M. and 6 P.M. if the temperature exceeds 75°F and the humidity exceeds 40 percent or if the temperature exceeds 70°F and the humidity exceeds 80 percent; or (2) between 6 P.M. and 7 A.M. if the temperature exceeds 80°F and the humidity exceeds 80 percent or if the temperature exceeds 85°F, regardless of the humidity.

Write a structured computer program that inputs temperature, humidity, and time of day and prints out a message specifying whether the air conditioner is on or off. Test the program with the following data:

(a) Time = 7:40 P.M., temperature = 81°F, humidity = 68 percent
(b) Time = 1:30 P.M., temperature = 72°F, humidity = 79 percent
(c) Time = 8:30 A.M., temperature = 77°F, humidity = 50 percent
(d) Time = 2:45 A.M., temperature = 88°F, humidity = 28 percent

11.4 Develop a structured program that uses the CASE construct to determine whether a number is positive, negative, or zero.

11.5 The American Association of State Highway and Transportation Officials provides the following criteria for classifying soils in accordance with their suitability for use in highway subgrades and embankment construction:

Grain Size, mm	Classification
>75	Boulders
2–75	Gravel
0.05–2	Sand
0.002–0.05	Silt
<0.002	Clay

Develop a structured program that uses the CASE construct to classify a soil on the basis of grain size. Test the program with the following data:

Soil Sample	Grain Size, mm
1	2×10^{-4}
2	10
3	0.6
4	120
5	0.01

11.6 Develop a structured program that uses the CASE construct to assign a letter grade to students on the basis of the following scheme: A (90–100), B (80–90), C (70–80), D (60–70), and F (<60). For the situation where the student's grade falls on the boundary (e.g., 80), have the program give the student the higher letter grade (e.g., a B).

11.7 Write a structured BASIC program that reads values of x and outputs absolute values of x without using the ABS(X) function.

11.8 Determine the line number to which the program will proceed according to the conditions of the following IF/THEN/ELSE statements. Show all the steps in the evaluation of the condition in a manner similar to Example 11.2. Note: $X = 10$, $Y = 20$, and $Z = 30$.

(*a*) IF X < = 10 OR Y = 20 THEN 200 ELSE 450

(*b*) IF X < = 20 OR Y > = 50 THEN 150 ELSE 210

(*c*) IF (Y = 20 AND Z = 30) AND (X > 10 OR Y > 30) THEN 45 ELSE 20

(*d*) IF (Y = 20 OR Z = 50) AND (X > = 10 AND Y < = 20) THEN 5 ELSE 7

(*e*) IF (Y = 50 AND X = 10) OR (Z < 40 OR Y > 30) THEN 2 ELSE 6

11.9 Reprogram Fig. 11.10 using the ON GOTO version of the CASE construct. Which would execute faster: the IF/THEN/ELSE or the ON GOTO version? Why?

CHAPTER 12

BASIC: REPETITION

Computers excel at performing tasks that are repetitive, boring, and time-consuming. Even very simple programs can accomplish huge tasks by executing the same chore over and over again. One of the earliest applications of computers was to tally and categorize the U.S. population in the census of 1890, a very repetitive task. Even today, one of the primary applications of computers is to maintain, count, and organize many pieces of information into smaller, more meaningful collections.

Nowhere is the repetitive power of the computer more valuable than in engineering. Many of our applications hinge on the ability of computers to implement portions of programs repetitively so that the expended effort is small compared to the amount of work produced. This is true in many areas of our discipline but is especially relevant to number-crunching computations. Before discussing the actual statements and constructs used for repetition, we will present a simple example to demonstrate the major concept underlying repetition—the loop.

12.1 GOTO, OR INFINITE, LOOPS

Suppose that you want to perform a statement or a group of statements many times. One way of accomplishing this is simply to write the set of statements over and over again. For example, a repetitive version of the simple addition program is

```
10 INPUT A, B
20 PRINT A + B
30 INPUT A, B
40 PRINT A + B
```

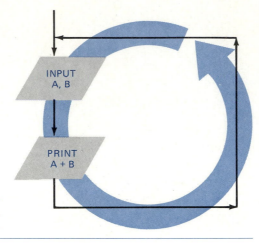

FIGURE 12.1
An example of a GOTO loop. Notice that the loop is infinite. That is, once it starts, it never stops.

```
50 INPUT A, B
60 PRINT A + B
       ·
       ·
       ·
```

Such a sequential approach is obviously inefficient. A much more concise alternative can be developed using a GOTO statement, as in

```
10 INPUT A, B
20 PRINT A + B
30 GO TO 10
```

As depicted in Fig. 12.1, the use of the GOTO statement directs the program to automatically circle or "loop" back and repeat statements 10 and 20. Such repetitive execution of a statement or a set of statements is called a *loop*. Because one of the primary strengths of computers is their ability to perform large numbers of repetitive calculations, loops are among the most important operations in programming.

A flaw of Fig. 12.1, called a *GOTO loop*, is that it is a *closed*, or *infinite, loop* (recall Fig. 6.7). Once it starts, it never stops. Consequently, provisions must be made so that the loop is exited after the repetitive computation is performed satisfactorily. One way to accomplish this is with the FOR/NEXT statements discussed in the next section.

12.2 FOR/NEXT LOOPS

Suppose that you wanted to perform a specified number of repetitions, or iterations, of a loop. One way, employing an IF/THEN/ELSE statement, is depicted in the flowchart and program fragment in Fig. 12.2a. The loop is designed to repeat five

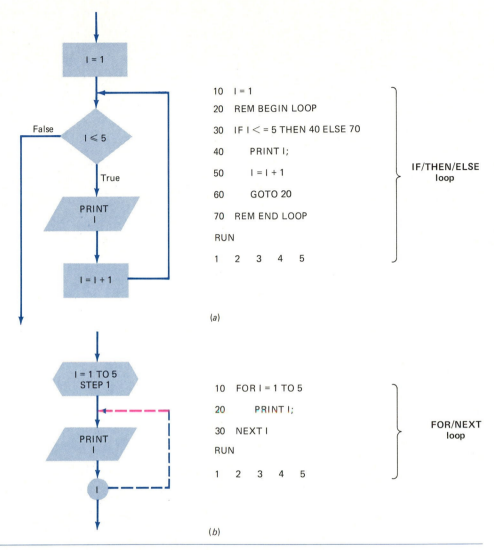

```
10   I = 1
20   REM BEGIN LOOP
30   IF I < = 5 THEN 40 ELSE 70
40      PRINT I;
50      I = I + 1
60      GOTO 20
70   REM END LOOP
RUN
1   2   3   4   5
```

IF/THEN/ELSE loop

(a)

```
10   FOR I = 1 TO 5
20      PRINT I;
30   NEXT I
RUN
1   2   3   4   5
```

FOR/NEXT loop

(b)

FIGURE 12.2 Two alternative methods for implementing a loop consisting of a set number of iterations. (a) An IF/THEN/ELSE and a counter I are used to determine when five passes have been implemented, (b) The FOR/NEXT loop offers a more efficient alternative to perform the same action.

times. The variable I is a counter that keeps track of the number of iterations. If I is less than or equal to 5, an iteration is performed. On each pass, the value of I is printed and I is incremented by 1. After the fifth iteration, I has a value of 6, therefore the ELSE condition of line 30 holds and control is transferred out of the loop to line 70.

Although the IF/THEN/ELSE loop is certainly a feasible option for performing a specified number of iterations, the operation is so common that a special set of statements is available in BASIC for accomplishing the same objective in a more efficient manner. Called the *FOR/NEXT loop*, its general format is

ln FOR *index* = *start* TO *finish* STEP *increment*

(Body of loop)

ln NEXT *index*

The FOR/NEXT loop works as follows. The *index* is a variable that is set at an initial value (*start*). The program then executes the body of the loop, moves to the NEXT statement, and then loops back to the FOR statement. Every time the NEXT statement is encountered, the *index* is automatically increased by the *increment*. Thus the *index* acts as a counter. Then, when the *index* is greater than the final value (*finish*), the computer automatically transfers control out of the loop to the line following the NEXT statement.

An example of the FOR/NEXT loop is provided in Fig. 12.2*b*. Notice how this version is much more concise than the IF/THEN/ELSE loop in Fig. 12.2*a*. Also notice that the terms "STEP *increment*" are omitted from line 10. If these terms are not included, the computer automatically assumes a default value of 1 for the *increment*.

As shown in Fig. 12.2*b*, all BASIC statements within a FOR/NEXT loop should be indented three spaces from that of the FOR statement. The NEXT statement should be indented the same number of spaces as its associated FOR statement. Any number of BASIC statements can be included within the loop.

EXAMPLE 12.1   Using a FOR/NEXT Loop to Input and Total Values

PROBLEM STATEMENT: A common application of repetition is to enter data from the keyboard. The FOR/NEXT loop is particularly well suited for cases where the user has foreknowledge of the number of values that are to be entered. Employ the FOR/NEXT loop to input a set number of values to a program. Also, design the code so that the values are totaled and their average determined.

SOLUTION: A program fragment to perform the above task is

```
10 LET TOTAL = 0
20 INPUT "NUMBER OF VALUES =";NUMVAL
30 FOR I = 1 TO NUMVAL
40    PRINT "VALUE#";I;" =";
50    INPUT VALUE
60    LET TOTAL = TOTAL + VALUE
```

```
70 NEXT I
80 AVERAG = TOTAL/NUMVAL
90 PRINT "AVERAGE = ";AVERAG
100 END
RUN
NUMBER OF VALUES = ? 3
VALUE #1 = ? 4
VALUE #2 = ? 10
VALUE #3 = ? 1
AVERAGE = 5
Ok
```

Line 20 is used to enter NUMVAL, the number of data values. NUMVAL then serves as the *finish* parameter for the loop. Because the *increment* is not specified, a step size of 1 is assumed. Therefore the loop repeats NUMVAL times while inputting one value per iteration.

On each pass, line 60 serves to sum the values. The variable TOTAL is used as an accumulator (recall Sec. 6.2.4). Before the loop is implemented, TOTAL is set to zero (line 10). On the initial pass, the first value is added to TOTAL and the result assigned to TOTAL on the left side of the equality (line 60). Therefore after the first pass TOTAL will contain the first value. On the next pass the second value is added, and TOTAL will then contain the sum of the first and second values. As the iterations continue, additional values are input and added to TOTAL. Thus after the loop is terminated, TOTAL will contain the sum of all the values. Line 80 is then used to compute the average.

The size of the *increment* can be changed from the default of 1 to any other value. The *increment* does not have to be an integer, nor does it even have to be positive. If a negative value is used, then the logic for stopping the loop is reversed. With a negative *increment*, the loop terminates when the value of the *index* is less than that of the *finish*. Step sizes such as -50, 10, -1, or 60 would all be possible.

Note that the *index* does not have to be included with the NEXT statement. However, it greatly improves the readability of the program because it reinforces, along with indentation, the connection between associated FOR and NEXT statements. Remember that FOR and NEXT statements must be used in pairs and that the FOR statement must always come before its NEXT statement.

12.2.1 Nesting of FOR/NEXT Loops

Loops may be enclosed completely within other loops. This arrangement, called *nesting*, is only valid if it follows the rule: If a FOR/NEXT loop contains either the FOR or the NEXT parts of another loop, it must contain both of them. Figure 12.3 presents some correct and incorrect versions. Notice how indentation is used to clearly delineate the extent of the loops.

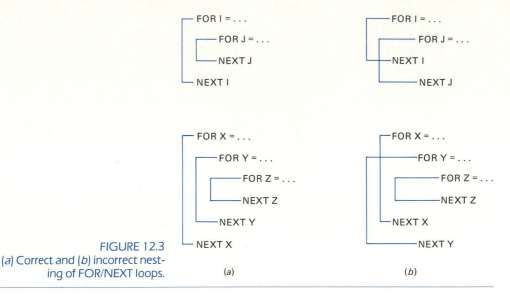

FIGURE 12.3
(a) Correct and (b) incorrect nesting of FOR/NEXT loops.

(a)   (b)

EXAMPLE 12.2   Using FOR/NEXT Loops

PROBLEM STATEMENT: Illustrate the use of FOR/NEXT loops with (a) an increment greater than 1, (b) a negative increment, and (c) several nested loops.

SOLUTION:
(a) The following program prints numbers from 2 to 22 with an increment of 4:

```
10 FOR COUNT = 2 TO 22 STEP 4
20    PRINT COUNT;
30 NEXT COUNT
40 PRINT: PRINT COUNT
50 END
RUN
   2 6 10 14 18 22
   26
Ok
```

Notice that upon exit from the loop, COUNT has been incremented by 4.
(b) The following program prints numbers from 20 down to −5, using an increment of −6:

```
200 FOR I = 20 TO -5 STEP -6
220    PRINT I;
240 NEXT I
260 END
RUN
  20 14 8 2 -4
Ok
```

(*c*) The following program illustrates both the nesting of loops and the associated indentation:

```
10 FOR I = 1 TO 3
20     FOR J = 1 TO 5
30         B = I * J
40         PRINT B;
50     NEXT J
60     PRINT
70 NEXT I
80 END
RUN
   1  2  3  4  5
   2  4  6  8  10
   3  6  9  12  15
Ok
```

12.3 STRUCTURED PROGRAMMING OF REPETITION

In Chap. 6, we mentioned that there are two fundamental types of loop constructs—the DOWHILE and the DOUNTIL structures. As depicted in Fig. 12.4*a*, the DOWHILE is used to create a loop that repeats as long as a condition is true. Because the condition precedes the event, it is possible that the action will not be implemented if the condition is false on the first pass. In contrast, as shown in Fig. 12.4*a*, the DOUNTIL is designed to have the event execute repeatedly until the condition is true. Then control is passed outside the loop. A key feature of this type of loop is that the event is always executed at least once. In the first part of this section, we will use IF/THEN/ELSE statements to devise these constructs. Then in a second part, we will turn to the WHILE/WEND statements that are available in Microsoft BASIC and that offer a superior means for expressing these constructs. Finally, we will mention some of the extended BASIC statements that are available for the advanced dialects of the language.

12.3.1 IF/THEN/ELSE Loops

In contrast to the FOR/NEXT loops described in the previous section, the IF/THEN/ELSE loops are designed for cases where the number of repetitions is not prespecified. Rather, as seen in Fig. 12.4*a*, termination of the loop hinges on a decision. Although some of the more popular BASIC dialects have statements expressly designed for the repetition constructs (such as the WHILE/WEND in the next section), the structures depicted in Fig. 12.4 can also be programmed with the more commonly available IF/THEN and IF/THEN/ELSE statements. Therefore, in order to ensure that most readers are able to program these constructs, we will first show how the IF/THEN/ELSE can be employed for this purpose.

The DOWHILE structure.   As seen in Fig. 12.4*c*, the DOWHILE version begins and ends with REM statements to document the extent of the loop. In addition, the first

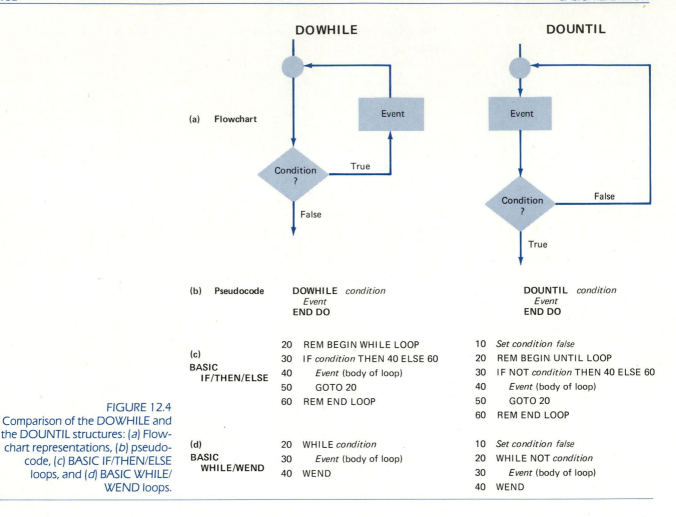

FIGURE 12.4
Comparison of the DOWHILE and the DOUNTIL structures: (*a*) Flowchart representations, (*b*) pseudocode, (*c*) BASIC IF/THEN/ELSE loops, and (*d*) BASIC WHILE/WEND loops.

REM is also included to clarify the unconditional GOTO transfer that is needed to loop back to the head of the construct. Note that the body of the loop is indented to make the structure more apparent.

The IF/THEN/ELSE version of the DOWHILE has many applications. In contrast to the FOR/NEXT loop, which is well suited to handle fixed numbers of repetitions, the IF/THEN/ELSE version is designed for cases where the number of iterations is not given. Rather, the repetitions depend on a condition that is tested every time the loop is implemented. As described in the following example, a situation where this sort of loop proves useful is when inputting data to a program.

EXAMPLE 12.3   Using a DOWHILE Loop to Input, Count, and Total Values

PROBLEM STATEMENT: As seen in Example 12.1, there are occasions when the user will have prior knowledge of the number of values that are to be input to a program. However, there are just as many instances when the user will either not know or will not desire to count the data beforehand. This is typically the case when large quantities of information are to be entered. In these situations, it is preferable to have the program count the data. Employ the IF/THEN/ELSE version of the DOWHILE loop to input, count, and total a set of values. Use the results to compute the average.

SOLUTION: The important issue that needs to be resolved in order to develop this program is how to signal when the last item is entered. A common way to do this is to have the user enter an otherwise impossible value to signal that data entry is complete. Such impossible data values are sometimes called *sentinel*, or *signal, values*. An IF/THEN/ELSE can then be used to test whether the sentinel value has been entered and whether to transfer control out of the loop. The sentinel value can be a negative number if all data values are known to be positive. A very large number, say 9999, might be employed if all valid data was small in magnitude. In the following code the number -999 is used as the sentinel value:

```
10 LET NUMVAL = 0
15 LET TOTAL = 0
20 INPUT "VALUE #1 = (ENTER -999 TO TERMINATE)";VALUE
25 REM BEGIN WHILE LOOP
30 IF VALUE <> -999 THEN 35 ELSE 60
35     LET NUMVAL = NUMVAL + 1
40     LET TOTAL = TOTAL + VALUE
45     PRINT "VALUE#";NUMVAL;" = (ENTER -999 TO
       TERMINATE)";
50     INPUT VALUE
55     GOTO 25
60 REM END LOOP
65 REM COMPUTE AVERAGE IF NUMVAL GREATER THAN ZERO
70 IF NUMVAL >0 THEN AVERG = TOTAL/NUMVAL
```

The input statement in line 20, which is called a *preparatory input,* occurs only once before the start of the loop. It is required so that the user has the option of entering no values at all. If the first entry were -999, then the body of the loop would not be implemented and NUMVAL would remain at zero. Thus the computation of the average value in line 70 is conditional in order to avoid division by zero for this case.

Notice also how the order of the statements within the loop differs from that in Example 12.1. For the present case, the input of data (line 50) is the last statement in the loop. This ordering prevents the sentinel value from being incorporated into TOTAL or counted in NUMVAL.

The DOUNTIL structure.   As seen in Fig. 12.4c, the DOUNTIL version employs REM statements and documentation in a fashion similar to the DOWHILE structure. However, there are two features that make the DOUNTIL somewhat more complicated. First, the condition in line 30 is formulated using the NOT logical operator. This is done so that, as in the flowchart, the loop occurs if the condition is not true. Although formulating the condition in this way is consistent with the DOUNTIL structure, it sometimes makes the condition a bit more difficult to formulate. Second, in order that the loop execute at least once, the condition must be artificially set false beforehand (line 10). The following example provides an illustration of how this is done.

EXAMPLE 12.4   *Using a DOUNTIL Loop to Check the Validity of Data*

PROBLEM STATEMENT: One situation where a DOUNTIL loop would be of use is in ensuring that a valid entry is made from the keyboard. For example, suppose that you developed a program that required nonzero data. Such would be the case where the data might serve as the divisor in a formula, and hence division by zero had to be avoided. It would be advisable to have the program check for zero values upon input. Use the DOUNTIL loop to accomplish this objective.

SOLUTION: The following program fragment inputs values and checks to ensure that they are not zero:

```
10 LET VALUE = 1
20 REM BEGIN UNTIL LOOP
30 IF NOT (VALUE = 0) THEN 40 ELSE 60
40    INPUT "VALUE = ";VALUE
50    GOTO 20
60 REM END LOOP
```

Line 10 forces the loop to execute at least once. This allows the input to occur prior to the test in line 30, as is proper for a DOUNTIL structure. Thus you would not be able to proceed beyond this set of statements unless a nonzero VALUE was entered.

12.4   WHILE/WEND STATEMENTS

The WHILE/WEND statements of Microsoft BASIC are of the general form of a DOWHILE loop in which the number of iterations is not necessarily known when the loop starts. The general form of the WHILE/WEND loop is

ln WHILE *condition*
     .
     .

↑
⟨*Body of loop*⟩
↑

↑

ln WEND

The WHILE and WEND statements must always be used in pairs. That is, each WHILE statement must have a corresponding WEND. The WHILE is placed at the start of the loop and the WEND the end. The *condition*, which determines when the loop stops repeating, must be on the same line as WHILE. The conditions that can govern the execution of the loop are the same as those used in the IF/THEN/ELSE statement (recall Secs. 11.3 and 11.4). Both relational and logical operators may be used in the conditional expressions.

Other control statements such as IF statements can be used within a WHILE/WEND loop. In fact, even a second (or third, or fourth, etc.) WHILE/WEND loop can be placed within a WHILE/WEND loop. Nesting of WHILE/WEND loops follows the same pattern and rules as for a single loop. The only limitation in nesting of WHILE/WEND loops is that an inner loop cannot cross the boundaries of an outer loop, since a WEND statement is always paired with the most recent WHILE.

The problem that develops if improper nesting of WHILE/WEND statements is used is that BASIC will not report a syntax error. The program will run but it will not function as the author intended. The level of indentation and comments then are in error and may mislead rather than clarify how the program works.

EXAMPLE 12.5   Using WHILE/WEND Loops

PROBLEM STATEMENT: Use WHILE/WEND loops to make the program fragments in Examples 12.3 and 12.4 more concise.

SOLUTION: Recall that the program fragment in Example 12.3 was designed to enter and count data. The fragment can be reprogrammed with a WHILE/WEND in place of the IF/THEN/ELSE, as in

```
10 LET NUMVAL = 0
15 LET TOTAL = 0
20 INPUT "VALUE#1 = (ENTER -999 TO TERMINATE)";VALUE
25 WHILE VALUE <> -999
30    LET NUMVAL = NUMVAL + 1
35    LET TOTAL = TOTAL + VALUE
40    PRINT "VALUE #";NUMVAL;" = (ENTER -999 TO
      TERMINATE)";
45    INPUT VALUE
50 WEND
55 REM COMPUTE AVERAGE IF NUMVAL GREATER THAN ZERO
60 IF NUMVAL >0 THEN AVERG = TOTAL/NUMVAL
```

Comparison with the fragment in Example 12.3 shows that the WHILE/WEND permits the DOWHILE construct to be implemented in a clearer and more concise fashion. Two lines, the REM in line 25 and the GOTO in line 55 from Example 12.3 can be omitted for this revised version.

The same simplification can be applied to the program fragment from Example 12.4 with the result

```
10 LET VALUE = 1
20 WHILE NOT (VALUE = 0)
30    INPUT "VALUE = "; VALUE
40 WEND
```

Again the use of the WHILE/WEND makes the construct clearer and more concise.

WHILE/WEND loops are certainly a tremendous asset in BASIC. This construct allows fully structured loops with obvious entry and exit points. The value is so great that extended dialects of BASIC offer both DOWHILE and DOUNTIL loops, as described in the next section.

12.5   DOWHILE AND DOUNTIL CONSTRUCTS IN EXTENDED BASICs

The extended forms of BASIC, such as True BASIC and BASIC-PLUS, include several versions of the DOWHILE and DOUNTIL constructs. Some of these are summarized in Fig. 12.5. The difference between the FOR/WHILE (Fig. 12.5a) or the FOR/UNTIL (Fig. 12.5b) and the WHILE (Fig. 12.5c) and UNTIL (Fig. 12.5d) loops is that the former include counters, whereas the latter do not.

FIGURE 12.5
BASIC-PLUS loops.

FOR/WHILE loop structure
For *index = start* STEP *increment* WHILE *condition*
  (Body of loop)
NEXT *index*

(a)

FOR/UNTIL loop structure
FOR *index = start* STEP *increment* UNTIL *condition*
  (Body of loop)
NEXT *index*

(b)

WHILE loop structure
WHILE *condition*
  (Body of loop)
NEXT

(c)

UNTIL loop structure
UNTIL *condition*
  (Body of loop)
NEXT

(d)

PROBLEMS

12.1 The exponential function can be evaluated by the following infinite series:

$$e^x = 1 + x + \frac{x^2}{2} + \frac{x^3}{3!} + \cdots$$

Write a program to implement this formula so that it computes the value of e^x as each term in the series is added. In other words, compute and print in sequence

$$e^x \cong 1$$

$$e^x \cong 1 + x$$

$$e^x = 1 + x + \frac{x^2}{2}$$

up to the order term of your choosing. For each of the above, compute the percent relative error as

$$\% \text{ error} = \frac{\text{true} - \text{series approximation}}{\text{true}} \times 100\%$$

Use the library function for e^x in your computer to determine the ''true value.'' Have the program print out the series approximations and the error at each step along the way. Employ loops to perform the analysis. As a test case, employ the program to evaluate $e^{1.5}$ up to the term $x^{10}/10!$.

12.2 Repeat Prob. 12.1 for the hyperbolic cosine which can be approximated by

$$\cosh x = 1 + \frac{x^2}{2!} + \frac{x^4}{4!} + \cdots$$

As a test case, employ the program to evaluate $\cosh 78°$ up to the term $x^{10}/10!$. Remember that both the cosh and the above formula use radians.

12.3 Repeat Prob. 12.1 for the sine, which can be approximated by

$$\sin x = x - \frac{x^3}{3!} + \frac{x^5}{5!} - \frac{x^7}{7!} + \cdots$$

As a test case, employ the program to evaluate $\sin 30°$. Remember, both the sine function and the above formula use radians.

12.4 Write a computer program to evaluate the following series:

$$\text{SUM} = \sum_{i=0}^{n} \frac{1}{2^i} \quad \text{for } n = 20$$

Print out SUM and n during each iteration. Does the series converge to a constant value or diverge to infinity?

12.5 Repeat Prob. 12.4, except with the following formula ($n = 20$):

$$SUM = \sum_{i=1}^{n} \frac{1}{i^2}$$

12.6 Repeat Prob. 12.4 given ($n = 20$)

$$SUM = \sum_{i=0}^{n} \left(\frac{2}{5}\right)^i$$

12.7 Consider the following formula from calculus:

$$\frac{1}{1-x} = 1 + x + x^2 + \cdots$$

Write a computer program to test the validity of this formula for values of x of 0.6, -0.3, 2, and $-3/2$.

12.8 The series

$$f(x) = \frac{4}{\pi}\left(\sin x + \frac{\sin 3x}{3} + \frac{\sin 5x}{5} + \cdots\right)$$

is an approximation of $f(x)$ as shown in Fig. P12.8. Write a computer program to evaluate the accuracy of the series as a function of the number of terms at $x = \pi/2$. The program should print out the error in the approximation and the number of terms in the series.

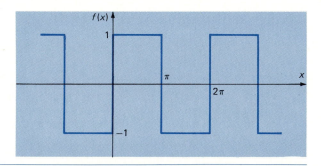

FIGURE P12.8

12.9 Repeat Prob. 12.8, but use Fig. P12.9, which is represented by the series

$$f(x) = \frac{\pi}{4} - \frac{2}{\pi}\left(\cos x + \frac{\cos 3x}{9} + \frac{\cos 5x}{25} + \cdots\right)$$

$$+ \left(\sin x - \frac{\sin 2x}{2} + \frac{\sin 3x}{3} - \cdots\right)$$

Check the approximation at $x = \pi/2$ and $x = 3\pi/2$.

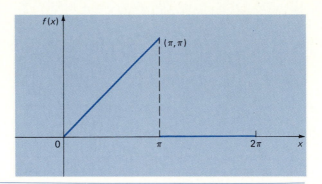

FIGURE P12.9

12.10 Repeat Example 9.6 but use loops to make the computation repetitive. Generate 10, 50, and 100 random coin tosses and tally how many heads and tails are tossed for each case. Comment on your results.

12.11 Reprogram Fig. 9.7 but use loops to make the code more efficient.

CHAPTER 13

BASIC:
SUBROUTINES

It is sometimes desirable to execute a particular set of statements several times and in different parts of the program. It would obviously be inconvenient and awkward to rewrite this set separately every time it is required. Fortunately, it is possible in BASIC to use a subroutine to write such sections once and then have them available at different locations within the program to use as many times as necessary. In addition, subroutines can be used to structure a program in a modular fashion.

13.1   SUBROUTINE STATEMENTS

The GOSUB statement is employed to transfer to the subroutine, and the RETURN statement is used to transfer back to the main program. The general forms of these statements are

$$ln_1 \ \ \texttt{GOSUB} \ \ ln_2$$

and

$$ln_3 \ \ \texttt{RETURN}$$

The GOSUB transfers control to line ln_2, which is the first line of the subroutine. This operation of transferring control is referred to as *calling* the subroutine. Within the subroutine, execution proceeds as normal until the RETURN is encountered, whereupon control is transferred back to the line immediately following ln_1. The process is depicted graphically in Fig. 13.1.

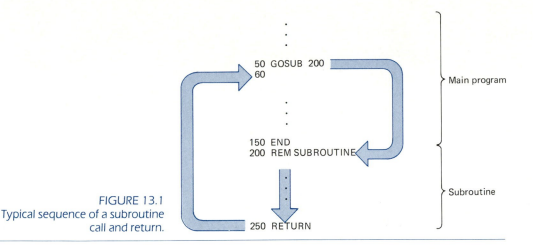

FIGURE 13.1
Typical sequence of a subroutine
call and return.

The line number referenced by the GOSUB, ln_2, must be a valid line number located at the start of the subroutine. Each RETURN must be paired with a GOSUB. A RETURN statement cannot be executed without having been reached via a GOSUB.

The effect of calling a subprogram is like having a GOTO in the main program that branches to another area of the program and a GOTO that then branches back to the statement immediately after the calling GOTO. The advantage of using GOSUB and RETURN instead of two GOTO statements is that the former conveniently allows the subroutine to be called from several locations. The RETURN assures that control is automatically transferred back to the correct line in the main program, whereas the use of a GOTO in the subprogram would require modification to return correctly for each particular case.

All subroutines should usually be located together at either the beginning or the end of the program. Because they are intended to be called by a GOSUB and never executed directly, they must be preceded by a GOTO or an END statement to prevent the program from reaching them without using a GOSUB. If the subroutines were all located at the beginning of the program, a GOTO would be required to detour around them to the main body of the program. In this book, we place BASIC subroutines at the end of the program. This organization reflects the top-down nature of program design, with the broad ideas in the main program followed by the detailed operations in the subroutines.

EXAMPLE 13.1   Subroutine to Compute Volume and Surface Area of a Rectangular Prism.

PROBLEM STATEMENT: Figure 13.2a summarizes formulas for the volume and surface area of a rectangular prism. Develop a subroutine to compute these values.

SOLUTION: Figure 13.3 shows BASIC code to solve this problem. Notice how both the input/output operations and the computations are sequestered in the subroutine.

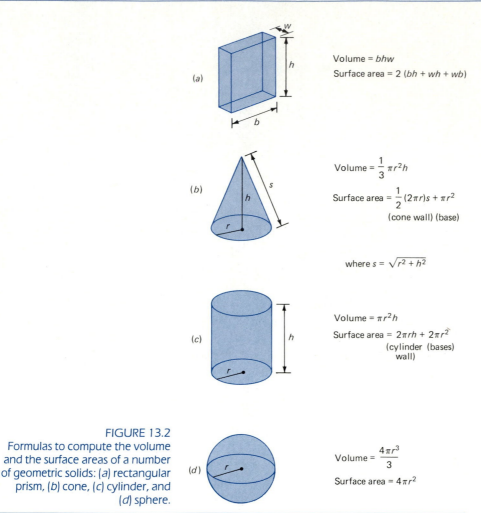

Volume = bhw

Surface area = $2\,(bh + wh + wb)$

Volume = $\dfrac{1}{3}\,\pi r^2 h$

Surface area = $\dfrac{1}{2}\,(2\pi r)s + \pi r^2$

(cone wall) (base)

where $s = \sqrt{r^2 + h^2}$

Volume = $\pi r^2 h$

Surface area = $2\pi rh + 2\pi r^2$

(cylinder (bases)
wall)

Volume = $\dfrac{4\pi r^3}{3}$

Surface area = $4\pi r^2$

FIGURE 13.2
Formulas to compute the volume
and the surface areas of a number
of geometric solids: (*a*) rectangular
prism, (*b*) cone, (*c*) cylinder, and
(*d*) sphere.

13.2 THE ORGANIZATION AND STRUCTURE OF SUBROUTINES

Just as chapters are the building blocks of a novel, subroutines are the building blocks
of a computer program. Each subroutine can have all the elements of a whole program,
but, in general, it is smaller and more specialized. Some computer experts suggest
that a reasonable size limit for a subroutine should be about 50 lines. Because this
size will fit on one or two monitor screens, it allows quick viewing of the whole
subroutine. It is a reasonable guideline because it facilitates visualizing the purpose
and the composition of the subroutine.

In addition to allowing you to easily view the subroutine, the 50-line limit will
also encourage you to restrict the subroutine to one or two well-defined tasks. It is
this characteristic of subroutines that makes them the ideal vehicle for expressing a
modular design in program form (recall Sec. 7.2).

```
100 REM    ***********************************************       300 REM **************** SUBROUTINE AREA ****************
110 REM    *          RECTANGULAR PRISM PROGRAM       *          310 REM
120 REM    *               (BASIC version)            *          320 REM
130 REM    *                   by  S. C. Chapra       *          330 REM    COMPUTES VOLUME AND AREA AND PRINT OUT RESULTS.
140 REM    *                                          *          340 REM
150 REM    *    This program calculates the volume and *         350 REM    DATA REQUIRED: NONE
160 REM    *  surface area of a rectangular prism.     *         360 REM    RESULTS RETURNED: VOLUME, AREA
170 REM    ***********************************************       370 REM
180 REM                                                          380 REM    LOCAL VARIABLES:
190 REM              .----------------------------------.        390 REM    ARB = BASE OF PRISM
200 REM              |   DEFINITION OF VARIABLES        |        400 REM    ARH = HEIGHT OF PRISM
210 REM              |                                  |        410 REM    ARW = WIDTH OF PRISM
220 REM              | VOLUME = VOLUME OF PRISM         |        420 REM
230 REM              | AREA   = SURFACE AREA OF PRISM   |        430 REM ----------------- BODY OF AREA  ----------------
240 REM              '----------------------------------'        440 CLS
250 REM                                                          450 PRINT "ENTER DIMENSIONS FOR A RECTANGULAR PRISM"
260 REM ***************** MAIN PROGRAM  *****************         460 PRINT:INPUT "BASE,LENGTH,WIDTH = ";ARB,ARH,ARW
270 REM                                                          470 VOLUME = ARB * ARH * ARW
280 GOSUB 300                                                    480 AREA = 2*(ARB*ARH + ARW*ARH + ARW*ARW)
290 END                                                          490 PRINT:PRINT "VOLUME = ";VOLUME;" AND AREA = ";AREA
                                                                 500 RETURN
```

FIGURE 13.3 A program with a subroutine to compute the volume and surface area of a rectangular prism.

The general organization of a subroutine is outlined in Fig. 13.4. Subroutines should always be considered major structures in the program and, consequently, should be delineated clearly using blank spaces and lines. Every subroutine should begin with a title line. It is the line number of this title line that should be called by the GOSUB statement. Additional labeling information—such as a short description of the purpose of the subroutine, the data required, the results returned, and definitions of local variables (that is, those used exclusively in the subroutine)—should be included after the title and before the body of the subroutine. Some of these characteristics are manifested in the program in Fig. 13.3.

FIGURE 13.4
The general organization of a subroutine.

```
REM
REM    ******************************************************
REM               SUBROUTINE name
REM
REM    Description of task . . .
REM
REM    Data required: Variable names and descriptions
REM
REM    Results returned: Variable names and descriptions
REM
REM    Local variables: Variable names and descriptions
REM
REM    Body of name
.
.
.
REM    RETURN
REM    ******************************************************
```

The statements in the body of the subroutine should be organized in a fashion similar to that of the main program. Structures should be identified by indentations and remarks as needed. Because a subroutine's task is, by definition, limited and focused, there should be few structures within each subroutine. Just as with the main program, each subroutine should start and end with no indentation.

The subroutine should always end with the RETURN statement. Strictly speaking, subroutines may have several RETURN statements for routing back to the main program. However, because of structured programming considerations, this practice is discouraged. Rather, each subroutine should have a single entrance at its beginning and a single exit at its end. In this way, the logical flow is kept simple and is easy to understand.

13.3   LOCAL AND GLOBAL VARIABLES

A two-way communication is required between the subroutine and the main program. The subroutine needs data to act on and must return its results for utilization in the main program. The data and results are said to be "passed" to and from the program. For many computer languages this exchange of information is a controlled process. That is, the information that is passed in and out of the subroutine must be specified explicitly by the programmer. These variables, along with the other variables in the main program, are commonly referred to as *global variables*. In contrast, variables that reside exclusively in the subroutine are called *local variables*.

The ability to have local variables in a subroutine can be a great asset. In essence, they serve to make each subroutine a truly autonomous entity. For example, changing a local variable will have no effect on variables outside the subroutine. In fact, true local variables can even have the same name as variables in other parts of the program yet still be distinct entities. One advantage of this is that it decreases the likelihood of "side effects" that occur when the value of a global variable is changed inadvertently in a subprogram. In addition, the existence of local variables also facilitates the debugging of a program. The subprograms can be debugged individually without fear that a working subroutine will cease functioning because of an error in another subroutine.

Unfortunately, Microsoft BASIC and other simpler dialects do not formally include local variables. That is, the language does not control the passage of information between subroutines and the main program. In BASIC, all variables are global and can be accessed by any part of a program. Extreme caution must therefore be exercised in naming variables to ensure that a subroutine does not inadvertently modify a variable of the same name used in another part of the program.

Although local variables are not explicitly a part of BASIC, there is still a need for subroutine variables that are unique and unrelated to those in the calling program. Therefore, it is good practice to define local variables on an ad hoc basis by choosing unique names for the variables in each subroutine. In the present book we do this by beginning all local variables with a prefix that is peculiar to that subroutine. The prefix must be short—at the maximum, two or three letters. Here we will use the first two

```
100 INPUT "TIME = ";T              100 INPUT "TIME = ";T
110 GOSUB 500                      110 GOSUB 500
120 PRINT "TIME = ";T              120 PRINT "TIME = ";T
130 END                           130 END
500 REM SUBROUTINE THERMO          500 REM SUBROUTINE THERMO
510 INPUT "TEMPERATURE = ";T       510 INPUT "TEMPERATURE = ";THT
520 IF T > 75 THEN 530 ELSE 550    520 IF THT > 75 THEN 530 ELSE 550
530      PRINT "AIR CONDITIONER ON"  530      PRINT "AIR CONDITIONER ON"
540        GOTO 570                540        GOTO 570
550    REM ELSE                    550    REM ELSE
560      PRINT "AIR CONDITIONER OFF" 560      PRINT "AIR CONDITIONER OFF"
570 REM ENDIF                      570 REM ENDIF
580 RETURN                         580 RETURN


RUN                               RUN
TIME = ? 1200                     TIME = ? 1200
TEMPERATURE = ? 83                TEMPERATURE = ? 83
AIR CONDITIONER ON                AIR CONDITIONER ON
TIME =  83                        TIME =  1200
Ok                                Ok
         (a)                               (b)
```

FIGURE 13.5

(a) A program illustrating the difficulty that can arise when a variable in a subroutine has the same name as a variable in the main program; (b) a program showing properly named variables.

letters of the subroutine title as the prefix (see Fig. 13.3). If no two subroutines begin with the same two letters and the global variables in the main program avoid using names starting with these letters, then each subroutine will have its own unique set of ''local'' variables. However, now the limitation of six-letter variable names can become restrictive. Once the first two letters are used to identify the particular subroutine, then the remaining four letters leave little room for variable names that are very descriptive. Herein we will continue to adhere to the six-letter limit, but you might consider using longer, more understandable names in your own programs.* Certainly, Microsoft BASIC allows very long names, but not all dialects of BASIC are so well endowed. Indeed, Applesoft BASIC is limited in this regard.

EXAMPLE 13.2   Global and Local Variables in BASIC

PROBLEM STATEMENT: The lack of local variables in BASIC subroutines can sometimes cause great difficulties. Problems often occur when different variables are assigned the same name in the main program and in a subroutine. This sometimes happens when dealing with variables such as time and temperature that are typically represented by the same letter, T.

The program in Fig. 13.5a illustrates the difficulty that can occur. The main program inputs the time of day. Then a subroutine is called to input the temperature and print a message signifying whether an air conditioner is to be turned on. Finally, control is returned to the main program, where the time is printed. However, because time and temperature have been assigned the same variable name, T, the value of time is erroneously printed as 83 (which is, in fact, the temperature) by line 120. This is because T was assigned this value when temperature was input in the subroutine.

*An alternative option for naming local variables is to affix an extension consisting of a period and the first three letters of the subroutine name. For example, ARB from Fig. 13.3 could alternatively be named B.ARE or, better still, BASE.ARE.

SOLUTION: The program in Fig. 13.5b circumvents this problem by defining a local variable, THT, for temperature in the subroutine THERMO. This local variable is a composite of the first two letters of the subroutine's name, TH, along with the T for temperature. Defining this local variable means that the variable in the subroutine is an entity distinct from those in other parts of the program. Consequently, the results of this version are correct.

It should be noted that some extended dialects, such as True BASIC (Kemeny, 1985) allow true local variables in subroutines. This represents a great improvement over Microsoft BASIC and is another reason why the extended dialects should become standard in the future.

13.4   THE ON GOSUB STATEMENT AND MENUS

Aside from the GOSUB statement, subroutines may be called with the ON GOSUB statement, which has the general form

$$ln \ \text{ON} \ expression \ \text{GOSUB} \ ln_1, ln_2, \ldots, ln_n$$

where the *expression* is a variable or formula that may take on values from 1 to *n*. If the value is 1, the program calls the subroutine at ln_1; if it is 2, the program calls the subroutine at ln_2; and so on. For example, for

```
10 LET A = 3
20 ON A GOSUB 50, 100, 140, 180
```

line 20 will call the subroutine at line 140. When the RETURN statement in this subroutine is encountered, control transfers back to the line number immediately following line 20 in the same manner as a GOSUB statement.

If the *expression* in the ON GOSUB is a noninteger, the fractional part is rounded. After it is rounded, if the *expression* is zero or greater than *n*, control transfers automatically to the next line. If the *expression* is less than zero or greater than 255, an error message results.

EXAMPLE 13.3   *Menu-Driven Program to Compute the Volume and Surface Area of a Number of Selected Geometric Solids*

PROBLEM STATEMENT: Expand the program from Example 13.1 (Fig 13.3) so that it computes the volumes and surface areas for all the geometric solids in Fig. 13.2.

SOLUTION: Figure 13.6 shows the BASIC code to solve this problem. Subroutines are employed to form well-defined modules to implement the computations for each solid. Notice how an ON GOSUB statement is used in conjunction with a menu to allow the user to select the solid for which the calculation is to be performed. Sample output from the program is shown in Fig. 13.7.

```
100 REM     ***************************************************
110 REM     *           VOLUME/AREA PROGRAM              *
120 REM     *             (BASIC version)                *
130 REM     *              by  S. C. Chapra              *
140 REM     *                                            *
150 REM     *  This program calculates the volume and    *
160 REM     *  surface area of different geometric solids. *
180 REM     ***************************************************
190 REM
200 REM **************** MAIN PROGRAM ****************
210 CLS
220 REM                     Create a Menu
230 REM
240 PRINT:PRINT:PRINT
250 PRINT "SELECT ONE OF THE FOLLOWING OPTIONS"
260 PRINT "TO DETERMINE THE VOLUME AND AREA FOR"
270 PRINT TAB(10);"1. RECTANGULAR PRISM"
280 PRINT TAB(10);"2. CONE"
290 PRINT TAB(10);"3. CYLINDER"
300 PRINT TAB(10);"4. SPHERE"
310 PRINT TAB(10);"5. TERMINATE THIS PROGRAM"
320 INPUT "ENTER YOUR SELECTION";SELECT
330 CLS
340 ON SELECT GOSUB 1000,2000,3000,4000,380
350 PRINT
360 INPUT "PRESS RETURN KEY TO CONTINUE"; DUMMY$
370 GOTO 210
380 END
1000 REM ************** SUBROUTINE PRISM **************
1010 REM
1020 REM      INPUTS DIMENSIONS, AND COMPUTES AND
1030 REM      PRINTS OUT VOLUME AND AREA FOR
1040 REM      A RECTANGULAR PRISM.
1050 REM
1060 REM      DATA REQUIRED: NONE
1070 REM      RESULTS RETURNED: NONE
1080 REM
1090 REM      LOCAL VARIABLES:
1100 REM
1110 REM         PRB = BASE
1120 REM         PRH = LENGTH
1130 REM         PRW = WIDTH
1140 REM
1150 REM --------------- BODY OF PRISM ---------------
1160 CLS
1170 PRINT "ENTER DIMENSIONS FOR A RECTANGULAR PRISM"
1180 PRINT:INPUT "BASE,HEIGHT,WIDTH = ";PRB,PRH,PRW
1190 PRVOL = PRB*PRH*PRW
1200 PRAREA = 2*(PRB*PRH+PRW*PRH+PRW*PRB)
1210 PRINT:PRINT "VOLUME = ";PRVOL;" AND AREA = ";PRAREA
1220 RETURN
1230 REM
2000 REM *************** SUBROUTINE CONE ***************
2010 REM
2020 REM      INPUTS DIMENSIONS, AND COMPUTES AND
2030 REM      PRINTS OUT VOLUME AND AREA FOR
2040 REM      A CONE.
2050 REM
2060 REM      DATA REQUIRED: NONE
2070 REM      RESULTS RETURNED: NONE
2080 REM
```

```
2090 REM      LOCAL VARIABLES:
2100 REM
2110 REM         COR = RADIUS
2120 REM         COH = HEIGHT
2130 REM         COSL = SLANT HEIGHT
2140 REM
2150 REM --------------- BODY OF CONE ---------------
2160 CLS
2170 PRINT "ENTER DIMENSIONS FOR A CONE"
2180 PRINT:INPUT "RADIUS,HEIGHT = ";COR,COH
2190 COSL = SQR(COR^2 + COH^2)
2200 COVOL = 3.14159*COR^2*COH/3
2210 COAREA = 3.14159*COR*COSL + 3.14159*COR^2
2220 PRINT:PRINT "VOLUME = ";COVOL;" AND AREA = ";COAREA
2230 RETURN
2240 REM
3000 REM ************** SUBROUTINE CYLINDER **************
3010 REM
3020 REM      INPUTS DIMENSIONS, AND COMPUTES AND
3030 REM      PRINTS OUT VOLUME AND AREA FOR
3040 REM      A CYLINDER.
3050 REM
3060 REM      DATA REQUIRED: NONE
3070 REM      RESULTS RETURNED: NONE
3080 REM
3090 REM      LOCAL VARIABLES:
3100 REM
3110 REM         CYR = RADIUS
3120 REM         CYH = HEIGHT
3130 REM
3140 REM --------------- BODY OF CYLINDER ---------------
3150 CLS
3160 PRINT "ENTER DIMENSIONS FOR A CYLINDER"
3170 PRINT:INPUT "RADIUS,HEIGHT = ";CYR,CYH
3180 CYVOL = 3.14159*CYR^2*CYH
3190 CYAREA = 2*3.14159*CYR*CYH + 2*3.14159*CYR^2
3200 PRINT:PRINT "VOLUME = ";CYVOL;" AND AREA = ";CYAREA
3210 RETURN
3220 REM
4000 REM *************** SUBROUTINE SPHERE ***************
4010 REM
4020 REM      INPUTS DIMENSIONS, AND COMPUTES AND
4030 REM      PRINTS OUT VOLUME AND AREA FOR
4040 REM      A SPHERE.
4050 REM
4060 REM      DATA REQUIRED: NONE
4070 REM      RESULTS RETURNED: NONE
4080 REM
4090 REM      LOCAL VARIABLES:
4100 REM
4110 REM         SPR = RADIUS
4120 REM
4130 REM --------------- BODY OF SPHERE ---------------
4140 CLS
4150 PRINT "ENTER DIMENSIONS FOR A SPHERE"
4160 PRINT:INPUT "RADIUS = ";SPR
4170 SPVOL = 4*3.14159*(SPR^3)/3
4180 SPAREA = 4*3.14159*SPR^2
4190 PRINT:PRINT "VOLUME = ";SPVOL;" AND AREA = ";SPAREA
4200 RETURN
```

FIGURE 13.6 Program to compute the volume and the surface area of the geometric solids from Fig 13.2.

13.5 YOUR SUBROUTINE LIBRARY

Subroutines facilitate the process of program development for two major reasons. First, they are discrete, task-oriented modules that are, therefore, easier to understand, write, and debug. Second, they can be used over and over both within a program and between programs. Testing a small subroutine to prove that it works properly is far easier than testing an entire program. Once a subroutine is thoroughly debugged and tested, it can be incorporated into other programs. Then if errors occur, the subroutine

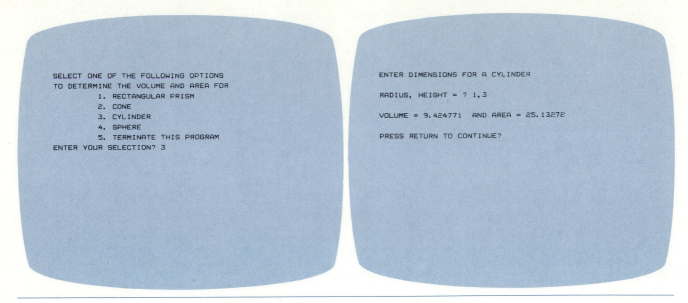

```
SELECT ONE OF THE FOLLOWING OPTIONS          ENTER DIMENSIONS FOR A CYLINDER
TO DETERMINE THE VOLUME AND AREA FOR
        1. RECTANGULAR PRISM                 RADIUS, HEIGHT = ? 1,3
        2. CONE
        3. CYLINDER                          VOLUME = 9.424771  AND AREA = 25.13278
        4. SPHERE
        5. TERMINATE THIS PROGRAM            PRESS RETURN TO CONTINUE?
ENTER YOUR SELECTION? 3
```

FIGURE 13.7 Output from the program in Fig. 13.6.

TABLE 13.1   Programs to Be Presented in the Remainder of the Book that Have General Applicability in Engineering. (You should set a goal to develop subroutines for each of these areas as the start of your own software library.)

Program	Application	Chapter
Plotter	Plot points or lines on cartesian coordinates	22,39
Shell sort	Sort numbers or words in ascending order	23
Histogram	Create a bar diagram depicting the shape of a data distribution	24
Statistics	Calculate a number of descriptive statistics to characterize data	24
Linear regression	Fit the best straight line through data plotted on cartesian coordinates	26
Lagrange interpolation	Use polynomials to estimate intermediate values between data plotted on cartesian coordinates	27
Bisection	Determine the root of a single algebraic or transcendental equation	28
Gauss elimination	Solve simultaneous linear algebraic equations	29
Numeric differentiation	Determine the derivative of data	30
Trapezoidal rule	Determine the integral of data	31
Euler's method	Solve a single, first-order differential equation	32

is unlikely to be the culprit, and the debugging effort can be focused on other parts of the program.

Many professional programmers accumulate a collection, or library, of useful subroutines that they access as needed. Thus instead of programming lines of code, they combine and manipulate larger, more powerful pieces of code—that is, subroutines. Programming with these custom modules saves overall effort and, more importantly, improves program reliability. The modules in a library can be thoroughly debugged under a number of different conditions so as to assure proper function. Such thorough testing can rarely be afforded for sections of code that are only used once.

In the remainder of the book we will be developing a number of programs that have general applicability in engineering (see Table 13.1). You should view these programs as the beginning of your own subroutine library with an eye to their utility in your future academic and professional efforts.

PROBLEMS

13.1 Write a program to determine the roots of a quadratic equation. Pattern the computation after Fig. 11.7. However, use two separate subroutines to (1) input the coefficients and (2) compute and print out the results. Organize the subroutines as discussed in Sec. 13.2 and employ local variables as discussed in Sec. 13.3.

13.2 Reprogram Fig. 9.7 so that it uses subroutines, is organized according to Sec. 13.2, and employs local variables as discussed in Sec. 13.3

13.3 Write a subroutine that accomplishes the same operation as the SGN function (recall Table 9.2). Organize the subroutine and employ local variables as described in Secs. 13.2 and 13.3.

13.4 Write a program to determine the values of all the trigonometric functions in Table 9.3 given an angle. Use a subroutine to input the angle and transform it to radians if it is entered in degrees. Employ another subroutine to determine the values of the trigonometric functions and print out the results. Organize the subroutine and employ local variables as described in Secs. 13.2 and 13.3. Test the program with an angle equal to 30°.

13.5 Write a program to determine the range of a projectile fired from the top of a cliff as described in Prob. 9.20. Employ a modular approach and use three subroutines to (1) input the data, (2) perform the computation, and (3) output the results. Design both the main program and the input subroutine so that they are menu-driven. The main program menu should include such choices as to input or modify data, to perform the computation and output results, and to terminate the program. Figure 13.6 can serve as the model for how this can be done. The input subroutine menu should include such choices as to input or modify (1) the initial velocity, (2) angle, and (3) cliff height and to (4) return to main menu. Use a CASE construct as the basis for the input subroutine.

13.6 Write a subroutine to input and convert temperatures from degrees Fahrenheit to degrees Celsius and Kelvin by the formulas

$$T_C = \frac{5}{9}(T_F - 32)$$

and

$$T_K = T_C + 273$$

Organize the subroutine and employ local variables as discussed in Secs. 13.2 and 13.3. Number this subroutine with line numbers between 1000 and 1999. Store the subroutine in ASCII form. Test it with the following data: $T_F = 20$, 203, and 98.

13.7 Write a subroutine to input and convert temperature from degrees Celsius to degrees Fahrenheit and Rankin by the formulas

$$T_F = \frac{9}{5}T_C + 32$$

and

$$T_R = T_F + 460$$

Follow Secs. 13.2 and 13.3 to organize and employ local variables in this subroutine. Number the subroutine with line numbers between 2000 and 2999. Store the subroutine in ASCII form. Test it with $T_C = -10$, 27, and 90.

13.8 Develop the subroutines in Probs. 13.6 and 13.7. Then merge them into a menu-driven program to perform general temperature conversions. The main program menu should include such choices as (1) to convert from Fahrenheit to Celsius and Kelvin, (2) to convert from Celsius to Fahrenheit and Rankin, and (3) to terminate the program. Figure 13.6 can be used as a model for how the menu can be programmed. Test the program with the following data: 10°C, 25°F, 120°C, and 800°F.

CHAPTER 14

BASIC:
LARGE AMOUNTS OF DATA

In our discussion of FORTRAN in Chap. 5, we mentioned that business typically deals with large amounts of data and simple computations, whereas engineering deals with small amounts of data and complicated computations. Although this is generally true, some engineering applications require large quantities of information. Here we present material that is particularly useful for these situations.

14.1 ARRAYS

To this point, we have learned that a variable name conforms to a storage location in primary memory. Whenever the variable name is encountered in a program, the computer can access the information stored in the location. This is an adequate arrangement for many small programs. However, when dealing with large amounts of information, it often proves inadequate.

For example, suppose that in the course of an engineering project you were using stream-flow measurements to determine how much water you might be able to store in a projected reservoir. Because some years are wet and some are dry, natural stream flow varies from year to year in an almost random fashion. Thus you would require about 25 to 50 years of measurements to accurately estimate the long-term average flow. Data of this type is listed in Table 14.1. A computer would come in handy in such an analysis. However, it would obviously be inconvenient to come up with a different variable name for each year's flow. Similar problems arise continually in engineering and other computer applications. For this season, subscripted variables, or arrays, have been developed.

Table 14.1 Thirty Years of Water-Flow Data for a River

Year	Flow*	Year	Flow*
1954	125	1969	63
1955	102	1970	115
1956	147	1971	71
1957	76	1972	106
1958	95	1973	53
1959	119	1974	99
1960	62	1975	153
1961	41	1976	112
1962	104	1977	82
1963	81	1978	132
1964	128	1979	96
1965	110	1980	102
1966	83	1981	90
1967	97	1982	122
1968	79	1983	138

*In billion gallons per year.

An *array* is an ordered collection of data. That is, it consists of a first item, second item, third item, etc. For the stream flow example, an array would consist of the first year's flow, the second year's flow, the third year's flow, etc.

In programming, all the individual items, or *elements*, of the array are referenced by the same variable name. In order to distinguish between the elements, each is given a subscript. This is a common practice in standard mathematics. For example, if we assign flow the variable name F, an index or subscript can be attached to distinguish each year's flow, as in

$$F_1 \quad F_2 \quad F_3 \quad F_4 \quad F_5$$

where F_i refers to the flow in year i. These are referred to as *subscripted variables*. In BASIC a similar arrangement is employed, but because subscripts are not allowed, the index is placed in parentheses, as in

F(1)   F(2)   F(3)   F(4)   F(5)

Note that for an array, the parentheses are part of the variable name. This makes the elements different from F1, F2, etc. In fact, F(1) and F1 are considered to be different variables and can be used in the same program. Each would be assigned a different storage location, as depicted in Fig. 14.1.

LET statements can be employed to assign values to the elements of an array in a fashion similar to assigning them to nonsubscripted variables. For example, using the data in Table 14.1,

```
10 LET F(1) = 125
20 LET F(2) = 102
```

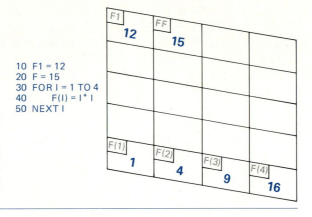

FIGURE 14.1
Visual depiction of memory locations for unsubscripted and subscripted variables. The program on the left assigns values to F1, FF, and F1(1) through F1(4), with the resulting constants being stored in the memory locations depicted on the right.

```
10 F1 = 12
20 F = 15
30 FOR I = 1 TO 4
40     F(I) = I * I
50 NEXT I
```

```
30 LET F(3) = 147
```

In addition, INPUT statements can also be used, as in

```
10 INPUT F(1)
20 INPUT F(2)
30 INPUT F(3)
```

Although the above methods can be employed, loops can be used to perform the same operation more efficiently. The following example illustrates the advantages of utilizing arrays and loops when dealing with large quantities of data.

EXAMPLE 14.1   Inputting Data to an Array with Loops

PROBLEM STATEMENT: Develop an efficient program to input the first 10 values from Table 14.1.

SOLUTION: Using LET or individual INPUT statements to enter the data would result in a program of 11 statements (including the END statement). A much more efficient arrangement is possible using a FOR/NEXT loop and an INPUT statement in tandem, as in

```
100 FOR I = 1 TO 10
150     INPUT FLOW(I)
200 NEXT I
250 END
```

Notice how we have used a more descriptive name, FLOW, for the array. In addition, notice how the index (I) of the loop is used as the subscript for FLOW. Thus, on the first pass through the loop, I is 1, and the program will prompt us to enter FLOW(1). Then, on subsequent passes, I will be incremented in order to input FLOW(2), FLOW(3), etc. Consequently, the loop allows us to enter the data using only 4 lines rather than 11.

The subscripts of an array are positive numbers ranging from 0 to 32,767.* However, when more than 10 subscripts are used, a special statement must be added, as described in the next section. Note that decimal subscripts may be employed but that the computer treats them as if they were rounded.

14.1.1   The DIM Statement

BASIC automatically sets aside 10 memory locations for any subscripted variable it encounters in a BASIC program. If more than 10 are needed, a DIM statement is required. Its general form is

$$ln \ \texttt{DIM} \ var_1(n_1), \ var_2(n_2), \ . \ . \ .$$

where var_1 and var_2 are the variable names of different arrays and n_1 and n_2 are the upper limits on the number of elements in each array. For example, if we wanted to perform an analysis on all the data in Table 14.1, we would have to include the following DIM statement in our program

```
50 DIM FLOW(30)
```

This would set aside 30 storage locations for the 30 flows in Table 14.1. Suppose that we also wanted to input the years as string variables. If we designate this variable as YEAR$, we would rewrite the DIM statement as

```
50 DIM FLOW(30), YEAR$(30)
```

The DIM statement must be placed before the first line where the subscripted variable is utilized. Otherwise it is automatically dimensioned to 10 when it is first used. Redimensioning after this point would result in an error message.

14.1.2   Size of Arrays

In the previous section, we showed how a DIM statement is required to define a 30-element array for the water-flow data in Table 14.1. We could just as well have set aside more locations, as in

```
50 DIM FLOW(100), YEAR$(100)
```

*Note that the minimum value for the subscript can be changed to 0 by using the statement OPTION BASE 0.

This might be done if we anticipated that at a later time additional data (say, from the years prior to 1954) would be obtained. Thus we do not have to specify the exact size of an array in the DIM statement; we merely have to make sure that there are enough elements available to store our data. Otherwise an error will occur.

The limit on the number of elements for subscripted variables depends on the amount of memory available on your computer. Therefore for large programs it is advisable to estimate the size of your array realistically so that you do not exceed your computer's memory capacity.

As noted on several occasions to this point, BASIC is an interpreted language. As a consequence, many systems allow you to use numerical expressions other than constants to specify the size of arrays. For example, you might read in the number of data points as a variable before dimensioning, as in

```
10 INPUT "NUMBER OF FLOWS = ";N
20 DIM FLOW(N), YEAR$(N)
```

This capability is extremely useful when you do not have a lot of available memory.

14.1.3 Two-Dimensional Arrays

There are many engineering problems where data is arranged in tabular or rectangular form. For example, temperature measurements on a heated plate (Fig. 14.2) are often taken at equally spaced horizontal and vertical intervals. Because such two-dimensional arrangements are so common, BASIC has the capability of storing data in rectangular arrays, or *matrices*. As shown in Figure 14.3, the horizontal sets of elements are called *rows*, whereas the vertical sets are called *columns*. The first subscript designates the number of the row in which the element lies. The second subscript designates the column. For example, element A(2,3) is in row 2 and column 3.

FIGURE 14.2
Temperature measurements on a heated plate.

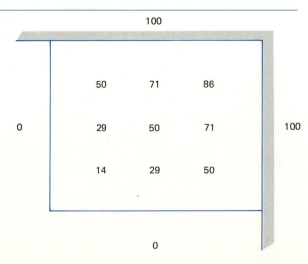

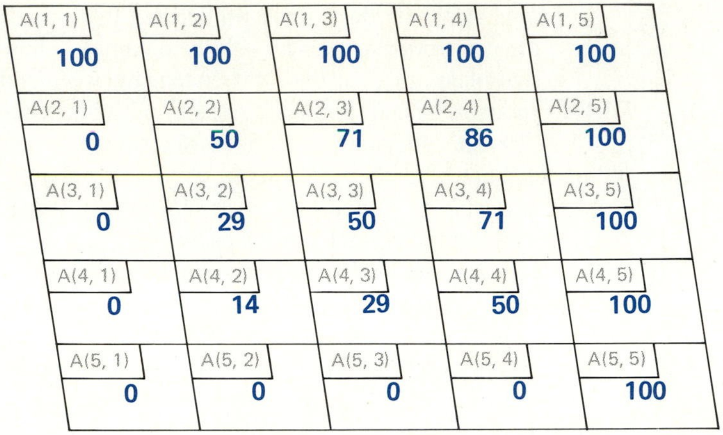

FIGURE 14.3
A two-dimensional array, or matrix, used to store the temperatures from Fig. 14.2. Note that the first and second subscripts designate the row and the column of the element, respectively.

EXAMPLE 14.2   Two-Dimensional Arrays

PROBLEM STATEMENT: Write a computer program to input the temperature data from Fig. 14.2. Then convert it to degrees Fahrenheit by the formula

$$T_F = \frac{9}{5} T_C + 32$$

Employ a user-defined function to make this conversion. Print out the results as integers.

SOLUTION: Figure 14.4 shows the program. Notice that a modular approach is used. The main, or calling, program consists of three GOSUB statements to call subroutines that input, convert, and output the data.

Note that, in Microsoft BASIC, arrays with more than two dimensions can be employed. In fact, up to 255 dimensions are allowed.

14.2 FILES

To this point, we have covered three methods for entering data into a program (see Table 14.2). Notice in the table that all of them have disadvantages related to large quantities of data and data for which frequent change might be required. In addition,

```
100 REM      *****************************************************
110 REM      *              HEATED PLATE PROGRAM                 *
120 REM      *                (BASIC version)                    *
130 REM      *                 by  S. C. Chapra                  *
140 REM      *                                                   *
150 REM      *   This program inputs temperatures for a          *
160 REM      *   heated plate. It converts temperatures from     *
170 REM      *   Celcius to Fahrenheit and transforms the        *
180 REM      *   results to integers. The results are output.    *
190 REM      *****************************************************
200 REM
210 REM           .-------------------------------------------.
220 REM           |          DEFINITION OF VARIABLES           |
230 REM           |                                           |
240 REM           |   NUMROW = NUMBER OF ROWS ON PLATE         |
250 REM           |   NUMCOL = NUMBER OF COLUMNS ON PLATE      |
260 REM           |   TEMP() = TEMPERATURE OF PLATE AT         |
270 REM           |            A PARTICULAR LOCATION           |
280 REM           `-------------------------------------------'
290 REM
300 DIM TEMP(20,20)
310 DEF FNCONVERT(T) = (9/5)*T + 32
320 REM
330 REM ******************  MAIN PROGRAM  ******************
340 REM
350 GOSUB 1000
360 GOSUB 2000
370 GOSUB 3000
380 END
390 REM
1000 REM **************  SUBROUTINE INPUT  ****************
1010 REM
1020 REM     INPUTS TEMPERATURES FOR THE ROWS AND
1030 REM     COLUMNS OF THE HEATED PLATE.
1040 REM
1050 REM     DATA REQUIRED: NONE
1060 REM     RESULTS RETURNED: NUMROW,NUMCOL,TEMP()
1070 REM
1080 REM     LOCAL VARIABLES:
1090 REM
1100 REM     INI, INJ = COUNTERS
1110 REM
1120 REM ------------------  BODY OF INPUT  ------------------
1130 CLS
1140 INPUT "NUMBER OF ROWS = ";NUMROW
1150 INPUT "NUMBER OF COLUMNS = ";NUMCOL
1160 FOR INI = 1 TO NUMROW
1170    FOR INJ = 1 TO NUMCOL
1180       PRINT "TEMPERATURE(";INI;",";INJ;") = ";
1190       INPUT TEMP(INI,INJ)
1200    NEXT INJ
1210 NEXT INI
1220 RETURN
1230 REM
2000 REM **************  SUBROUTINE CONVERT  ****************
2010 REM
2020 REM     CONVERTS TEMPERATURES FROM CELCIUS
2030 REM     TO FAHRENHEIT AND CONVERTS THE RESULT
2040 REM     TO AN INTEGER.
2050 REM
2060 REM     DATA REQUIRED: NUMROW,NUMCOL,TEMP()
2070 REM     RESULTS RETURNED: TEMP()
2080 REM
2090 REM     LOCAL VARIABLES:
2100 REM
2110 REM     COI, COJ = COUNTERS
2120 REM
```

```
2130 REM ---------------  BODY OF CONVERT  ------------------
2140 FOR COI = 1 TO NUMROW
2150    FOR COJ = 1 TO NUMCOL
2160       TEMP(COI,COJ) = FNCONVERT(TEMP(COI,COJ))
2170       TEMP(COI,COJ) = INT(TEMP(COI,COJ))
2180    NEXT COJ
2190 NEXT COI
2200 RETURN
2210 REM
3000 REM *************  SUBROUTINE OUTPUT  ***************
3010 REM
3020 REM     PRINTS OUT THE RESULTS IN TABULAR
3030 REM     FORM.
3040 REM
3050 REM     DATA REQUIRED: NUMROW,NUMCOL,TEMP()
3060 REM     RESULTS RETURNED: NONE
3070 REM
3080 REM     LOCAL VARIABLES:
3090 REM
3100 REM     OUI, OUJ = COUNTERS
3110 REM
3120 REM ---------------  BODY OF OUTPUT  ------------------
3130 CLS
3140 FOR OUI = 1 TO NUMROW
3150    FOR OUJ = 1 TO NUMCOL
3160       PRINT USING "####    ";TEMP(OUI,OUJ);
3170    NEXT OUJ
3180    PRINT
3190 NEXT OUI
3200 RETURN

RUN

NUMBER OF ROWS = ? 5
NUMBER OF COLUMNS = ? 5
TEMPERATURE( 1 , 1 ) = ? 100
TEMPERATURE( 1 , 2 ) = ? 100
TEMPERATURE( 1 , 3 ) = ? 100
TEMPERATURE( 1 , 4 ) = ? 100
TEMPERATURE( 1 , 5 ) = ? 100
TEMPERATURE( 2 , 1 ) = ? 0
TEMPERATURE( 2 , 2 ) = ? 50
TEMPERATURE( 2 , 3 ) = ? 71
TEMPERATURE( 2 , 4 ) = ? 86
TEMPERATURE( 2 , 5 ) = ? 100
TEMPERATURE( 3 , 1 ) = ? 0
TEMPERATURE( 3 , 2 ) = ? 29
TEMPERATURE( 3 , 3 ) = ? 50
TEMPERATURE( 3 , 4 ) = ? 71
TEMPERATURE( 3 , 5 ) = ? 100
TEMPERATURE( 4 , 1 ) = ? 0
TEMPERATURE( 4 , 2 ) = ? 14
TEMPERATURE( 4 , 3 ) = ? 29
TEMPERATURE( 4 , 4 ) = ? 50
TEMPERATURE( 4 , 5 ) = ? 100
TEMPERATURE( 5 , 1 ) = ? 0
TEMPERATURE( 5 , 2 ) = ? 0
TEMPERATURE( 5 , 3 ) = ? 0
TEMPERATURE( 5 , 4 ) = ? 0
TEMPERATURE( 5 , 5 ) = ? 100

 212     212     212     212     212
  32     122     159     186     212
  32      84     122     159     212
  32      57      84     122     212
  32      32      32      32     212
```

FIGURE 14.4 Program to input, convert, and output a two-dimensional array for the temperatures from Fig. 14.3.

TABLE 14.2   The Advantages and Disadvantages of Four Ways in Which Information Is Entered into BASIC Programs

Method	Advantages	Disadvantages	Primary Applications
LET (assignment) statement	Is simple and direct. Values are "wired" into the program and do not have to be entered on each run.	Takes up a line per assignment statement. If a new value is to be assigned, the user must retype the LET statement.	Appropriate for assigning values to true constants (such as pi or the acceleration of gravity) that are used repeatedly throughout a program and never change from run to run.
INPUT statement	Permits different values to be assigned to variables on each run. Allows interaction between user and program.	Values must be entered by user on each run.	Appropriate for those variables that take on different values on each run.
READ/DATA statements	Provide concise assignment of values. The DATA statement can be accessed more than once by different READ statements in the program.	If a new value is to be assigned, the user must retype the DATA statement.	Same as for the LET statement, but for cases where many constants are to be assigned.
FILE statements	Provides concise way to handle large amounts of input data. Allows data to be tranferred between programs.	Is relatively complicated	Appropriate for large quantities of data.

all the methods deal exclusively with the information requirements of a single program. If the same data is to be used for another program, it must be retyped.

Another option is to store information on an external memory device such as a magnetic disk. This is conceptually similar to the way we have been saving our programs on disks so that they are not destroyed when the computer is shut off. Along with our programs, we can store data in external memory files just as office information is stored in file cabinets for later use (Fig. 14.5).

A number of advantages result from storing data on files. First, data entry and modification become easier. In addition, files make it possible to run several programs using the data from the same file and, alternatively, to run the same program from several different files. Before proceeding to the specifics of manipulating files, we will introduce some new terminology.

14.2.1   File Terminology

In its most general sense, a *file* can be defined as a section of external storage that has a name. However, a more precise definition is possible if we define the component parts that make up a file.

As seen in Fig. 14.5, each individual piece of information on a file is called an *item*. Examples might be a name, a social security number, a flow rate, or a price. A *record* is a group of items that relate to the same object or individual. For example,

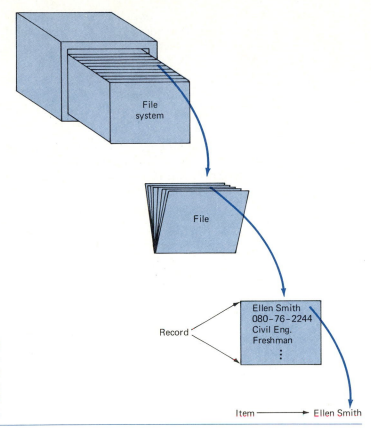

FIGURE 14.5
The various components associated with data files. The figure uses the analogy of an office file system and a file system employed on the personal computer.

in your university's mainframe computer there is undoubtedly a record that refers to your academic standing. It probably consists of a number of items, including your name, social security or identification number, class, department, grades in individual courses, credit hours, etc. This record is in turn a part of a larger entity called a *file*, which contains the academic records of all the students at your school. Finally, this file is one of many files that are maintained in the *storage* or *file system* of the university's mainframe computer.

Now we can redefine a *file* as a collection of records. In many, but not all, cases the records are related. In the case of related records, each usually has a *key item* that is used for identification purposes. For the example given above, your social security or ID number would be the logical choice for the key item. Your name would probably be a bad choice because at an institution of any size there would likely be names that would duplicate each other—for example, there is a good chance that more than one John Smith would be enrolled.

There are two types of files used on personal computers: sequential and direct-access files. *Sequential files* are those in which items are entered and accessed in sequence. That is, there is a first item, a second item, etc. The key concept is that when we want to access one of these items, we have to start at the beginning of the file and make our way through each item until we come across the one we want. Sequential files are very much like READ/DATA statements in this regard. (Recall the way the pointer moves from beginning to end of the DATA statements of Fig. 10.5.) In contrast, for *direct-access,* or *random-access*, *files*, we can access a record directly without having to make our way through all previous records. In this sense these files are similar to arrays. If we desire to use the data in the fifth row and eleventh column of a two-dimensional array, we merely use the subscripts to designate our choice, as in A(5,11). We do not have to sequentially go through each row, column by column to arrive at the value we want. Random access works in a similar fashion by using some identifier, such as its key item or a designation of its position on a file, to directly retrieve a record.

Direct-access files have a number of advantages, not the least of which is the fact that they retrieve data more quickly and more efficiently than sequential files. They are especially attractive for large-scale business applications such as a bank's billing or an airline's ticketing program.

Although sequential files are inefficient, they have their place in data processing. They are especially well suited for data that is to be used sequentially anyway, such as mailing labels. In engineering, data is often in sequential form—for example, the flows that were ordered by year in Table 14.1. Consequently, many engineering applications are adequately handled by sequential files. Finally, and most important in the present context, they are easier to explain and are more uniform than random-access files. Because of the scope of this text, we have chosen to devote this chapter to sequential files. You can consult your computer's user's manual in the event that you desire or need to employ direct-access files.

14.2.2 BASIC File Statements

Before describing how sequential files are used, it must be noted that they are among the least standardized concepts of BASIC. Fortunately, most of the differences relate to nomenclature—that is, the statement labels that are used to designate the fundamental file manipulations. In general, the actual operations themselves are fairly similar from system to system. Therefore in the following discussion we have tried to keep description of the procedures as general as possible. Where we do use specific nomenclature, we have opted to use statements compatible with the IBM PC's MS-DOS operating system. Chances are that some of the specific names that we employ will differ somewhat (in some cases considerably) from those used on your system. You may therefore have to obtain details from your system's user's manual in order to implement the material effectively. In order to accommodate your needs, we have included Appendix F that summarizes comparable statements for the other major personal computer system available today—the Apple II.

In the following pages, we will first introduce the fundamental statements that are used. Then we will show how they are employed to execute a number of standard operations. At every step of the way you should consult your user's manual to determine the proper commands for your own personal computer. In this way, you will successfully master the fundamentals of file manipulation and maintenance for your system.

The following types of statements are fairly standard on sequential file systems:

1. Opening a file—that is, establishing communication between a program and a file
2. Transferring data to a file from a program
3. Transferring data from a file to a program
4. Closing a file—that is, terminating communication between a program and a file

Before describing these statements, we must first introduce the conventions for identifying or labeling a file in a program.

Identifying a file.   The *file specification* or *filespec* is used to identify the file in auxiliary storage. Its general form is

$$device\ name\ \textbf{:}\quad file\ name$$

where the *device name* specifies the storage device on which the file is located and the *file name* is the actual name by which the file is identified. The file name itself consists of a *name* and an *extension*. Therefore a more complete representation of the file specification is

$$device\ name\ \textbf{:}\quad name\ \textbf{,}\ extension$$

The most commonly used device names are given below.

Device Name	Type of I/O Device
A:	First disk drive
B:	Second disk drive
CAS1:	Cassette recorder
KYBD:	Keyboard
LPT1:	Printer
SCRN:	Screen

If the device name is omitted (as will be the case for all the following material), Microsoft BASIC assumes that the device is the first drive, or, as in the above, drive A.

The *name* of the file consists of up to eight characters followed (optionally) by a period and an *extension* of up to three characters. The *name* should describe the contents of the file. It must start with a letter followed by letters, numbers, or select special symbols. (@, $, #, %, $\wedge$, !, and − are among the acceptable symbols; check your user's manual for a complete list.) Note that, because of its employment as a delimiter between the *name* and the *extension*, a period cannot be used. In addition, blank spaces cannot be used.

The *extension* usually serves to identify the file type. For example, BAS is used to designate a BASIC file. If you do not supply an extension, the disk operating system will assume an extension of BAS when you store or retrieve a program from the disk in BASIC. The extensions COM and EXE are used by the system to designate files that are in machine language. Aside from these, you can concoct your own extensions to identify the contents of a file. For example, the extension DAT could be used to designate that data is stored in a file, whereas TXT could be used if a word-processing text is stored. Extensions are particularly handy to discriminate between files with the same name.

In the present part of the book, we will create a file to store the stream-flow data from Table 14.1. The name we will use for this file is FLOW.DAT.

Opening a file.

Some dialects of BASIC do not require that you formally open a file. For these cases, the first line that reads or writes to a file automatically links the program to the file. However, many more dialects require such a statement. For Microsoft BASIC, two formats can be used. The first is typical of OPEN statements in other dialects. It is

ln OPEN $mode_1$,#*file number* ,*filespec*

where $mode_1$ designates the mode in which the file is to be opened. The string ''O'' is used if information is to be output from your program to the file, ''I'' is used if information is to be input to your program from the file, and ''A'' is used if new information is to be appended to a presently existing file. The *file number* is a positive integer that is employed to identify the file within the BASIC program. This is required because it is common to utilize several files in a program. For such cases, it is good practice to number them in ascending order, starting at 1. For many dialects, up to 15 can be opened, but you should consult your user's manual to check how many are permitted with your dialect.

It is important to realize that the *file number* identifies the file *within the program*, whereas the *filespec* identifies it *on the external storage device*. Therefore, if a file is used in two different programs, it could have different file numbers, but the same filespec would be required in both instances. For example, for the first program the file might be opened for output of data from the program with

```
50 OPEN "O",#1,"GRADE1.BAS"
```

whereas for the second, it could be opened for input of data to the program with

```
70 OPEN "I",#2,"GRADE1.BAS"
```

An alternative format for the OPEN statement is also available. It is

ln OPEN *filespec* FOR $mode_2$ AS #*file number*

where all the terms are the same as for the first format with the exception of $mode_2$. Instead of string variables such as ''I'' or ''O,'' the mode is written out in full, without quotation marks, as INPUT, OUTPUT, or APPEND. Thus the two examples in this format would be

```
50 OPEN "GRADE1.BAS" FOR OUTPUT AS #1
```

and

```
70 OPEN "GRADE1.BAS" FOR INPUT AS #2
```

Transferring data from a file.   Although it would probably be more logical to first explain how data is transferred to a file, there is a good reason for discussing the reverse process first. This is because the data must be in a particular form in the file in order for it to be transferred properly from the file to the program. By first understanding how information is transferred from the file, we can then appreciate some of the problems associated with the reverse process.

The general form of the statement to transfer data from a file to the program is

ln INPUT #*file number*,*item list*

where the *file number* is the same one designated in the OPEN statement and the *item list* is the list of values that are being transferred from the file. An example of such a statement, along with an appropriate OPEN statement, is

```
10 OPEN "FLOW.DAT" FOR INPUT AS #1
20 INPUT#1,YEAR,FLOW
```

The INPUT# statement works just like the INPUT statement that is used to accept data from the keyboard. Thus, if a number of items are to be input in a row, they must be delimited by commas. Similarly, string values must be enclosed in quotations. The presence or absence of these characters on the file is related to the way in which they were originally output to the file, as described in the next section.

Transferring data to a file.   The PRINT#, PRINT# USING, and WRITE# statements can be employed to transfer data to a file. They differ in the way the data is formatted as it is output to the file. In general, the PRINT# and the PRINT# USING statements are employed to create reports and text files for later printing, whereas the WRITE# is most commonly used to create files for subsequent processing by other programs.

The general forms of the PRINT# and the PRINT# USING statements are

ln PRINT #*file number*, *item list*

and

ln PRINT #*file number* USING *format string*; *item list*

where the *file number* corresponds to the one defined by the corresponding OPEN statement and the *item list* is the values that are being transferred to the file.

The PRINT# and the PRINT# USING statements output data to the file in a fashion identical to the way that the corresponding PRINT and PRINT USING statements will display it on the monitor screen. Thus comma and semicolon delimiters work the same way in PRINT# statements as they do in PRINT statements. Similarly, the *format string* of the PRINT# USING controls the format to the file in a fashion identical to the one by which the PRINT USING controls the format to the screen.

Although the PRINT# and PRINT# USING statements are suited to store a report or text file, they are less convenient when creating a file that is to be input to another program. This is due to the fact that, as described in the preceding section, the statements used to input the data from the file to another program require that the data be delimited by commas and that string values be enclosed in quotation marks. Although this can be done with either the PRINT# or the PRINT# USING statement, it cannot be accomplished very conveniently.

The WRITE# statement is designed to avoid these inconveniences. Its general form is

> *ln* `WRITE` #*file number*, *item list*

For this case, the *item list* can be simply delimited with commas, as in

> `10 WRITE #1, NAME$, NUMBER`

The WRITE# statement automatically sets such data in the proper format for subsequent input. That is, as the data is output to the file, commas are automatically inserted between items, and string values are enclosed in quotation marks.

Closing a file.   Before terminating a run, it is good practice (and sometimes mandatory) to close all files. Among other things, this will prevent the computer from writing other information onto the files and possibly destroying valuable data. The command to close a file is

> *ln* `CLOSE` #*file number*

14.2.3   Standard File Manipulations

Now that we have introduced some of the BASIC file commands, we can show how they are used to perform some of the standard file manipulations that are needed for routine engineering applications. These are

1. Creating a file
2. Printing a file
3. Adding new records to a file
4. Modifying or deleting records that are already on a file

Creating a file.   Figure 14.6 shows a simple program to transfer the data from Table 14.1 into a file. Notice that the file name is input as the string variable FILNAM$. This generalized format allows the program to be used to store other information

```
1110 REM      CREATE A SEQUENTIAL FILE
1120 INPUT "CREATE A FILE NAMED";FILNAM$
1130 OPEN FILNAM$ FOR OUTPUT AS #1
1140     WHILE LAST <> -999
1150         INPUT "YEAR = ";YEAR
1160         INPUT "FLOW = ";FLOW
1170         WRITE #1,YEAR,FLOW
1180         INPUT "LAST RECORD (ENTER -999)";LAST
1190     WEND
1200 CLOSE #1
1210 LAST = 0
```

FIGURE 14.6
A program to output data from a program to create a file.

consisting of pairs of string [YEAR$(I)] and numeric [FLOW(I)] constants. For example, if we had up to 50 years of flow data from another stream, we could use this program to store it in a file. However, to accomplish this correctly, we would have to give the file a name other than "FLOW.DAT". Otherwise the new data would be transferred onto the file and supplant (and hence destroy) the old data.

Printing the contents of a file. Figure 14.7 shows a program to transfer the flow data back from the file to the program. This version also prints out the data on the monitor or printer so that we can check that the data transfer has been performed as expected. Notice that in this example, we use a file number of 2 instead of 1 as in the previous example. As long as we are consistent within a program either number could be used. However, also notice that both examples would employ the name "FLOW.DAT", which is how the file is referenced on the disk.

As the data is being input, we must develop a way to detect when there is no more data to be read from the file. Otherwise, if we try to read data that is not present, we will obtain an error and the program will stop.

There are several ways to detect that we are at the end of a file. One way is with the end-of-file function. When a file is created, an end-of-file mark is placed after the last item of data. The end-of-file function can detect this mark. The function can be then used in the WHILE/WEND loop, as seen in Fig. 14.7. As long as we have not read all the data from the file, EOF will be false and control will transfer to the next line. However, when the last item is read, the EOF becomes true and control transfers out of the loop. Thus, if we are looping to read the data, this statement allows us to transfer control out of the loop before we attempt to read a nonexistent record.

Appending records to a file. Just as we have used "INPUT" and "OUTPUT" to designate input and output modes, "APPEND" can be used to append or add data to

FIGURE 14.7
A program to transfer data from the disk back to the program.

```
2100 REM      PRINT OUT FILE CONTENTS
2110 INPUT "LIST A FILE NAMED";FILNAM$
2120 CLS
2130 PRINT "YEAR";TAB(20);"FLOW"
2140 OPEN FILNAM$ FOR INPUT AS #2
2150 WHILE EOF(2) = 0
2160     INPUT #2,YEAR,FLOW
2170     PRINT YEAR;TAB(20);FLOW
2180 WEND
2190 CLOSE #2
```

FIGURE 14.8
A program to append additional
data to a file.

```
3110 REM      APPEND A SEQUENTIAL FILE
3120 INPUT "APPEND DATA TO A FILE NAMED";FILNAM$
3130 OPEN FILNAM$ FOR APPEND AS #1
3140    WHILE LAST <> -999
3150        INPUT "YEAR = ";YEAR
3160        INPUT "FLOW = ";FLOW
3170        WRITE #1,YEAR,FLOW
3180        INPUT "LAST RECORD (ENTER -999)";LAST
3190    WEND
3200 CLOSE #1
3210 LAST = 0
```

FIGURE 14.8
A program to append additional data to a file.

the end of a file. Figure 14.8 shows a program to append additional years of flow data to the file created for Table 14.1.

The same procedure can be used to insert data into the middle of a file. For this case, we would read all the records up to the point where the insertion is to be made and write these onto a new file. Then we could add the new information onto the new file. Finally, the remaining records could be read onto the new file.

Modifying or deleting records from a file.   Once we have stored data on a file, there will be occasions when we will have to return and modify or delete some of the records. Both these operations can be performed in a fashion similar to the previous method for inserting new data into an existing file. Figure 14.9 shows a program to modify the flow for a particular year.

14.2.4   A SIMPLE DATABASE MANAGEMENT PROGRAM

Each of the capabilities embodied in Figs. 14.6 through 14.9 can be incorporated into the simple database management program seen in Fig. 14.10. Notice that the program is menu-driven and modular. Each file manipulation is treated as a subroutine that is called by the menu-driven main program. In this way, the user can choose whatever option is required to manipulate the files.

The file created by a program such as that shown Fig. 14.10 is conventionally referred to as a *database*. Although Fig. 14.10 is adequate for introducing you to how

FIGURE 14.9
A program to modify an existing record on a file.

```
4130 REM      CORRECT A VALUE
4140 OPEN "FLOW.DAT" FOR INPUT AS #1
4150 OPEN "FLOW.DUM" FOR OUTPUT AS #2
4160 REM      INPUT YEAR TO BE CORRECTED
4170 INPUT "YEAR TO BE CORRECTED = ";YRCORR
4180 WHILE EOF(1) = 0
4190    INPUT #1,YEAR,FLOW
4200    IF YRCORR <> YEAR THEN 4210 ELSE 4230
4210        WRITE #2,YEAR,FLOW
4220        GOTO 4270
4230      REM ELSE
4240        PRINT "CURRENT VALUE = ";FLOW
4250        INPUT "CHANGE TO ";FLOW
4260        WRITE #2,YEAR,FLOW
4270    REM ENDIF
4280 WEND
4290 CLOSE 1
4300 CLOSE 2
4310 KILL "FLOW.DAT"
4320 NAME "FLOW.DUM" AS "FLOW.DAT"
4330 END
```

```
100 REM     ***************************************************
110 REM     *             SEQUENTIAL FILE PROGRAM             *
120 REM     *                 (BASIC version)                 *
130 REM     *                by  S. C. Chapra                 *
140 REM     *                                                 *
150 REM     *   This program creates, prints, appends and     *
160 REM     *   corrects data on a sequential file.  This     *
170 REM     *   particular version is set up for river        *
180 REM     *   flows and their corresponding years.          *
190 REM     ***************************************************
200 REM
210 REM           .---------------------------------------------.
220 REM           |          DEFINITION OF VARIABLES            |
230 REM           |                                             |
240 REM           |   FILNAM$ = NAME TO BE GIVEN TO FILE        |
250 REM           '---------------------------------------------'
260 REM
270 REM ***************** MAIN PROGRAM  *****************
280 REM
290 CLS
300 INPUT "NAME THE FILE";FILNAM$
310 CLS
320 REM                         Create a Menu
330 REM
340 PRINT:PRINT:PRINT
350 PRINT "SELECT ONE OF THE FOLLOWING OPTIONS"
360 PRINT "TO MANIPULATE THE FILE ";FILNAM$:PRINT
370 PRINT TAB(10);"1.  CREATE THE FILE"
380 PRINT TAB(10);"2.  PRINT THE FILE"
390 PRINT TAB(10);"3.  APPEND THE FILE"
400 PRINT TAB(10);"4.  CORRECT THE FILE"
410 PRINT TAB(10);"5.  TERMINATE THIS PROGRAM"
420 INPUT "ENTER YOUR SELECTION";SELECT
430 CLS
440 ON SELECT GOSUB 1000,2000,3000,4000,470
450 INPUT "PRESS RETURN KEY TO CONTINUE"; DUMMY$
460 GOTO 310
470 END
480 REM
1000 REM ***************** SUBROUTINE CREATE  *****************
1010 REM
1020 REM     CREATES A SEQUENTIAL FILE.
1030 REM
1040 REM     DATA REQUIRED:  FILNAM$
1050 REM
1060 REM     LOCAL VARIABLES:
1070 REM
1080 REM     CRYEAR = YEAR WRITTEN TO FILNAM$
1090 REM     CRFLOW = FLOW WRITTEN TO FILNAM$
1100 REM
1110 REM ------------------ BODY OF CREATE  ----------------
1120 REM
1130 OPEN FILNAM$ FOR OUTPUT AS #1
1140     WHILE CRLAST () -999
1150        INPUT "YEAR = ";CRYEAR
1160        INPUT "FLOW = ";CRFLOW
1170        WRITE #1,CRYEAR,CRFLOW
1180        INPUT "LAST RECORD (ENTER -999)";CRLAST
1190     WEND
1200 CLOSE 1
1210 CRLAST = 0
1220 RETURN
1230 REM
2000 REM ***************** SUBROUTINE PRINT  *****************
2010 REM
2020 REM     PRINTS OUT THE DATA STORED IN FILNAM$.
2030 REM
2040 REM     DATA REQUIRED: FILNAM$
2050 REM
2060 REM     LOCAL VARIABLES:
2070 REM
2080 REM     PRYEAR = YEAR INPUT FROM FILNAM$
```

```
2090 REM     PRFLOW = FLOW INPUT FROM FILNAM$
2100 REM
2110 REM ------------------ BODY OF PRINT  ----------------
2120 CLS
2130 PRINT "YEAR";TAB(20);"FLOW"
2140 OPEN FILNAM$ FOR INPUT AS #1
2150 WHILE EOF(1) = 0
2160     INPUT #1, PRYEAR, PRFLOW
2170     PRINT PRYEAR;TAB(20);PRFLOW
2180 WEND
2190 CLOSE #1
2200 RETURN
2210 REM
3000 REM *************** SUBROUTINE APPEND  *************
3010 REM
3020 REM     ADDS DATA TO THE END OF THE FILE.
3030 REM
3040 REM     DATA REQUIRED:    FILNAM$
3050 REM
3060 REM     LOCAL VARIABLES:
3070 REM
3080 REM     APYEAR = YEAR ADDED TO FILNAM$
3090 REM     APFLOW = FLOW ADDED TO FILNAM$
3100 REM
3110 REM ------------------ BODY OF APPEND  ---------------
3120 REM
3130 OPEN FILNAM$ FOR APPEND AS #1
3140     WHILE APLAST () -999
3150        INPUT "YEAR = ";APYEAR
3160        INPUT "FLOW = ";APFLOW
3170        WRITE #1,APYEAR,APFLOW
3180        INPUT "LAST RECORD (ENTER -999)";APLAST
3190     WEND
3200 CLOSE 1
3210 APLAST = 0
3220 RETURN
3230 REM
4000 REM *************** SUBROUTINE CORRECT  *************
4010 REM
4020 REM     ALLOWS THE USER TO CORRECT A FLOW ENTRY.
4030 REM
4040 REM     DATA REQUIRED: FILNAM$
4050 REM
4060 REM     LOCAL VARIABLES:
4070 REM
4080 REM     COYR   = YEAR FOR FLOW TO BE CORRECTED
4090 REM     COYEAR = YEAR INPUT FROM / WRITTEN TO FILNAM$
4100 REM     COFLOW = FLOW INPUT FROM / WRITTEN TO FILNAM$
4110 REM
4120 REM ------------------ BODY OF CORRECT  ----------------
4130 REM
4140 OPEN FILNAM$ FOR INPUT AS #1
4150 OPEN "FLOW.DUM" FOR OUTPUT AS #2
4160 REM     INPUT YEAR TO BE CORRECTED
4170 INPUT "YEAR TO BE CORRECTED = ";COYR
4180 WHILE EOF(1) = 0
4190     INPUT #1,COYEAR,COFLOW
4200     IF COYR () COYEAR THEN 4210 ELSE 4230
4210         WRITE #2,COYEAR,COFLOW
4220         GOTO 4270
4230     REM ELSE
4240         PRINT "CURRENT VALUE = ";COFLOW
4250         INPUT "CHANGE TO ";COFLOW
4260         WRITE #2,COYEAR,COFLOW
4270     REM ENDIF
4280 WEND
4290 CLOSE 1
4300 CLOSE 2
4310 KILL FILNAM$
4320 NAME "FLOW.DUM" AS FILNAM$
4330 RETURN
```

FIGURE 14.10 An example of a primitive database management system, a menu-driven BASIC program to maintain a sequential file system.

the computer can be used to create and manage a database, it is not very comprehensive or efficient. Sequential or random-access files can be employed to write much more effective and elaborate database management programs in BASIC. In addition, canned software packages called *database management systems*, or *DBMS*, are available for the same purpose. We will discuss DBMS in more detail when we return to the topic in Chap. 41.

PROBLEMS

14.1 Reprogram Fig. 10.6, employing loops and subscripted variables to make the code more efficient.

14.2 Repeat Prob. 10.1 but employ loops and subscripted variables to make the program more efficient.

14.3 Repeat Prob. 10.4 but employ loops and subscripted variables to make the program more efficient. Add columns for n = 10,000 and 15,000. Use a two-dimensional array to store the oxygen concentrations.

14.4 Write a short program to input a series of numbers and compute their average. Use the following algorithm as the basis for the program:

   • **Step 1**   Use a loop to input, count, and sum the numbers. Example 12.3 might serve as a model for inputting and counting. Employ an array to store the values as subscripted variables.
   • **Step 2**   Compute the average value as the sum divided by the total number of values.
   • **Step 3**   Print out the individual values in tabular form.
   • **Step 4**   Print out the average.

Test the program by determining the average flow for the data in Table 14.1.

14.5 Repeat Prob. 14.4 but rather than inputting the data via the keyboard, use READ/DATA statements to incorporate the data into the program. Employ subscripted variables to store the values for the flows.

14.6 Repeat Prob. 14.4 but employ a sequential file to store the flow data on a disk; use the database management program from Fig. 14.10. Utilize subscripted variables to store the values of the flows.

14.7 Which of the following external storage devices allows direct-access files: magnetic disk or cassette tape? Why?

14.8 Using Fig. 14.9 as a model, develop a program fragment to delete a record from a file.

14.9 Program Fig. 14.10 for your computer. Utilize it to create a file for the first 10 years of data from Table 14.1. Print this file. Append the next 10 years of data to the file. Print this file. Change the value for 1963 from 81 to 89. Print this file.

14.10 Employ arrays and files to develop a modular computer program to determine your grade for this course.

14.11 Employ arrays and files to create a modular program to determine the statistics for a sports team. See Prob. 10.2 for additional information.

PART FIVE

PROGRAMMING IN FORTRAN

After the development of a sound and comprehensive plan, the next step of the program-development process is to compose the program using a high-level language. This part of the book shows how FORTRAN is used for this purpose.

Chapter 15 is devoted to the most elementary information regarding FORTRAN programming. This material is designed to get you started by providing information sufficient to successfully write and execute simple programs. *Chapters 16 through 21* cover additional material that will allow you to develop more complicated and powerful programs.

CHAPTER 15
THE FUNDAMENTALS
OF FORTRAN

15.1   WHY FORTRAN?

There are a variety of reasons for the popularity of FORTRAN. In particular, it was a language that was in the right place at the right time. Prior to FORTRAN, assembly language had few arithmetic operators, and programming computations were difficult and often unreliable. In the late 1950s, IBM developed FORTRAN to fill this void and met the growing demand of engineers and scientists for a high-level language that could easily handle arithmetic operations. Consequently it grew in popularity.

Beyond making computations easy to formulate, FORTRAN also makes them very efficient and accurate. Because FORTRAN is almost always a compiled language, FORTRAN programs execute quite rapidly in comparison to programs written in interpreted languages such as BASIC. In addition, because FORTRAN is most commonly implemented on mainframe computers, its representations of numbers afford the high precision that is available on such machines.

Aside from its utility for computations, the availability of FORTRAN also contributed to its popularity. Because of IBM's dominance in the computer marketplace, other manufacturers developed FORTRAN compilers to remain competitive with IBM. Thus FORTRAN compilers have been readily available for all types and sizes of computers. Even microcomputers, which were originally too limited to handle a full FORTRAN compiler, have evolved to the point that today most popular machines can accommodate FORTRAN.

As the result of the widespread use of FORTRAN, a wealth of programs have been written in the language. These programs are so ubiquitous that it behooves every engineer to learn the language in order to tap this body of knowledge.

TABLE 15.1   *Major Enhancements of FORTRAN 77 over FORTAN IV*

New Statements for FORTRAN 77
  Block IF, ELSE IF, ELSE, ENDIF statements
  CHARACTER data type
  PROGRAM statement
Changes in FORTRAN IV statements
  DIMENSION statement
    • Index values of an array can be negative or zero.
    • Index range can be specified by a constant or variable.
  DO statement
    • The control variable for the loop can be nonpositive, can be of real or double-precision data type, and is defined on normal completion of loop.
    • Testing of the loop termination criteria occurs prior to the first pass through the loop so that if the criteria are met prior to the first pass through the loop the contents of the loop are never executed.
    • The value of the parameters of the DO loop (initial value, stop value, and increment) may be changed within the loop without affecting the function of the loop.
  END statement
    • It may have a label and be branched to.

15.2   WHICH FORTRAN?

Interestingly, the movement of FORTRAN to new and different computer systems did not spawn a lot of FORTRAN dialects. Most versions adhere to a common set of statements and syntax. This is probably a result of two factors: the role of a dominant company like IBM in its creation and development and the efforts of the American National Standards Institute (ANSI) to promote uniform dialects.

The dialect originally introduced in 1957 came to be known as FORTRAN I. It was followed by a number of improved versions that culminated in 1962 in FORTRAN IV. In about 1966 a standard FORTRAN was adopted by ANSI. This dialect, which has come to be known as FORTRAN 66, essentially corresponds to FORTRAN IV. Even today, many compilers use this version.

In 1978 a new ANSI standard called FORTRAN 77 was adopted. The compiler used in this book is the VS FORTRAN which runs on IBM mainframes and compatible equipment.* VS FORTRAN is based on FORTRAN 77, with a few IBM-specific features. All of the programs contained here utilize standard FORTRAN 77 and use none of the IBM-specific features. Should you have only a FORTRAN IV compiler at your disposal, you should have few problems. Most of our examples will be directly compatible with the FORTRAN IV compiler. Major differences between FORTRAN IV and FORTRAN 77 are summarized in Table 15.1. You will appreciate these differences as we describe the language in detail.

It should be noted that other dialects of FORTRAN exist. One version, which is common at many universities, is WATFIV. This dialect was developed at the Uni-

---

*VS is short for virtual storage and refers to the way that programs are stored (recall Sec. 4.3.1).

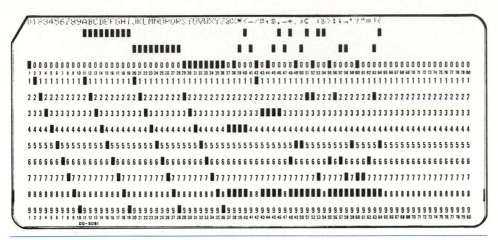

FIGURE 15.1 A punch card.

versity of Waterloo in Canada. Advanced versions of WATFIV are very similar to FORTRAN 77. In addition, the dialect has some structured programming features that are unavailable in FORTRAN 77. We will mention these special features where appropriate.

15.3   "TALKING" TO THE COMPUTER IN FORTRAN

How you actually go about interacting with the computer to enter and run a FORTRAN program depends on the system you employ. In the early years of computing, most machines operated as *batch-processing systems*. This mode allowed little or no direct interaction between the user and the machine. Each program was typed by the user on punched cards (Fig. 15.1) with a card-punch machine called a *keypunch*. Together with the program, data and job control commands (recall Sec. 4.3.1) were also typed on cards. The entire deck was assembled into what was called a FORTRAN package, or "job" (Fig. 15.2). Each package would usually be submitted to a computer operator who would run it and then return the job along with printed output. The key to this mode of operation was the separation between the user and the execution of the program. Often there was a significant delay between the time that a batch job was submitted and the time that the output was received. If an error was detected, the user would not know it until he or she received the output; in that event then, the program would have to be corrected and the job resubmitted.

Although batch systems are still used today, an alternative mode called a *conversational system* is also available. As the name implies, this mode allows more interaction between the user and the machine. Usually the system operates as a *time-sharing* system, where many individuals access the computer simultaneously. Information is often entered into the computer via a *terminal*. This is typically a video display unit consisting of a cathode-ray tube and a keyboard. The terminal is either

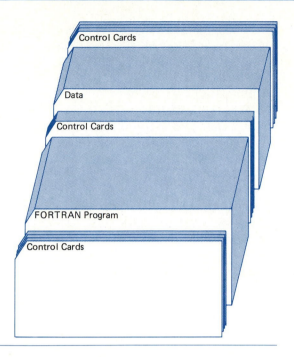

Control Cards

Data

Control Cards

FORTRAN Program

Control Cards

FIGURE 15.2
A FORTRAN package, or "job."

directly wired into the computer or connected by a telecommunications hookup. A set sign-on procedure, which varies from system to system, is usually required to establish communication with the computer. During sign-on, you will typically be required to provide the computer with information regarding your needs for the session (for example, the language and peripheral equipment you require). In addition, you must usually type in an identification code and a password to gain access to the system. These are needed for billing and to prevent unauthorized individuals from gaining access and tampering with your programs and data.

Once you have gained access to the computer, you are ready to type in your program. In contrast to the punching of cards, the typing of a program is a much more interactive process. Most terminals have editing capabilities that allow you to modify lines in an efficient manner. Once the program is to your liking, it is submitted to the computer by typing appropriate commands. We will discuss these commands in further detail in a later section. For the time being, it is sufficient to understand that the running of the program for a compiled language such as FORTRAN proceeds in stages. Whereas in batch mode, you lose contact once the process begins, conversational systems allow you to maintain some control. Thus as each phase is completed, you can be apprised of the progress. If a problem occurs in the first phase, you can be alerted immediately so that you can correct the mistakes and run the program again. It is this level of control that usually makes conversational systems preferable to batch-mode systems.

The operation of a FORTRAN program on a microcomputer is quite similar to operation on a timesharing system except that you obviously will not use a password or identification code to gain access. As with the timesharing system, the user types and edits the program until it is acceptable. A number of commands are then employed to take the program through the stages needed to come up with the final result. (Appendix H of this book includes the procedures for accomplishing this with Microsoft FORTRAN on the IBM PC.)

Now that you know something about the manner in which you will interact with the computer, we will take a closer look at the FORTRAN program itself. As mentioned previously, a program is a set of instructions to tell the computer what to do. In FORTRAN each instruction is usually written as a single line, and each instruction is typically aligned in specific columns. The conventions regarding the alignment stem from the use of punch cards in the early years of computing. Although cards are employed infrequently today, the spacing conventions have usually been maintained. It should be noted that some systems do not employ these conventions and use numbered lines in an interactive mode that is very similar to BASIC. Obviously, if your system employs such a scheme, you will not need to observe the spacing conventions. However, it is not a bad idea to acquaint yourself with the following practices because they are still employed in many versions of FORTRAN.

The 80 columns of a punch card are called the card's *field*. Parts of the field are set aside for specific purposes. These are illustrated on the coding form shown in Fig. 15.3. A coding form is a piece of paper on which the program can be written and checked for mistakes prior to being entered in the computer. Note that the form has an 80-column field just like the punch card. Also observe that specific parts of the field are used for particular purposes.

As shown in Fig. 15.3, FORTRAN accepts programs in lines of 80 characters or less. Usually all characters beyond column 72 are ignored. The last eight columns were traditionally set aside for labeling each card, usually by numbering them in ascending order. This make sense when one thinks in terms of cards rather than terminal lines. When using cards, there was always the distinct possibility that you would drop a large deck on the floor. Because card-sorting machines were available, the numbers in the last columns could serve as the basis for putting a shuffled deck back into correct order.

Notice that some program lines are numbered using columns 1 through 5. The number can start in any of the five columns but cannot extend into the sixth. Numbers can therefore range from 1 to 99999 and do not have to be in any particular order. However, each line number can only be used once. Thus its sole function is to provide a distinct identity to a particular line within the program.

The reasons for numbering lines will become clearer as we learn more about programming. A first example is provided in Fig. 15.4. A computer executes a program by starting at the first line and then progressing from one line to the next unless instructed to do otherwise. Therefore, one reason for numbering the lines in columns 1 to 5 is to override normal sequence. To do this, you could type a statement directing the computer to jump or ''go to'' the line of your choice. The only way that the computer would know where it was supposed to go would be if the destination statement had a line number (Fig. 15.4).

Such jumping around is not done in the simple program in Fig. 15.5. For this example, each statement is executed in a straightforward sequence. The first is a comment (as described in Fig. 15.3) that provides a title for the program. Then values

DATA PROCESSING CENTER
Texas A&M University

PROGRAM: INTRODUCTION TO COMPUTING FOR ENGINEERS
PROGRAMMER: S.C. CHAPRA

	Statement / Comments	Identification
C	THIS CODING FORM IS USED TO DEMONSTRATE THE PURPOSE OF THE VARIOUS	LINE 1
C	FIELDS ON A PUNCHED CARD.	LINE 2
C	COMMENT CARDS, OF WHICH THE PRESENT LINE IS AN EXAMPLE, ARE	LINE 3
C	DESIGNATED BY A "C" IN COLUMN 1	LINE 4
C	OTHERWISE, COLUMNS 1 THROUGH 5 ARE RESERVED FOR STATEMENT NUMBERS.	LINE 5
C	FOR EXAMPLE, THE FOLLOWING LINES ARE STATEMENTS 1, 83, AND 2006.	LINE 6
	1 FORMAT(I8)	LINE 7
	83 Y=A+(B-X)**2.-5.*c/D	LINE 8
	2006 Y = A + (B-X)**2. - 5.*c/D	LINE 9
C	COLUMNS 7 THROUGH 72 ARE RESERVED FOR THE ACTUAL STATEMENTS.	LINE 10
C	THE COMPUTER IGNORES BLANK SPACES WITHIN THIS FIELD AND, THEREFORE,	LINE 11
C	SPACING CAN BE USED TO FACILITATE READING. FOR INSTANCE, STATEMENTS	LINE 12
C	83 AND 2006 ACCOMPLISH THE SAME ALGEBRAIC MANIPULATIONS BUT	LINE 13
C	STATEMENT 2006 IS MORE READABLE DUE TO THE INCLUSION OF BLANKS.	LINE 14
C	IF A STATEMENT IS TOO LONG FOR ONE CARD, IT CAN BE CONTINUED ON	LINE 15
C	THE FOLLOWING CARD BY PLACING ANY NUMBER OTHER THAN ZERO IN COLUMN 6	LINE 16
	CONC = CONC + (THETA -XMAX) / (THETA - XMIN) * POPUL - (THETA -	LINE 17
	1 YMAX) / (THETA -YMIN) * POPUL * POPUL	LINE 18
C	COLUMNS 73 THROUGH 80 CAN BE USED TO NUMBER THE LINES IN A PROGRAM	LINE 19
C	OR FOR ANY OTHER LABELLING PURPOSE.	LINE 20

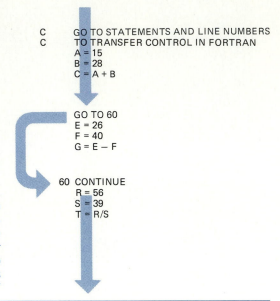

```
C       GO TO STATEMENTS AND LINE NUMBERS
C       TO TRANSFER CONTROL IN FORTRAN
        A = 15
        B = 28
        C = A + B


        GO TO 60
        E = 26
        F = 40
        G = E − F


    60  CONTINUE
        R = 56
        S = 39
        T = R/S
```

FIGURE 15.4
A FORTRAN program starts at the top line and progresses downward unless instructed to do otherwise. One way to circumvent the normal sequence is with a GO TO statement that directs the computer to jump or "go to" the line of your choice. One reason for numbering a FORTRAN line is to let the computer know where it is "going to."

```
C Simple Addition Program
       A = 28
       B = 15
       C = A + B
       WRITE(6,*) C
       STOP
       END
```

FIGURE 15.5
A simple FORTRAN program to add two numbers.

```
** Output Results **
43.0000000
```

are assigned to the variables A and B. The following line sums these variables and assigns the answer to C. Next the computer is instructed to print out or "write" the answer. When the program is executed this will make the computer print the result on an output device. The parenthetical term (6, *) tells the computer how this is to be done. For this case, the 6 specifies that the result is to be output on a printer.† Other characters could have been used to specify alternative output devices. The asterisk tells the computer the manner, or *format*, in which the answer is to be printed. The asterisk instructs the computer that we want the answer to be *list-directed*. That is, we will not try to control exactly how the result is output and will leave it up to the computer to print it out in an acceptable fashion. In a later chapter, we will learn how to control the formatting of both input and output. For the time being, the simple approach embodied by the asterisk will suffice.

Before moving on to learn how to run a program, we should mention a number

†The convention of using a 6 to designate printer output is widespread but not universal. You should check to determine the numbers employed by your computer to specify the printer and other input or output devices.

of points. First, notice that all the letters in Fig. 15.5 (except the comments) are capitalized. Most FORTRAN compilers require that programs be typed in uppercase. Second, certain FORTRAN statements are called *executable* because they represent instructions to the computer to carry out tasks during the program's execution. All others are called *nonexecutable*. In Fig. 15.5 the comments and the END statement are nonexecutable. Finally, notice that the last lines in the program are a STOP and an END. The STOP is an executable statement that causes the computer to terminate execution. Although more than one STOP may appear in a program, the first one to be reached causes the program to terminate. The END is a nonexecutable statement that identifies the last physical line in a FORTRAN program or subprogram. There is only one END statement in a program or subprogram. It informs the compiler where it is to terminate the translation of the program into machine language. Consequently every program or subprogram must have END as its last line.

15.4 HOW TO PREPARE AND EXECUTE FORTRAN PROGRAMS

Every computer system has its own protocol for preparing and executing a FORTRAN program. Consequently we are somewhat limited in the guidance we can provide regarding these tasks. However, although the details differ, the general steps are the same. In the present section, we will describe these general steps as background information that you may find useful when you learn the specific procedures for your system. Note that the commands for implementing Microsoft FORTRAN on the IBM PC are outlined in Appendix H.

The four general steps for preparing and running a FORTRAN program on any conversational system are outlined in Fig. 15.6. (A somewhat different, but related, process is employed for batch systems.) The following sections describe these steps in detail.

FIGURE 15.6 Schematic of the four steps involved in running a FORTRAN program. The boxes indicate the type of program generated by each of the system programs, which are designated as ovals. External arrows indicate other information required or generated by the system programs.

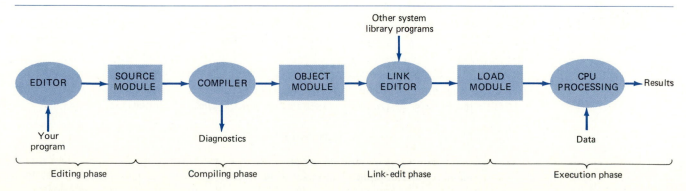

15.4.1 Creating and Editing Programs

After the program has been composed, the first step in implementing it is to type it and save it in the computer. This involves two phases: allocating the storage space for the program and then typing the program into this space. Allocation of storage is often automatically accomplished by simply naming the program. However, in some cases the amount of storage must be specified.

The actual typing of the program is performed on the terminal. The typing process employs the system's editor program, which is designed to facilitate the creation of files. Some editors are line-oriented and others are organized around pages. Most will have capabilities for inserting or deleting characters, words, or lines anywhere in a program. Most will also have the ability to move or copy one section of a program to any other location. This is facilitated by the fact that FORTRAN programs do not require sequentially numbered lines.

All the above capabilities should allow you to create a file containing your FORTRAN program. When you are satisfied that the program is correct, the file is saved. At this point, it is referred to as the *source module* or the *source code*.

15.4.2 Compiling Programs

In the compilation step, the FORTRAN version of the program stored as the source module is translated into a machine-usable form—the *object module* or the *object code*. If the computer detects errors in your program, *diagnostics*, or *error messages*, will be printed out. The errors can be corrected by returning to the editing step and then resubmitting the corrected source module.

Once the translation is successfully completed, the compiled object module can be used over and over again without repetition of the translation process. This is why compiled programs are much faster to run than interpreted programs, which must be translated line by line every time they are executed.

15.4.3 Linking Programs

Before they can be run, almost all programs will require other special programs from the system library. For example, most computers cannot compute the square root directly, and so a special library program is needed for this purpose. Other programs may contain the routines to print on the terminal screen or on a hard-copy device such as a printer. These system library programs are combined with your object module in the link step. This process may occur automatically, or it may require special instructions on your part.

In addition to merging programs, linking also specifies the *memory map* for your program. That is, the link program takes care of all details necessary to place your program in the computer's memory so that the CPU can process it.

15.4.4 Executing or Running Programs

After the object module is linked to the system library programs it is called a *load module*. At this point, it is ready to be executed. This is usually accomplished by a simple command. On some systems this entails merely typing the program's name and then striking the RETURN key on the terminal. In other cases, you may have to type the word RUN followed by the program name. Whatever the procedure, the effect is the same. For the simple addition program of Fig 15.5, the result of the addition—43.0000000—would be printed on the output device.

15.5 A SIMPLE FORTRAN PROGRAM

Now that you have been introduced to how a program is executed, we can begin to discuss the actual mechanics of developing a FORTRAN computer program. The remainder of this chapter will be devoted to expanding Fig. 15.5 into a quality FORTRAN program. Because of the simple nature of the algorithm underlying Fig. 15.5, the following material does not represent a detailed description of the capabilities of FORTRAN. A more in-depth discussion of the language will be deferred to subsequent chapters. For the time being, we will focus on the fundamentals of FORTRAN programming. In addition, we will introduce some practices that will contribute to the quality of your finished product.

15.5.1 Adding Descriptive Comments: The Comment Statement

As we will reiterate time and time again, an important attribute of a high-quality computer program is how easy it is to read and to understand. One deficiency of the program in Fig. 15.5 is that it is almost devoid of documentation. That is, it consists merely of a series of instructions to get a job done. Although this is not a major problem for such a simple program, inadequate documentation can detract immensely from the utility of larger and more complicated programs. If the user has to expend a great deal of time and energy to decipher your program's operation and logic, it will diminish the impact and usefulness of your work.

One of our primary tools for the internal documentation of a program is the comment statement. The general form of this statement is

C *comment*

where the letter C is in the first column and the *comment* is any descriptive information you would like to include. For example, the *comment* included at the beginning of Fig. 15.5 provides a title for the program, and thus the user can immediately identify the general subject of the program.

Beyond giving titles comments can be employed to incorporate additional documentation. Figure 15.7 is an example of how this can be done for the simple addition program. Notice that a group of comments has been used to form a title module at the program's beginning. Along with the title, this module includes the author's name and a concise description of what the program is to accomplish.

After the title module, comments can be used to interject descriptive statements as well as to make the program more readable. As seen in Fig. 15.7, this latter objective can be accomplished by using comments to clearly delineate the program's

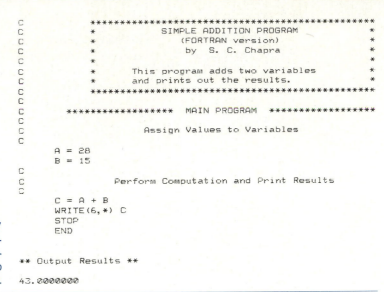

```
C       ********************************************************
C       *                                                      *
C       *              SIMPLE ADDITION PROGRAM                 *
C       *                 (FORTRAN version)                    *
C       *                  by  S. C. Chapra                    *
C       *                                                      *
C       *           This program adds two variables            *
C       *           and prints out the results.                *
C       ********************************************************
C
C       *****************  MAIN PROGRAM  *****************
C
C                    Assign Values to Variables
C
        A = 28
        B = 15

C
C                Perform Computation and Print Results
C
        C = A + B
        WRITE(6,*) C
        STOP
        END

** Output Results **

43.0000000
```

FIGURE 15.7
A revised version of the simple addition program from Fig. 15.5, including comments that serve to document the program.

major modules. Notice how the body of each comment is centered. This is one style that makes it easier to pick out the comments. Also notice how several blank comments and a line of asterisks are included to set off and separate some of the comments and modules. The judicious use of lines and blank spaces is one effective tool for making the program more readable.

While the use of comments is optional in FORTRAN, it is strongly encouraged. But do not overuse comment statements. If the action of a line or lines is obvious, do not add excess verbiage in the form of remarks. A program can quickly become cluttered with unneeded comments that actually make it harder to read and follow. As described in the next section, careful selection of variable names helps to minimize the need for comments.

15.6 NUMERIC CONSTANTS, VARIABLES, AND ASSIGNMENT

As an engineer, one of the primary uses you will have for a computer is to manipulate numbers. Consequently, it is important to understand how numeric quantities are represented on the computer.

There are two fundamental ways in which numeric data is expressed in FORTRAN: directly as constants and indirectly as variables. *Constants* are values that remain fixed during program execution, whereas *variables* are symbolic names to which a value is assigned that can be changed during execution.

Numeric constants.   Numeric constants are numbers that remain fixed throughout program execution. They may be either positive or negative and may be expressed in integer or real form. *Integers* have no decimal point, as in

$$508 \quad 67 \quad -1507 \quad 883$$

where the standard integer values for a 32-bit computer can range from $-2,147,483,648$*
to $2,147,483,647$. (Note that some compilers have different precision.)

The *real variables* are those with decimal points:

$$896. \quad -4.78 \quad 0.0056 \quad -100.4$$

or those that are greater than $2,147,483,647$ or less than $-2,147,483,648$.

In addition, real constants can be expressed in exponential notation. *Exponential constants* are similar in spirit to scientific notation and are handy for expressing very large or very small numbers. To review, a number is expressed in scientific notation by converting it to a number between 1 and 10 multiplied by an appropriate integral power of 10. This can be done by moving the decimal point to the right or left until there remains a number between 1 and 10. This remaining number must then be multiplied by 10 raised to a power equal to the number of places the decimal point was moved. If the decimal has been moved right, the power is negative, and if it has been moved left, the power is positive. The general form is

$$b \times 10^n$$

where b is the value that remains after the number has been converted to a value between 1 and 10 and n is the integer number of places the decimal was moved. For example, the number $8,056,000$ would be converted to 8.056 by moving the decimal six places to the left. Therefore in scientific notation it would be expressed as 8.056×10^6.

FORTRAN employs the same concept, using the exponential notation

$$b\mathrm{E}n$$

Thus in FORTRAN the number $8,056,000$ is expressed as 8.056E6. Note that in FORTRAN the b must contain a decimal point. In addition, as we will discuss shortly, a double-precision exponential form is also available, one using a D in place of the E.

Letters (other than E and D) and symbols such as dollar signs and commas are not allowed in numeric constants. Therefore \$56.23 and $1,760,453$ are illegal numeric constants and would constitute a syntax error in FORTRAN. In addition, only one minus sign may be used at the front of a constant to designate that it is negative. Minus signs cannot be used as delimiters (that is, separators), as is commonly done

---

*These ranges relate to the fact that $2^{31} = 2,147,483,648$. The power is 31 because one bit is used to specify the sign. The positive limit of the range is 1 less than 2^{31} in order to allow for zero.

in social security numbers. For example, $120-44-5605$ is an illegal constant in FORTRAN. However, a minus sign can be used to designate negative exponents. For example, $-3.2E-4$ is a legal FORTRAN representation of -0.00032.

It should be noted that real numbers in FORTRAN can be expressed in either single or double precision. *Single-precision numbers* are usually stored with seven significant digits, whereas *double-precision numbers* are stored with sixteen (different machines range from 14 to 18). If a number is written with more than seven digits, it will automatically be truncated when stored unless it is written in the double-precision version of the exponential form. This form uses a D rather than an E, as in

$$3.1415926536D0 \qquad 2.56001753D-6 \qquad -2.050000768D9$$

You may be wondering at this point why so many different forms are needed and why we would want to force the computer to store a constant in a particular form. There are a number of reasons why we might want to go to the trouble. For example, an integer takes up less space in memory and can be processed faster during computations than a real constant. Similarly, single-precision numbers have the same advantages relative to double-precision constants. Thus, if we use integers rather than single precision, and single precision rather than double precision, the resulting programs will run faster and take up less memory. On the other hand, double-precision constants are advantageous when a computation must be performed with a high degree of accuracy. As discussed previously in Chap. 4, such situations often occur in engineering.

Variables and assignment.   In addition to constant quantities, it is useful to have variable quantities that can change during the execution of a program. These variable quantities are designated by symbolic names. For example, in the simple addition programs from Figs. 15.5 and 15.7, we used three variables—A, B, and C. (It is common practice to refer to the variable name as the variable itself.)

To understand how variables work, it is convenient to think of the computer memory as consisting of a number of cells that are similar to post-office boxes (Fig. 15.8). The variable is the name that the computer assigns to each of these cells. For example, in the simple addition problem, each of the variables—A, B, and C—would correspond to one of the cells. When we use the line A = 28 to assign a value of 28 to A, the computer stores the value in the memory location for A (Fig. 15.8*a*). Then later in the program when we perform the addition C = A + B, the computer "remembers" that A is equal to 28 by referring to the memory cell where it is stored. Now suppose that after the addition we redefine A by another assignment statement such as

A = 80

When this line is executed, the computer replaces the old value by this new one. Using our analogy of the postal boxes, this act would be akin to placing the new number in the box and pushing the old one out the back and into oblivion (Fig. 15.8*b*). This ability to change value is the reason symbolic names are called *variables*.

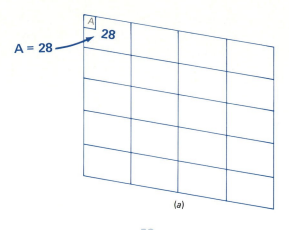

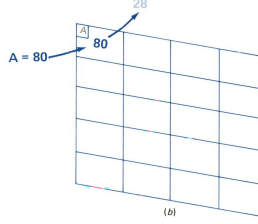

FIGURE 15.8
Analogy between computer memory locations and post-office boxes. (*a*) The FORTRAN statement A = 28 assigns the variable name A to one of the memory locations and stores a value of 28 in that location. (*b*) A subsequent assignment statement, A = 80, places a new value of 80 in the location, and the process destroys the prior value of 28.

Because lines such as A = 28 and B = 15 assign constant values to variables, they are called *assignment statements*. They can be written in general form as

variable = expression

where *variable* stands for the variable name and *expression* can be a constant, another variable, or an arithmetic expression. An example of the use of an arithmetic expression in an assignment statement is the line

```
C = A + B
```

When such a statement is executed, the numerical value of the expression is automatically stored in the storage location for C. Note that variables can also be subtracted, multiplied, and divided, as in

```
C = A - B
C = A * B
C = A/B
```

As shown in Fig. 15.3, all the above equations must lie within columns 7 through 72. In general, most compilers ignore blank spaces between terms in such equations. For example, the addition equation could be written as

```
C  =  A  +  B
```

Algebraic vs. computer equations.   The equal sign in the above statements can be thought of as meaning "is replaced by." This meaning is very significant because it suggests that computer equations differ in a very important way from standard algebraic equations. That is, computer equations must always have a single variable on the left-hand side of the equal sign. Thus, because it violates this rule, the following legitimate algebraic equation is illegal in FORTRAN:

```
A  +  B  =  C
```

In the same sense, although

```
X  =  X  +  1
```

is not a proper algebraic equation, it is a perfectly acceptable FORTRAN assignment statement. This is evident when the words "is replaced by" are substituted for the equal sign. For example, if X had a prior value of 5, the new value, after execution of $X = X + 1$, would be 6.

Variable names.   The symbolic name for the variable should reflect, as simply as possible, what information is stored by the variable. Mnemonic names that are easy to remember are best. Names that are short enough to type quickly and accurately are helpful, so long as the name is long enough to convey the meaning or purpose of the variable. While all of the following are valid FORTRAN variable names, those in the first row are inferior because their meanings are more obscure than for those in the second row:

```
SXEQW    Z19943    ABC4DE    I1948J    S
PROD23   FLUX      TOTAL     NAME1     SUM
```

Aside from being communicative, the actual symbolic names used to designate variables must follow rules (Table 15.2). Names must begin with a letter followed by up to five additional letters and numbers. Therefore, names cannot contain symbols, nor may they be longer than six characters. FORTRAN reserved words or keywords may not be used as variable names. For example, the reserved word END has a special meaning to the compiler and can never be used as a variable name. However, the name may include imbedded keywords—for example ENDMAS would be acceptable even though it includes the keyword END. The following are some examples of unusable FORTRAN names:

```
100001  99VAR  LONGVARNAME  A+VAR  IF  FORMAT
```

TABLE 15.2　The Fundamental Rules for Naming Variables in FORTRAN 77

Rule	Correct	Incorrect
Every name must begin with a letter of the alphabet.	FUEL VELOC1	1MASS 2001
Names must consist of only letters and numbers.	AREA3 TEMP1	AREA#3 TEMP.1
Names must consist of six or fewer characters.	SUM3 RESULT	VELOCITY1 TOOLONG
FORTRAN reserved words cannot be used as names (but they may be imbedded in names).	WRITE1 END5	WRITE END

100001 and 99VAR do not begin with a letter. The name A + VAR provides a good example why names cannot include symbols. Because the symbol + connotes addition, the computer would not recognize A + VAR as a single variable name but would perceive it as the sum of two separate variables, A and VAR.

FORTRAN has six types of variables (Table 15.3). In other words, the storage location represented by the variable can contain six types of information. These are integers, real single-precision numbers, real double-precision numbers, complex numbers with an imaginary portion, character(s), and logical data that is either true or false.

Using the analogy from Fig. 15.8, the different variable types relate to different types of mailboxes for storing information. Three means are available for telling the computer the type of data that a variable will contain.

TABLE 15.3　Types of Variables Used in FORTRAN 77

Type	Declaration	Description
Integer	INTEGER	Contains an integer; unless otherwise declared, variables starting with the letters I through N are assumed to be of this type.
Real (single precision)	REAL	Contains a single-precision, floating-point or decimal number; unless otherwise declared, variables starting with the letters A through H and O through Z are assumed to be of this type.
Real (double precision)	DOUBLE PRECISION	Contains a double-precision real number.
Complex number	COMPLEX	Contains a complex number with an imaginary portion.
Characters	CHARACTER	Contains character or alphanumeric information.
Logical	LOGICAL	Contains values that are either true or false.

1. *Naming conventions* The simplest means of specifying variable type is by the first letter in the variable name. Unless instructed otherwise, FORTRAN assumes that all names starting with the letters I, J, K, L, M, or N are integers, whereas names beginning with any other letter of the alphabet are real single-precision variables. For example, X, SUM, PRODUC, ONE, TWO, and THREE would represent real variable, and N, I, ICOUNT, LAST, MEAN, and MAX would represent integers.

This is the most common way to declare a variable type. Because the convention is built into the language, the type of variable can be defined conveniently and unambiguously. There are two shortcomings to this approach. First, sometimes the best name for one type of variable may start with a letter corresponding to the other type. For example, the maximum value of a group of numbers would be nicely named as MAXVAL. However, this label automatically limits the value to integers. Second, the naming convention does not extend to the last four variable types from Table 15.3.

2. *The IMPLICIT declaration* A second way to inform the FORTRAN compiler of variable type is by using the IMPLICIT statement, which is of the general form

IMPLICIT *type (range of letters), ..., type (range of letters)*

where *type* is chosen from the names of the different kinds of variables—INTEGER, REAL, DOUBLE PRECISION, COMPLEX, CHARACTER, or LOGICAL—and *range of letters* is a list of the first letters that you want to designate the corresponding type— for example,

```
IMPLICIT REAL (A,B,C), INTEGER (D-F), LOGICAL (L)
IMPLICIT DOUBLE PRECISION (Q,S,U)
IMPLICIT CHARACTER (M-P)
IMPLICIT CHARACTER*80 (T)
```

The IMPLICIT statement overrides the FORTRAN naming conventions so that, in the first example, all variables beginning with the letter L will be logical variables and not integer variables (as would be the case by convention). If the type of variable is not declared by an IMPLICIT statement, then the original FORTRAN naming conventions hold. For example, the letters I and J are not specified by the above IMPLICIT statements, and so names beginning with these letters are integers.

The last IMPLICIT statement declares that all variables beginning with T be of type CHARACTER. This means that the variable may contain string constants—that is, letters, numbers, and symbols for purposes such as labeling. The *80 signifies that the string constant may be up to 80 characters long. If the length is not specified (as in the third example), the variable may include a string of only one character. The other variable types can also have their maximum sizes specified in this way, but this is not commonly done.

The IMPLICIT statement is a convenient way to declare variable type. However, it can lead to confusion if the reader of the program does not remember which first letters correspond to which variable types.

3. *The explicit declaration* The final way to declare the type of a variable is by explicitly giving the variable type and then the variable names corresponding to the type. This is accomplished with an explicit statement that is of the general form

$$type \quad variable_1, \quad variable_2, \quad \ldots, \quad variable_n$$

where the *type* is as in the IMPLICIT statement (that is, REAL, INTEGER, etc.) and *variable*$_i$ is the name of the *i*th variable of a particular *type*. Explicit declaration should always be done early in the program and prior to the first executable statement. Some examples of explicit declaration are

```
REAL SUM, NUMVAL, ONE, TWO, MEAN
DOUBLE PRECISION DIFF
INTEGER ICOUNT, YVAR, GRADE
CHARACTER LETTER
CHARACTER*80 STRING
LOGICAL OK, DONE
```

An explicit statement of variable type overrides both the naming conventions and any IMPLICIT statement declarations. Aside from its expressed purpose of declaring the type of each variable, the explicit declaration has an added benefit. That is, the names and types of all variables can be located in one area of the program for easy reference by the user.

In the FORTRAN programs in the remainder of the book, we will primarily use naming conventions to establish numeric variable types. Explicit declarations will be employed for other variables and wherever the naming conventions must be circumvented to obtain a better name for a variable. The declarations will immediately follow the portion of each module where the use of the variables is described.

Figure 15.9 is a revised version of the simple addition program illustrating a more communicative use of variable names. Notice how a module has been incorporated to define the meaning of the variables at a single location at the program's beginning and that more descriptive names have been used to specify the variables. FIRST, SECOND, and TOTAL are much more suggestive of the function of the variables than the original names, A, B, and C.

15.7 INTRODUCTION TO INPUT/OUTPUT

''Input/output'' refers to the means whereby information is transmitted to and from a program. You have already been introduced to output with the WRITE command in the simple addition program. Now we will elaborate on the topic.

The READ statement. One disadvantage of the assignment statement is that it assigns a specific value to a variable. Thus, in the simple-addition example program, we use assignment statements to set FIRST equal to 28 and SECOND equal to 15. If we

```
C      ****************************************************
C      *                                                  *
C      *           SIMPLE ADDITION PROGRAM                *
C      *              (FORTRAN version)                   *
C      *              by  S. C. Chapra                    *
C      *                                                  *
C      *        This program adds two variables           *
C      *        and prints out the results.               *
C      ****************************************************
C
C      .--------------------------------------------.
C      |              DEFINITION OF VARIABLES        |
C      |                                             |
C      |  FIRST,SECOND = VARIABLES TO BE ADDED        |
C      |  TOTAL = SUM OF THE NUMBERS                  |
C      `--------------------------------------------'
C
C      ****************  MAIN PROGRAM  ****************
C
C              Obtain Numbers From User
C
       READ(5,*) FIRST
       READ(5,*) SECOND
C
C              Perform Computation and Print Results
C
       TOTAL = FIRST + SECOND
       WRITE(6,*) TOTAL
       STOP
       END


** Output Results **

43.0000000
```

FIGURE 15.9
A revision of the simple addition program from Fig. 15.7. A module has been included to define the variables used in the program. Notice that more-descriptive variable names are employed.

want to add two different numbers, say, FIRST = 8 and SECOND = 33, we have to retype the two lines in order to assign new values to the variables.

The READ statement is an alternative that allows new data to be entered every time the program is run. Its general form is

$$\text{READ} \quad (\textit{unit number, format number}) \quad var_1, \ldots, var_n$$

where the *unit number* specifies the hardware device on which the information is being input, the *format number* specifies the manner in which the information is being input, and *var_i* is the name of the *i*th variable to which the information is being assigned.

For example, rather than assign a value to FIRST in the simple addition program, we could alternatively input a value with

$$\text{READ} \quad (5,*) \quad \text{FIRST}$$

The first character within the parentheses is the unit number. On most systems, the number 5 instructs the computer that a default device is to be employed to input the data. By "default" we mean that we leave it to the computer to select the device. In most cases, it automatically chooses a common input device such as a terminal keyboard or a card reader. You should consult your computer's user's manual to ascertain what unit numbers are employed for the input devices on your system. In Chap. 21 we will discuss how other input media such as files can be used to enter data to a program.

The second character within the parentheses is the format number. This is employed to tell the computer the way in which the data will be entered. For the time being, our use of the asterisk instructs the computer that we are employing *free format*. For input this means that the data is unformatted. If one value is to be input, it is merely typed on the keyboard or a card and entered. For example, if device 5 is the terminal, the computer would execute the simple addition program line by line until it encountered the statement

```
READ (5,*) FIRST
```

Upon reaching this statement, the computer would stop and wait for you to enter a value for FIRST. Suppose that you want to set FIRST equal to 15. Then by merely typing in a 15 and striking RETURN you would have instructed the computer that FIRST = 15. It then would continue executing, reading in a value for SECOND (say 28) and then printing out the correct answer, TOTAL = 15 + 28 = 43.

More than one value may be entered with a READ statement. For example, to input values for both FIRST and SECOND, you could type

```
READ(5,*) FIRST, SECOND
```

Then, when the computer came to this line, you would merely type the two values separated by a comma—for example,

```
15,28
```

and strike return to enter values for FIRST and SECOND. Thus a value of 15 would be stored in FIRST, 28 would be stored in SECOND, and the program would sum the two values and output the result, 43.0000000.

Some examples of READ statements along with acceptable entries are shown in Fig. 15.10. In all but the second of these examples, the number of data entered is consistent with the READ statement. If more entries are included on an input line than required by the READ statement, the extra values are ignored. The values 99 and 17 would be ignored by READ(5,*)N and any subsequent READ statements. If the RETURN key is pressed without entering enough data items (as is the case in Fig. 15.10), the READ will just keep looking for data until all the variables have been assigned a value.

The data entered must match the variable type specified in the READ statement. Therefore FINAME and LANAME would have to have been previously defined as character variables, using, for example,

```
CHARACTER*30 LANAME,FINAME
```

```
C
C Program to Demonstrate Read Statement
C
      CHARACTER*30 FINAME, LANAME
      READ (5,*) ONE, TWO, THREE
      READ (5,*) N
      READ (5,*) FINAME, LANAME, GRADE
      READ (5,*) A, B, C
      STOP
      END
//$DATA
1.0E2,19,33.29
89,99,17
'JOE','SMITH',74.5
17.9
21
0.11234E-3
```

FIGURE 15.10
A short program showing some READ statements along with the accompanying data.

If this has not been done, an error message results and execution is terminated when the computer attempts to read the character data.

The WRITE statement. The WRITE statement is the major vehicle for outputting data. Its general form is

$$\text{WRITE} (unit\ number,\ format\ number)\ \ exp_1,\ ...,\ exp_n$$

where the *unit number* specifies the hardware device on which the information is being output, the *format number* specifies the manner in which the information is being output, and exp_i is the *i*th value that is to be output.

The form of the WRITE statement in the simple addition program is

$$\text{WRITE (6,*) TOTAL}$$

On many systems, the number 6 is used to specify a typical output device. In most cases, this will either be your terminal screen or printer. You should consult your system's user's manual to ascertain what *unit numbers* are employed on your system. In Chap. 21, we will discuss how other output media such as files can be used to store data generated by a program.

The second character within the parentheses is the format number. For the time being, our use of the asterisk instructs the computer to print out the results in a free format. Integer numbers will be displayed without a decimal and real numbers will be displayed in either decimal or scientific notation.

Some examples of WRITE statements along with possible (depending on the value of the variables) output are shown in Fig. 15.11. The *expressions* following the WRITE statement can be either a constant, a variable, or a mathematical expression. The data enclosed between apostrophes will be displayed in the output without the quotation marks but otherwise unchanged. The WRITE statement alone causes a blank line to be output.

String constants and variables. Now that we have introduced the READ and WRITE

```
C
C Program to Demonstrate Write Statement
C
        CHARACTER*30 NAME
        SUM = 199.5
        NAME = 'TOM'
        I = 2
        N = 7
        WRITE (6,*) 10, SUM, NAME
        WRITE (6,*) I + N
        WRITE (6,*)
        WRITE (6,*) 'TOTAL = '
        STOP
        END
```

FIGURE 15.11
A short program showing some
WRITE statements along with the
resulting output.

```
** Output Results **

        10      199.50000000      TOM
         9

TOTAL =
```

statements, we will show how they can be used to input and output data in a more descriptive fashion. To do this, we have to elaborate more on the concept of string constants and variables.

Aside from numbers, computers are also designed to handle alphanumeric information (that is, groups of letters, numbers, and symbols) such as names, addresses, and other miscellaneous labels. *Character or string constants* are specific groupings of alphanumeric information. In FORTRAN they are enclosed in single quotation marks, or apostrophes, as in

```
'JOHN DOE'    '8/5/48'    'THE FIRST RESULT IS'
```

String variables work in the same way as numeric variables in the sense that they are names that designate a specific location in the computer's memory. As described previously, a CHARACTER statement must be used to declare these variables.

String variables can be given constant values in the same manner as numeric variables. One way is via an assignment statement, as in

```
NAME = 'JOHN DOE'
DATE = 'JULY 4, 1976'
SUM1 = 'THE FIRST RESULT IS'
```

In addition, a READ statement can be used for the same purpose. For example,

```
READ(5,*) NAME1
```

would result in a prompt appearing on the monitor during execution. In response to the prompt you could type in the appropriate value, as in

```
?  'JOHN DOE' ®
```

After you have struck the RETURN (designated by ®), the storage location corresponding to NAME1 would hold the string constant JOHN DOE.

Before describing how character data is used to enhance input and output, we must mention that, prior to FORTRAN 77, most dialects could not accommodate character variables. However, character constants (that is, messages enclosed in quotes) were available. As shown in the next section, these are sufficient to devise descriptive input and output messages.

Descriptive input.   The following material on descriptive input is only relevant to those computer systems where a FORTRAN program can be run interactively. That is, once the program is executed, intermediate results are output to the user's terminal. For batch systems employing input media such as punched cards, the following material does not apply. However, the subsequent section on descriptive output has relevance to both interactive and noninteractive systems.

One problem with the READ statements described to this point is that the user has been provided no clue as to what information is supposed to be input. For example,

```
READ (5,*) FIRST
```

would cause the computer to stop and wait for you to enter a value. In most cases, the terminal screen would provide no explicit indication that the program required data. Unless the user knew beforehand that a value of FIRST was to be entered, the system provides no clue as to what variables are to be input. One way to remedy this situation is to use a WRITE statement and string constants to print a message that alerts the user to what is required, for example,

```
WRITE (6,*) 'FIRST ='
READ (5,*) FIRST
```

When these lines are executed, the computer will print

```
FIRST =
```

where the message signifies that the computer is prompting you for information. Thus the WRITE statement lets the user know what is required.

Descriptive output.   Just as input can be made more user-friendly, output can also be made more descriptive using character strings. For example, in the simple addition program the statement

```
WRITE (6,*) TOTAL
```

merely results in the answer, 43, being printed out. A message can easily be incorporated into the statement, as in

```
WRITE(6,*) 'TOTAL =', TOTAL
```

```
C           **************************************************
C           *              SIMPLE ADDITION PROGRAM           *
C           *                 (FORTRAN version)              *
C           *                 by  S. C. Chapra               *
C           *                                                *
C           *        This program adds two variables         *
C           *        and prints out the results.             *
C           **************************************************
C
C           .------------------------------------------.
C           |            DEFINITION OF VARIABLES       |
C           |                                          |
C           | FIRST,SECOND = VARIABLES TO BE ADDED     |
C           | TOTAL = SUM OF THE NUMBERS               |
C           '------------------------------------------'
C
C       ****************  MAIN PROGRAM  *****************
C
C                        Identify Program
C
        WRITE(6,*)
        WRITE(6,*) 'PROGRAM TO ADD TWO NUMBERS'
C
C                   Obtain Numbers From User
C
        READ(5,*) FIRST
        READ(5,*) SECOND
C
C                Perform Computation and Print Results
C
        TOTAL = FIRST + SECOND
        WRITE(6,*)
        WRITE(6,*) 'SUM OF ',FIRST,' PLUS ',SECOND
        WRITE(6,*) 'IS EQUAL TO ',TOTAL
C
C                Signal that Program was Successful
C
        WRITE(6,*)
        WRITE(6,*) 'COMPUTATION COMPLETED'
        STOP
        END
```

FIGURE 15.12 A final revision of the simple addition program. Descriptive output are incorporated in this version.

When the program is run, this results in

```
TOTAL = 43.0000000
```

Figure 15.12 is a final revised version of the simple addition program. We have added a variety of descriptive WRITE statements to make the program more readable and easier to use. A run of the program is shown in Fig. 15.13. We will return to the topic of input and output in more detail in Chap. 17.

15.8   QUALITY CONTROL

To this point you have acquired sufficient information to compose a simple FORTRAN computer program. Using the information from the previous section, you should now be capable of writing a program that will generate some output. However, the fact

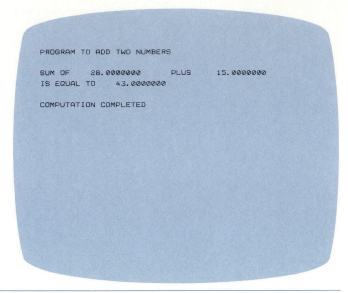

```
PROGRAM TO ADD TWO NUMBERS

SUM OF    28.0000000     PLUS      15.0000000
IS EQUAL TO   43.0000000


COMPUTATION COMPLETED
```

FIGURE 15.13
A run of the program from
Fig. 15.12.

that a program prints out information is no guarantee that these answers are correct. Consequently, we will now turn to some issues related to the reliability of your program. We are emphasizing this subject at this point to stress its importance in the program-development process. As problem-solving engineers, most of our work is used directly or indirectly by clients. For this reason, it is essential that our programs be reliable—that is, that they do exactly what they are supposed to do. At best, lack of reliability can be embarrassing; at worst, for engineering programs involving public safety, it can be tragic.

15.8.1 Errors or "Bugs"

During the course of preparing and executing a program of any size, it is likely that errors will occur. These errors are sometimes called *bugs* in computer jargon. This label was coined by Capt. Grace Hopper, one of the pioneers of computer languages. In 1945 she was working with one of the earliest electromechanical computers when it went dead. She and her colleagues opened the machine and discovered that a moth had become lodged in one of the relays. Thus the label ''bug'' came to be synonymous with problems and errors associated with computer operations, and the term *debugging* came to be associated with their solution.

When developing and running a program, three types of bugs can occur:

1. *Syntax errors* violate the rules of the language such as spelling, number formation, line numbering, and other conventions. These errors are often the result of mistakes such as typing REAA rather than READ. Errors of this type are often detected when your source code is being compiled or is being converted to equivalent object code. They usually prompt the computer to print out an error message. Such messages are called *diagnostics* because the computer is helping you to ''diagnose'' the problem.

2. *Run-time errors* are those that occur during execution of the object code—for example, when a statement involves division by zero or when you try to take the logarithm of a negative number. For such situations a diagnostic message will be printed, and you must then reenter the data properly. For other run-time errors the computer may simply terminate the run or print a message informing you of the location where the error occurred. In any event, you will be cognizant of the fact that an error has occurred.

3. *Logic errors*, as the name implies, are due to faulty program logic. These are the worst types of errors because no diagnostics will flag them. Thus your program will appear to be working properly in the sense that it executes and generates output. However, the output will be incorrect.

All the above types of errors must be removed before a program can be employed for engineering applications. We divide the process of quality control into two categories: debugging and testing. As mentioned above, *debugging* involves correcting known errors. In contrast, *testing* is a broader process intended to detect errors that are not known. Additionally, testing should also determine whether the program fills the needs for which it was designed.

As will be clear from the following discussion, debugging and testing are interrelated and are often carried out in tandem. For example, we might debug a module to remove the obvious errors. Then, after a test, we might uncover some additional bugs that require correction. This two-pronged process is repeated until we are convinced that the program is absolutely reliable.

15.8.2 Debugging

As stated above, debugging deals with correcting known errors, those usually detected by the FORTRAN compiler. There are three ways in which you will become aware of errors. First, an explicit diagnostic message can provide you with the exact location and the nature of the error. Second, a diagnostic may be printed, but the exact location of the error is unclear. For example, the computer may indicate that an error has occurred in a line representing a long equation including some user-defined functions (we will discuss these in the next chapter). This error could be due to syntax errors in the equation, or it could stem from errors in the statements defining the functions. Third, no diagnostics are printed, but the program does not operate properly. These errors are usually due to faulty logic such as infinite loops or mathematical mistakes that do not yield syntax errors. Corrections are easy for the first type of error because the FORTRAN compiler will do most of the work. However, for the second and third types, some detective work is required in order to identify the errors and their exact locations.

Unfortunately, the analogy with detective work is all too apt. That is, ferreting out errors is often difficult and involves a lot of a "legwork," collecting clues and running up some blind alleys. In particular, as with detective work, there is no clear-cut methodology involved. However, there are some ways of going about it that are more efficient than others.

One key to finding bugs is to print out intermediate results. With this approach it is often possible to determine at what point the computation went bad. Similarly, the use of a modular approach will obviously help to localize errors in this way. You could put a different WRITE statement at the start of each module and in this way focus on the module in which the error occurs. In addition, it is always good practice to debug (and test) each module separately before integrating it into the total package.

Although the above procedures can help, there are often times when your only option will be to sit down and read the code line by line, following the flow of logic and performing all the computations with a pencil, paper, and pocket calculator. Just as a detective sometimes has to think like the criminal in order to solve a case, there will be times you will have to "think" like a computer to debug a program.

Once you have discovered an error, there are several ways to make corrections. If the line has already been typed and entered, merely use the computer's editor to retype it and strike RETURN again. The new version automatically takes the place of the older, incorrect line. On the other hand, if you notice the error before you strike RETURN, most computers have a key that permits you to backspace and then type over the line correctly. In addition, most operating systems have an editing feature to correct a line without retyping it. The program source code must be recompiled after all the bugs have been corrected.

15.8.3 Testing

One of the greatest misconceptions of the novice programmer is the belief that if a program runs and prints out results, it is correct. Of particular danger are those cases where the results appear to be reasonable but are in fact wrong. To ensure that such cases do not occur, the program must be subjected to a battery of tests for which the correct answer is known beforehand.

As mentioned in the previous section, it is good practice to debug and test modules prior to integrating them into the total program. Thus testing usually proceeds in phases. These phases include module tests, development tests, whole-system tests, and operational tests.

Module tests, as the name implies, deal with the reliability of individual modules. Because each is designed to perform a specific, well-defined function, the modules can be run in isolation to determine that they are executing properly. Sample input data can be developed for which the proper output is known. This data can be used to run the module, and the outcome can be compared with the known result to verify successful performance.

Development tests are implemented as the modules are integrated into the total program package. That is, a test is performed after each module is integrated. An effective way to do this is with an approach that starts with the first module and progresses down through the program in execution sequence. If a test is performed after each module is incorporated, problems can be isolated more easily, since new errors can usually be attributed to the latest module added.

After the total program has been assembled, it can be subjected to a series of *whole-system tests*. These should be designed to subject the program to (1) typical data, (2) unusual but valid data, and (3) incorrect data to check the program's error-handling capabilities.

Finally, *operational tests* are designed to check how the program performs in a realistic setting. A standard way to do this is to have a number of independent subjects (including the client) implement the program. Operational tests sometimes turn up bugs. In addition, they provide feedback to the developer regarding ways to improve the program so that it more closely meets the user's needs.

Although the ultimate objective of the testing process is to ensure that programs are error-free, there will be complicated cases where it is impossible to subject a program to every possible contingency. Also, it should be recognized that the level of testing depends on the importance and magnitude of the problem context for which the program is being developed. There will obviously be different levels of rigor applied to testing a program to determine the averages of your intramural softball team as compared with a program to regulate the operation of a nuclear reactor. In each case, however, adequate testing must be performed, testing consistent with the liabilities connected with a particular problem context.

15.9 DOCUMENTATION

Documentation is the addition of English-language descriptions to allow the user to implement the program more easily. It is an ongoing process that begins when the program is first being written and is only completed after final debugging and testing. Remember, along with other people who might employ your software, you are also a user. Although a program may seem clear and simple when you first compose it, after the passing of time the same code may seem inscrutable. Therefore sufficient documentation must be included to allow you and other users to immediately understand your program and how it can be implemented. This task has both internal and external aspects.

15.9.1 Internal Documentation

Picture for a moment a text with just words and no paragraphs, subheadings, or chapters. Such a book would be more than a little intimidating to study. The way the pages of a text are structured—that is, all the devices that serve to break up and organize the material—serve to make the text more effective and enjoyable. In the same sense, internal documentation of computer code can greatly enhance the user's understanding of a program and how it works.

We have already broached the subject of internal documentation when we introduced the comment statement in Sec. 15.5.1. The following are some general suggestions for documenting programs internally:

1. Include a module at the head end of the program, giving the program's title and your name and address. This is your signature and marks the program as your work.
2. Include a second module to define each of the key variables.
3. Select variable names that are reflective of the type of information the variables are to store.

PROGRAM: The Simple Addition Program

PROGRAMMER: S.C. Chapra
                                  Civil Engineering Department
                                  Texas A & M University
                                  College Station, Texas 77843

DESCRIPTION: This program adds two numbers and prints out the results.

SPECIAL REQUIREMENTS: None

REFERENCE: Chapra, S.C. and R.P. Canale; *Introduction to Computing for Engineers.* (McGraw-Hill, New York). 1986. Fig. 15.12.

EXAMPLE RUN:

(Insert example
run here.)

FIGURE 15.14 A simple one-page format for program documentation. This page could be stored in a binder along with a printout of the program.

4. Insert spaces within statements to make them easier to read.

5. Use comment statements liberally throughout to program for purposes of labeling and clarifying.

6. In particular, use comment statements to clearly label and set off all the modules.

7. Use indentation to clarify the programs structures. In particular, indentation can be used to set off loops and decisions.

15.9.2   External Documentation

External documentation refers to instructions in the form of output messages and supplementary printed matter. The messages occur when the program runs and are intended to make the output attractive and user-friendly. This involves the effective use of spaces, blank lines, and special characters to illuminate the logical sequence and structure of the program output. Attractive output simplifies the detection of errors and enhances the communication of program results.

The supplementary printed matter can range from a single sheet to a comprehensive user's manual. Figure 15.14 is an example of a simple documentation form that we recommend you prepare for every program you develop. These forms can be maintained in a notebook to provide a quick reference for your program library. The user's manual for your computer is an example of comprehensive documentation. This manual tells you how to run your computer system and disk operating programs. Note that for well-documented programs, the title information in Fig. 15.14 is unnecessary because the title module is included in the program.

15.10   STORAGE AND MAINTENANCE

In order to retain a program for later use, you must save it to a secondary storage device such as a file before signing off the computer. Every system has its own set of commands for accomplishing this.

Aside from this physical act of retaining the program, storage and maintenance consist of two major tasks: (1) upgrading the program in light of experience and (2) ensuring that the program is safely stored. The former is a matter of communicating with users to gather feedback regarding suggested improvements to your design.

The question of safe storage of programs is especially critical for those who implement FORTRAN on microcomputers and use diskettes as their secondary storage media. Although these devices offer a handy means for retaining your work and transferring it to others, they can be easily damaged. Table 8.7 outlines a number of tips for effectively maintaining your programs on diskettes. Follow the suggestions on this list carefully in order to avoid losing your valuable programs. In addition, as discussed in Sec. 15.9.2, printouts of all programs and important results should be maintained in notebooks.

PROBLEMS

15.1   Describe some of the strengths of FORTRAN.

15.2   Learn how to run a FORTRAN program on your system. Write up the procedure in a step-by-step form.

15.3   The arithmetic operator for FORTRAN multiplication is the asterisk. Therefore, a simple multiplication program can be developed by reformulating the addition in Fig. 15.5 as

```
C = A * B
```

Develop a version of this program patterned after Fig. 15.12. Generate a paper printout or hard copy of the output. Document this program using the format specified in Fig. 15.14. Use values of FIRST = 20 and SECOND = 16 for your example run.

15.4 Develop, test and document a program to input the length, width, and height of a box. Have the program compute the volume and the surface area of the box. Also, allow the user to input the units (for example, feet) for the dimensions of the box as character constants and have the program print out the correct units with the answers. Document this program using the format specified in Fig. 15.14. Use values of length, width, and height equal to 5, 1.5, and 1.25 m, respectively, for your example run.

15.5 In composing and entering a FORTRAN program, there are certain conventions that must be observed. For example, because FORTRAN programs were originally typed on punched cards, program lines must be contained in an 80-column field. This field is further subdivided into subfields that can contain only certain types of information. Use Fig. 15.3 and information in the text to identify all the subfields and their functions.

15.6 When are FORTRAN programs not executed sequentially from top to bottom?

15.7 Describe the entire process of creating and executing a FORTRAN program on your particular computer system.

15.8 Note the unit numbers that your system employs to specify the conventional modes whereby information is input and output to FORTRAN programs.

15.9 What are the naming conventions used for variables on your compiler?
(*a*) What is the acceptable length?
(*b*) With what characters can names begin? Of which characters can they be composed?
(*c*) Based on its first letter, what does FORTRAN assume about a variable?

15.10 What are the three ways in which variable type can be specified? Provide examples and comment on when each might be appropriate.

15.11 How many characters can a string variable contain? How is this specified?

15.12 Write a program that accepts your name and address as input and then outputs this information with appropriate labels. For example, the output might look like

NAME: (*your name as input*)
ADDRESS: (*your address as input*)

CHAPTER 16

FORTRAN: COMPUTATIONS

As stated on several occasions to this point, one of the primary engineering applications of computers is to manipulate numbers. In the previous chapter, you were introduced to some fundamental concepts that lie at the heart of this endeavor: constants, variables, and assignment. Now we will expand on these concepts and introduce some additional material that will allow you to exploit the further potential of FORTRAN to perform computations.

16.1 ARITHMETIC EXPRESSIONS

Obviously, in the course of engineering problem solving, you will be required to perform more complicated mathematical operations than the simple process of addition described in the previous chapter. A key to understanding how these operations are accomplished is to realize that a primary distinction between computer and conventional equations is that the former are constrained to a single line. In other words, a multilevel, or "built-up," equation such as

$$X = \frac{A + B}{Y - Z^2} + \frac{33.2}{X^3}$$

must be reexpressed so that it can be written as a single-line expression in FORTRAN.

The operators that are used to accomplish this are listed in Table 16.1. Notice how the two operations that are intrinsically multilevel—that is, division and exponentiation—can be expressed in single-line form using these operators. For example, A divided by B, which in algebra is

$$\frac{A}{B}$$

TABLE 16.1   Arithmetic Symbols and Their Associated
Priorities

Symbol	Meaning	Priority
( )	Parentheses	1
**	Exponentiation	2
−	Negation	3
*	Multiplication	4
/	Division	4
+	Addition	5
−	Subtraction	5

is expressed in FORTRAN as A/B. Similarly, the cube of A, which in algebra demands that the 3 be raised as a superscript, as in A^3, is expressed in FORTRAN as A**3. Note that the power to which the number is raised can be a positive or negative integer or decimal. However, a negative number can only be raised to an integer power.

In order to construct more complicated expressions, a number of rules must be followed. These involve the priorities that are given to the operations. As seen in Table 16.1, parentheses are given top priority in the sense that the calculations within parentheses are implemented first. For example, in

```
C * (A + B)
```

the addition is performed first and the result is then multiplied by C. If several sets are nested within each other the operations in the innermost are performed first.

Note that parentheses cannot be used to imply multiplication as is done in conventional arithmetic. For example, the above multiplication could not be expressed as C(A + B) in FORTRAN. The asterisk must always be included to indicate multiplication.

Parentheses must always come in pairs. A good practice when writing expressions is to check that there is always an equal number of open and closed parentheses. For example, the expression

```
((X + 2 * (3 - B - (4 * A/B) - 7))
```

is incorrect because there is one too many open parentheses. Checking the number of open and closed parentheses does not ensure that the expression is correct, but if the number of parenthesis are not equal, the expression is definitely erroneous.

After operations within parentheses, the next highest priority goes to exponentiation. If multiple exponentiations are required, as in X**Y**Z, they are implemented from right to left. This is followed by negation—that is, the placement of a minus sign before an expression to connote that it is negative. Next, all multiplications and divisions are performed from left to right. Finally, the additions and subtractions are performed from left to right. The use of these priorities to construct mathematical expressions is best shown by example.

EXAMPLE 16.1   Arithmetic Expressions

PROBLEM STATEMENT: Determine the proper values of the following arithmetic expressions in FORTRAN. Although decimals are not shown, assume that all the numbers used are real numbers.

(a) $5 + 6 * 7 + 8$ (f) $5 ** 2 ** 3$
(b) $(5 + 6) * 7 + 8$ (g) $(5 ** 2) ** 3$
(c) $(5 + 6) * (7 + 8)$ (h) $8/2 * 4 - 3$
(d) $-2 ** 4$ (i) $(10/(2 + 3) * 5 ** 2) ** 0.5$
(e) $(-2) ** 4$

SOLUTION
(a) Because multiplication has a higher priority than addition, it is performed first to yield

   $5 + 42 + 8$

Then the additions can be performed from left to right to yield the correct answer, 55.

(b) Although multiplication has a higher priority than addition, the parentheses have even higher priority, and so the parenthetical addition is performed first:

   $11 * 7 + 8$

followed by multiplication

   $77 + 8$

and finally addition to give the correct answer, 85.

(c) First evaluate the additions within the parentheses:

   $11 * 15$

and then multiply to give the correct answer, 165.

(d) Because exponentiation has a higher priority than negation, the 2 is first raised to the power 4 to give 16. Then negation is applied to give the correct answer, -16.

(e) In contrast to the expression in (d), here the parentheses force us to first transform 2 into -2. Then the -2 is raised to the power 4 to give the correct answer, 16.

(f) Because the order of exponentiation is from right to left, 2 is first cubed to give 8. The 5 is then raised to the power 8 to yield the correct answer, 390625.

(g) In contrast to the expression in (f), here the normal order of exponentiation is overidden, and 5 is first squared to give 25, which in turn is raised to the power 3 to yield the correct answer, 15625.

(h) Because multiplication and division have equal priority, they must be performed from left to right. Therefore 8 is first divided by 2 to give

   $4 * 4 - 3$

Then 4 is multiplied by 4 to yield

$$16 - 3$$

which can be evaluated for the correct answer, 13. Note that if the left-to-right rule had been disobeyed and the multiplication performed first, an erroneous result of -2 would have been obtained.

(*i*) Because the parentheses in this expression are nested (that is, one pair is within the other), the innermost is evaluated first:

$$(10/5*5 ** 2) ** 0.5$$

Next the expression within the parentheses must be evaluated. Because it has highest priority, the exponentiation is performed first:

$$(10/5 * 25) ** 0.5$$

Then the division and multiplication are performed from left to right to yield

$$(2 * 25) ** 0.5$$

and

$$50 ** 0.5$$

which can be evaluated to give the correct answer, 7.071.

When dealing with very complicated equations, it is often good practice to break them down into component parts. Then the components can be combined to come up with the end results. This practice is especially helpful in locating and correcting typing errors that frequently occur when entering a long equation.

EXAMPLE 16.2   Complicated Arithmetic Expressions

PROBLEM STATEMENT: Write the comparable FORTRAN version for the algebraic expression

$$y = \frac{\dfrac{x^2 + 2x + 3}{\sqrt{5 + z}} - \sqrt{\dfrac{15 - 77x^3}{z^{3/2}}}}{x^2 - 4xz - 5x^{-4/5}}$$

Treat each major term of the equation as a separate component in order to facilitate programming the equation in FORTRAN.

SOLUTION: The two terms in the numerator and the denominator can be programmed separately as

```
PART1 = (X**2+2*X+3)/((5+Z)**0.5)
PART2 = ((15-77*X**3)/(Z**(3./2.)))**0.5
DENOM = X**2-4*X*Z-5*X**(-4./5.)
```

Then the total equation could be formed from the component parts, as in

```
Y = (PART1 - PART2)/DENOM
```

Notice that we included unnecessary pairs of parentheses in the equations for PART1 and PART2. We could just as correctly have written these expressions as

```
PART1 = (X**2+2*X+3)/(5+Z)**0.5
PART2 ((15-77*X**3)/Z**(3./2.))**0.5
```

We included the extra parentheses because in our judgment they make the equation easier to read. As long as they are matched, extra parentheses are allowable. As in this example, it is a good stylistic practice to include them in order to make the expression more readable. In addition, spaces can be employed to make the expressions clearer. For example, the equation for the denominator could be written as

```
DENOM = X**2 - 4*X*Z - 5+X**(-4./5.)
```

16.1.1   Integer and Mixed-Mode Arithmetic

Integer arithmetic.   All of the computations to this point have involved real variables and constants. Expressions with only integers behave similarly. However, integer division presents a special case that demands care. Intermediate results and the final assignment that involve integer division are truncated—that is, any digits following the decimal point are dropped. Unlike most other computer languages, FORTRAN never uses the decimal portion of integer expressions in the intermediate stages of a calculation. This can have a pronounced effect on the final result, as shown in the following example.

EXAMPLE 16.3   *Integer Arithmetic*

PROBLEM STATEMENT: Determine the values of the following operations involving integers: (*a*) 3/4 * 4, (*b*) 5/3 − 17, (*c*) 7 − 19/5, (*d*) 4 ** (5/10).

SOLUTION
(*a*) Because precedence is from left to right, the division is performed first to give 0.75. However, because the division involves two integers, the decimal portion of this result is dropped to give 0. Therefore the final result is 0 * 4 = 0.

(b) First 5/3 = 1.6667, which is truncated to 1. Then 1 − 17 = − 16.
(c) The division is implemented first to give 19/5 = 3.8, which is truncated to 3.
Then 7 − 3 = 4.
(d) The parenthetical division is performed to give 0.5, which is truncated to 0. Then
the exponentiation 4 ** 0 gives 1.

Mixed-mode arithmetic.   Expressions involving both real and integer parts are called
mixed-mode expressions. They follow these rules:

1. If an operation involves an integer and a real part, the integer is converted to
a real number prior to the operation, and the result is real.
2. If an operation involves two integers, the result is an integer. For the case
where division of two integers gives an outcome with a decimal portion, the result is
truncated to give the final integer value.
3. Assignment transforms the result on the right-hand side of the equality into
the type of the variable on the left-hand side.

EXAMPLE 16.4   *Mixed-Mode Arithmetic*

PROBLEM STATEMENT   Evaluate the following mixed-mode expressions: (a) X =
3/4 + 1.0; (b) J = 3.0/4 + 1.0; (c) X = 3.0/4 * 4; (d) X = 3/4 * 4; (e) I =
7. − 19/5; (f) J = 7 − 19/5.

SOLUTION
(a) 3/4 = 0.75, but because integers are involved, this result is truncated to 0.
Therefore the final result is 1.0.
(b) 3.0/4 = 0.75, and because a real number is involved this result is not truncated
but is added to 1.0 to give 1.75. However, because the result is assigned to the integer
J, the final value is truncated to give an integer of 1.
(c) 3.0/4 = 0.75, which is multiplied by 4 to give 3.0.
(d) 3/4 = 0.75, which is truncated to 0 because the operation involves integers
exclusively. This result is multiplied by 4 to give 0.
(e) 19/5 = 3.8, which is truncated to 3 because the operation involves integers. This
value is subtracted from 7. to give 4.0. However, the final result is transformed to an
integer 4 when it is assigned to I.
(f) 19/5. is equal to 3.8. This value is subtracted from 7 to give 3.2. However, the
final result is transformed to an integer value of 3 when it is assigned to J.

The easiest way to avoid confusion in arithmetic is to avoid mixed-mode expressions unless you actually desire the results of an expression to be truncated. Most
engineering results require real results, and integer calculations are used almost exclusively for counting items.

16.1.2 Complex Variables

Complex variables can be included in FORTRAN. As mentioned previously in Sec. 15.6, a type-specification statement must be employed to declare complex variables. For example,

COMPLEX var_1, var_2, . . ., var_n

tells the computer that the variables, var_1 through var_n are complex. Two constants are associated with each complex variable, one for the real part and the other for the imaginary part. These can be assigned to the variable as in

```
COMPLEX C1, C2
      .
      .
      .
C1 = (1.0,5.0)
C2 = (0.4,2.5)
```

Note that the values must be separated by a comma as shown. The values can also be entered via input as in

```
READ(5,*) C1, C2
```

where the input data could be entered as

```
1.0,5.0,0.4,2.5
```

Arithmetic operations can be performed on complex variables resulting in a complex answer. Remember that the variable to which the answer is assigned must also be declared as complex by a type-specification statement. If an operation involves a complex and a real or integer variable, the latter is automatically converted to a complex variable whose imaginary part is zero. Operations between complex- and double-precision variables are not allowed. Note that special intrinsic functions (as described in the next section) are available to manipulate complex variables. Some of these are listed in Appendix G.

16.2 INTRINSIC, OR BUILT-IN, FUNCTIONS

In engineering practice many mathematical functions appear repeatedly. Rather than require the user to develop special programs for each of these functions, manufacturers include them directly as part of the computer's software library. These *built-in* (or *intrinsic* or *library*) *functions* are of the general form

name (argument)

TABLE 16.2   Commonly Available Built-in Functions in FORTRAN (The argument of the function X can be a constant, variable, another function, or an algebraic expression.)

Function	Description
SQRT(X)	Calculates the square root of X where X must be positive.
ABS(X)	Determines the absolute value of X.
ALOG10(X)	Calculates the common logarithm of X where X must be positive.
ALOG(X)	Calculates the natural logarithm of X where X must be positive.
EXP(X)	Calculates the exponential of X.
SIN(X)	Calculates the sine of X where X is in radians.
COS(X)	Calculates the cosine of X where X is in radians.
TAN(X)	Calculates the tangent of X where X is in radians.
ASIN(X)	Calculates the inverse sine of X.
ACOS(X)	Calculates the inverse cosine of X.
ATAN(X)	Calculates the inverse tangent of X.
SINH(X)	Calculates the hyperbolic sine where X is in radians.
COSH(X)	Calculates the hyperbolic cosine where X is in radians.
TANH(X)	Calculates the hyperbolic tangent where X is is radians.
INT(X)	Calculates the largest integer less than or equal to X.
IFIX(X)	Calculates the truncated value of X, ignoring the decimal point.

where the *name* designates the particular intrinsic function (Table 16.2) and the *argument* is the value that is to be evaluated. The argument can be a constant, a variable, another function, or an algebraic expression.

There are a great variety of built-in functions available for FORTRAN 77. In the present chapter, we will discuss the more commonly available functions listed in Table 16.2. Additional functions including those to manipulate complex variables are compiled in Appendix G.

16.2.1   General Mathematical Functions: SQRT, ABS, ALOG10, ALOG, and EXP

The first group of functions deals with some standard mathematical operations that have broad application in engineering.

The square root function (SQRT).   To this point we have learned one way to determine the square root of a number by raising the number to the 0.5 power, as in

```
Y = X**(0.5)
```

After this statement is executed, Y will have the value of the square root of X. Because this is such a common operation, the built-in function named SQRT performs the same task. An example is

```
Y = SQRT(X)
```

The absolute value function (ABS).   There are many situations where you will be interested in the value of a function but not its sign. The ABS function,

```
Y = ABS(X)
```

provides the value of X without any sign. If X is greater than or equal to 0, ABS(X) will equal X. If it is less than 0, ABS(X) will equal X times -1.

EXAMPLE 16.5   Using SQRT and ABS Functions to Determine the Complex Roots of a Quadratic Equation

PROBLEM STATEMENT: Given the quadratic equation

$$f(x) = ax^2 + bx + c$$

the roots may be determined by the formula

$$x = \frac{-b \pm \sqrt{b^2 - 4ac}}{2a}$$

If $b^2 - 4ac \geq 0$, this equation can be used directly. However, if $b^2 - 4ac < 0$, then the roots are complex and a revised solution is

$$x = \text{Re} \pm \text{Im} \; i$$

where Re is the real part of the root, which can be calculated as

$$\text{Re} = \frac{-b}{2a}$$

Im is the imaginary part, which is calculated as

$$\text{Im} = \frac{\sqrt{|b^2 - 4ac|}}{2a}$$

and $i = \sqrt{-1}$. Use the ABS and SQRT functions to determine the complex roots for the case where $b^2 - 4ac < 0$.

SOLUTION: The code to determine the complex roots of a quadratic equation is listed in Fig. 16.1. Notice how the ABS function is located within the argument of the SQRT function.

FIGURE 16.1
A FORTRAN program to determine the complex roots of a quadratic equation.

```
REAL IMAGIN
READ(5,*) A,B,C
REAL = -B/(2*A)
IMAGIN = SQRT(ABS(B**2 - 4*A*C))/(2*A)
WRITE(6,*) REAL, IMAGIN
STOP
END
```

```
** Output Results **

-0.500000000      1.50000000
```

The common logarithm function (ALOG10).   Logarithms have numerous applications in engineering. Recall that the general form of the logarithm is

$$y = \log_a x$$

where a is called the *base* of the logarithm. According to this equation, y is the power to which the base must be raised to give x. The most familiar and easy-to-understand form is the *base 10*, or *common, logarithm*, which corresponds to the power to which 10 must be raised to give x; for example,

$$\log_{10} 10,000 = 4$$

because 10^4 is equal to 10,000.

Common logarithms define powers of 10. For example, common logs from 1 to 2 correspond to values from 10 to 100. Common logs from 2 to 3 correspond to values from 100 to 1000 and so on. Each of these intervals is called an *order of magnitude*. The common logarithm is computed in FORTRAN as

```
Y = ALOG10(X)
```

The natural logarithm function (ALOG).   Besides the base 10 logarithm, the other important form in engineering is the *base e logarithm* where $e = 2.71828. \ldots$. The base e logarithm, which is also called the *natural logarithm*, is extremely important in calculus and other areas of engineering mathematics. For this reason, FORTRAN includes an ALOG function to compute the natural logarithm, as in

```
Y = ALOG(X)
```

The exponential function (EXP).   An additional mathematical function that has widespread engineering applications is the exponential function, or e^x, where e is the base of the natural logarithm. In FORTRAN, it is computed as

```
Y = EXP(X)
```

16.2.2   Trigonometric Functions

Another set of mathematical functions that have great relevance to engineering is the trigonometric functions. In FORTRAN the sine (SIN), cosine (COS), and tangent (TAN) functions are commonly available. As outlined in Fig. 16.2, other trigonometric functions can be derived by inverting these three. In addition, the inverse trigonometric functions are available on most systems. For example, if the value of the tangent is known, the arctangent (ATAN) can be used to determine the corresponding angle. Finally, functions are available to compute the hyperbolic sine (SINH), cosine (COSH), and tangent (TANH).

For all of these functions, the angle is expressed in radians (Fig. 16.3). A *radian* is the measure of an angle at the center of a circle that corresponds to an arc that is equal to the circle's radius (Fig. 16.3*b*). The conversion from degrees to radians is (Fig. 16.3*c*)

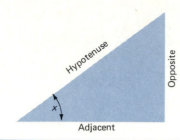

$$\sin x = \frac{\text{opposite}}{\text{hypotenuse}} \qquad \csc x = \frac{1}{\sin x}$$

$$\cos x = \frac{\text{adjacent}}{\text{hypotenuse}} \qquad \sec x = \frac{1}{\cos x}$$

$$\tan x = \frac{\text{opposite}}{\text{adjacent}} \qquad \cot x = \frac{1}{\tan x}$$

FIGURE 16.2
Definitions of the trigonometric functions.

FIGURE 16.3
Relationship of radians to degrees: (a) the angle x of a right triangle placed in standard position at the center of a circle; (b) the definition of a radian; (c) π radians = 180°.

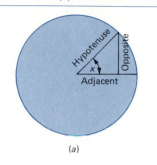

(a)

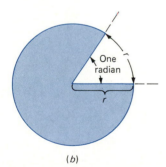

(b)

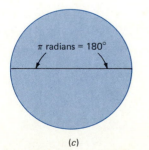

(c)

π radians $= 180°$

Therefore, if X is expressed originally in degrees, the sine of X can be computed in FORTRAN as

```
Y = SIN(X * 3.14159/180)
```

Aside from the standard trigonometric functions, other trigonometric functions that can be evaluated on the basis of intrinsic functions are listed in Table 16.3.

16.2.3   Miscellaneous Functions: INT, IFIX, and RND

Aside from general mathematical and trigonometric functions, there are other built-in functions that are usually available and that have utility to engineers.

Functions to determine integer values (INT and IFIX).   The *integer function* gives the greatest integer (that is, whole number) that is less than or equal to the argument. Because of this definition, it is sometimes referred to as the *greatest integer function*. For example,

```
X = 3.9
Y = INT(X)
```

will give Y equal to 3. Also, it is important to realize that

```
X = -2.7
Y = INT(X)
```

will give Y equal to −3, because it is the greatest integer that is less than −2.7.

The *IFIX function* discards the decimal part of a number and returns the integer part. For positive numbers, this gives the same result as does the INT function. For example,

```
X = 3.9
Y = IFIX(X)
```

TABLE 16.3 Derived Trigonometric Functions that Can Be Developed from the Intrinsic Functions of FORTRAN

Function	Equivalent*
Secant	1/COS(X)
Cosecant	1/SIN(X)
Cotangent	1/TAN(X)
Inverse secant	ATAN(SQRT(X * X − 1)) + (SGN(X) − 1) * 1.5708
Inverse cosecant	ATAN(1/SQRT(X * X − 1)) + (SGN(X) − 1) * 1.5708
Inverse cotangent	−ATAN(X) + 1.5708
Hyperbolic secant	1/COSH(X)
Hyperbolic cosecant	1/SINH(X)
Hyperbolic cotangent	1/TANH(X)
Inverse hyperbolic sine	LOG(X + SQRT(X * X + 1))
Inverse hyperbolic cosine	LOG(X + SQRT(X * X − 1))
Inverse hyperbolic tangent	LOG((1 + X)/(1 − X))/2
Inverse hyperbolic secant	LOG((SQRT(−X * X + 1)+1)/X)
Inverse hyperbolic cosecant	LOG((SGN(X) * SQRT(X * X + 1)+1)/X)
Inverse hyperbolic cotangent	LOG((X + 1)/(X − 1))/2

*SGN(X) = ABS(X)/X for X ≠ 0.

will give Y equal to 3. However, in contrast to the INT function,

```
X = -2.7
Y = IFIX(X)
```

will give Y equal to −2, because the IFIX simply truncates or chops off the 0.7 to attain its result.

EXAMPLE 16.6   Using the INT function to Truncate or Round Positive Numbers

PROBLEM STATEMENT: It is often necessary to round numbers to a specified number of decimal places. One very practical situation occurs when performing calculations pertaining to money. Whereas an economic formula might yield a result of $467.9587, in reality we know that such a result must be rounded to the nearest cent. According to rounding rules, because the first discarded digit is greater than 5 (that is, it is 8), the last retained digit would be rounded up, giving a result of $467.96 (Fig. 16.4). The INT function provides a way to accomplish this same manipulation with the computer.

FIGURE 16.4
Illustration of the retained and discarded digits for a number rounded to two decimal places. The result after rounding is 467.96.

Last retained digit   First discarded digit

467.95 87

Retained digits   Discarded digits

SOLUTION: Before developing the rounding formula, we will first derive a formula to truncate, or "chop off," decimal places. First, we can multiply the value by 10 raised to the number of decimal places we desire to retain. For example, in the present situation we desire to retain two decimal places, and therefore we multiply the value by 10^2, or 100, as in

$$467.9587 \times 100 = 46795.87 \qquad \text{(E16.6.1)}$$

Next we can apply the INT function to this result to chop off the decimal portion:

```
INT(46795.87) = 46795
```

Finally we can divide this result by 100 to give

$$46795/100 = 467.95$$

Thus we have retained two decimal places and truncated the rest. The entire process can be represented by the FORTRAN statement

```
X = INT(X * 10**D)/10**D                                              (E16.6.2)
```

where X is the number to be truncated and D is the number of decimal places to be retained.

Now it is a very simple matter to modify this relationship slightly so that it will round rather than truncate. This can be done by looking more closely at the effect of the multiplication in Eq. (E16.6.2) on the last retained and the first discarded digit in the number being rounded (Fig. 16.4). Multiplying the unknown by 10 ** D essentially places the last retained digit in the unit position and the first discarded digit in the first decimal place. For example, as shown in Eq. (E16.6.1), the multiplication moves the decimal point between the last retained digit (5) and the first discarded digit (8). Now it is simple to see that if the first discarded digit is above 5, adding 0.5 to the quantity will increase the last retained digit by 1. For example, for the result of Eq. (E16.6.1)

$$46795.87 + 0.5 = 46796.37 \qquad \text{(E16.6.3)}$$

If the first discarded digit were less than 5, adding 0.5 would have no effect on the last retained digit. For instance, suppose the number were 46795.49. Adding 0.5 yields 46795.99, and therefore the last retained digit (5) is not changed.

To complete the process, the INT function is applied to Eq. (E16.6.3)

```
INT(46796.37) = 46796
```

and the result divided by 100 to give the correct answer, 467.96. The entire process is represented by

```
X = INT(X * 10**D + 0.5)/10**D                    (E16.6.4)
```

It should be noted that numbers can be rounded to an integer value by setting D = 0, as in

```
X = INT(467.9587 * 10**0 + 0.5)/10**0 = 468
```

or to the tens, hundreds, or higher places by setting D to negative numbers. For example, to round to the 10s place, D = −1 and Eq. (E16.6.4) becomes

```
X = INT(467.9587 * 10**(-1) + 0.5)/10**(-1) = 470
```

Notice that we did not round negative numbers in Example 16.6. Although Eq. (E16.6.4) works properly for most negative numbers, there is one special case where it yields an erroneous result. If the first discarded digit is a 5 and all other discarded digits are zeros, Eq. (E16.6.4) rounds negative numbers improperly. For example, rounding −467.95 to one decimal place gives

```
X = INT(-467.95 * 10**1 + 0.5)/10**1 = -467.9
```

Thus, instead of rounding to the correct answer of −468.0, it rounds to −467.9. Therefore a slight adjustment of the equation using the IFIX function and a relationship to determine the sign can be made to correct this problem:

```
X = IFIX(X * 10**D + ABS(X)/X * 0.5)/10**D
```

where ABS(X)/X will be −1 if X is negative and +1 if X is positive. For the example

```
X = IFIX(-467.95*10**1 + ABS(-467.95)/(-467.95)*0.5)/10**1
```

which when evaluated yields the correct result of −468.0. Thus this equation is superior to Eq. (E16.6.4) in that it adequately rounds both negative and positive numbers.

The random-number function (RND).* In later chapters, we will introduce you to the subjects of statistics and simulation. One important aspect of these subjects is the random number. To understand what this means, suppose that you toss an honest coin over and over again and record whether it comes up heads (H) or tails (T). Your result

---

*Because the random-number generator is not universally available on FORTRAN compilers, it is not listed in Table 16.2. However, it is of sufficient interest that we describe the function in detail here. If your system does not have a random-number generator, Appendix I includes a short program that you can use for this purpose.

for 20 tosses might look like

HTTHHTHHHHTHTTTHTHTT

This is called a *randomly generated sequence* because each letter represents the result of an independent toss of the coin. If we repeated the experiment, we would undoubtedly come up with a different sequence because for an honest coin the outcome of each toss is random. That is, there is an equal likelihood that each toss will yield either a head or a tail, and there is no way to predict the outcome of an individual toss beforehand.

The same sort of experiment could be performed with a die (that is, one of a pair of dice). For this case, every toss would result in either 1, 2, 3, 4, 5, or 6, and the sequence of 20 tosses might look like

53122341626444153566

Random-number sequences are very important in engineering because many of the systems that you will study have aspects that vary in a random fashion. We will investigate some of these processes in depth in later chapters. For the time being, it is useful to know that some FORTRAN compilers have a random-number function. One common version is represented as

variable = RND(*seed*)

where the *seed* is typically a number greater than zero that is used to start the random-number generation. Every time this function is implemented, a random decimal fraction somewhere between 0 and 1 will be returned.

EXAMPLE 16.7   *Simulating a Random Coin Toss with the Computer*

PROBLEM STATEMENT: Use the RND function to simulate the random toss of a coin.

SOLUTION: The following short program can be used to simulate the random toss of a coin:

```
X = RND(1000)
Y = INT(X * 10**0 + 0.5)/10**0
WRITE(6,*)X,Y
STOP
END
```

The first line generates a random number between 0 and 1. The second line rounds the number to an integer. If X is less than 0.5, the rounded value will be 0, and we will consider the toss a tail. If X is greater than or equal to 0.5, the rounded value will be 1, and we will consider the toss a head. A run yields

```
RUN
   0.9794235        1
```

Therefore a head has been tossed. To obtain additional values, the statements can be repeated. For example, for three tosses a simple program would be

```
X = RND(1000)
Y = INT(X * 10 ** 0 + 0.5)/10 ** 0
WRITE(6,*) X,Y
X = RND(1000)
Y = INT(X * 10 ** 0 + 0.5)/10 ** 0
WRITE(6,*) X,Y
X = RND(1000)
Y = INT(X * 10 ** 0 + 0.5)/10 ** 0
WRITE(6,*) X,Y
STOP
END
RUN
   0.9794235        1
   0.8679581        1
   0.3929367        0
```

Thus for this case the second and third tosses come out heads and tails, respectively.

Every time the program at the end of Example 16.7 is run on a particular computer, the same series of random numbers will be generated. This is because the computer generates the random number on the basis of a starting value called a seed. Unless the seed is changed, a particular computer will generate the same sequence of random numbers. We will return to the subject of random-number generation when we discuss simulation in Chap. 25.

16.3　STATEMENT FUNCTIONS

Aside from built-in functions, you might be interested in defining functions of your own. The *statement function* is used for this purpose and has the general form

$$variable\,(\,argument\,) \; = \; expression$$

where *variable* is the name of the function (any valid variable name), *argument* is the value at which the function is to be evaluated, and *expression* is the actual arithmetic

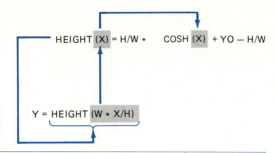

FIGURE 16.5
An example of a statement
function.

expression. For example, in the simple addition program we could have defined a function for the addition of A and B as

```
ADD(X,Y) = X + Y
```

Then the addition could have been rewritten as

```
TOTAL = ADD(FIRST,SECOND)
```

In FORTRAN several arguments may be passed into a statement function. Other variables in the statement have the same value as they had in the main program at the point the statement function was invoked (Fig. 16.5). Statement functions must always be placed prior to the first executable statement in a program.

EXAMPLE 16.8   Statement Functions

PROBLEM STATEMENT: The height of a hanging cable is given by the formula (Fig. 16.6)

$$y = \frac{H}{w} \cosh\left(\frac{w}{H} x\right) + y_0 - \frac{H}{w}$$

where H is the horizontal force at $x = a$, y_0 is the height at $x = a$, and w is the cable's weight per unit length. Compose a computer program to calculate the height of a cable at equally spaced values of x from 0 to a maximum x, given w and H. Employ a statement function to compute the height. Test the program for the following case: $w = 10$, $H = 2000$, and $y_0 = 10$ for $x = 0$ to 30. Print out values of y at 3 equally spaced increments from $x = 0$ to 30.

SOLUTION: Figure 16.7 shows a program to compute the height of a hanging cable. Notice how a statement function has been set up to calculate the height. The Y assignment statements access this function to compute the cable height. Notice how the argument of HEIGHT in these lines is an arithmetic expression W * X/H that is passed to the statement function. There its value is ascribed to the dummy variable X, which is used to evaluate the height. This result is then passed back and substituted

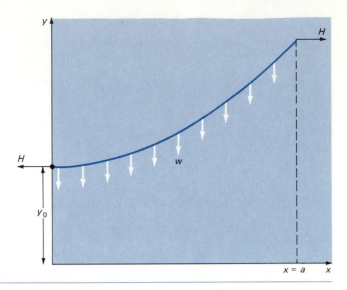

FIGURE 16.6
A section of a hanging cable.

for HEIGHT (W * X/H). X is called a *dummy variable* because its use as the argument in the statement function does not affect its value in the rest of the program. In fact, in the present example, it has a value of W * X/H.

FIGURE 16.7 A program to determine the height of a hanging cable.

```
C       ***************************************************
C       *       PROGRAM TO COMPUTE THE HEIGHT OF A        *
C       *                 HANGING CABLE                   *
C       *               (FORTRAN version)                 *
C       *              by   S. C. Chapra                  *
C       *                                                 *
C       *       This program computes the height of a     *
C       *       hanging cable at four equally-spaced      *
C       *       points along the X-dimension.             *
C       ***************************************************
C
C       .-------------------------------------------------.
C       |                                                 |
C       |   DEFINITION OF VARIABLES                       |
C       |                                                 |
C       |   W     = WEIGHT PER LENGTH                     |
C       |   H     = HORIZONTAL FORCE                      |
C       |   Y0    = INITIAL HEIGHT                        |
C       |   XMAX  = MAXIMUM LENGTH                        |
C       |   XINC  = INCREMENT OF X WHERE Y COMPUTED       |
C       |   HEIGHT = FUNCTION TO COMPUTE CABLE HEIGHT     |
C       '-------------------------------------------------'
C
        HEIGHT(X) = H/W*COSH(X) + Y0 - H/W
C
C                   Initialzation of Variables
C
        X = 0
```

```
C
C       ******************  MAIN PROGRAM  ******************
C
        READ(5,*) W,H,Y0,XMAX
        XINC = XMAX/4.
        WRITE(6,*) 'DISTANCE          HEIGHT'
        WRITE(6,*)
C                       Print Initial Values
C
        WRITE(6,*) X,Y0
C
C              Calculate and Print Remaining Values
C
        X = X + XINC
        ARG = W*X/H
        Y = HEIGHT(ARG)
        WRITE(6,*) X,Y
        X = X + XINC
        Y = HEIGHT(W*X/H)
        WRITE(6,*) X,Y
        X = X + XINC
        Y = HEIGHT(W*X/H)
        WRITE(6,*) X,Y
        STOP
        END
```

```
** Output Results **

DISTANCE              HEIGHT

   0.0000000        10.0000000
  10.0000000        10.2500458
  20.0000000        11.0007782
  30.0000000        12.2541046
```

Although the foregoing example shows how a user-defined function is employed, it does not really illustrate its true value. This is seen when the function is extremely long and complicated and is used numerous times throughout the program. In these instances, the concise representation of the user-defined function is a potent asset.

PROBLEMS

Which of the following are valid assignment statements in FORTRAN? State the errors for those that are incorrect.

16.1 R1 = −4 ** 1.5

16.2 C = C + 22

16.3 A + B = C + D

16.4 X = Y = 7

16.5 4 = X + Y

If R = 3., S = 5., and T = −2., evaluate the following. Perform the evaluation step by step, following the priority order of the individual operations.

16.6 −S * R + T

16.7 S + R/S * T

16.8 S + R ** T * S ** T

16.9 R * (R + (R + 1) + 1

16.10 T + 10 * S − S ** R ** R

Write FORTRAN statements for the following algebraic equations.

16.11

$$x = a \div c + \frac{c - d^2}{g} \frac{r^5}{2f^2 - 4}$$

16.12

$$r = \frac{cb^3}{r^2} (3d + 8x) - \frac{4}{9} x$$

16.13 In civil engineering the vertical stress under the center of a loaded circular area is given by

$$s = q \left[1 - \frac{1}{(r^2/z^2 + 1)^{3/2}} \right]$$

Write this equation in FORTRAN.

16.14 In engineering economics the annual payment A on a debt that increases linearly with time at a rate G for n years is given by the equation

$$A = G \left[\frac{1}{n} - \frac{n}{(1 + i)^n - 1} \right]$$

Write this equation in FORTRAN. *Hint*: Be careful with your variable names to avoid some of the problems associated with mixed-mode arithmetic.

16.15 In chemical engineering the residence time of five equal-sized reactors in series is given by

$$t = \frac{5}{k} \left[\left(\frac{c_0}{c_5} \right)^{1/5} - 1 \right]$$

Write this equation in FORTRAN.

16.16 In fluid mechanics the pressure of a gas at the stagnation point is given by

$$p = p_0 \left(1 + m^2 \frac{k - 1}{2} \right)^{k/(k-1)}$$

Write this equation in FORTRAN.

16.17 In electrical engineering the electric field at a distance x from the center of a ring-shaped conductor is given by

$$E = \frac{1}{4\pi\epsilon_0} \frac{Qx}{(x^2 + a^2)^{3/2}}$$

Write this equation in FORTRAN.

16.18 In chemical engineering van der Waals equation gives pressure of a gas as

$$p = \frac{RT}{(V/n) - b} - a \left(\frac{n}{V} \right)^2$$

Write this equation in FORTRAN.

16.19 Evaluate the following: (*a*) X = 6/9 − 4; (*b*) J = 5./2 − 4.6; (*c*) I = 5/2 − 4.6; (*d*) Y = 6/9 − 4.0.

16.20 The saturation value of dissolved oxygen, c_s, in fresh water is calculated by the equation

$$c_s = 14.652 - (4.1022 \times 10^{-1})T_C + (7.9910 \times 10^{-3})T_C^2 \\ - (7.7774 \times 10^{-5})T_C^3$$

where T_C is temperature in degrees Celsius. Saltwater values can be approximated by multiplying the freshwater result by $1 - (9 \times 10^{-6})n$, where n is salinity in milligrams per liter. Finally, temperature in degrees Celsius is related to temperature in degrees Fahrenheit, T_F, by

$$T_C = \frac{5}{9} (T_F - 32)$$

Develop a computer program to compute c_s, given n and T_F. Test the program for $n = 18,000$ and $T_F = 70$.

16.21 A projectile is fired with a velocity v_0, at an angle of θ with the horizontal, off the top of a cliff of height y_0 (Fig. P16.21). The following formulas can be used to compute how far horizontally (x) the projectile will carry prior to impacting on the plane below.

$$(v_y)_0 = v_0 \sin \theta$$

$$(v_x)_0 = v_0 \cos \theta$$

$$t = \frac{-(v_y)_0 - \sqrt{(v_y)_0^2 - 2gy_0}}{g}$$

$$x = (v_x)_0 t$$

where $(v_y)_0$ and $(v_x)_0$ are the vertical and horizontal components of the initial velocity, t is the time of flight, and g is the gravitational constant (or -9.81 m/s^{-2}), where the negative sign designates that we are assigning a negative value to downward acceleration). Prepare a program to compute x, given v_0, θ, and y_0. If $v_0 = 180$ m/s, $\theta = 30°$, and $y_0 = 150$ m, x will be equal to 3100 m. Use this result to test your program. Then employ the program to determine what value of θ results in the greatest value of x. Use trial and error to determine θ.

FIGURE P16.21

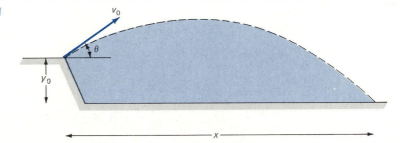

16.22 Evaluate the following expressions. Note which ones involve mixed-mode arithmetic:

(a) 3./2 * 2   (c) 10 ** 3/2   (e) 8 ** (2/10)
(b) 30/9 * 9   (d) 30/(10 * 1.5)   (f) (2 * 4)/10/2.

16.23 In FORTRAN the intrinsic logarithm function is used to evaluate exponentiation if the power is expressed as a real number. For instance, if X = 1.5 ** 2.5, the computer will take the logarithm of 1.5 and multiply the result by 2.5.

In contrast, if an integer value is employed for the exponent, the computer calculates the expression by multiplying the value by itself the number of times represented by the value of the exponent. Therefore $Y = X ** 4$ will be evaluated by the computer as $Y = X * X * X * X$. Given this information, describe how the following expressions will be evaluated, and give the numerical result for Y:

(*a*) Y = 10.5 ** 2   (*c*) Y = SQRT(16.) ** 1/2

(*b*) Y = ALOG(17.5) ** 3.   (*d*) Y = 2.5 * (3.5) ** (−2)

16.24 Given X and Y, the coordinates of a point on the circumference, write a statement function that calculates the radius of a circle centered at the origin. Use Fig. 16.3*a* to develop the proper formula.

CHAPTER 17

FORTRAN: INPUT/OUTPUT

Because it represents the interface between the user and the program, input/output is a critical aspect of software design. You have been introduced to two statements that are employed for this purpose: the READ and the WRITE statements. In Chap. 15 you were shown how these statements can be used in conjunction with string constants to devise descriptive input and output. Now we turn to some additional statements and concepts that will extend your capabilities in this area.

17.1 TABULAR OUTPUT WITH FREE FORMAT

There are many cases in engineering where you would want to control the format whereby information is displayed on the screen or on printed output. This is particularly important when the program is to be used for report generation.

The simplest way to produce tabular output of data is to use the *free-format* or *list-directed* WRITE statement, with the variables separated by commas. The output device, such as the monitor screen or printer, can then be thought of as consisting of vertical columns, or print zones, that are an equal number of characters wide (Fig. 17.1). The first numeric value in a WRITE statement is placed at the far right of the first print zone, and subsequent values, delimited by commas, are placed sequentially at the far right of the subsequent zones. When numbers are printed in this way they are said to be *right-justified*. String values are handled differently. They are merely printed end to end, starting at the left side of the screen or printed page.

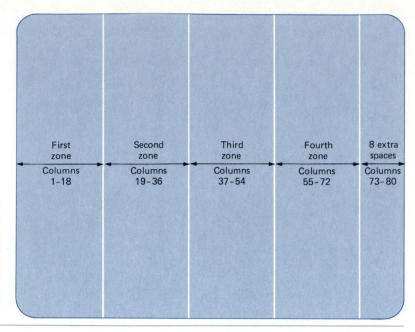

First zone	Second zone	Third zone	Fourth zone	8 extra spaces
Columns 1–18	Columns 19–36	Columns 37–54	Columns 55–72	Columns 73–80

FIGURE 17.1
When values are delimited with commas in a WRITE statement, the output device can be thought of as consisting of equal-width print zones. For the above case, the zones are 18 characters wide. Other systems use somewhat different widths.

EXAMPLE 17.1   Outputting a Table with the Free-Format or List-Directed WRITE

PROBLEM STATEMENT: The students in a class have obtained the following grades:

Student Name	Average Quiz	Average Homework	Final Exam
Smith, Jane	82.00	83.90	100
Jones, Fred	96.55	92.00	77

Their final grade (FG) is computed with the following weighted average:

```
FG = AQ/3 + AH/3 + FE/3
```

where AQ, AH, and FE are the average quiz, average homework, and final exam grades, respectively. Develop a computer program to determine the final grade for each student. Print out the students' grades in a tabular format as shown above, with the final grade added as the last column.

SOLUTION: The computer code to determine the final grade and to output the table is shown in Fig. 17.2. The results illustrate some of the idiosyncrasies of free-format output. First, the character strings used for the title of the table are all jammed together. Second, although the numbers are output in nice, orderly 18-character zones, the fact that each line of numbers exceeds the total width of the output device (in this case,

```
CHARACTER*14 NAME1,NAME2
READ(5,*) NAME1,AQ1,AH1,FE1
READ(5,*) NAME2,AQ2,AH2,FE2
FG1 = AQ1/3 + AH1/3 + FE1/3
FG2 = AQ2/3 + AH2/3 + FE2/3
WRITE(6,*) 'STUDENT','AVERAGE','AVERAGE','FIN','FIN'
WRITE(6,*) 'NAME','QUIZ','HOMEWORK','EXAM','GRADE'
WRITE(6,*)
WRITE(6,*) NAME1,AQ1,AH1,FE1,FG1
WRITE(6,*) NAME2,AQ2,AH2,FE2,FG2
STOP
END
```

FIGURE 17.2
The computer code to determine the final grade and output the results in tabular form for Example 17.1. Because these results were output on an 80-column printer, each line of numbers extends, or wraps around, to the next line.

```
** Output Results **

STUDENTAVERAGEAVERAGEFINFIN
NAMEQUIZHOMEWORKEXAMGRADE

SMITH, JANE                 82.0000000        83.8999939        100.000000
              88.6333300
JONES, FRED                 96.5500000        92.0000000        77.000000
              88.5166700
```

an 80-column screen) means that the last numbers extend to the next line. This tendency to "wrap around" means that for this case we are limited to printing an 18-character name and three zones of numerical information.

If the output device were a 132-column printer, we would be able to output up to seven 18-character-width zones. For this case, the four zones of numbers in the present example would fit nicely. A program to do this is shown in Fig. 17.3, along with the resulting output. Notice that an attempt has been made to correct the table heading by entering the string constants used for this purpose as 18-character strings. Unfortunately, doing this with assignment statements results in the headings being *left-justified*—that is, they are placed at the far left of the field. Therefore, although the headings and the numbers are each aligned nicely, they are not aligned with each other. One way to correct this is to rewrite the WRITE statements for the headings as

```
WRITE (6,*)'          ',HEADT1,HEADT2,HEADT3,HEADT4,HEADT5
WRITE (6,*)'          ',HEADB1,HEADB2,HEADB3,HEADB4,HEADB5
```

By interjecting a character string of 10 blanks at the beginning of each line, the heading is shifted over so that it is centered above the numbers.

17.2 FORMATTED INPUT AND OUTPUT

The free format of FORTRAN for input and output is often quite satisfactory. Free-format input is especially useful for programs that use the keyboard for data entry. Free format, with entries separated by commas, is probably the manner that would

```
CHARACTER*18 NAME1,NAME2
CHARACTER*18 HEADT1,HEADT2,HEADT3,HEADT4,HEADT5
CHARACTER*18 HEADB1,HEADB2,HEADB3,HEADB4,HEADB5
HEADT1 = 'STUDENT'
HEADT2 = 'AVERAGE'
HEADT3 = 'AVERAGE'
HEADT4 = 'FINAL'
HEADT5 = 'FINAL'
HEADB1 = 'NAME'
HEADB2 = 'QUIZ'
HEADB3 = 'HOMEWORK'
HEADB4 = 'EXAM'
HEADB5 = 'GRADE'
READ(5,*) NAME1,AQ1,AH1,FE1
READ(5,*) NAME2,AQ2,AH2,FE2
FG1 = AQ1/3 + AH1/3 + FE1/3
FG2 = AQ2/3 + AH2/3 + FE2/3
WRITE(6,*) HEADT1,HEADT2,HEADT3,HEADT4,HEADT5
WRITE(6,*) HEADB1,HEADB2,HEADB3,HEADB4,HEADB5
WRITE(6,*)
WRITE(6,*) NAME1,AQ1,AH1,FE1,FG1
WRITE(6,*) NAME2,AQ2,AH2,FE2,FG2
STOP
END
```

FIGURE 17.3
The computer code to determine the final grade and output the results in tabular form for Example 17.1. These results were output on a 132-column printer. The headings and the columns of numbers are not aligned because the headings are left-justified and the numbers are right-justified.

STUDENT NAME	AVERAGE QUIZ	AVERAGE HOMEWORK	FINAL EXAM	FINAL GRADE
SMITH,JANE	82.0000000	83.8999939	100.000000	88.6333160
JONES,FRED	96.5500031	92.0000000	77.0000000	88.5166473

be employed by most individuals in the absence of any other guidelines. In addition, free formatting can also be used when simple tables consisting of a few columns are to be output. However, for larger and more complex tables, additional control of the output is required. The FORMAT statement of FORTRAN provides such control.

The FORMAT statement specifies an explicit form for inputting or outputting data. The form of numeric real data is specified as to the number of digits, the number of decimal places, location of data on a line, and whether scientific notation is to be used. Integer and character data is handled in a somewhat simpler fashion because no decimal or scientific notation is possible. Therefore only the number and position of digits or characters need be specified. All this information is included in the FORMAT statement. Each READ or WRITE statement in the program references one of these FORMAT statements. In this way, the input and output are under rigid control.

The reasons for formatting output should be obvious; formatted output allows us to control the way that information is displayed or printed. However, the formatting of input may not seem justified. In fact, when data is entered via a keyboard, free format is superior to formatted input. As stated at the beginning of this section, free-formatted entry with commas is probably the way most people would naturally enter data at a keyboard. In contrast, a FORMAT statement would require that numbers begin at specific columns on a line. Trying to keep track of such spacing on a monitor screen is very inconvenient. Thus inputting formatted data on a keyboard is usually inefficient. Formatting requirements for input made sense when data was entered on punch cards that had clearly marked columns. In addition, formatted entry of data was employed to put as much information on a punched card as possible. Free format wastes a certain amount of space for the commas. By specifying exactly where one number begins and another ends, a FORMAT statement can cause numbers to be placed end to end with no space in between.

But formatted input, which had validity when punch cards were prevalent, has less merit in today's primarily keyboard environment. Consequently, although formatted input is treated in the following sections, our emphasis will be on output.

17.3 THE FORMAT STATEMENT

Recall that the general form of the READ statement is (Sec. 15.7)

READ *(unit number, format number) var$_1$, ..., var$_n$*

and that the general form of the WRITE statement is

WRITE *(unit number, format number) exp$_1$, ..., exp$_n$*

To this point, you have learned that an asterisk for the format number causes data to be input or output in free format.

The FORMAT statement provides an alternative to the asterisk. Its general form is

format number FORMAT *(format code$_1$, ..., format code$_n$)*

where the *format number* is a unique, positive whole number that is employed to reference the FORMAT statement to a particular READ or WRITE statement. For example, as seen in Fig. 17.4, the FORMAT statement provides explicit instructions as to the manner in which each variable is to be output.

The format number must be located in the first five columns of a line. It serves to link each READ or WRITE with its corresponding FORMAT statement. Note that several READ or WRITE statements may access the same FORMAT statement.

The *format codes* supply a symbolic description of the form of the data. As seen in Fig. 17.4, they always follow the word FORMAT and are enclosed in parentheses. Each format code is separated from the next by a comma. The codes themselves have several forms. Some examples are

```
10 FORMAT (10X,2I6,T40,E20.3)
20 FORMAT (5X,F5.2)
30 FORMAT (F7.0,F8.3)
40 FORMAT (A30)
50 FORMAT (E16.8)
```

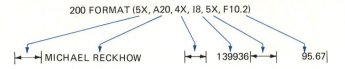

FIGURE 17.4
An example of how a FORMAT statement provides the information to guide the output of data by a WRITE statement. Each variable has a corresponding format code that specifies how the variable is to be output.

In the above, the X and T are used to skip over column spaces in the line, while the rest of the letters specify how the numbers or characters are to be displayed. For example, the letter X skips spaces on the output device, whereas I6 tells the computer that the data is an integer (I) that can be up to six spaces long. Several of the codes have numbers in front of them. This merely tells the computer how many times to repeat the code. For example, 2I6 tells the computer that there will be two integers that are each six spaces long. Thus

```
80 FORMAT (I6,I6)
```

and

```
80 FORMAT (2I6)
```

are equivalent. If a FORMAT statement has more format codes than are needed by a READ or WRITE statement, then the extra codes are ignored. For example, for

```
     WRITE (6,100) NUM1,NUM2
100 FORMAT (3I8)
```

because there are only two variables to be written, the third I8 is unnecessary and is ignored. The statement 100 FORMAT(2I8) would work in an identical fashion for this case.

FORMAT statements are not executable FORTRAN statements. That is, the compiler makes use of them to obtain information regarding how READ and WRITE statements are to be implemented, but they are not executed by the computer when the program is run. As a consequence, they can be placed anywhere in the program prior to the END statement. Logical locations are immediately after the input/output (or I/O) statement that uses them or at the beginning or end of the module where they are used. The convention employed here will be to place FORMAT statements that are used once next to the I/O statement that references them. If a FORMAT statement is employed by more than one I/O statement, then it will be placed at the end of the major module where it occurs. Because FORMAT statements must precede the END statement, we will place them between the STOP and the END statements at the end of the program. This makes them easy to locate.

Now that you have assimilated some preliminary information on FORMAT statements, we can look in detail at the format codes that are at the heart of their operation. The codes can be broken into different categories: (1) codes to control spacing, (2) codes for character data, (3) codes for integers, and (4) codes for real numbers.

17.3.1 FORMAT Codes to Control Spacing

The codes to control spacing deal with two types of spacing: between lines and between characters. Spacing between lines for a WRITE statement can be determined by placing a special control code in the first column on a line. This first character, called a *carriage-control character,* is not actually printed but instead controls the line spacing. The various codes are listed in Table 17.1. Note that if the first character is not one of the five given in Table 17.1, the results will be unpredictable—that is, they are system-dependent. Consequently, a carriage-control character should start each format specification that relates to printed output.

For example, if we desire to skip a line and print the message ENTER MORE DATA, the following WRITE and FORMAT statements would be used:

```
    WRITE (6,1)
    WRITE (6,2)
  1 FORMAT (' ')
  2 FORMAT (' ENTER MORE DATA')
```

which causes a line to be skipped and the message printed. Because the zero is in column 1, it is not printed. A single WRITE and FORMAT could also be used to perform the same action:

```
    WRITE (6,1)
  1 FORMAT ('ØENTER MORE DATA')
```

TABLE 17.1 First Column FORMAT Codes to Control Spaces Between Lines

First Character	Effect	Spacing
' '	Advances one line and then prints	Single
'+'	No line feed before printing	Overstrike
'0'	Skips to second line and then prints	Double
'−'	Skips to third line and then prints	Triple
'1'	Skips to top of next page and then prints	New page

In addition to carriage-control characters, the slash (/) can also be used to control line spacing. When a slash is encountered, it directs the computer to terminate output on the current line. How the slash affects the output depends upon whether the output is via line printer or some other device. For a line printer, the code

```
     WRITE(6,800)
 800 FORMAT(' GET',/,' DOWN')
```

results in the output

```
 GET
 DOWN
```

Note that, in contrast to other FORMAT codes, the commas before and after the slash are optional. Thus the FORMAT statement above could also be written as

```
 800 FORMAT(' GET'/' DOWN')
```

The spaces before the GET and the DOWN in FORMAT statement 800 are carriage-control characters that are required if a line printer is being utilized. They are needed because the slash simply directs the computer to terminate printing on the current line and does not specify that output is to be continued on the next line. In contrast, if the output is not to a line printer, the carriage-control characters are not required. In that case, the following FORMAT statement results in the same printout:

```
 800 FORMAT('GET'/'DOWN')
```

Consecutive slashes can be employed to skip several lines, but their effect is dependent on their position in the line. If n consecutive slashes appear at the beginning or end of a FORMAT statement, n lines will be skipped. However, if they are in the interior of the statement, $n - 1$ lines will be skipped.

Spacing between characters is determined by special codes, as outlined in Table 17.2. Remember that the first column is not printed, and so the location specified by the first format code in the FORMAT statements in Table 17.2 is 1 greater than the actual column for output.

The T and the X codes are especially useful for creating tabular output. The T code is analogous to a typewriter tab setting in that it specifies the starting position of the following value as measured from the left margin. In contrast, the X code is analogous to a typewriter space bar in that it inserts spaces between values.

EXAMPLE 17.2   Table Headings with the T and X Codes

PROBLEM STATEMENT: Use both the T and the X codes to create headings for the table previously developed in Example 17.1.

TABLE 17.2 Format Codes to Control Spaces Between Characters

Format Code	Effect	Example
*n*X	Skip *n* spaces	10 FORMAT (10X, 'Name', 25X, 'Grade')
T*n*	Tab to column *n*	20 FORMAT (T11, 'Name', T30, 'Grade')
/	Begin new line	30 FORMAT (1X, 'NAME', /, ' Grade')

Output from FORMAT statements 10, 20, and 30:
```
12345678901234567890123456789012345678901234567890
            Name                            Grade
            Name                Grade
    Name
    Grade
```

SOLUTION: A program to create the headings with the T code is shown in Fig. 17.5. The comparable FORMAT statements using the X code are

```
100 FORMAT (' STUDENT',16X,'AVERAGE', 4X,'AVERAGE',4X,'FINAL',6X,'FINAL')
200 FORMAT(' NAME',19X,'QUIZ',7X,'HOMEWORK',3X,'EXAM',7X,'GRADE',/)
```

Statement 100 for both cases is displayed in Fig. 17.6 with the output. This diagram clearly demonstrates the difference between the T and the X codes.

FIGURE 17.5
A FORTRAN program to create headings using the T code to set the spacing.

```
      CHARACTER*14 NAME1,NAME2
      READ(5,*) NAME1,AQ1,AH1,FE1
      READ(5,*) NAME2,AQ2,AH2,FE2
      FG1 = AQ1/3 + AH1/3 + FE1/3
      FG2 = AQ2/3 + AH2/3 + FE2/3
      WRITE(6,100)
      WRITE(6,200)
      WRITE(6,*)
      WRITE(6,*) NAME1,AQ1,AH1,FE1,FG1
      WRITE(6,*) NAME2,AQ2,AH2,FE2,FG2
      STOP
100   FORMAT(' STUDENT',T25,'AVERAGE',T36,'AVERAGE',T47,'FINAL',T58,
     + 'FINAL')
200   FORMAT(' NAME',T25,'QUIZ',T36,'HOMEWORK',T47,'EXAM',T58,
     + 'GRADE')
      END
//$DATA
'SMITH,JANE',82.00,83.90,100.
'JONES,FRED',96.55,92.00,77.


** Output Results **

STUDENT                 AVERAGE   AVERAGE    FINAL     FINAL
NAME                    QUIZ      HOMEWORK   EXAM      GRADE

SMITH,JANE    82.0000000          83.8999939    100.000000    88.6333160
JONES,FRED    96.5500031          92.0000000    77.0000000    88.5166473
```

100 FORMAT (' STUDENT', T25,'AVERAGE', T36,'AVERAGE', T47, 'FINAL',T58,'FINAL')

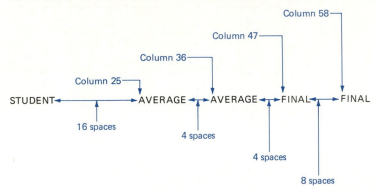

FIGURE 17.6
Comparison of FORMAT statements using the T and the X codes to implement identical output.

100 FORMAT (' STUDENT', 16X,'AVERAGE', 4X,'AVERAGE', 4X, 'FINAL',8X,'FINAL')

17.3.2 FORMAT Codes for Character Data

The input or output of character data to or from a variable is accomplished with the alphanumeric, or A, format code. The A code can be preceded by an integer number to indicate the repetitions of the A format code (see Table 17.3), for example,

```
50 FORMAT (4A30,A40)
```

The 4A30 would allow for four 30-character-long strings and the A40 for one 40 characters long. If a string variable contains more characters than the FORMAT statement permits, then only those characters described in the FORMAT statement will be printed. Figure 17.7 shows an example. Note how the strings are left-justified.

String constants can be output using a format code, but usually it is best to print

TABLE 17.3   General Forms of the Format Codes to Input and Output Letters and Numbers

Format Code	Name	Interpretation
nAw	Alphanumeric	This defines n alphanumeric fields, each w columns wide, each holding left-justified character data.
nIw	Integer	This defines n integer fields, each w columns wide, each holding a right-justified integer value.
$nFw.d$	Floating point	This defines n real fields, each w columns wide, each containing a right-justified value with a decimal point and with d decimal places.
$nEw.d$	Exponential	This defines n real fields, each w columns wide, each containing a right-justified value in scientific notation. The exponent is placed at the far right of the field preceded by the letter E. The decimal point of the mantissa is d columns to the left of the E.

FIGURE 17.7
A program fragment illustrating
the use of the A code for outputting character data.

```
Program fragment:
      CHARACTER*80 STRING
      STRING = 'Mary J.'
      WRITE(6,60) STRING
      WRITE(6,70) STRING
      STOP
   60 FORMAT(' First Initial =', 1A)
   70 FORMAT(' Name = ', 7A)
      END
Output:
First Initial =M
Name = Mary J.
```

FIGURE 17.7
A program fragment illustrating the use of the A code for outputting character data.

them within apostrophes, as shown in Fig. 17.7. String constants cannot appear in FORMAT statements used by a READ. To output an apostrophe that appears within a string constant use two apostrophes together, as in 'STUDENT''S GRADE'.

17.3.3 FORMAT Codes for Integers

Integer numbers are formatted with the I format code (Table 17.3). The I is always followed by an integer number indicating the allocated width.

An example of the use of the I code is provided in Fig. 17.8. The integer values are printed right-justified in the field. For I4 only four columns are allocated, whereas for 2I10 two areas of 10 columns each are set aside for the numbers. If a number does not fill the entire field, the empty spaces are merely left blank. On the other hand, if the number exceeds the field, asterisks are printed to signify that the space allocated was insufficient. Such is the case for 32,000 in the third WRITE statement

FIGURE 17.8
A program fragment illustrating the use of the I code for outputting integer data.

```
Program fragment:
      1234567890123456789012345678 90
            WRITE(6,10) 20, 199
            WRITE(6,10) 4300, 7, 39
            WRITE(6,10) 32000, -2786, -32000
   10 FORMAT ( ' ', I4, 2I10)

Results:
      1234567890123456789012345678 90
        20       199
      4300         7         39
      ****     -2786     -32000
```

in Fig. 17.8. When the I format codes are used for READ statements, the entries have to be right-justified in the field when they are typed.

17.3.4 FORMAT Codes for Real Numbers

Real numbers may be displayed or entered in either scientific or exponential notation (E format code) or as decimal numbers (F, or floating point, format code). For these cases, both the width of the field *and* the number of decimal places must be specified. The general forms are shown in Table 17.3. The number of decimal places must not exceed the width allocated. If the number to be printed is too large, then the space will be filled with asterisks to indicate the program's inability to print the number.

Figure 17.9 presents examples of both the F and E codes. These examples illustrate why these codes are especially well suited for devising tables. Because the decimal of a number is always placed in a specific location, when the same FORMAT statement is used repeatedly (as in Fig. 17.9), the decimals automatically line up.

In Fig. 17.9 the semicolons are inserted between numbers to highlight where one number ends and another begins. If these semicolons are left out, the numbers will

FIGURE 17.9
A program fragment illustrating the use of the F and E codes to output real numbers in decimal and exponential form.

```
                            F Format Codes
Program fragment:
        WRITE(6,10) 387.945, 387.945, 387.945
        WRITE(6,10) -4.876, -592.333, 0.269E7
        WRITE(6,10) -0.521E-2, 0.000722, -0.099
10      FORMAT ( ' ', F12.4, ';', F8.2, ';', F6.0)


Results:
    12345678901234567890123456789012345678901234567890
        387.9451;  387.95;  388.
        -4.8760; -592.33;******
        -0.0052;    0.00;   -0.
    ----------------------------------------------------------
                            E Format Codes
Program fragment:
        WRITE(6,10) 387.945, 387.945, 387.945
        WRITE(6,10) -4.876, -592.333, 2.69E+6
        WRITE(6,10) -0.521E-2, 0.000722, -0.099
10      FORMAT ( ' ', E12.4, ';', E9.3, ';', E14.1)

Results:
    12345678901234567890123456789012345678901234567890
        0.3879E+03;0.388E+03;      0.4E+03
        -0.4876E+01;-.592E+03;      0.3E+07
        -0.5210E-02;0.722E-03;     -0.1E-00
```

run together. For example, the last WRITE (without semicolons) looks like

$$-0.5210E-020.722E-03 \qquad -0.9E-01$$

which is not very easy to read. Another way (besides using semicolons) to make such numbers clearer is to leave spaces between the numbers, using the X code.

The row of asterisks occurs when the allotted field cannot accommodate a number. The value 0.269E7 exceeds the width of the F6.0 format code. The value -592.333 is printed with the E9.3 format code even though after the four places for the exponent and the five places for the number and the decimal are set aside, there is no room for the minus sign. This occurred because the FORTRAN compiler used to prepare this output automatically dropped the zero before the decimal in order to make room for the minus sign. Another compiler might not have done this, and a row of asterisks would have been printed. You should experiment with your compiler to learn how it handles such situations. In any event always leave adequate space for the fields in your output.

EXAMPLE 17.3   Using FORMAT Statements to Output a Table

PROBLEM STATEMENT: Employ FORMAT statements to output the table from Example 17.1.

SOLUTION: Figure 17.10 shows the program and the output for the table. Notice how for the FORMAT statement labeled 300 a space is used to avoid having a character output in column 1. Also, compare this program with that in Fig. 17.3 to demonstrate to yourself why the formatted output is superior.

FIGURE 17.10
A computer program to print out a table of student grades.

```
      CHARACTER*14 NAME1,NAME2
      READ(5,*) NAME1,AQ1,AH1,FE1
      READ(5,*) NAME2,AQ2,AH2,FE2
      FG1 = AQ1/3 + AH1/3 + FE1/3
      FG2 = AQ2/3 + AH2/3 + FE2/3
      WRITE(6,100)
      WRITE(6,200)
      WRITE(6,*)
      WRITE(6,300) NAME1,AQ1,AH1,FE1,FG1
      WRITE(6,300) NAME2,AQ2,AH2,FE2,FG2
      STOP
100   FORMAT(' STUDENT',T25,'AVERAGE',T36,'AVERAGE',T47,'FINAL',T58,
     + 'FINAL')
200   FORMAT(' NAME',T25,'QUIZ',T36,'HOMEWORK',T47,'EXAM',T58,
     + 'GRADE')
300   FORMAT(' ',A17,6X,4(F6.2,5X))
      END
```

```
** Output Results **

STUDENT        AVERAGE      AVERAGE       FINAL       FINAL
NAME           QUIZ         HOMEWORK      EXAM        GRADE

   SMITH,JANE  82.00        83.90         100.00      88.63
   JONES,FRED  96.55        92.00         77.00       88.52
```

17.4   THE DATA STATEMENT

Data can be entered to a FORTRAN program in a variety of ways. For example, the READ statement allows the user to enter different values of a variable on each run without modifying the program itself. Although this capability has its advantages, there will be times when it will be inconvenient to enter all or a part of the data from the keyboard. This is particularly true when entering large amounts of data.

One remedy might be to assign values to the variables within the body of the program using assignment statements. Although this circumvents the problem of re-typing the data for each run, it can add quite a few lines to your program. DATA statements provide a more concise and efficient alternative for incorporating large amounts of information into the program code. They allow the data to be kept together, apart from the body of the program so that it is not cluttered. Having data in a discrete place may also facilitate program maintenance, especially if the data is to be changed. The general form of the DATA statement is

$$\text{DATA } var_1/ \text{ } const_1/, \text{ ..., } var_n/ \text{ } const_n/$$

where $const_i$ is the ith numeric or string constant that is being entered and var_i is the ith numeric or string variable name to which $const_i$ is being assigned.

Functionally, the DATA statement

```
DATA NUMBER/7/
```

is equivalent to the assignment statement NUMBER = 7. Because it is not executable the DATA statement can appear anywhere in the program. In our programs we place it just after the data declaration statements at the start of the program. Examples of some acceptable DATA statements are shown in Fig. 17.11.

EXAMPLE 17.4   DATA Statements

PROBLEM STATEMENT: Employ DATA statements to input values to the program developed in Examples 17.1 through 17.3.

FIGURE 17.11 Examples of acceptable DATA statements. These examples assume that variable type follows the FORTRAN naming convention, where I and N are integers and GRADE PI, and E are real. LETTER and STUNAM should have been explicitly declared as character variables.

```
DATA N / 17 /
DATA I / 7 /, GRADE / 89.3 /
DATA STUNAM / 'Mary J' /
DATA LETTER / 'Y' /, PI / 0.31416E1 /, E / 2.71828 /
```

```
        CHARACTER*14 NAME1,NAME2
        DATA NAME1/'SMITH,JANE'/,AQ1/82.00/,AH1/83.90/,FE1/100./
        DATA NAME2/'JONES,FRED'/,AQ2/96.55/,AH2/92.00/,FE2/77./
        FG1 = AQ1/3 + AH1/3 + FE1/3
        FG2 = AQ2/3 + AH2/3 + FE2/3
        WRITE(6,100)
        WRITE(6,200)
        WRITE(6,*)
        WRITE(6,300) NAME1,AQ1,AH1,FE1,FG1
        WRITE(6,300) NAME2,AQ2,AH2,FE2,FG2
        STOP
100     FORMAT(' STUDENT',T25,'AVERAGE',T36,'AVERAGE',T47,'FINAL',T58,
      + 'FINAL')
200     FORMAT(' NAME',T25,'QUIZ',T36,'HOMEWORK',T47,'EXAM',T58,
      + 'GRADE')
300     FORMAT(' ',A17,6X,4(F6.2,5X))
        END
```

** Output Results **

FIGURE 17.12
A revised program to input student grades using the DATA
statement.

STUDENT NAME	AVERAGE QUIZ	AVERAGE HOMEWORK	FINAL EXAM	FINAL GRADE
SMITH,JANE	82.00	83.90	100.00	88.63
JONES,FRED	96.55	92.00	77.00	88.52

SOLUTION: The revised program is shown in Fig. 17.12. The inclusion of DATA statements means that the names and grades do not have to be input on each run, as was the case with the programs in Examples 17.1 through 17.3. This feature would be particularly useful for large amounts of data.

PROBLEMS

17.1 Members of a sales force receive commissions of 5 percent for the first $1000, 7 percent for the next $1000, and 10 percent for anything over $2000. Compute and print out the commissions and social security numbers for the following salespeople: John Smith, who has $5325.76 in sales and a social security number of 080-55-6642, Priscilla Holottashakingoinon, who has $8050.63 in sales and a social security number of 306-77-5520, and Bubba Schumacker, who has $6643.96 in sales and a social security number of 707-00-1122. Output the results in a neat, tabular format. Use READ statements to input the information.

17.2 Write, debug, and document a computer program to determine statistics for your favorite sport. Pick anything from softball to bowling to basketball. Design the program so that it is user-friendly and provides valuable and interesting information to anyone (for example, a coach or player) who might use it to evaluate athletic performance. In particular, print out the results in a neat, tabular format. Use DATA statements to input the information.

17.3   Expand the program in Examples 17.1 through 17.4 so that it also includes a class participation grade (PG). The final grade would be recalculated as

```
FG = 0.3 * (AQ + AH + FE) + 0.1 * PG
```

Give Jane a PG equal to 95 and give Fred a 75. Include this new information on the table along with their identification numbers (Jane, 739936E; and Fred, 440169S).

17.4   In Prob. 16.20, you were given equations to predict the saturation value of oxygen for various values of temperature and salinity. Use those formulas and the capabilities you have gathered in the present chapter to compute and print out the following table of saturation concentrations:

$$n = 0 \quad n = 5000$$

$T = 40°F$
$T = 50°F$

17.5   What output would be produced if the following statements were executed?
(a)
```
      X = 3.5
      R = 3
      WRITE(6,301) X,R
  301 FORMAT(' THE VALUE OF X IS',F10.3/'0THE VALUE OF R IS',F10.0)
```
(b)
```
      SUM = 35.5
      NUM = 35
      WRITE(6,302) NUM,SUM
  302 FORMAT(' THE SUM OF THE ',I3,' NUMBERS = ',F12.5)
```
(c)
```
      X = 10E02
      Y = 12.5E-2
      WRITE(6,303) X,Y
  303 FORMAT(' ',F10.2/'+',T20,E12.1)
```

17.6   What is the difference between printing numeric and character data in correctly specified fields?

17.7   Determine the WRITE and FORMAT statements needed to produce the following output. Include the top-line column index as a part of the output.

```
12345678901234567890123456789012345678901234567890
      NUMBER               X'S       X - AVERAGE
        1                  2.0          -3.5
        2                  3.0          -2.5
        3                  4.0          -1.5
        4                  7.0           1.5
        5                  8.0           2.5
        6                  9.0           3.5
        7                  7.0           1.5
```

```
         8                    6.0              0.5
         9                    5.0             -0.5
        10                    4.0             -1.5
              AVERAGE  =         5.5
```

17.8   Write a short program that employs DATA statements to input the following character constants into six-character variables:

```
'ANGLE','DEGREES','RADIANS','  X','  Y','  Z'
```

Then output the information so that each character constants are the headings for a six-zone table. Each zone should be 10 spaces wide, with the character constants printed right-justified in the 10-space zone.

17.9   Select 10 of your favorite issues from the New York Stock Exchange. Write, debug, and document a FORTRAN program that reads the name of the stock, its selling price today, and its 52-week low. The program should then print out the name of the stock, its selling price, and the percentage gain over the 52-week low. Output the results in a neat, tabular format.

CHAPTER 18

FORTRAN: SELECTION

To this point we have written programs that perform instructions sequentially. That is, the program statements are executed line by line starting with the first line and ending with the last. However, a strict sequence is only one of the ways in which a program can be executed. As discussed in Chap. 6, the flow of logic can branch (selection) or loop (repetition). The present chapter deals with the statements required to branch, or to route the flow of logic to a line other than the next one.

Although selection can greatly enhance your power and flexibility, its indiscriminate and undisciplined application can introduce enormous complexity to a program. Because this greatly detracts from program maintenance, computer scientists have developed a number of conventions to impose order on selection. Some of these structured-programming conventions involve the way in which the statements are applied—that is, they are voluntary and involve the "style" with which the programmer chooses to construct the code. Others actually involve specialized FORTRAN statements that provide very clear and coherent ways to perform selection. Unfortunately, some of these specialized types of statements are unavailable in certain FORTRAN dialects. Because of their great utility, these innovations will undoubtedly be universally available in the future. For the time being, however, their absence complicates our description of selection. On the one hand, we want to introduce you to the best and most highly advanced capabilities of FORTRAN. On the other hand, we do not want you to be frustrated by your inability to implement advanced constructs on your machine. To resolve this dilemma, we have chosen to include the advanced constructs in the following sections. However, in order that all readers will be capable of using them, we have included ways to reformulate the constructs employing simpler statements available in all forms of FORTRAN. Although this may sometimes involve a little extra effort, the long-term, positive impact on the quality of your programs will more than justify the investment.

18.1 NONSEQUENTIAL PROGRAM FLOW: THE GO TO STATEMENT

The GO TO is the simplest of all transfer statements. It allows you to directly override the numeric sequence in which the program is executed. The general form is

```
GO TO ln
```

where *ln* is the line number that is executed next, for example,

```
    READ(5,*) A,B,C
    GO TO 40
10  D = A * B * C
    WRITE(6,*) A,B,C
40  READ(5,*) X,Y,Z
```

Under normal circumstances, the program would advance to line number 10 after inputting values for A, B, and C. However, because of the GO TO statement, rather than transferring to line 10, the program here would jump ahead or "go to" line 40.

By allowing transfer to any other line, the GO TO greatly increases the program's flexibility. However, a word of caution is in order. All the jumping around that results from the indiscriminate use of GO TO statements can lead to programs that are extremely difficult to understand. For this reason, structured programming considerations mandate that they be employed only when absolutely necessary.

The negative feeling about GO TO statements is so strong that in some programming languages (Pascal for one) the GO TO construct is discouraged and is rarely seen. Languages that shun the GO TO statement usually have other constructs that control the flow of program execution in a more coherent and structured manner. As mentioned at the beginning of this chapter, many dialects of FORTRAN do not presently include a wealth of constructs for controlling program flow in this fashion. For this reason, we will use the GO TO to compensate for the lack of certain types of structured control statements in most FORTRANs. Our use of GO TOs will be limited according to the following guidelines:

1. The GO TO statement will be used only when it functions as a part of a structured control construct.

2. The line to which the GO TO transfers control will always be accompanied by a comment to clarify the reason for the transfer.

3. The GO TO statement will not cause a program loop to have more than one exit point.

The impact and importance of these limitations will be clearer once we begin to show how selection can be expressed in a structured manner. Before doing this, we must mention another statement that is sometimes employed in conjunction with the GO TO statement.

The CONTINUE statement.   The CONTINUE statement performs no specific operation but is sometimes used as the target of a GO TO statement. Once it is reached, the program merely "continues" on to the program's next executable line. Its general form is *ln* CONTINUE

Because of the deemphasis of the GO TO statement, the CONTINUE statement is employed far less than in the past. However, as described in the next chapter, it is still used in the construction of loops.

18.2   CONDITIONAL TRANSFER: THE LOGICAL IF STATEMENT

The logical IF statement represents the diamond-shaped decisions in the flowcharts shown in Fig. 18.1. It is called a *conditional transfer* because it only transfers control if a particular condition is true. The fundamental format for the logical IF statement is

IF ⟨*condition*⟩ *statement*

where *condition* represents a logical variable or a relation between expressions that is either true or false and *statement* represents a FORTRAN statement that is to be implemented if the relation is true. If the relation is false, implementation of the *statement* does not occur, and control passes automatically to the line following the logical IF statement.

The logical IF statement can control which path a program takes at a branch point. The chosen path could theoretically lead anywhere in the program. However, this could lead to the same type of confusion that follows from the indiscriminate application of GO TO statements. Consequently, the paths should always lead to statements immediately following the logical IF statement. This concise branching is exhibited by the flowchart representations for the two types of selection—the single- and double-alternative decision patterns—illustrated in Fig. 18.1.

An example of a form of the logical IF that branches to another statement is

IF (VAR .EQ. 1.) GO TO 120

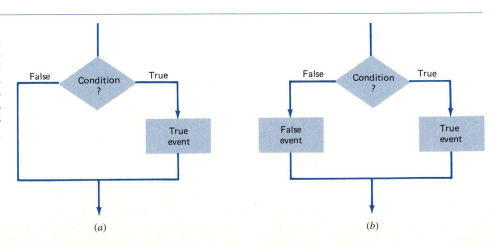

FIGURE 18.1
Two kinds of constructs, the single- and double-alternative decision patterns, are the fundamental building blocks for all types of selection. The combination of these two patterns allows any possible selection process to be implemented.

(a)   (b)

For this case, if the value of VAR is equal to 1, control will transfer to line 120. If not, control passes to the next line. If the number following the GO TO is not an existing line number in your program then an error message results.

Another example of the logical IF statement is

```
IF (NUMBER .GT. 0) AVERG = SUM/NUMBER
```

For this example, if the value of the variable NUMBER is greater than zero (greater than signified by .GT.), then the variable AVERG will be computed as the SUM divided by the NUMBER. After the computation is implemented, control will transfer to the next line. If NUMBER is equal to or less than zero, the statement AVERG = SUM/NUMBER is ignored, and control transfers immediately to the next line.

In both examples, the condition to be tested is located after the IF and before the *statement*. Now we will take a closer look at how these conditions can be formulated.

18.3 FORMULATING THE CONDITION OR DECISION OF LOGICAL IF STATEMENT

The *condition* in a logical IF statement is always a quantity that is either true or false. The two simplest cases are a single logical variable or the relation between two expressions.

Logical variable. In Table 15.3, we introduced the concept of a logical variable. Logical variables may take on only two constants: .TRUE. or .FALSE.. Note that the periods at the beginning and end of each of these words are a part of the constant. An example of using a logical variable as the *condition* of a logical IF statement is

```
      LOGICAL CHECK
      CHECK = .TRUE.
      IF (CHECK) WRITE(6,*)'TEST IS TRUE'
  40 CONTINUE
```

The first line is an explicit declaration that is required to tell the computer that the variable CHECK is of the logical type. Then, the second line assigns a value of .TRUE. to CHECK. Consequently, the logical IF in the third line will test true and, therefore, the message will be printed. If CHECK had been set equal to .FALSE., the *condition* would have tested false and the program would proceed to line 40 without printing the message.

Relation between expressions. Aside from logical variables, a more common example of a *condition* is the comparison of two expressions, which can be represented generally by

expression$_1$ *relation expression*$_2$

TABLE 18.1   Relational Operators

Operator	Relationship	Example
.EQ.	Equal	100 IF (VAR .EQ. 0.) VAR = 99
.NE.	Not equal	110 IF (C .NE. '9') WRITE (6,*) C
.LT.	Less than	120 IF (VAR .LT. 0.) VAR = −VAR
.GT.	Greater than	130 IF (X .GT. Y) Z = Y/X
.LE.	Less than or equal to	140 IF (3.9 .LE. X/3) GO TO 200
.GE.	Greater than or equal to	150 IF (VAR .GE. 0.) GO TO 160

where the *expressions* can be variables, constants, or formulas. The *relation* is any one of the relational operators in Table 18.1. These include equality, inequality, and testing whether one expression is greater or less than another.

The equality test of line 100 is true when VAR = 0. Thus, when VAR is exactly equal to zero, the value of 99 will be assigned to VAR. If VAR is not exactly zero, the assignment statement VAR = 99 will not be executed and control passes to the next line.

The inequality tests whether the expression on the right side of the .NE. is different from that on the left. According to line 110 (Table 18.1), the value of C will be printed only when C is not equal to the character 9.

Testing whether one value is less than another is accomplished with the .LT. operator. Likewise, the .GT. determines if the value on the left side of .GT. is greater than that on the right side. The operator .GE. allows the left-hand value to be either exactly equal to or greater than the right-hand value. Similarly, the operator .LE. will be true if the two are equal or if the left-hand value is less than the right-hand value. Remember that the rules for working with inequalities with these operators require the operator to change if the values on the left and right side of inequality are exchanged. X .GT. Y is the same test as Y .LE. X.

The order of evaluation of the operations in the logical IF statement is important to note. First the condition is tested. Then, if the condition is true, the *statement* is carried out. In line 100, VAR is tested to determine whether it is equal to zero prior to performing the assignment. Otherwise VAR can never equal zero. It will always equal 99.

Arithmetic and relational operators can be combined in a condition being tested. The arithmetic operations are always implemented before testing the relationship. In line 140 of Table 18.1 X is first divided by 3, and then 3.9 is tested to see if it is equal or less than the quotient.

Strings and character data can also be tested for their relationship to other string and character data. Most frequently the test for this type of data will be for equality or inequality, as in line 110. Character data cannot be directly compared to numeric data. That is why the 9 in line 110 has to be enclosed in apostrophes. By enclosing the 9 in apostrophes, we cause it to be treated as a string character, like A or B, and ignores its numeric value. If the 9 is not set in apostrophes, FORTRAN will stop processing the line and report an error. Just as string variables cannot have numbers assigned to them or vice versa, relational operations require that both sides of the relation have either numeric data or string data, but not both in one operation.

Be cautious when testing for the equality or inequality of two real expressions. As mentioned in Sec. 4.1, real numbers are stored in the computer as a series of binary digits and an exponent. If the value to be tested cannot be exactly represented by the seven digits of a single-precision real variable then the test may fail, even though it is, in fact, approximately true. For example, the following condition would test as false:

```
1/3 .EQ. 0.333
```

Even though 0.333 is approximately equal to 1/3, the computer must determine an exact equality for the condition to be true. More subtle differences can occur, as illustrated in the following example.

EXAMPLE 18.1   *Problems with Testing Equalities*

PROBLEM STATEMENT: Divide 1 by 82 and then test whether the result multiplied by 82 is equal to 1.

SOLUTION: The following code was written to investigate the testing of equalities between real numbers:

```
        A = 1./82.
        B = A * 82.
     30 IF (B .EQ. 1.) GO TO 60
        WRITE(6,*)'EQUALITY DOES NOT HOLD'
        GO TO 70
     60 WRITE(6,*)'EQUALITY HOLDS'
     70 CONTINUE
   C OUTPUT VALUE FOR B
        WRITE(6,*)B
        STOP
        END
        RUN
        EQUALITY DOES NOT HOLD
        0.999999940
```

Although we know that 1/82 multiplied by 82 should be exactly equal to 1, such is not the case for this code. As seen above, the computer represents the result as being 0.999999940. Because of the approximate way that the computer stores real numbers, the test in line 30 is false.

One way to circumvent this problem is to reformulate it. Rather than testing whether the two numbers are equal, you can test whether their difference is small.

For example, line 30 could be reformulated as

```
30 IF (ABS(B - 1) .LT. 0.00001) GO TO 60
```

This statement tests whether the absolute value of the difference between B and 1 is less than 0.00001—that is, less than 0.001 percent. Formulated in this way, line 30 would test true, as desired. Although you should be aware of this problem, testing whether one value is less than (.LT. or .LE.) or greater than (.GT. or .GE.) another can usually always be performed without difficulty.

18.4　COMPOUND LOGICAL IF STATEMENTS AND LOGICAL OPERATORS

Many dialects of FORTRAN allow the testing of more than one logical condition in the same logical IF statement. These *compound logical IF statements* employ the logical operators AND, OR, and NOT to simultaneously test conditions (Table 18.2). For example, the AND is employed when both conditions must be true, as in the general form

```
IF (condition₁ .AND. condition₂) statement
```

If both conditions are true then the *statement* will be implemented. If either or both are false, the program advances to the next without implementing the *statement*.

In contrast, the OR is employed when either one or the other or both of the conditions are true, as in

```
IF (condition₁ .OR. condition₂) statement
```

If either or both are true, the *statement* will be executed.

Finally, the NOT is used for the case where the transfer is completed if a condition is false, as in

```
IF (.NOT. condition) statement
```

TABLE 18.2　Logical Operators

Operator	Meaning	Example
.NOT.	This negates a condition. In the example, the condition is true if X ≤ 0 is false, and vice versa.	100 IF (.NOT. (X .LE. 0.)) WRITE(6,*) 'X IS POSITIVE'
.AND.	Both conditions must be true, otherwise the whole operation is considered false.	110 IF (X .GT. 0 .AND. X .LT. 5) WRITE(6,*) '0 < X < 5'
.OR.	If either or both of the conditions are true, the whole operation is true.	120 IF ((L .EQ. 'NO') .OR. (L .EQ. 'N')) WRITE(6,*) 'NO'

TABLE 18.3   Complements for the Relational Operators

Operator	Complement	Identities
.EQ.	.NE.	(A .EQ. B) is equivalent to .NOT. (A .NE. B)
.LT.	.GE.	(A .LT. B) is equivalent to .NOT. (A .GE. B)
.GT.	.LE.	(A .GT. B) is equivalent to .NOT. (A .LE. B)
.LE.	.GT.	(A .LE. B) is equivalent to .NOT. (A .GT. B)
.GE.	.LT.	(A .GE. B) is equivalent to .NOT. (A .LT. B)
.NE.	.EQ.	(A .NE. B) is equivalent to .NOT. (A .EQ. B)

The complement of a condition is true when the condition is false and vice versa. For example, the complement of $X \leq 0$ is $X > 0$. Thus, line 100 from Table 18.2 is equivalent to

```
100 IF (X .GT. 0) WRITE(6,*)'X IS POSITIVE'
```

The complements for the relational operators are summarized in Table 18.3. Note that, in most cases, it is clearer and more convenient to use the complement of a relational operator than to employ the NOT.

Logical expressions can contain more than one of the above operators, for example,

```
IF ((A.GT.0).OR.(B.LT.0).AND.(C.EQ.0)) GO TO 50
```

However, just as with arithmetic expressions, the order in which the operations are carried out must follow rules. The priorities for the operators (along with all the other priorities established to this point) are specified in Table 18.4. Thus, all other things being equal, the NOT would be implemented prior to the AND and the AND prior to the OR.

For complicated logical expressions, parentheses and position also dictate the order. Thus operations within parentheses are implemented first. Then the priorities

TABLE 18.4   The Priorities of Operators Used in FORTRAN

Type	Operator	Priority
Arithmetic operators	Parentheses, ( )	Highest
	Exponentiation, **	
	Negation, −	
	Multiplication, *, and division, /	
	Addition, +, and subtraction, −	
Relational operators	.EQ., .LT., .GT., .LE., .GE., .NE.	
Logical operators	.NOT.	
	.AND.	
	.OR.	Lowest

in Table 18.4 are used to determine the next step. Finally, within the same priority class, operations are executed from left to right.

EXAMPLE 18.2   *Priorities in Compound Logical Expressions*

PROBLEM STATEMENT: Determine whether the following compound logical expressions are true or false.

(*a*) (A .EQ. 0) .AND. (X .GT. 7) .OR. (Y .LT. 9)
(*b*) (A .EQ. 0) .AND. ((X .GT. 7) .OR. (Y .LT. 9))

where A = 1, X = 2, and Y = 3.

SOLUTION
(*a*) Using the priorities from Table 18.4, we can determine that the expression is true by the following evaluations:

```
(A .EQ. 0) .AND. (X .GT. 7) .OR. (Y .LT.9)

     F                   F                T
     └────────┬────────┘
              F
     └─────────────────────┬─────────────────┘
                           T
```

The AND comparison is implemented first because it has a higher priority than OR. It is false because both A is not equal to zero and X is not greater than 7. The OR is then implemented with the result that the entire expression is deemed true because Y is less than 9.

(*b*) Because of the introduction of the parentheses, we are forced to evaluate the OR first. As a result, the entire operation yields a false result:

```
(A .EQ. 0) .AND. ((X .GT. 7) .OR. (Y .LT. 9))

    F                 F              T
                      └──────┬──────┘
                             T
    └──────────────────┬──────────────────┘
                       F
```

18.5   THE IF/THEN/ELSE STATEMENTS

An alternative to the logical IF is the IF/THEN/ELSE construct. In these statements an alternative branch is executed when the condition expression is false. As such, it is designed expressly for the programming of the double-alternative selection structure of Fig. 18.1. Its general form is

```
IF (condition) THEN
    statements representing the true alternative
ELSE
    statements representing the false alternative
ENDIF
```

For the single-alternative selection (Fig. 18.1*a*), the general form is

```
IF (condition) THEN
    statements representing the true alternative
ENDIF
```

For the double-alternative selection, the program automatically transfers over the false-alternative statements, if the *condition* is true. Conversely, if the *condition* is false, the program skips over the true-alternative statements and implements the false-alternative statements. As such, it functions in an identical fashion to the structure in Fig. 18.1*b*.

Unfortunately, FORTRAN IV does not include these constructs. For this dialect, the standard logical IF with a GO TO must be employed as described in Sec. 18.6.1.

18.6   STRUCTURED PROGRAMMING OF SELECTION

The statements introduced in the previous sections can be orchestrated to introduce all sorts of complicated logic into your computer programs. However, as we cautioned at the beginning of the chapter, their indiscriminate application can lead to code so complex that it is practically indecipherable. In order to avoid this situation, computer scientists have developed a series of rules and suggested practices that when followed can lead to easy-to-understand code. In the present section, we will show how the two fundamental selection constructs shown in Fig. 18.1 can be programmed according to these rules. We will first use the simple logical IF statement to express these constructs. Then we will show how the block-structured IF/THEN/ELSE provides a more efficient vehicle for the same purpose. We do this for two reasons. First, it illustrates the logic underlying the more advanced block-structured version. Second, FORTRAN IV does not allow a block-structured IF/THEN/ELSE construct. Therefore we will show how a logical IF and a few GO TOs can be used to simulate a block-structured IF/THEN/ELSE construct so that they may be employed by everyone using this book.

18.6.1   Selection with the Logical IF Statement

There are several ways that the logical IF statement can be employed to program a double-alternative decision pattern. Figure 18.2 shows a flowchart that will serve to illustrate this point. Figure 18.2*a* through *d* presents four sets of FORTRAN code that all implement the algorithm in the flowchart. Figure 18.2*a* and *b* both employ the form of the condition shown in the flowchart—that is, they test whether A is less than zero. They differ in that Fig. 18.2*a* has the false alternative first, whereas Fig. 18.2*b*

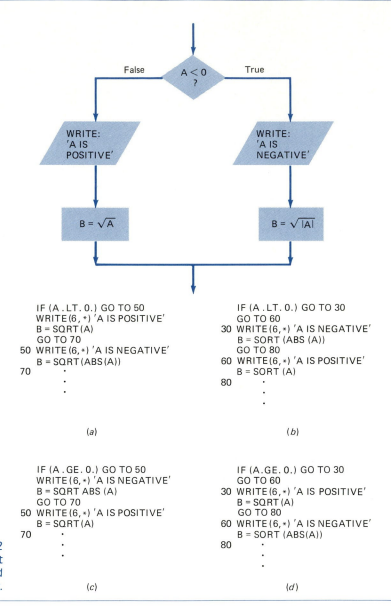

FIGURE 18.2
Four sets of FORTRAN code that
implement the algorithm specified
by the flowchart.

has the true alternative first. Figure 18.2c and d are alternatives that result from utilizing the complement of the original condition—that is, they test whether A is greater than or equal to zero. They also differ in that Fig. 18.2c has the false alternative first, whereas Fig. 18.2d has the true alternative first. Notice how, in all cases, GO TO statements are required to transfer control properly. As mentioned in Sec. 18.1, this is the only way in which GO TO statements will be employed in this book.

The fact that four separate versions can be developed to perform the same double-alternative decision is one example of how selection can introduce flexibility and an associated complexity into a program. The result is that the logic of such statements is often very difficult to decipher. In order to reduce this potential confusion, computer scientists have suggested that selection be limited to the forms shown in Fig. 18.2*b* and *d*. That is, they suggest that the true alternative always be placed first. Notice that this means that two GO TOs are required. However, this is more than compensated for by the clarity introduced in having the true alternative directly following the logical IF. Thus every time you see a logical IF, you will automatically expect that the true alternative will be first. The construct can be made even clearer by adding comment and CONTINUE statements to document its logic. For example, for Fig. 18.2*b*,

```
        IF (A .LT. 0) GO TO 30
        GO TO 60
   30 CONTINUE
C       THEN
        WRITE(6,*) 'A IS NEGATIVE'
        B = SQRT(ABS(A))
        GO TO 90
   60 CONTINUE
C       ELSE
        WRITE(6,*) 'A IS POSITIVE'
        B = SQRT(A)
   90 CONTINUE
C       ENDIF
```

Line 60 has been added to mark the beginning of the false alternative. In addition, the ENDIF comment is included to designate the end of the logical IF construct. The documentation can now be taken a final step by using indentation to visually express the structure:

```
        IF(A .LT. 0) GO TO 30
            GO TO 60
   30       CONTINUE
C           THEN
            WRITE(6,*) 'A IS NEGATIVE'
            B = SQRT(ABS(A))
            GO TO 90
   60     CONTINUE
C         ELSE
            WRITE(6,*) 'A IS POSITIVE'
            B = SQRT(A)
   90 CONTINUE
C       ENDIF
```

Note how the indentation clearly sets off the true and the false alternatives. The same style can also be applied to the single-alternative pattern, as will be described in the next section.

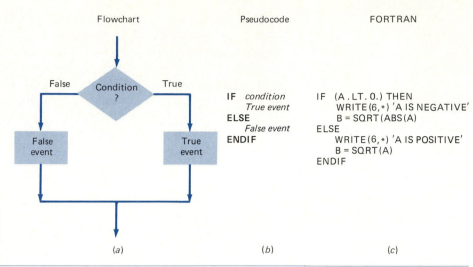

FIGURE 18.3
A block-structured IF/THEN/ELSE
construct for a double-alternative
decision pattern.

18.6.2 The Block-Structured IF/THEN/ELSE Construct

The logical IF construct in the previous section can be simplifed further if the block-structured IF/THEN/ELSE statement (previously introduced in Sec. 18.5) is available for your computer. The result is shown in Fig. 18.3. The same style can also be developed for the single-alternative pattern. For example, the IF/THEN/ELSE construct in Fig. 18.4c can be developed to characterize the single-alternative pattern in Fig. 18.4a. Note that because there is no false alternative, an ELSE is not used. If the condition tests to be false, control passes to the ENDIF.

The codes shown in Figs. 18.3 and 18.4 are referred to as *block-structured IF/*

FIGURE 18.4
(a) A flowchart for a single-alternative decision pattern; (b) the pseudocode; and (c) the corresponding
block-structured IF/THEN/ELSE
construct in FORTRAN.

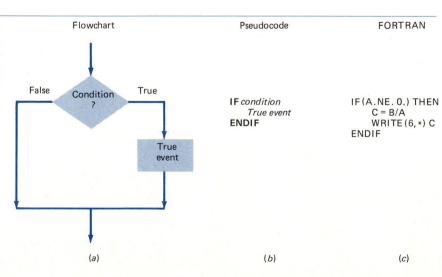

TABLE 18.5 Rules for Writing Block-Structured IF/THEN/ELSE Constructs

Rule 1	The last statement is always ENDIF.
Rule 2	The true alternative is always placed first.
Rule 3	The false condition follows the ELSE clause. If there is no separate action to be taken, as in the single-decision pattern, the ELSE clause is not employed.
Rule 4	For a double-alternative pattern, the last statement of the true alternative is executed, and then the execution of the program continues with the first statement following the ENDIF.
Rule 5	The ENDIF is indented to the same degree as the start of its IF statement.
Rule 6	The ELSE statement is indented three spaces more than the IF statement.
Rule 7	The FORTRAN statement(s) within the THEN and ELSE clause(s) is indented three spaces from the level of the ELSE or the IF statement to which it refers, whichever is greater.
Rule 8	The indentation of the conditional statements, the ELSE, and the ENDIF is always determined by the IF/THEN/ELSE statements to which they refer.

THEN/ELSE constructs. The rules for building these constructs are summarized in Table 18.5. The benefits of using these constructs are great. Each IF/THEN/ELSE construct has only one entry point and one exit. The FORTRAN statements that are a part of the IF/THEN/ELSE construct are immediately apparent, and the true and false alternatives are clearly delineated. The primary objective here is to make the resulting code coherent and easily understandable to someone other than the original author of the program. These conventions help standardize the writing of programs so that others may more readily grasp the organization of an unfamiliar program.

All but the simplest logical statements should be constructed as block structures for easier readability. Figure 18.5, which reprints Figs. 18.2*a* and 18.3*b*, should make this evident. Even though this is a very simple example, the superiority of the structured approach is obvious. For more complicated code, the benefits become even more pronounced.

18.6.3 Nesting of Block IF/THEN/ELSE Constructs

An IF/THEN/ELSE construct can have another IF/THEN/ELSE statement as part of its THEN or ELSE statements. The inclusion of a construct within a construct is called *nesting*. For the case of the block IF/THEN/ELSE construct, nesting provides an

FIGURE 18.5
A side-by-side comparison of Figs. 18.2*a* and 18.3*b*. Both sets of code accomplish the same objective, but the structured version in (*b*) is much easier to read and decipher.

```
      IF (A .LT. 0) GO TO 50
      WRITE(6,*)'A IS POSITIVE'
      B = SQRT(A)
      GO TO 70
50    WRITE(6,*)'A IS NEGATIVE'
      B = SQRT(ABS(A))
70    CONTINUE
               (a)
```

```
      IF (A .LT. 0) THEN
         WRITE (6,*)'A IS NEGATIVE'
         B = SQRT(ABS(A))
      ELSE
         WRITE(6,*)'A IS POSITIVE'
         B = SQRT(A)
      ENDIF
               (b)
```

alternative to having involved conditional expressions with many AND or OR clauses. By using simpler conditional expressions it may be easier to follow the logic of the decisions.

The more complex nested IF/THEN/ELSE structures are much easier to understand when proper rules for indentation are followed. The rules for forming and indenting the nested IF/THEN/ELSE statements are essentially the same as those for a single IF/THEN/ELSE structure. Just remember that indentation of the conditional statement ELSE and the ENDIF is always determined by the IF/THEN/ELSE statement to which they refer. An example using nested IF/THEN/ELSE statements will demonstrate the use of nesting and proper identation.

EXAMPLE 18.3   Nesting of IF/THEN/ELSE Constructs

PROBLEM STATEMENT: Develop a program to determine the real and complex roots of the quadratic equation.

FIGURE 18.6
A structured flowchart to determine the real and complex roots of a quadratic equation. Note that the first selection prevents division by zero in the equation if $a = 0$.

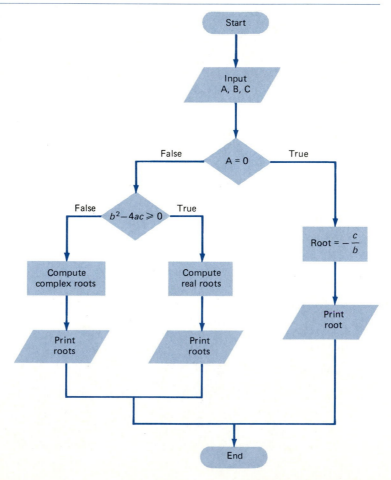

```
C    ************************************************************
C    *                ROOTS OF A QUADRATIC EQUATION             *
C    *                     (FORTRAN version)                    *
C    *                     by  S. C. Chapra                     *
C    *                                                          *
C    *            This program determines the real             *
C    *            and complex roots of a quadratic.            *
C    ************************************************************
C
C            .------------------------------------.
C            |         DEFINITION OF VARIABLES     |
C            |                                     |
C            | A = COEFFICIENT OF X SQUARED        |
C            | B = COEFFICIENT OF X                |
C            | C = CONSTANT                        |
C            '------------------------------------'
C
C    ***********************  MAIN PROGRAM  *****************
C
     READ(5,*) A,B,C
     IF (A .EQ. 0) THEN
         ROOT = -C/B
         WRITE(6,*) 'SINGLE ROOT = ',ROOT
     ELSE
         IF (B**2 - 4*A*C .GE. 0) THEN
C            COMPUTE REAL ROOTS
             ROOT1 = (-B + SQRT(B**2 - 4*A*C))/(2*A)
             ROOT2 = (-B - SQRT(B**2 - 4*A*C))/(2*A)
             WRITE(6,*) 'REAL ROOT = ',ROOT1
             WRITE(6,*) 'REAL ROOT = ',ROOT2
         ELSE
C            COMPUTE COMPLEX ROOTS
             REAL = -B/(2*A)
             IMAGIN = SQRT(ABS(B**2 - 4*A*C))/(2*A)
             WRITE(6,*) 'COMPLEX ROOT = ',REAL,' + (',IMAGIN,' I)'
             WRITE(6,*) 'COMPLEX ROOT = ',REAL,' - (',IMAGIN,' I)'
         END IF
     END IF
     STOP
     END
```

FIGURE 18.7
A structured FORTRAN program to implement the algorithm from Fig. 18.6. Notice how indentation clearly sets off the nested loops.

```
** Output Results **

REAL ROOT =   -1.0000000
REAL ROOT =   -3.0000000
```

SOLUTION: A structured flowchart for this problem is shown in Fig. 18.6. Note that there are two nested selections. The first selection is designed to avoid division by zero in the quadratic equation for the case where $a = 0$. The second selection branches to compute the real or complex roots, depending on whether the term within the square root is positive or negative.

Structured code to implement the flowchart is listed in Fig. 18.7. Indentation is used to clearly show the true and false options of each IF/THEN/ELSE.

At this point you may wonder why all FORTRAN programs are not written in a structured form? What are the disadvantages of using this type of program organization? For one thing, the amount of typing is increased in the structured code. In addition, obtaining results with the structured code may be slightly slower than with the nonstructured version because the structured code has more lines of FORTRAN to compile and execute. Thus there is a trade-off here. The structured program is easier to understand and change later, but more time is consumed developing it initially. However, in the overall scheme of program development and use, the time and effort are almost always a worthwhile investment. As FORTRAN 77 (which has block-structured IF/THEN/ELSE constructs formally built into it) becomes more widely available, the use of these constructs will become routine.

18.7   THE CASE CONSTRUCT

The CASE construct is a special type of IF/THEN/ELSE block. The CASE construct selects one path from a group of alternative paths. This is a very useful construction in a situation where, for example, the user is presented with a menu and makes a selection, after which the program takes an action that depends upon the choice of the user.

As seen in Fig. 18.8, the CASE statement tests to find which of the *values,* 1 through *n*, matches the *variable*. Only the statements associated with the matched *value* are then executed. The statements following ELSE would be carried out if none of the *values* matched the contents of the *variable*.

Although it is a part of many advanced languages such as Pascal, the CASE statement is not formally a part of FORTRAN. However, it can be simulated with a series of IF/THEN/ELSE statements. In addition, FORTRAN 77 has a feature that facilitates building a CASE structure—the ELSE IF. Instead of each IF statement having a paired ENDIF, only the last IF in a group has an ENDIF and all intermediate IFs end with ELSE IFs.

To make this construction more obvious to the reader, slightly different indenting and branching in the IF/THEN/ELSE structure is helpful. The approach is depicted

FIGURE 18.8
Generalized form of the CASE construct.

```
CASE variable
    value₁
        statements
    value₂
        statements
        •
        •
        •
    valueₙ
        statements
    ELSE
        statements
ENDCASE
```

```
IF (variable .EQ. value₁) THEN
    FORTRAN statements
        ·
        ·
        ·
ELSE IF (variable .EQ. value₂) THEN
    FORTRAN statements
        ·
        ·
        ·
ELSE IF (variable .EQ. valueₙ) THEN
    FORTRAN statements
        ·
        ·
        ·
ELSE
    FORTRAN statements
ENDIF
```

FIGURE 18.9
Generalized format for the CASE construct using IF/THEN/ELSE statements.

in Fig. 18.9. Again statements following the condition are executed when the *variable* equals the *value₁*.

The rules governing the forming of a CASE construction using the IF/THEN/ELSE statements of FORTRAN, which are different from the rules for the usual IF/THEN/ELSE block structure, are summarized in Table 8.6.

EXAMPLE 18.4   The CASE Construct

PROBLEM STATEMENT: Suppose that each class at a university is to go to a different room to register for courses. The classroom assignments are

Class	Identification Code	Classroom
Freshmen	1	108
Sophomores	2	306
Juniors	3	214
Seniors	4	215
Graduates	5	107

The identification code is a number that signifies the student's class. Design a program that uses the CASE construct to print out the proper classroom number in response to the student's identification code.

TABLE 18.6   *Rules Governing the Formation of the CASE Construct Using the IF/THEN/ELSE Statements of FORTRAN that Are Different from the Rules for the Conventional Block-Structured IF/THEN/ELSE.*

Rule 1	The ELSE IF statement is used so that the group of logical IF statements has only one ENDIF at the end of the construct.
Rule 2	The logical IF and ELSE IF statements are all at the same level of indentation.

SOLUTION: The FORTRAN program to accomplish the objective is listed in Fig. 18.10. Notice that the CASE construct is used to print out the correct message on the basis of the student's code.

Computed GO TO form of the CASE construct. The *computed GO TO* statement offers an alternative to the IF/THEN/ELSE statements for forming the CASE construct. The general form of the computed GO TO statement is

$$GO\ TO\ (ln_1, ln_2, \ldots, ln_n)\ variable$$

where the *variable* is an integer that may take on values from *1* to *n*. If the value is *1,* the program goes to ln_1; if it is *2*, the program goes to ln_2; and so on.

The use of the computed GO TO statement is discouraged by some advocates of structured programming because its undisciplined use can lead to code that is very confusing. However, if employed for a structured construct such as the CASE, it is

FIGURE 18.10
A program using the CASE construct.

```
C Program Fragment to Illustrate
C CASE Decisions   (FORTRAN)
C *************************************************
      READ(5,*) CLASS
      IF (CLASS.EQ.1) THEN
          WRITE(6,*) 'PROCEED TO ROOM 108'
      ELSE IF (CLASS.EQ.2) THEN
          WRITE(6,*) 'PROCEED TO ROOM 306'
      ELSE IF (CLASS.EQ.3) THEN
          WRITE(6,*) 'PROCEED TO ROOM 214'
      ELSE IF (CLASS.EQ.4) THEN
          WRITE(6,*) 'PROCEED TO ROOM 215'
      ELSE IF (CLASS.EQ.5) THEN
          WRITE(6,*) 'PROCEED TO ROOM 107'
      ELSE
          WRITE(6,*) 'NO SUCH CLASS'
          WRITE(6,*) 'PLEASE RUN PROGRAM AGAIN'
      END IF
      STOP
      END


** Output Results **

PROCEED TO ROOM 214
```

perfectly acceptable. In fact, because it allows direct transfer to particular lines, it is actually somewhat more efficient than the IF/THEN/ELSE construct of Fig. 18.9. Problem 18.10 at the end of this chapter involves reprogramming Fig. 18.10 so that it employs the computed GO TO version of the CASE.

PROBLEMS

18.1 The following code is designed to print out a message to a student according to his or her final exam grade (FINAL) and total points accrued in a course (POINTS).

```
IF (FINAL .LE. 60) WRITE(6,*) 'FALLING GRADE'
IF (FINAL .GT. 60) WRITE(6,*) 'PASSING GRADE'
IF (FINAL .GT. 60 .AND. POINTS .LE. 200) WRITE(6,*) 'POOR'
IF (FINAL .GT. 60 .AND. POINTS .GE. 200 .AND. POINTS .LE. 400) WRITE(6,*) 'GOOD'
IF (FINAL .GT. 60 .AND. POINTS .GT. 400) WRITE(6,*) 'EXCELLENT'
```

Reformulate this code using block-structured IF/THEN/ELSE constructs. Test the program for the following cases
(a) FINAL = 50, POINTS = 250
(b) FINAL = 70, POINTS = 321
(c) FINAL = 90, POINTS = 561

18.2 As an industrial engineer you must develop a program to accept or reject a rod-shaped machine part according to the following criteria: Its length must be not less than 7.75 cm nor greater than 7.85 cm; its diameter must be not less than 0.335 cm nor greater than 0.346 cm. In addition, under no circumstances may its mass exceed 5.6 g. Note that the mass is equal to the volume (cross-sectional area times length) multiplied by the rod's density (7.8 g/cm^3). Write a structured computer program to input the length and diameter of a rod and print out whether it is accepted or rejected. Test it for the following cases. If it is rejected, print out the reason (or reasons) for rejection.
(a) Length = 7.71, diameter = 0.338
(b) Length = 7.81, diameter = 0.341
(c) Length = 7.83, diameter = 0.343
(d) Length = 7.86, diameter = 0.344
(e) Length = 7.63, diameter = 0.351

18.3 An air conditioner both cools and removes humidity from a factory. The system is designed to turn on (1) between 7 A.M. and 6 P.M. if the temperature exceeds 75°F and the humidity exceeds 40 percent or if the temperature exceeds 70°F and the humidity exceeds 80 percent; or (2) between 6 P.M. and 7 A.M. if the temperature exceeds 80°F and the humidity exceeds 80 percent or if the temperature exceeds 85°F, regardless of the humidity.

Write a structured computer program that inputs temperature, humidity, and time of day and prints out a message specifying whether the air conditioner is on or off. Test the program with the following data:

(*a*) Time = 7:40 P.M., temperature = 81°F, humidity = 68 percent
(*b*) Time = 1:30 P.M., temperature = 72°F, humidity = 79 percent
(*c*) Time = 8:30 A.M., temperature = 77°F, humidity = 50 percent
(*d*) Time = 2:45 A.M., temperature = 88°F, humidity = 28 percent

18.4 Develop a structured program that uses the CASE construct to determine whether a number is positive, negative, or zero.

18.5 The American Association of State Highway and Transportation Officials provides the following criteria for classifying soils in accordance with their suitability for use in highway subgrades and embankment construction:

Grain Size, mm	Classification
>75	Boulders
2–75	Gravel
0.05 − 2	Sand
0.002 − 0.05	Silt
<0.002	Clay

Develop a structured program that uses the CASE construct to classify a soil on the basis of grain size. Test the program with the following data:

Soil Sample	Grain Size, mm
1	2×10^{-4}
2	10
3	0.6
4	120
5	0.01

18.6 Develop a structured program that uses the CASE construct to assign a letter grade to students on the basis of the following scheme: A (90–100), B (80–90), C (70–80), D (60–70), and F (<60). For the situation where the student's grade falls on the boundary (e.g., 80), have the program give the student the higher letter grade (e.g., a B).

18.7 Write a structured FORTRAN program that reads values of x and outputs absolute values of x without using the ABS(X) function.

18.8 How many times will line 40 be executed and what is the value of TOTAL when the ENDIF in the following code is reached?

```
LOGICAL CHECK
CHECK = .TRUE.
I = 4
TOTAL = 20
```

```
40 IF (CHECK .AND. I .GT. 0) THEN
      TOTAL = TOTAL + I
      I = I - 1
      GO TO 40
   ENDIF
```

18.9 In pipelines certain valves can be set in only two positions, open or closed. Write a nested IF/THEN/ELSE construct that will print out which pipes are flowing based upon the configuration shown in Fig. P18.9. Use logical variables to represent the status of valves A through H. Test it with B and D closed and all others open. Also test it with A closed.

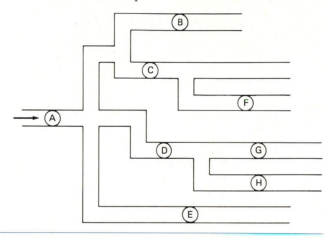

FIGURE P18.9

18.10 Reprogram Fig. 18.10 so that it employs the computed GO TO version of the CASE construct.

CHAPTER 19

FORTRAN: REPETITION

Computers excel at performing tasks that are repetitive, boring, and time-consuming. Even very simple programs can accomplish huge tasks by executing the same chore over and over again. One of the earliest applications of computers was to tally and categorize the U.S. population in the census of 1890, a very repetitive task. Even today, one of the primary applications of computers is to maintain, count, and organize many pieces of information into smaller, more meaningful collections.

Nowhere is the repetitive power of the computer more valuable than in engineering. Many of our applications hinge on the ability of computers to implement portions of programs repetitively so that the expended effort is small compared to the results produced. This is true in many areas of our discipline but is especially relevant to number-crunching computations. Before discussing the actual statements and constructs used for repetition, we will present a simple example to demonstrate the major concept underlying repetition—the loop.

19.1 GO TO, OR INFINITE, LOOPS

Suppose that you want to perform a statement or a group of statements many times. One way of accomplishing this is simply to write the set of statements over and over again. For example, a repetitive version of the simple addition program is

```
READ(5,*) A, B
WRITE(6,*) A + B
READ(5,*) A, B
WRITE(6,*) A + B
```

```
READ(5,*) A, B
WRITE(6,*) A + B
        .
        .
        .
```

Such a sequential approach is obviously inefficient. A much more concise alternative can be developed using a GO TO statement, as in

```
10 READ(5,*) A, B
   WRITE(6,*) A + B
   GO TO 10
```

As depicted in Fig. 19.1, the use of the GO TO statement directs the program to automatically circle or "loop" back and repeat the READ and WRITE statements. Such repetitive execution of a statement or a set of statements is called a *loop*. Because one of the primary strengths of computers is their ability to perform large numbers of repetitive calculations, loops are among the most important operations in programming.

A flaw of Fig. 19.1, called a *GO TO loop*, is that it is a *closed*, or *infinite*, loop (recall Fig. 6.7). Once it starts, it never stops. Consequently, provisions must be made so that the loop is exited after the repetitive computation is performed satisfactorily. One way to accomplish this is with the DO statements discussed in the next section.

19.2 DO LOOPS

Suppose that you wanted to perform a specified number of repetitions, or iterations, of a loop. One way, employing a logical IF statement, is depicted in the flowchart and program fragment in Fig. 19.2a. The loop is designed to repeat three times. The

FIGURE 19.1
An example of a GO TO loop. Notice that the loop is infinite. That is, once it starts, it never stops.

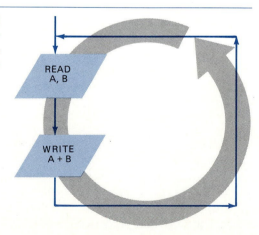

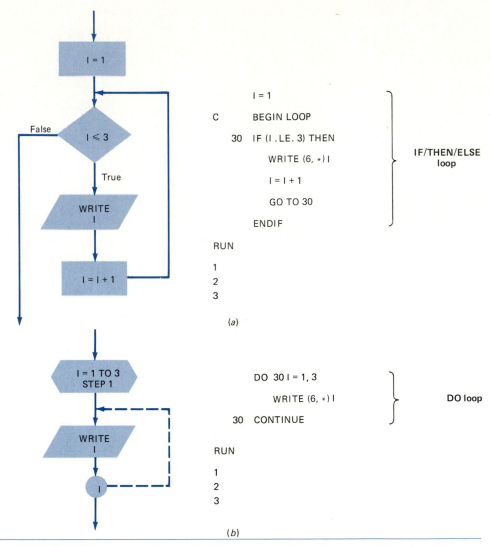

FIGURE 19.2 Two alternative methods for implementing a loop consisting of a set number of iterations: (a) An IF/THEN/ELSE and a counter I are used to determine when three passes have been implemented. (b) The DO loop offers a more efficient alternative to perform the same action.

variable I is a counter that keeps track of the number of iterations. If I is less than or equal to 3, an iteration is performed. On each pass, the value of I is printed and I is incremented by 1. After the third iteration, I has a value of 4, and therefore the THEN condition of line 30 is not executed and control is transferred out of the loop to the line following the ENDIF statement.

Although the logical IF loop is certainly a feasible option for performing a specified number of iterations, the operation is so common that a special set of statements is available in FORTRAN for accomplishing the same objective in a more efficient manner. Called the *DO loop*, its general format is

```
DO  ln index  =  start, finish, increment
     ↓

     ↓

     ↓
  (Body of loop)
     ↓

     ↓

     ↓
 ln CONTINUE
```

The DO loop works as follows. The *index* is a variable that is set at an initial value (*start*). The program then executes the body of the loop, moves to the CONTINUE statement, and then loops back to the DO statement. Every time the CONTINUE statement is encountered, the *index* is automatically increased by the *increment*. Thus the *index* acts as a counter. Then, when the *index* is equal to or greater than the final value (*finish*), the computer automatically transfers control out of the loop to the line following the CONTINUE statement. Note that the last line of a DO loop does not have to be a CONTINUE statement, but it is good practice to program DO loops in this way.

An example of the DO loop is provided in Fig. 19.2*b*. Notice how this version is much more concise than the logical IF loop in Fig. 19.2*a*. Also notice that the term "*increment*" is omitted from the DO statement. If this term is not included, the computer automatically assumes a default value of 1 for the increment.

As seen in Fig. 19.2*b*, all FORTRAN statements within a DO loop should be indented three spaces from that of the DO statement. The CONTINUE statement should be indented the same number of spaces as its associated DO statement. Any number of FORTRAN statements can be included within the loop.

EXAMPLE 19.1   Using a DO Loop to Input and Total Values

PROBLEM STATEMENT: A common application of repetition is to enter data from the keyboard. The DO loop is particularly well suited for cases where the user has foreknowledge of the number of values that are to be entered. Employ the DO loop to input a set number of values to a program. Also, design the code so that the values are totaled and their average determined.

SOLUTION: A program fragment to perform the above task is

```
   TOTAL = 0
   READ(5,*)NUMVAL
20 DO 45 I = 1, NUMVAL
      READ(5,*)VALUE
      TOTAL = TOTAL + VALUE
```

```
45 CONTINUE
   AVERAG = TOTAL/NUMVAL
   WRITE(6,*)'AVERAGE =',AVERAG
   STOP
   END
```

First the number of data values, NUMVAL, is entered. NUMVAL then serves as the finish parameter for the loop. Because the increment is not specified, a step size of 1 is assumed. Therefore the loop repeats NUMVAL times while inputting one value per iteration.

On each pass, the values are summed. The variable TOTAL is used as an accumulator (recall Sec. 6.2.4). Before the loop is implemented, TOTAL is set to zero. On the initial pass, the first value is added to TOTAL and the result assigned to TOTAL on the left side of the equal sign. Therefore after the first pass TOTAL will contain the first value. On the next pass the second value is added, and TOTAL will then contain the sum of the first and second values. As the iterations continue, additional values are input and added to TOTAL. Thus after the loop is terminated, TOTAL will contain the sum of all the values. It can then be used to compute the average.

The size of the *increment* can be changed from the default of 1 to any other value except zero. For FORTRAN 77 the *increment* does not even have to be positive. If a negative value is used, the logic for stopping the loop is reversed. With a negative

FIGURE 19.3
(a) Correct and (b) incorrect
nesting of DO loops.

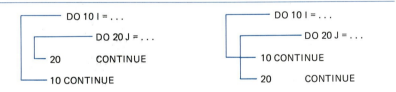

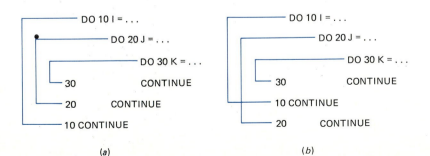

(a)   (b)

increment, the loop terminates when the value of the *index* is less than that of the *finish*. Step sizes such as −50, 10, −1, or 60 would all be possible.

19.2.1 Nesting of DO Loops

Loops may be enclosed completely within other loops. This arrangement, called *nesting*, is only valid if it follows the rule: If a DO loop contains either the DO or the CONTINUE parts of another loop, it must contain both of them. Figure 19.3 presents some correct and incorrect versions. Notice how indentation is used to clearly delineate the extent of the loops.

EXAMPLE 19.2   DO Loops

PROBLEM STATEMENT: Illustrate the use of DO loops with (*a*) an increment greater than 1 (*b*) a negative increment, and (*c*) several nested loops.

SOLUTION
(*a*) The following program prints numbers from 2 to 22 with an increment of 4:

```
      DO 30 COUNT = 2, 22, 4
          WRITE(6,*)COUNT
   30 CONTINUE
      WRITE(6,*)'VALUE AFTER EXIT =',COUNT
      STOP
      END
      RUN
       2
       6
      10
      14
      18
      22
      VALUE AFTER EXIT = 26
```

Notice that upon exit from the loop, COUNT has been incremented by 4.
(*b*) The following program prints numbers from 20 down to −5, using an increment of −6:

```
      DO 240 I = 20, -5, -6
          WRITE(6,*)I
  240 CONTINUE
      STOP
      END
      RUN
      20
      14
       8
       2
      -4
```

(c) The following program illustrates both the nesting of loops and the associated indentation:

```
      DO 70 I = 1, 2
          DO 50 J = 1, 3
              K = I * J
              WRITE(6,*)K
50        CONTINUE
          WRITE(6,*)' *'
70    CONTINUE
      STOP
      END
      RUN
      1
      2
      3
      *
      2
      4
      6
      *
```

19.2.2   DO Loops in FORTRAN 77 and in FORTRAN IV

In FORTRAN 77 the index variable and the start and finish values can be either integers or real numbers. In FORTRAN IV these must all be integer numbers and the start and finish values must be unsigned integers greater than zero. The index variable of FORTRAN 77 is compared with the finish value prior to the first execution of the DO loop. If the index variable is initially greater than the finish value, the contents of the DO loop will never be executed. In contrast, in FORTRAN IV the index variable is not compared to the finish value until one pass through the loop. Thus the FORTRAN IV DO loop is always executed at least once. To have FORTRAN IV behave in the same way as FORTRAN 77, the FORTRAN IV DO loop has to be preceded by an IF statement that explicitly tests whether the index variable initially exceeds the finish value. If so, the program will branch around (past the end of) the DO loop.

19.3   STRUCTURED PROGRAMMING OF REPETITION

In Chap. 6, we mentioned that there are two fundamental types of loop constructs— the DOWHILE and the DOUNTIL structures. As depicted in Fig. 19.4a, the DO-WHILE is used to create a loop that repeats as long as a condition is true. Because the condition precedes the event, it is possible that the action will not be implemented if the condition is false on the first pass. In contrast, as seen in Fig. 19.4a, the DOUNTIL is designed to have the event execute repeatedly until the condition is true.

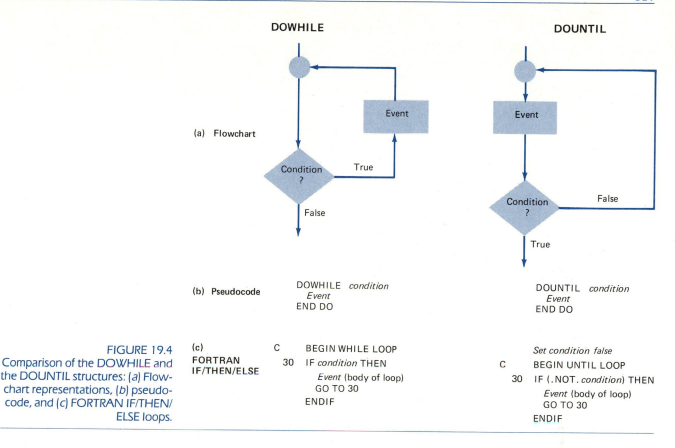

Then control is passed outside the loop. A key feature of this type of loop is that the event is always executed at least once.

Unfortunately, FORTRAN 77 has no structure equivalent to the general form of the DOWHILE or DOUNTIL loops. However, as described in the next section, the IF/THEN/ELSE and a GO TO can be used to simulate the structure. Then at the end of the chapter we will mention some structured dialects of FORTRAN that do include explicit statements for both the DOWHILE and DOUNTIL structures.

19.3.1 IF/THEN/ELSE Loops

In contrast to the DO loops described in the previous section, the IF/THEN/ELSE loops are designed for cases where the number of repetitions is not prespecified. Rather, as seen in Fig. 19.4a, termination of the loop hinges on a decision. Although some FORTRAN dialects have statements expressly designed for the repetition constructs (such as the WHILE and UNTIL loops described in the next section), the structures depicted in Fig. 19.4 can also be programmed with the more commonly available IF/THEN/ELSE statements. Therefore, in order to ensure that all readers be able to program these constructs, we will show how this is done.

The DOWHILE structure.   As seen in Fig. 19.4c, the DOWHILE version begins with a comment statement to document the start of the loop. In addition, a GO TO is used to loop back to the head of the construct. Note that the body of the loop is indented to make the structure more apparent.

The IF/THEN/ELSE version of the DOWHILE has many applications. In constrast to the DO loop, which is well suited to handle fixed numbers of repetitions, the IF/THEN/ELSE version is designed for cases where the number of iterations is not given. Rather, the repetitions depend on a condition that is tested every time the loop is implemented. As described in the following example, a situation where this sort of loop proves useful is when inputting data to the program.

EXAMPLE 19.3   Using a DOWHILE Loop to Input, Count, and Total Values

PROBLEM STATEMENT: As seen in Example 19.1, there are occasions when the user will have prior knowledge of the number of values that are to be input to a program. However, there are many instances when the user will either not know or will not desire to count the data beforehand. This is typically the case when large quantities of information are to be entered. In these situations, it is preferable to have the program count the data. Employ the IF/THEN/ELSE version of the DOWHILE loop to input, count, and total a set of values. Use the results to compute the average.

SOLUTION: The important issue that needs to be resolved in order to develop this program is how to signal when the last item is entered. A common way to do this is to have the user enter an otherwise impossible value to signal that data entry is complete. Such impossible data values are sometimes called *sentinel*, or *signal*, *values*. An IF/THEN/ELSE can then be used to test whether the sentinel value has been entered and thus whether to transfer control out of the loop. The sentinel value can be a negative number if all data values are known to be positive. A very large number, say 9999, might be employed if all valid data is small in magnitude. In the following code the number -999 is used as the sentinel value:

```
        NUMVAL = 0
        TOTAL = 0
        READ(5,*)VALUE
C DOWHILE LOOP
    30 IF (VALUE .NE. -999) THEN
            NUMVAL = NUMVAL + 1
            TOTAL = TOTAL + VALUE
            READ(5,*) VALUE
            GO TO 30
        ENDIF
C COMPUTE AVERAGE IF NUMVAL GREATER THAN ZERO
    70 IF (NUMVAL .GT. 0) AVERG = TOTAL/NUMVAL
```

The READ statement prior to the loop, which is called a *preparatory input*, occurs only once before the start of the loop. It is required so that the user has the option of entering no values at all. If the first entry were -999, then the body of the loop would not be implemented and NUMVAL would remain at zero. Thus the computation of the average value in line 70 is conditional in order to avoid division by zero for this case.

Notice also how the order of the statements within the loop differs from that in Example 19.1. For the present case, the input of data is at the end of the loop. This ordering prevents the sentinel value from being incorporated into TOTAL or counted in NUMVAL.

The DOUNTIL structure. As seen in Fig. 19.4c, the DOUNTIL version employs indentation and comments in a fashion similar to the DOWHILE structure. However, there are two features that make the DOUNTIL somewhat more complicated. First, the condition in line 30 is formulated using the NOT logical operator. This is done so that, as in the flowchart, the loop occurs if the condition is not true. Although formulating the condition in this way is consistent with the DOUNTIL structure, it sometimes makes the condition a bit more difficult to formulate. Second, in order that the loop execute at least once, the condition must be artificially set false beforehand. The following example provides an illustration of how this is done.

EXAMPLE 19.4   *Using a DOUNTIL Loop to Check the Validity of Data*

PROBLEM STATEMENT: One situation where a DOUNTIL loop would be of use is in ensuring that a valid entry is made from the keyboard. For example, suppose that you developed a program that required nonzero data. Such would be the case where the data might serve as the divisor in a formula, and hence division by zero had to be avoided. Therefore it would be advisable to have the program check for zero values upon input. Use the DOUNTIL loop to accomplish this objective.

SOLUTION: The following program fragment inputs values and checks to ensure that they are not zero:

```
   10 VALUE = 1
 C DOUNTIL LOOP
   30 IF (.NOT. (VALUE .EQ. 0)) THEN
   40    WRITE(6,*)'VALUE = '
         READ(5,*)VALUE
   50    GO TO 30
      ENDIF
   60 CONTINUE
```

Line 10 forces the loop to execute at least once. This allows the input to occur prior to the test in line 30, as is proper for a DOUNTIL structure. Thus, you are not able to proceed beyond this set of statements unless a nonzero VALUE is entered.

```
WHILE condition              LOOP
    (Body of loop)               (Body of loop)
END WHILE                    UNTIL condition
    (a)                          (b)
```

FIGURE 19.5
The (a) WHILE and (b) UNTIL constructs of WATFIV FORTRAN.

19.4 WHILE AND UNTIL STATEMENTS

Some extended forms of FORTRAN, such as WATFIV, include several versions of the DOWHILE and DOUNTIL constructs. Some of these are summarized in Fig. 19.5.

These are much more concise than the simulated DOWHILE and DOUNTIL constructs presented in the previous section. In particular, they do not include the GO TO statements that were needed in the simulated constructs. Because of their economy and elegance, statements such as the WHILE and UNTIL constructs will undoubtedly be incorporated into the major FORTRAN dialects in the years to come.

PROBLEMS

19.1 The exponential function can be evaluated by the following infinite series:

$$e^x = 1 + x + \frac{x^2}{2} + \frac{x^3}{3!} + \cdots$$

Write a program to implement this formula so that it computes the value of e^x as each term in the series is added. In other words, compute and print in sequence

$$e^x \cong 1$$

$$e^x \cong 1 + x$$

$$e^x = 1 + x + \frac{x^2}{2}$$

up to the order term of your choosing. For each of the above, compute the percent relative error as

$$\% \text{ error} = \frac{\text{true} - \text{series approximation}}{\text{true}} \times 100\%$$

Use the library function for e^x in your computer to determine the "true value." Have the program print out the series approximations and the error at each step along the way. Employ loops to perform the analysis. As a test case, employ the program to evaluate $e^{1.5}$ up to the term $x^{10}/10!$.

19.2 Repeat Prob. 19.1 for the hyperbolic cosine, which can be approximated by

$$\cosh x = 1 + \frac{x^2}{2!} + \frac{x^4}{4!} + \cdots$$

As a test case, employ the program to evaluate cosh 78° up to the term $x^{10}/10!$. Remember that both the cosh and the above formula use radians.

19.3 Repeat Prob. 19.1 for the sine, which can be approximated by

$$\sin x = x - \frac{x^3}{3!} + \frac{x^5}{5!} - \frac{x^7}{7!} + \cdots$$

As a test case, employ the program to evaluate sin 30°. Remember, both the sine function and the above formula use radians.

19.4 Write a computer program to evaluate the following series:

$$\text{SUM} = \sum_{i=0}^{n} \frac{1}{2^i} \qquad \text{for } n = 20$$

Print out SUM and n during each iteration. Does the series converge to a constant value or diverge to infinity?

19.5 Repeat Prob. 19.4, except with the following formula ($n = 20$):

$$\text{SUM} = \sum_{i=1}^{n} \frac{1}{i^2}$$

19.6 Repeat Prob. 19.4 given ($n = 20$)

$$\text{SUM} = \sum_{i=0}^{n} \left(\frac{2}{5}\right)^i$$

19.7 Consider the following formula from calculus:

$$\frac{1}{1 - x} = 1 + x + x^2 + \cdots$$

Write a computer program to test the validity of this formula for values of x of 0.6, -0.3, 2, and $-3/2$.

19.8 The series

$$f(x) = \frac{4}{\pi}\left(\sin x + \frac{\sin 3x}{3} + \frac{\sin 5x}{5} + \cdots\right)$$

is an approximation of $f(x)$ as shown in Fig. P19.8. Write a computer program to evaluate the accuracy of the series as a function of the number of terms at $x = \pi/2$. The program should print out the error in the approximation and the number of terms in the series.

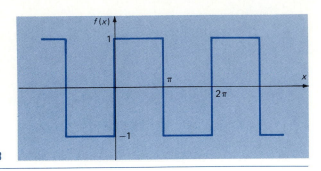

FIGURE P19.8

19.9 Repeat Prob. 19.8, but use Fig. P19.9, which is represented by the series

$$f(x) = \frac{\pi}{4} - \frac{2}{\pi}\left(\cos x + \frac{\cos 3x}{9} + \frac{\cos 5x}{25} + \cdots\right)$$

$$+ \left(\sin x - \frac{\sin 2x}{2} + \frac{\sin 3x}{3} - \cdots\right)$$

Check the approximation at $x = \pi/2$ and $x = 3\pi/2$.

FIGURE P19.9

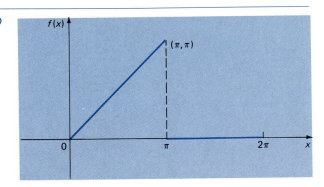

19.10 Repeat Example 16.7 but use loops to make the computation repetitive. Generate 10, 50, and 100 random coin tosses and tally how many heads and tails are tossed for each case. Comment on your results.

19.11 Reprogram Fig. 16.7 but use loops to make the code more efficient.

CHAPTER 20

FORTRAN:
SUBPROGRAMS

It is sometimes desirable to execute a particular set of statements several times and at different points in a program. It would obviously be inconvenient and awkward to rewrite this set separately every time it is required. Fortunately, it is possible in FORTRAN to employ a subroutine or a user-defined function subprogram to write such sections once. You can then have them at your disposal within the program to use as many times as necessary. In addition, subroutines can be used to structure a program in a modular fashion.

20.1 SUBROUTINE STATEMENTS

The CALL statement is employed to transfer to the subroutine and the RETURN statement is used to transfer back to the main program. The general form of this statement is

CALL *name* ⟨*arg₁*, *arg₂*, ..., *argₙ*⟩

where *name* is the name of the subroutine and *argᵢ* is the *i*th argument that is passed between the main program and the subroutine.

The subroutine itself begins with a statement of the form

SUBROUTINE *name* ⟨*arg₁*, *arg₂*, ..., *argₙ*⟩

where *name* is identical to the name in the CALL statement and each of the arguments corresponds to arguments listed in the CALL. These arguments are not required to

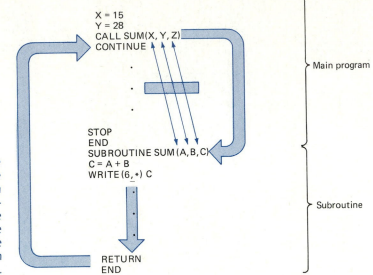

FIGURE 20.1
Typical sequence of a subroutine call and return. Variables are passed between the main program and the subroutine via the arguments. After the execution of the subroutine, the variable Z in the main program will contain the value of the sum as computed in the subroutine.

have the same name given in the CALL statement. Only their position in the argument list determines which values get passed to which variables. Although the names may differ, the variable types must match. The subroutine ends with a RETURN and an END statement. The CALL statement transfers control to the first line of the subroutine. This operation of transferring control is referred to as *calling* the subroutine. Within the subroutine, execution proceeds as normal until the RETURN is encountered, whereupon control is transferred back to the line immediately following the CALL statement. The process is depicted graphically in Fig. 20.1.

The effect of calling a subroutine is like having a GO TO that branches from the middle of the main program to another area of the program and a GO TO that then branches back to the statement immediately after the calling GO TO. The advantage of using CALL and SUBROUTINE statements instead of two GO TO statements is that the former conveniently allows the subroutine to be called from several locations. The RETURN assures that control is automatically transferred back to the correct line in the main program, whereas the GO TO would have to be modified to return correctly for each particular case.

All subroutines should be located together after the END statement of the main program. This organization reflects the top-down nature of program design, with the broad ideas in the main program followed by the detailed operations in the subroutines.

EXAMPLE 20.1   Subroutine to Compute Volume and Surface Area of a Rectangular Prism

PROBLEM STATEMENT: Figure 13.2a summarizes formulas for the volumes and surface areas of a rectangular prism. Develop a subroutine to compute these values for this geometric solid.

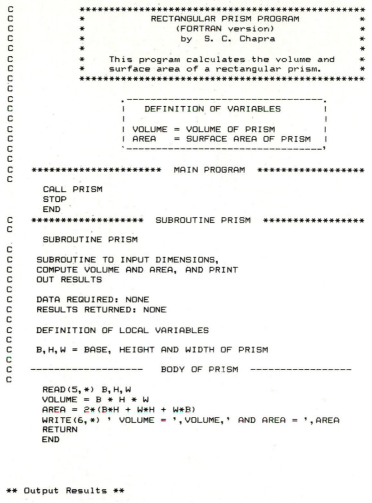

```
C     ****************************************************
C     *                                                  *
C     *            RECTANGULAR PRISM PROGRAM             *
C     *               (FORTRAN version)                  *
C     *                 by  S. C. Chapra                 *
C     *                                                  *
C     *       This program calculates the volume and     *
C     *       surface area of a rectangular prism.       *
C     ****************************************************
C
C
C            .------------------------------------.
C            I                                    I
C            I       DEFINITION OF VARIABLES      I
C            I                                    I
C            I  VOLUME = VOLUME OF PRISM          I
C            I  AREA   = SURFACE AREA OF PRISM    I
C            '------------------------------------'
C
C
C     ********************* MAIN PROGRAM  *****************
C
      CALL PRISM
      STOP
      END
C     ****************** SUBROUTINE PRISM  ***************
C
      SUBROUTINE PRISM
C
C     SUBROUTINE TO INPUT DIMENSIONS,
C     COMPUTE VOLUME AND AREA, AND PRINT
C     OUT RESULTS
C
C     DATA REQUIRED: NONE
C     RESULTS RETURNED: NONE
C
C     DEFINITION OF LOCAL VARIABLES
C
C     B,H,W = BASE, HEIGHT AND WIDTH OF PRISM
C
C     ------------------ BODY OF PRISM ----------------
C
      READ(5,*) B,H,W
      VOLUME = B * H * W
      AREA = 2*(B*H + W*H + W*B)
      WRITE(6,*) ' VOLUME = ',VOLUME,' AND AREA = ',AREA
      RETURN
      END
```

FIGURE 20.2
A FORTRAN program to compute the volume and surface area of a rectangular prism.

```
** Output Results **

VOLUME =    18.0000000    AND AREA =    48.0000000
```

SOLUTION: Figure 20.2 shows FORTRAN code to solve this problem. Notice how both the input/output operations and the computations are sequestered in the subroutine.

20.2   THE ORGANIZATION AND STRUCTURE OF SUBROUTINES

Just as chapters are the building blocks of a novel, subroutines are the building blocks of a computer program. Each subroutine can have all the elements of a whole program, but, in general, it is smaller and more specialized. Some computer experts suggest that a reasonable size limit for a subroutine should be about 50 lines. Because this size will fit on one or two monitor screens, it allows quick viewing of the whole subroutine. It is a reasonable guideline because it facilitates visualizing the purpose and the composition of the subroutine.

In addition to allowing you to easily view the subroutine, the 50-line limit will also encourage you to restrict the subroutine to one or two well-defined tasks. It is this characteristic of subroutines that makes them the ideal vehicle for expressing a modular design in program form.

The general organization of a subroutine is outlined in Fig. 20.3. Subroutines should always be considered major structures in the program and, consequently, should be delineated clearly using blank spaces and lines. Every subroutine should begin with a short description of the purpose of the subroutine, the data required, the results returned, and definitions of local variables (that is, those used exclusively in the subroutine) should be included after the title and before the body of the subroutine. Some of these characteristics are mainfested in the program in Fig. 20.2.

The statements in the body of the subroutine should be organized in a fashion similar to that of the main program. Structures should be identified by indentations and remarks as needed. Because a subroutine's task is, by definition, limited and focused, there should be few structures within each subroutine. Just as with the main program, each subroutine should start and end with no indentation.

The subroutine should always terminate with the RETURN and END statements. Strictly speaking, subroutines may have several RETURN statements for routing back to the main program. However, because of structured programming considerations, this practice is discouraged. Rather, each subroutine should have a single entrance at its beginning and a single exit at its end. In this way, the logical flow is kept simple and is easy to understand.

20.3 LOCAL AND GLOBAL VARIABLES

A two-way communication is required between the subroutine and the main program. The subroutine needs data to act on and must return its results for utilization in the

FIGURE 20.3
The general organization of a subroutine.

```
SUBROUTINE name (arg₁, arg₂, ..., argₙ)
Type declaration of arguments
Comments describing the action of the subroutine
Definition of global variables
Type definition of local variables
Body of the subroutine

     .
     .
     .

RETURN
END
```

main program. The data and results, which are called *arguments*, are said to be "passed" to and from the program. For FORTRAN, this exchange of information is a controlled process. That is, the information that is passed in and out of the subroutine must be specified explicitly by the programmer; the arguments must be of the same type and must be declared in both the CALL and the SUBROUTINE statements. These variables, along with the other variables in the main program, are commonly referred to as *global variables*. In contrast, variables that reside exclusively in the subroutine are called *local variables*.

The ability to have local variables in a subroutine can be a great asset. In essence, they serve to make each subroutine a truly autonomous entity. For example, changing a local variable will have no effect on variables outside the subroutine. In fact, true local variables can even have the same name as variables in other parts of the program yet still be distinct entities. One advantage of this is that it decreases the likelihood of "side effects" that occur when the value of a global variable is changed inadvertently in a subprogram. In addition, the existence of local variables also facilitates the debugging of a program. The subprograms can be debugged individually without fear that a working subroutine will cease functioning because of an error in another subroutine.

20.3.1   Global Variables in FORTRAN Programs: COMMON

Although local variables are a great benefit, there will be occasions when you will want your subprograms to have direct access to certain variables in the calling program. Any type of variable may be made available for use by a subprogram by using the COMMON statement, which allows variables to be defined as global variables. The COMMON statement directs the compiler to set aside a certain area in the computer's memory which the main program and subprograms may access to obtain values for the variable. COMMON statements can be unnamed blank COMMONs, or they can be given explicit names. For an explicitly named COMMON the general form is

        COMMON /name/ var$_1$, ..., var$_n$

For a blank COMMON the general form is

        COMMON var$_1$, ..., var$_n$

or

        COMMON // var$_1$, ..., var$_n$

Both the main program and the subprogram must contain the COMMON statement in order for both parts to have access to a variable.

Within a subprogram the number of variables in a COMMON statement can be fewer than in the main program, and the variables do not have to have the same name as in the main program. The correspondence between a variable in the main program and in a subprogram is by position in the COMMON statement. If the main program has a COMMON statement

```
COMMON /SUBSET/ ARR, X, VEL, NAME
```

the subprogram could have a COMMON statement such as

```
COMMON /SUBSET/ XMATR, Y
```

or

```
COMMON /SUBSET/ ARR, Y, VAL, LNAME
```

The values of ARR in the main program would be found in XMATR in the first COMMON statement and ARR in the second. In the first statement VEL and NAME would not be available to the subprogram. In fact, these same names could be used by the subprogram for local variable names with entirely different values from those in the main program. In the subprogram, Y would refer to the same data as X in the main program.

The COMMON statement should follow type declaration statements, but it should be placed prior to the executable statements of the program or subprograms.

20.4   USER-DEFINED FUNCTION SUBPROGRAMS

The major difference between a subroutine and a user-defined function subprogram is that the latter returns a result that can be used directly, just as if it were a constant. To this point, we have covered two types of statements that behave in a similar fashion—the intrinsic functions and statement functions discussed in Chap. 16. Recall how an intrinsic function such as SIN(X) can be used as the part of, say, a mathematical expression. The user-defined function subprograms can be employed in a similar fashion. However, they differ from the intrinsic and the statement functions in some fundamental ways. In contrast to the intrinsic function, which is supplied by the system, the user-defined function subprogram (as the name implies) is concocted by the user. In contrast to the user-defined statement function, which is limited to a single line, the function subprogram can consist of many lines. Thus, as seen in Fig. 20.4, the function subprogram lies midway between statement functions and the subroutines.

The general form of the function subprogram is

$type$ FUNCTION $name(arg_1, arg_2, ..., arg_n)$
    .
    .
    .
$name$ =

```
        +
        +
        +
RETURN
END
```

Intrinsic or Library Functions: Built-in functions that perform commonly employed mathematical or trigonometric operations; generate a single value that can be used as part of a mathematical expression.

```
Definition:        Defined by system
Application:       Y = SIN(X)
```

STATEMENT FUNCTIONS: User-defined functions that can be employed as part of a mathematical expression; the function is defined by a single line, which must be placed prior to the first executable statement.

```
Definition:        ROUND(X) = INT(X + 0.5)
Application:       Y = ROUND(X)
```

FUNCTION SUBPROGRAMS: User-defined miniprograms that consist of several lines. which must be placed after the main program; generate a single value that can be used as part of a mathematical expression.

```
Application:       Y = AREA(R)
Definition:        FUNCTION AREA(R)
                   AREA = 3.14159 * R ** 2
                   RETURN
                   END
```

SUBROUTINES: User-defined miniprograms that consist of several lines, which must be placed after the main program. Several results may be generated and passed back to the main program as variables. The results are passed between the subroutine and the main program via the subroutine's arguments or via COMMON statements.

```
Application:       CALL SUM(X,Y,Z)
Definition:        SUBROUTINE SUM(A,B,C)
                   C = A + B
                   WRITE(6,*)C
                   STOP
                   END
```

where *type* declares the data type of the result returned (this can be omitted if the *type* is REAL), the *name* must be a valid variable name that is consistent with the *type*, and *arg*$_i$ is the *i*th argument passed to the subprogram. Notice that one assignment statement (without arguments) must appear in the body of the subprogram to assign a value to *name*. It is this value that is then given to the *name* when it is invoked (with arguments) in the calling program.

The organization of a user-defined function subprogram follows that of other elements of the program. The subprogram should, therefore, be set apart by lines and spaces and should include comments and indentation as necessary. Notice also that the last lines must be RETURN and END statements. In general, these subprograms are usually placed with the subroutines at the end of the program.

EXAMPLE 20.2   *User-Defined Function Subprograms*

PROBLEM STATEMENT: Employ a user-defined function subprogram to determine cartesian (*x*, *y*) coordinates, given the radius *r* and the angle θ in polar coordinates. Allow the angle to be input in either degrees or radians. Enter the radius in inches and have the main program output the result in feet.

SOLUTION: The program to solve this problem is shown in Fig. 20.5. Notice how the user-defined function subprogram consists of several lines. If necessary, the value of the angle is first converted to radians. Then the *x* coordinate is computed. This result is returned directly to an equation in the main program where it is converted from inches to feet.

FIGURE 20.5
A FORTRAN program to calculate the *x* coordinate on the basis of the polar cordinates.

```
C -----------------------------------------------------
      READ(5,*) RADIUS,THETA,NUNIT
      X = XCOORD(RADIUS,THETA,NUNIT)/12
      WRITE(6,*) 'THE X COORDINATE IS ',X
      STOP
      END
C -----------------------------------------------------
      FUNCTION XCOORD(R,T,N)
      IF (N.EQ.1) THEN
         T = T * 3.14159/180
      END IF
      XCOORD = R*COS(T)
      RETURN
      END



** Output Results **

THE X COORDINATE IS    1.66666889
```

20.5 YOUR SUBROUTINE LIBRARY

Subroutines facilitate the process of program development for two major reasons. First, they are discrete, task-oriented modules and are therefore easier to understand, write, and debug. Second, they can be used over and over both within a program and between programs. Testing a small subroutine to prove that it works properly is far easier than testing an entire program. Once a subroutine is thoroughly debugged and tested, it can be incorporated into other programs. Then if errors occur, the subroutine is unlikely to be the culprit, and the debugging effort can be focused on other parts of the program.

Many professional programmers accumulate a collection, or library, of useful subroutines that they access as needed. Thus instead of programming lines of code, they combine and manipulate larger, more powerful pieces of code—that is, subroutines. Programming with these custom modules saves overall effort and, more importantly, improves program reliability. The modules in a library can be thoroughly debugged under a number of different conditions so as to assure proper function. Such thorough testing can rarely be afforded for sections of code that are only used once.

In the remainder of the book we will be developing a number of programs that have general applicability in engineering (see Table 20.1). You should use these programs as the beginning of your own subroutine library with an eye to their utility in your future academic and professional efforts.

TABLE 20.1 *Programs to Be Presented in the Remainder of the Book That Have General Applicability in Engineering (You should set a goal to develop subroutines for each of these areas as the start of your own software library.)*

Program	Application	Chapter
Plotter	Plot points or lines on cartesian coordinates	22,39
Shell sort	Sort numbers or words in ascending order	23
Histogram	Create a bar diagram depicting the shape of a data distribution	24
Statistics	Calculate a number of descriptive statistics to characterize data	24
Linear regression	Fit the best straight line through data plotted on cartesian coordinates	26
Lagrange interpolation	Use polynomials to estimate intermediate values between data plotted on cartesian coordinates	27
Bisection	Determine the root of a single algebraic or transcendental equation	28
Gauss elimination	Solve simultaneous linear algebraic equations	29
Numerical differentiation	Determine the derivative of data	30
Trapezoidal rule	Determine the integral of data	31
Euler's method	Solve a single, first-order rate equation	32

PROBLEMS

20.1 Write a program to determine the roots of a quadratic equation. Pattern the computation after Fig. 18.7. However, use two separate subroutines to (1) input the coefficients and (2) compute and print out the results. Organize the subroutines as discussed in Sec. 20.2.

20.2 Develop a function subprogram to round a number if it is positive and to truncate it if it is negative. Design the subprogram so that it rounds and truncates to a specified number of decimals.

20.3 Write a subroutine that accomplishes the same operation as the SGN function of BASIC. (See Table 9.2.) Organize the subroutine as described in Sec. 20.2.

20.4 Write a program to determine the values of all the trigonometric functions in Table 16.3 given an angle. Use a subroutine to input the angle and transform it to radians if it is entered in degrees. Employ another subroutine to determine the values of the trigonometric functions and print out the results. Organize the subroutine as described in Sec. 20.2. Test the program with an angle equal to 30°.

20.5 Write a program to determine the range of a projectile fired from the top of a cliff as described in Prob. 16.21. Employ a modular approach and use three subroutines to (1) input the data, (2) perform the computation, and (3) output the results.

20.6 Write a subroutine to input and convert temperatures from degrees Fahrenheit to degrees Celsius and Kelvin by the formulas

$$T_C = \frac{5}{9}(T_F - 32)$$

and

$$T_K = T_C + 273$$

Organize the subroutine as discussed in Sec. 20.2. Test it with the following data; $T_F = 20$, 203, and 98.

20.7 Write a subroutine to input and convert temperature from degrees Celsius to degrees Fahrenheit and Rankin by the formulas

$$T_F = \frac{9}{5}T_C + 32$$

and

$$T_R = T_F + 460$$

Organize the subroutine as discussed in Sec. 20.2. Test it with $T_C = -10$, 27, and 90.

20.8 What are the differences between a user-defined-function subprogram, a subroutine, and a statement function? How many arguments can each one have in its argument list?

20.9 What is the function of the COMMON statement?

20.10 Write a subroutine to compute the volume and surface area of a sphere. See Fig. 13.2 for the formulas. Employ this subroutine in conjunction with the main program from Fig. 20.2. However, modify the main program so that it reads the value for the radius and writes out the values for the volume and surface area. *Note:* This requires that the subroutine pass the values as arguments. The subroutine definition line might, therefore, look something like

```
SUBROUTINE SPHERE(R,VOL,AREA)
```

20.11 Write a function subprogram to linearly interpolate between two points: x_1,y_1 and x_2,y_2. Place the interpolated value as the value of the function. Specifically, the user-defined function will have five values in its argument list: x_1, y_1, x_2, y_2, and the value of x at which the interpolation is to be performed. Note also that your function should check if the value of x is within the interval from x_1 to x_2 and should print out an error message if not. *Hint:* Note that the function name must always be assigned a value in the subprogram. Guard against the possibility that it might not.

20.12 Write a function subprogram that checks whether the first value in the argument list is equal to one of the four remaining values. If the first value is identical to one of the other arguments, assign a value of .TRUE. to the function's name, otherwise assign it a value of .FALSE. *Note:* This requires that the function name itself be declared a logical variable in the main program. For example, if the function is named CHECK, then the main program must declare CHECK to be logical with the statement

```
LOGICAL CHECK
```

Additionally, at the start of the subprogram, the function name must be declared as a logical function. This can be done with the statement

```
LOGICAL FUNCTION CHECK(TEST,X1,X2,X3,X4)
```

Test your main program and function by assigning different values to the arguments; after the function is executed, have the program write a message stating whether a match was found. Note that on some compilers, the above procedure may yield an error message. If this occurs on your system, determine how you would circumvent the error.

CHAPTER 21
FORTRAN: LARGE AMOUNTS OF DATA

In our discussion of FORTRAN in Chap. 5, we mentioned that business typically deals with large amounts of data and simple computations, whereas engineering deals with small amounts of data and complicated computations. Although this is generally true, some engineering applications require large quantities of information. Here we present material that is particularly useful for these situations.

21.1 ARRAYS

To this point we have learned that a variable name conforms to a storage location in primary memory. Whenever the variable name is encountered in a program, the computer can access the information stored in the location. This is an adequate arrangement for many small programs. However, when dealing with large amounts of information, it often proves inadequate.

For example, suppose that in the course of an engineering project you were using stream-flow measurements to determine how much water you might be able to store in a projected reservoir. Because some years are wet and some are dry, natural stream flow varies from year to year in an almost random fashion. Thus you would require about 25 to 50 years of measurements to accurately estimate the long-term average flow. Data of this type is listed in Table 21.1. A computer would come in handy in such an analysis. However, it would obviously be inconvenient to come up with a different variable name for each year's flow. Similar problems arise continually in engineering and other computer applications. For this reason subscripted variables, or arrays, have been developed.

TABLE 21.1   Thirty Years of Water-Flow Data for a River

Year	Flow*	Year	Flow*
1954	125	1969	63
1955	102	1970	115
1956	147	1971	71
1957	76	1972	106
1958	95	1973	53
1959	119	1974	99
1960	62	1975	153
1961	41	1976	112
1962	104	1977	82
1963	81	1978	132
1964	128	1979	96
1965	110	1980	102
1966	83	1981	90
1967	97	1982	122
1968	79	1983	138

*In billion gallons per year.

An *array* is an ordered collection of data. That is, it consists of a first item, second item, third item, etc. For the stream flow example, an array would hold data from the first year's flow, the second year's flow, the third year's flow, etc.

In programming, all the individual items, or *elements*, of the array are referenced by the same variable name. In order to distinguish between the elements, each is given a subscript. This is a common practice in standard mathematics. For example, if we assign flow the variable name F, an index, or subscript, can be attached to distinguish each year's flow, as in

$$F_1 \quad F_2 \quad F_3 \quad F_4 \quad F_5$$

where F_i refers to the flow in year i. These are referred to as *subscripted variables*. In FORTRAN a similar arrangement is employed, but because subscripts are not allowed, the index is placed in parentheses, as in

F(1)   F(2)   F(3)   F(4)   F(5)

Note that for an array the parentheses are part of the variable name. This makes the elements different from F1, F2, etc. In fact, F(1) and F1 are considered to be different variables by FORTRAN and can be used in the same program. Each would be assigned a different storage location, as depicted in Fig. 21.1.

Assignment statements can be employed to assign values to the elements of an array in a fashion similar to assigning them to nonsubscripted variables. For example, using the data in Table 21.1,

```
F(1) = 125
F(2) = 102
```

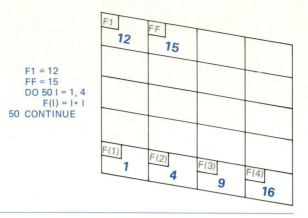

FIGURE 21.1
Visual depiction of memory locations for unsubscripted and subscripted variables. The program on the left assigns values to F1, FF, and F(1) through F(4), with the resulting constants being stored in the memory locations depicted on the right.

```
F(3) = 147
     *
     *
     *
```

In addition, READ statements can also be used, as in

```
READ(5,*) F(1)
READ(5,*) F(2)
READ(5,*) F(3)
     *
     *
     *
```

Although the above methods can be employed, loops can be used to perform the same operation more efficiently. The following example illustrates the advantages of utilizing arrays and loops when dealing with large quantities of data.

EXAMPLE 21.1   Inputting Data to an Array with Loops

PROBLEM STATEMENT: Develop a program fragment to input the first 10 values from Table 21.1.

SOLUTION: Using assignment or individual READ statements to enter the data would result in a program fragment of 10 statements. A much more efficient arrangement is possible using a DO loop and a READ statement in tandem, as in

```
        DO 200 I = 1,10
            READ(5,*) FLOW(I)
    200 CONTINUE
```

Notice how we have used a more descriptive name, FLOW, for the array. In addition, notice how the index (I) of the loop is used as the subscript for FLOW. Thus, on the first pass through the loop, I is 1, and the program fragment will prompt us to enter FLOW(1). Then on subsequent passes I will be incremented in order to input FLOW(2), FLOW(3), etc. Consequently, the loop allows us to enter the data using only 3 lines rather than 10.

The subscripts of an array can be any positive whole number, including zero. However, the special DIMENSION statement must be added to declare the variable names as being subscripted, as described in the next section.

21.1.1 The DIMENSION Statement

It is necessary to set aside memory locations for any subscripted variable in a FORTRAN program. A DIMENSION statement is required to do this. Its general form is

```
DIMENSION var₁(n₁), var₂(n₂), ..., varₙ(nₙ)
```

where var_1 and var_2 are the variable names of different arrays and n_1 and n_2 are the upper limits on the number of elements in each array. For example, if we wanted to perform an analysis on all the data in Table 21.1, we would have to include the following DIMENSION statement in our program:

```
DIMENSION FLOW(30)
```

This would set aside 30 storage locations for the 30 flows in Table 21.1. Suppose that we also wanted to input the years as string variables. If we declared this as a character variable YEAR, we would rewrite the DIMENSION statement as

```
DIMENSION FLOW(30), YEAR(30)
```

The DIMENSION statement must be placed at the beginning of the module where the subscripted variable is utilized. Otherwise the compiler will have no idea whether a subscripted name is a subscripted variable or a function defined by the user. Also, in the main program you must use integer constants, not variables, to dimension variables.

21.1.2 Size of Arrays

In the previous section we showed how a DIMENSION statement is required to define a 30-element array for the water-flow data in Table 21.1. We could just as well have set aside more locations, as in

```
DIMENSION FLOW(100), YEAR(100)
```

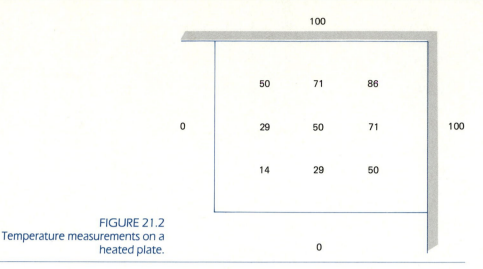

FIGURE 21.2
Temperature measurements on a
heated plate.

This might be done if we anticipated that at a later time additional data (say, from the years after 1983) would be obtained. Thus we do not have to specify the exact size of an array in the DIMENSION statement; we merely have to make sure that there are enough elements available to store our data. Otherwise an error will occur.

The limit on the number of elements for subscripted variables depends on the amount of memory available on your computer. Therefore for large programs it is advisable to estimate the size of your array realistically so that you do not exceed your computer's memory capacity.

21.1.3 Two-Dimensional Arrays

There are many engineering problems where data is arranged in tabular or rectangular form. For example, temperature measurements on a heated plate (Fig. 21.2) are often taken at equally spaced horizontal and vertical intervals. Because such two-dimensional arrangements are so common, FORTRAN has the capability of storing data in rectangular arrays or *matrices*. As seen in Fig. 21.3, the horizontal sets of elements are called *rows*, whereas the vertical sets are called *columns*. The first parenthetical numeral designates the number of the row in which the element lies. The second parenthetical numeral designates the column. For example, element A(2,3) is in row 2 and column 3.

EXAMPLE 21.2   Two-Dimensional Arrays

PROBLEM STATEMENT: Write a computer program to input the temperature data from Fig. 21.2. Then convert it to degrees Fahrenheit by the formula

$$T_F = \frac{9}{5}(T_C + 32)$$

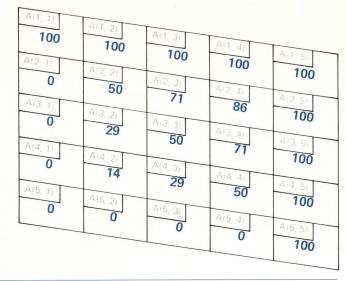

FIGURE 21.3
A two-dimensional array, or matrix, used to store the temperatures from Fig. 21.2. Note that the parenthetical numerals designate the row and column of the element, respectively.

Employ a user-defined function to make this conversion. Print out the results in a tabular format.

SOLUTION: Figure 21.4 shows the program. Notice that a modular approach is used. The main, or calling, program consists of three CALL staements for subroutines that input, convert, and output the data.

Note that in FORTRAN arrays with more than two dimensions can be used. The maximum number of dimensions allowed in FORTRAN 77 is seven.

21.1.4 Dimension, Type, Specification Statements in Program Modules

Because each major module (that is, the main program, subroutines, and function subprograms) is compiled separately, each must contain any type declarations, specification statements (for example, DIMENSION, COMMON), and DATA statements that are required in that module.

In each module, these statements should be placed in a particular order. The order is

1. Title statements
2. IMPLICIT type statements
3. EXPLICIT type statements
4. DIMENSION statements
5. COMMON statements

```
C      *******************************************************
C      *                                                     *
C      *            HEATED PLATE PROGRAM                     *
C      *              (FORTRAN version)                      *
C      *               by  S. C. Chapra                      *
C      *                                                     *
C      *  This program inputs temperatures for a             *
C      *  heated plate. It converts temperatures from        *
C      *  Celsius to Fahrenheit and transforms the           *
C      *  results to integers. The results are output.       *
C      *******************************************************
C
C          .-------------------------------------------.
C          |                                           |
C          |         DEFINITION OF VARIABLES           |
C          |                                           |
C          |  NUMROW = NUMBER OF ROWS ON PLATE         |
C          |  NUMCOL = NUMBER OF COLUMNS ON PLATE       |
C          |  TEMP() = TEMPERATURE OF PLATE AT         |
C          |           A PARTICULAR LOCATION           |
C          |                                           |
C          '-------------------------------------------'
C
C
C      *******************  MAIN PROGRAM  *******************
C
       DIMENSION TEMP(7,7)
C
       CALL INPUT(NUMROW,NUMCOL,TEMP)
       CALL CONVER(NUMROW,NUMCOL,TEMP)
       CALL OUTPUT(NUMROW,NUMCOL,TEMP)
       STOP
       END
C
C      **************  SUBROUTINE INPUT  **************
C
       SUBROUTINE INPUT(NUMROW,NUMCOL,TEMP)
C
C
C          INPUTS TEMPERATURES FOR THE ROWS AND
C          COLUMNS OF THE HEATED PLATE.
C
C          DATA REQUIRED: NONE
C          RESULTS RETURNED: NUMROW,NUMCOL,TEMP()
C
C          LOCAL VARIABLES:
C
C          I, J = COUNTERS
C
       DIMENSION TEMP(7,7)
C
C      ------------------ BODY OF INPUT ------------------
C
       READ(5,*) NUMROW,NUMCOL
       DO 10 I = 1,NUMROW
          READ(5,*) (TEMP(I,J),J = 1,NUMCOL)
   10  CONTINUE
       RETURN
       END
C
```

```
C      ***************  SUBROUTINE CONVER  ***************
C
       SUBROUTINE CONVER(NUMROW,NUMCOL,TEMP)
C
C          CONVERTS TEMPERATURES FROM CELSIUS
C          TO FAHRENHEIT AND CONVERTS THE RESULT
C          TO AN INTEGER.
C
C          DATA REQUIRED: NUMROW,NUMCOL,TEMP()
C          RESULTS RETURNED: TEMP()
C
C          LOCAL VARIABLES:
C
C          I, J = COUNTERS
C
       DIMENSION TEMP(7,7)
       CONVRT(T) = (9./5.)*T + 32.
       ROUND(T) = INT(T+.5)
C
C      ----------------- BODY OF CONVER -----------------
C
       DO 30 I = 1,NUMROW
          DO 20 J = 1,NUMCOL
             TEMP(I,J) = CONVRT(TEMP(I,J))
             TEMP(I,J) = ROUND(TEMP(I,J))
   20     CONTINUE
   30  CONTINUE
       RETURN
       END
C
C      **************  SUBROUTINE OUTPUT  **************
C
       SUBROUTINE OUTPUT(NUMROW,NUMCOL,TEMP)
C
C          PRINTS OUT THE RESULTS IN TABULAR
C          FORM.
C
C          DATA REQUIRED: NUMROW,NUMCOL,TEMP()
C          RESULTS RETURNED: NONE
C
C          LOCAL VARIABLES:
C
C          I, J = COUNTERS
C
       DIMENSION TEMP(7,7)
C
C      ----------------- BODY OF OUTPUT -----------------
C
       DO 40 I = 1,NUMROW
          WRITE(6,100) (TEMP(I,J),J = 1,NUMCOL)
   40  CONTINUE
       RETURN
  100  FORMAT(7(3X,F7.1))
       END
//$DATA
5,5
100.,100.,100.,100.,100.
0.,50.,71.,86.,100.
0.,29.,50.,71.,100.
0.,14.,29.,50.,100.
0.,0.,0.,0.,100.

** Output Results **

   212.0     212.0     212.0     212.0     212.0
    32.0     122.0     160.0     187.0     212.0
    32.0      84.0     122.0     160.0     212.0
    32.0      57.0      84.0     122.0     212.0
    32.0      32.0      32.0      32.0     212.0
```

FIGURE 21.4 A program to input, convert, and output a two-dimensional array.

6. DATA statements
7. Statement functions
8. The body of the program
9. END statement

Aside from the proper ordering of statements, some comments on dimensioning of variables are in order. Dimensioned variables that are exclusively employed in a subroutine or function should be dimensioned in the subprogram in exactly the same manner as they would be in a main program.

For variables that are used in both the main program and in a subprogram, a number of approaches can be taken. In the case of one-dimensional arrays, the method is simple. That is, it does not matter how the array is dimensioned in the subprogram so long as it is dimensioned. For example

```
C MAIN PROGRAM
        DIMENSION A(100)
            .
            .
            .
        CALL ADD(A)
            .
            .
            .
        SUBROUTINE ADD(X)
        DIMENSION X(1)
            .
            .
            .
```

Notice that when the dimensioned variable is passed as an argument, its name alone is passed. The purpose of the DIMENSION statement in the subroutine merely informs the compiler that the variable X is an array. Therefore, it can be dimensioned as X(1) [or X(20) or X(1000) for that matter] and the value in parentheses is disregarded.

However, for two- and higher-dimensioned arrays, the DIMENSION declaration must be precisely stated. One way to do this is to declare it in exactly the same fashion in both the main and the subprogram. Although this is acceptable, it represents a limited approach because it means that every array passed to the subprogram must be of the same size. A way around this dilemma is to pass the dimension size through the subprogram's argument as in the following example

```
C MAIN PROGRAM
        DIMENSION B(100,200),C(50,50)
        MB = 100
        NB = 200
        MC = 50
        NC = 50
            .
            .
            .
```

```
CALL PROD(B,MB,NB)
CALL PROD(C,MC,NC)
       .
       .
       .
SUBROUTINE PROD(Y,M,N)
DIMENSION Y(M,N)
       .
       .
       .
```

Thus, every time the subroutine is called the dimension of the subprogram array is automatically adjusted to match the particular array that is being passed from the main program.

21.2 FILES

To this point we have covered three methods for entering data into a program. Notice in Table 21.2 that all the methods have disadvantages related to large quantities of data and/or data for which frequent change might be required. In addition, all the methods deal exclusively with the information requirements of a single program. If the same data is to be used for another program, it must be retyped.

TABLE 21.2 The Advantages and Disadvantages of Four Ways in Which Information is Entered into FORTRAN Programs

Method	Advantages	Disadvantages	Primary Applications
Assignment statement	Is simple and direct. Values are "wired" into the program and do not have to be reentered on each run.	Takes up a line per assignment statement. If a new value is to be assigned, the user must retype the assignment statement.	Appropriate for assigning values to true constants (such as pi or the acceleration of gravity) that are used repeatedly throughout a program and never change from run to run.
READ statement	Permits different values to be assigned to variables on each run. Allows interaction between user and program.	Values must be entered by user on each run.	Appropriate for those variables that take on different values on each run.
DATA statement	Permits concise assignment of values.	If a new value is to be assigned, the user must retype the DATA statement.	Same as for the assignment statement, but for cases where many constants are to be assigned.
Files	Permit concise way to handle large amounts of input data. Allows data to be transferred between programs.	Are relatively complicated. Involves other devices in the computer.	Appropriate for large quantities of data.

System command:	Effect:
`RUN Grade1`	Begin executing program Grade1.
`RUN Grade1 7=ResultFile`	Attach UNIT 7 to the file called ResultFile instead of the default device.
`RUN Grade1 5=DataFile 6=PrintFile`	Have DataFile associated with UNIT 5 and PrintFile with UNIT 6.
`RUN Grade1 9=InFile`	All statements that reference UNIT 9 will use InFile.

FIGURE 21.5 Example of system commands to attach the unit numbers in a program to devices other than the default options.

Another available option is to store the data on an external memory device such as a magnetic disk or tape. This is conceptually similar to the way we have been saving our programs on disks or tapes so that they are not destroyed when the personal computer is shut off or when we sign off the mainframe. Along with your programs, we can store data in external memory files just as office information is stored in file cabinets for later use.

A number of advantages result from storing data on files. First, data entry and modification become easier. In addition, files make it possible to run several programs using the data from the same file and, alternatively, to run the same program from several different files.

Creating and using a file in FORTRAN is very easy. Recall from Chap. 17 that the READ and WRITE statements include a unit number that instructs the computer as to the devices for input and output. To this point we have focused on common input/output devices such as the keyboard, card reader, monitor, and printer. By using the proper unit number, we can READ or WRITE data to a file in an analogous fashion. Caution should be exerted when doing this because any data on a particular file could be lost when the WRITE statements from a current program are implemented. This is due to the fact that when you write on a file, the new information will supplant the old information on the file.

Although FORTRAN files are easy to access, it should be noted that special system commands are needed to tell the computer that a unit number refers to a particular file. Some examples from a typical mainframe computer system are shown in Fig. 21.5. Most mainframe computers and minicomputers will use commands that are similar to those seen in Fig. 21.5. However, microcomputers often use other means because their operating systems may not provide for unit specification at the time the program is run. Consult your computer's FORTRAN manual for guidance on how your system operates in this regard.

PROBLEMS

21.1 Reprogram Figure 17.10, employing loops and subscripted variables to make the code more efficient.

21.2   Repeat Prob. 17.1 but employ loops and subscripted variables to make the program more efficient.

21.3   Repeat Prob. 17.4 but employ loops and subscripted variables to make the program more efficient. Add columns for n = 10,000 and 15,000. Use a two-dimensional array to store the oxygen concentrations.

21.4   Write a short program to input a series of numbers and compute their average. Use the following algorithm as the basis for the program:

   • **Step 1**   Use a loop to input, count, and sum the numbers. Example 19.3 might serve as a model for inputting and counting. Employ an array to store the values as subscripted variables.
   • **Step 2**   Compute the average value as the sum divided by the total number of values.
   • **Step 3**   Print out the individual values in tabular form.
   • **Step 4**   Print out the average.

Test the program by determining the average flow for the data in Table 21.1.

21.5   Repeat Prob. 21.4 but rather than inputting the data via the keyboard, use DATA statements to incorporate the data into the program. As with Prob. 21.4 employ subscripted variables to store the values for the flows.

21.6   Write a short program fragment which swaps values between two variables. For example, before the program fragment is entered, X(1) = 2.75 and X(2) = 3.43; after the program fragment is executed, X(1) = 3.43 and X(2) = 2.75.

21.7   Check your system to determine how many bytes are used to represent integer values and then determine how many bytes are being reserved by the following FORTRAN statements:

```
INTEGER SUM
DIMENSION SUM(50,10)
```

21.8   Write a short program that reads a 3 × 4 array row by row and then sums the rows. This program should output the array, interchanging the columns for rows so that the output will be a 4 × 3 matrix with the sums of each column printed below.

21.9   Repeat Prob. 21.8 but design the program so that the original array is read from a file and the final array is read to another file.

21.10   Write a sorting subroutine that will arrange a one-dimensional array in ascending order. That is, the first value of the array, A(1), will contain the smallest value, A(2) will contain the second smallest, and so on. Assign values to the array in the main program and then call the subroutine to sort the values. Two items should be noted:

1. Arrays have to be dimensioned in the subprogram but can be given a dummy argument such as 1. This is because when an array is passed as an argument, the values are not actually passed, only the location of the array.

2. There are many different ways to sort but one of the easiest is similar to the method you might use if you were performing the sort without a computer. You would probably compare the first number with the second and pick the smaller. Then you would compare this one with the next number on the list and determine which of these was smaller. After making your way to the end of the list you would know the smallest number, which you would then bring to the head of the list. At that point you would repeat the process, starting at the second number, and so on. *Hint:* This procedure involves nested DO loops and the process of switching treated in Prob. 21.6.

PART SIX

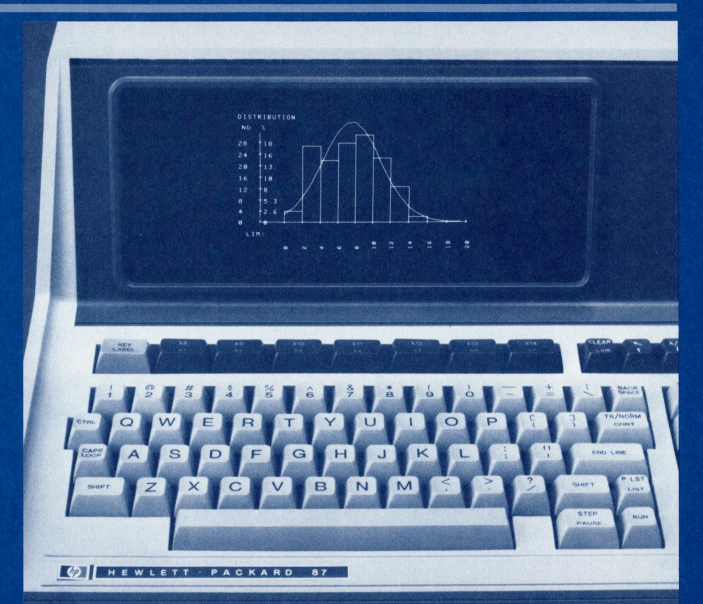

DATA ANALYSIS

Now that you have mastered the rudiments of programming, we turn to the ways in which this capability can be applied in engineering practice. Most of the remainder of this book is devoted to skills and tools that contribute to the problem-solving process. In all cases we emphasize the ways in which computers are used to facilitate the process. The first topic we cover is data analysis.

Problem solving with the computer is a futile exercise if it is based on inadequate information. In computer jargon, this is sometimes referred to as GIGO, or "garbage in, garbage out." Consequently, data analysis is one of the most important skills of the practicing engineer. Although many projects rely on existing data and formulas, there are a great many more situations where information is not available in these convenient forms. Special cases and new problem contexts often require that you measure your own data and develop your own predictive relationships.

The present part of the book introduces a number of specific tools and techniques that can prove useful in these tasks. Chapter 22 is devoted to graphical methods that provide a visual expression of data and relationships. Chapter 23 deals with algorithms for sorting information. Chapter 24 deals with quantifying a measurement's reliability with statistics. Finally, Chap. 25 describes how a physical process can be simulated on a computer.

CHAPTER 22
GRAPHICS

It is often remarked that "one picture is worth a thousand words." This saying holds true whether we are assimilating information or conveying it to others. Graphical depictions are invaluable communication devices for conveying data and concepts to an audience (Fig. 22.1). In addition, it is well known that visualization is a key prerequisite for our individual efforts to understand otherwise abstruse concepts and phenomena. Because engineers must continually deal with abstract information and must communicate findings to clients, we make extensive use of graphics in our profession. We have already seen one way in which visual devices can serve our needs. In Chap. 6, flowcharts were employed to map out the logic of a computer algorithm. Other examples abound throughout engineering.

The microelectronic revolution adds an exciting new dimension to these endeavors. Before computers, hand drafting was a time-consuming and, hence, a costly process. Even after computers were invented, the cost and logistical problems connected with utilizing mainframe computers discouraged the widespread application of computer graphics. Because of their speed, memory, low cost, and accessibility, personal computers offer a significant increase in our capability to produce and modify drawings. As such, they will not only make us more efficient but will also broaden and modify the role of graphics in our work.

The present chapter is devoted to ways in which engineers present information visually and how computers can facilitate the process. The first sections deal with precomputer methods for developing graphs. This is necessary because many characteristics of well-designed computer graphs are related to the characteristics of graphs that are concocted by hand. This review is followed by an introduction to computer graphics.

FIGURE 22.1 An engineer using personal computer graphics to present technical information to clients.

22.1 LINE GRAPHS

A routine task in engineering practice is to establish the relationship between two variables. Line graphs are the most common visual displays employed for this purpose. As seen in Fig. 22.2, line graphs display the relationship on a two-dimensional space. They are useful because the pattern on the graph illustrates qualitatively how one variable changes with respect to another—that is, whether it is increasing, decreasing, or oscillating.

Figure 22.3 illustrates the two primary coordinate systems used for line drawings in engineering: cartesian and polar. *Cartesian coordinate systems* define position in terms of distance along a vertical and a horizontal axis (Fig. 22.3*a*). In contrast, *polar coordinate systems* define position in terms of an angle and a distance from the center of the space (Fig. 22.3*b*).

Cartesian coordinate systems are further distinguished by the manner in which distance along each axis is delineated. The type shown in Fig. 22.3*a*, called a *rectilinear* line graph, divides each axis into intervals of equal length. In contrast, the *semilogarithmic* and *logarithmic* line graphs in Fig. 22.4*a* and *b* use logarithmic distances for one or both of the axes.

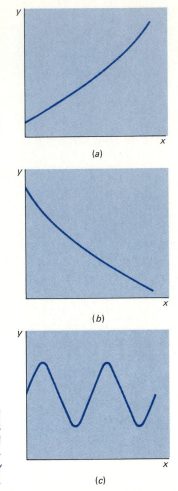

FIGURE 22.2
Three examples of line graphics (a) increasing, (b) decreasing, and (c) oscillating relationships between the dependent variable y and the independent variable x.

22.2 RECTILINEAR GRAPHS

For cartesian coordinate systems, it is conventional to plot the *dependent variable* on the vertical axis, or *ordinate*, and the *independent variable* on the horizontal axis, or *abscissa* (Fig. 22.5). As the names imply, these variables usually characterize a cause-effect relationship, where the dependent variable (the effect) "depends" on the value of the independent, or "given," variable (the cause). Thus the dependent variable is often said to be plotted "versus," or "as a function of," the independent variable.

As seen in Fig. 22.6, several curves can be plotted. In Fig. 22.6a, several dependent variables are plotted versus a single independent variable. However, because this necessitates including additional ordinates and labels, the number of dependent variables plotted in this fashion is usually limited to two or three. As shown

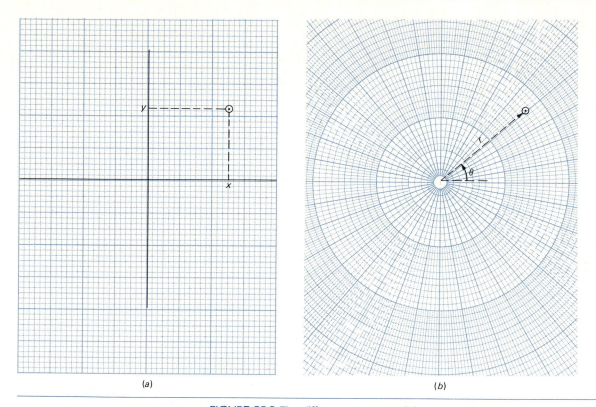

(a)

(b)

FIGURE 22.3 The difference between (a) cartesian and (b) polar coordinate systems. In both cases, two values are required to specify location on the two-dimensional space.

FIGURE 22.4
(a) Semilog and (b) log-log graph paper.

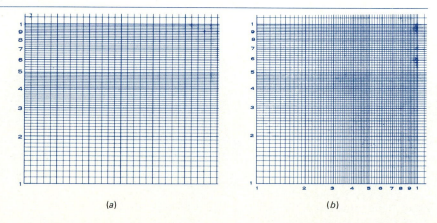

(a)

(b)

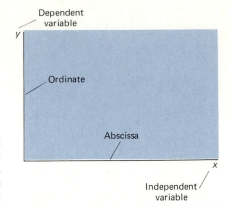

FIGURE 22.5
Conventions regarding the placement of the dependent and independent variables for cartesian coordinate systems.

in Fig. 22.6*b*, several curves can also be plotted to relate a single dependent variable to a single independent variable. Each curve corresponds to the relationship between the variables for a fixed parameter value. In the present context, a *parameter* is a constant or variable that characterizes or identifies a particular curve on a line graph. For example, each curve in Fig. 22.6*b* corresponds to a unique value of the parameter, temperature (T).

The procedure for constructing a graph can be reduced to the following steps:

1. Tabulating the data
2. Selecting the correct type of graph paper
3. Positioning the axes
4. Graduating and calibrating the axes
5. Plotting the points and drawing the lines
6. Adding labels and titles

22.2.1 Tabulating Data

It is good practice to tabulate the data before plotting it on a graph. Two situations normally arise. When the data originates from an experiment or a test, it is usually

FIGURE 22.6
Two cases where more than one curve is plotted on a graph. (*a*) Three dependent variables—y_1, y_2, and y_3—versus an independent variable x. Note how the extra ordinates may be added to the left or on the opposite side from the primary ordinate. (*b*) Graphs of a dependent variable y versus an independent variable x for the three values of the parameter T.

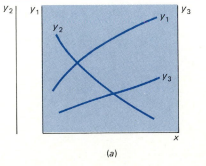

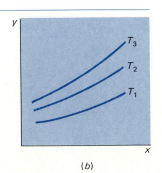

in the form of a set of readings that can be directly assembled in tabular form. By convention, the independent variable is usually listed in the first column and dependent variables in the adjacent columns. For cases where there are several dependent variables conforming to different parameter values, the parameters should be clearly specified in the table's heading.

A second source of data for plotting is an equation. Because the equation allows the determination of the value of the dependent variable on the basis of any value of the independent variable, an unlimited number of points can be generated in order to plot the continuous curve represented by the equation. Thus the equation is used to generate a table of values that can be used to plot the curve. The actual data values in the generated table are not plotted as points that are visible on the final graph but are used as the basis for sketching a smooth, continuous curve. The number of data points needed for this purpose depends on the equation. A smooth equation would require few points, whereas a complicated equation with oscillations would require many.

EXAMPLE 22.1   Tabulating Data from Experiments and Equations

PROBLEM STATEMENT: The fall velocity of a 68.1-kg parachutist is measured as a function of time; the results are summarized in Table 22.1. Notice that two experiments were performed using jumpsuits with different air resistance. This effect is designated by a parameter called the drag coefficient c, which has units of kilograms per second. In addition to these measurements, a theoretical equation has been developed to predict fall velocity as a function of time:

$$v = \frac{gm}{c} (1 - e^{-(c/m)t}) \qquad\qquad\text{(E22.1.1)}$$

where v = velocity, m/s
$\quad\ \ g$ = acceleration of gravity, 9.8 m/s^2
$\quad\ m$ = mass, kg
$\quad\ \ t$ = time, s

Use this equation to generate a table of values to plot and compare with the experimental data.

TABLE 22.1   The Measured Fall Velocity of a
Parachutist versus Time for Two
Different Drag Coefficients

Time, s	Fall Velocity, m/s	
	c = 12 kg/s	c = 18 kg/s
3	22	19
6	34	27
9	43	33
12	47	34

TABLE 22.2 The Computed Fall Velocity of a
Parachutist versus Time for Two
Different Drag Coefficients

| | Fall Velocity, m/s | |
Time, s	c = 12 kg/s	c = 18 kg/s
0	0.0	0.0
2	16.5	15.2
4	28.1	24.2
6	36.3	29.5
8	42.0	32.6
10	46.1	34.4
12	48.9	35.5
14	50.9	36.2

SOLUTION: The equation can be used to compute a velocity for each of the times and each of the drag coefficients in Table 22.1. For example, for $t = 6$ s and $c = 12$ kg/s,

$$v = \frac{9.8(68.1)}{12}(1 - e^{-(12/68.1)6}) = 36.3 \text{ m/s}$$

This process is repeated for the other values in the table. However, in order to more effectively characterize the curve, additional points must be generated. As seen in Table 22.2, we compute values from $t = 0$ to $t = 14$ using increments of 2 s. Notice that this extends several seconds beyond either side of the measured data range. This amount of detail is required to effectively characterize this curve. If the curve oscillates, even more detail is required.

22.2.2 Choosing Graph Paper

Printed graph paper is available in a variety of colors, paper types, and grid sizes. The choice of the proper kind of paper is predicated on a number of factors, including the purpose of the graph and the range and the precision of the data. For example, semilog and log-log paper is often required when one or more of the variables range over many orders of magnitude. Additionally, such paper is often used for curve fitting. Obviously, logarithmic scales cannot be employed for variables with both positive and negative values. We will return to these types of graphs and their applications in later sections of this chapter.

For rectilinear graphs, a number of different grid spacings are available. Figure 22.7 shows three examples, ranging from a very fine to a very coarse grid. The very fine grids are usually avoided in cases where the graph is to be reproduced or photographed. However, fine grids are required when the data must be plotted to a high degree of precision. Because the data in Example 22.1 is not very precise, we would use a fairly coarse grid, one with perhaps 10 divisions per inch, for a plot of the fall velocity versus time.

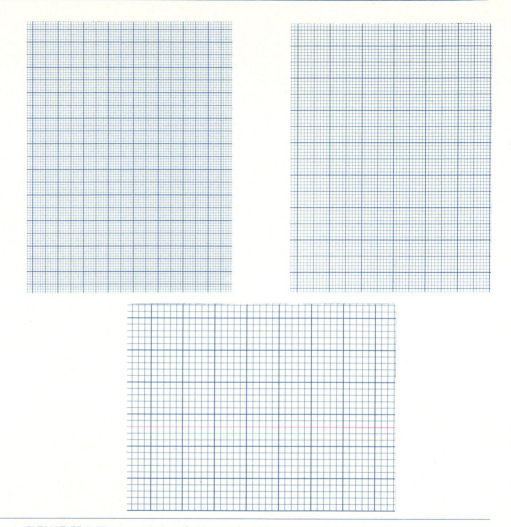

FIGURE 22.7 Three examples of grid spacing for rectilinear graph paper.

22.2.3 Positioning the Axes

Because data often contains both positive and negative values for the variables, rectilinear graphs may fall within any or all of the four quadrants depicted in Fig. 22.8. The positioning of the axis is determined by the range of the data and the objective of the graph. For situations where all the values are positive, we can limit the plot to the first quadrant of Fig. 22.8. In such cases it is usually desirable to place the origin (that is, the values of zero for the variables) at the lower left corner of the graph. An exception is for situations where such positioning results in the data being compressed drastically (Fig. 22.9a). If the object of the graph is to depict the pattern in which the points fall with respect to each other, such compression will make it difficult to

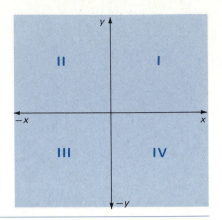

FIGURE 22.8
The four quadrants of a rectilinear graph.

perceive the configuration of the points. For such cases, the minimum and maximum values of the axes can be chosen so that the points fill the space of the graph (Fig. 22.9b). However, if the object of the graph is, in fact, to show that the data is grouped in a cluster far from the origin, the original version (Fig. 22.9a) is perfectly valid. For situations where both positive and negative values are to be plotted, the origin should be positioned so that all values can be shown.

22.2.4   Graduating and Calibrating the Axes

After the axes are properly positioned, they must be *scaled*. That is, a series of marks, called *graduations*, must be marked off in order to specify distances along each axis. It is convenient when the graduations are chosen so that the smallest divisions are a power of 10 times 1, 2, or 5. This is done to facilitate *interpolation*—that is, estimation of values falling between graduations (Fig. 22.10).

FIGURE 22.9
(a) An example of how placement of the origin at the lower left corner can lead to a compressed data set. (b) A remedy for this situation can be developed by using axis breaks so that the points fill the space of the graph.

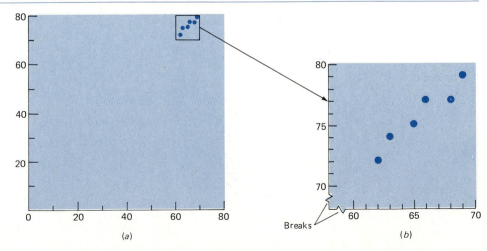

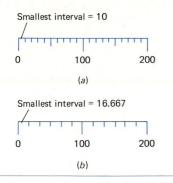

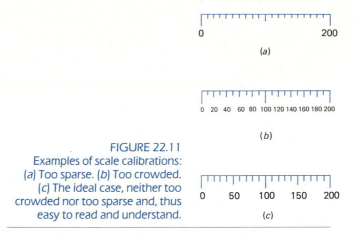

FIGURE 22.10
Examples of graduations: (a) Follows the rule of making the graduations a power of 10 times 1, 2, or 5; hence interpolation is facilitated. In contrast, (b) violates the rule and interpolation is difficult.

FIGURE 22.11
Examples of scale calibrations: (a) Too sparse. (b) Too crowded. (c) The ideal case, neither too crowded nor too sparse and, thus easy to read and understand.

After the graduations are added, numerical values are included to *calibrate* the scale. As seen in Fig. 22.11, calibrations should be neither too crowded nor too spread out. The object is to make the graph as easy to understand as possible.

22.2.5 Plotting Points and Drawing Lines

Data to be plotted on rectilinear graphs may consist of points or lines, or both. Points are usually employed to represent experimental data, whereas curves are typically used to represent theoretical or empirical equations. Points for a particular set of data are identified by using one of the distinctive shapes shown in Fig. 22.12. The exact location of each point is specified by the dot in the center of each symbol. The dot may be omitted, provided the symbol is centered on the data point. Note that for graphs involving several sets of data, a different symbol can be used to designate each set.

Lines are employed to represent continuous trends, such as those specified by an equation. As with points, several lines may be included on the same graph. When these lines overlap, it is often convenient to use different types, as shown in Fig. 22.13, in order to visually distinguish between the various curves.

There are several instances where lines and points appear on the same curve. One example is the case where a line is used to trace out, or "fit," the trend suggested

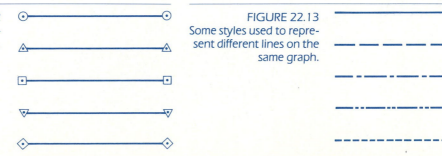

FIGURE 22.12
Some distinctive shapes used to locate data points on graphs.

FIGURE 22.13
Some styles used to represent different lines on the same graph.

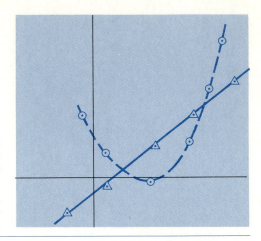

FIGURE 22.14
Example of lines and points on the
same graph. In this example, the
lines are intended to trace out the
trend suggested by the points.
Notice that the lines are not allowed to penetrate the points.

by a set of points. Notice in Fig. 22.14 that the lines do not penetrate the points. Another example occurs when the lines and points represent independent measures of the same quantity. This was the case for the theoretical and empirical data recorded for the velocity of the falling parachutist in Tables 22.1 and 22.2. As shown in Fig. 22.15, plotting both points and lines on the same graph provides a visual comparison of the theoretical and empirical results. Different symbols are used to distinguish between the two cases. The same solid line is employed for the equations because it is clear which line is associated with which data. However, if the lines overlapped it would be preferable to use a dashed line for one of them. Notice that Fig. 22.15 indicates that the theoretical relationships (the lines) consistently overestimate the empirical data (the points). Also notice that the points from Table 22.2 are not shown explicitly on the figure but, rather, are used as the basis for sketching the curves.

FIGURE 22.15
Plot of data from Tables 22.1 and
22.2 showing graduations, calibrations, points, and lines.

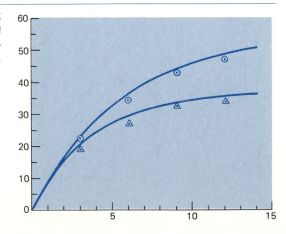

22.2.6   Adding Labels, an Identification Key, and a Title

The final phase in producing a graph is to add words to identify the plot and its elements. Your general objective should be to make the graph as easy as possible to understand. The process involves adding labels, an identification key, and a title for the plot.

The axes are labeled with the variable names and the units of measurement. In certain cases, the mathematical symbols [such as t for time or $f(x)$ for a function of x] may be used in place of the variable names. However, because your intent should be to communicate easily, this practice is usually not recommended. Figure 22.16 provides examples of some acceptable ways to label axes.

Where several curves and types of points are included, labels can be added to designate their identities. This can be done in two ways. Where the curves are sufficiently separate, the identifier (such as a run number, parameter value, date, etc.) may be placed adjacent to the curve (Fig. 22.17a). An alternative is to include a key. As seen in Fig. 22.17b, the key is usually in the bottom half of the graph. Of course, where it is positioned depends on the available space on the plot.

The final element to be added is the title. This is a short description that is usually placed in the top half of the plot. For experimental data, it is common practice to add

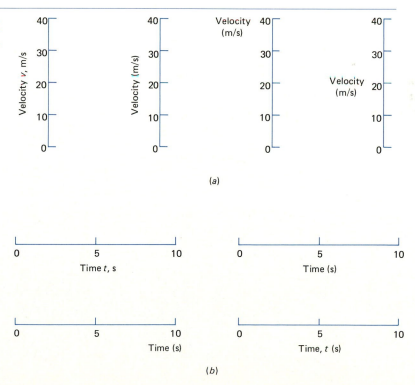

FIGURE 22.16
Ways to label axes. (a) examples of ordinate labels and (b) examples of abscissa labels.

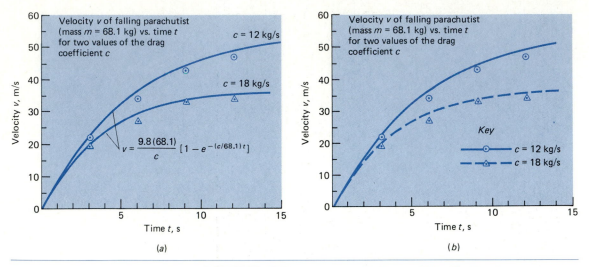

(a)   (b)

FIGURE 22.17 Two ways to add titles and identification keys.

the name of the experimenter and the date and location of the experiment. For plots that are reproduced in reports, papers, or books, the title is typically expressed as a caption placed below or alongside the plot. Figures in this book use this approach.

22.3   OTHER TYPES OF GRAPHS

The rectilinear graphs described in the previous section are but one, albeit the most common, of the graphs you will use as an engineer. Other important types are logarithmic and polar graphs, nomographs, and charts.

FIGURE 22.18
(a) Untransformed data ranges over several orders of magnitude. For this case, the smaller values tend to be bunched together. (b) When such data is transformed logarithmically, the points are spread out. Sometimes, as in this example, a predictable relationship such as a straight-line, or linear, trend emerges.

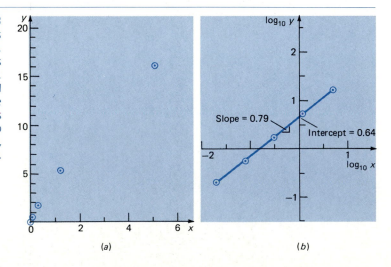

(a)   (b)

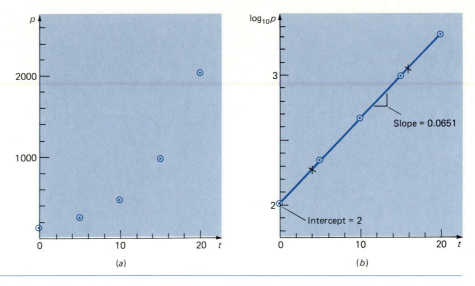

FIGURE 22.19
(a) A curving plot of population versus time. (b) When this data is graphed on a semilog plot it is transformed to a straight line. This line in (b) can be used to determine a mathematical equation or model to predict values of the curved function in (a). The additional points designated by x's in (b) are employed to determine the slope.

22.3.1 Logarithmic Graphs

Two types of logarithmic graphs are commonly used in engineering: semilog and log-log plots. These are sometimes employed when one or more of the variables range over many orders of magnitude. For such cases, using logarithms permits a wide range of values to be graphed without compressing the smaller values (Fig. 22.18). Another reason, as described next, is to develop mathematical relationships between the variables.

Semilogarithmic plots. A graph in which the ordinate is plotted with a logarithmic scale and the abscissa with a linear scale is referred to as a *semilogarithmic*, or *semilog, plot*. Oftentimes data that is curved when plotted on rectilinear graph paper (Fig. 22.19a) will plot as a straight line on a semilog plot (Fig. 22.19b). The intercept and the slope of the straight line can be used to formulate the following equation:

$$\log_{10} y = \log_{10} y_0 + (slope)x \tag{22.1}$$

where y_0 is the value of the dependent variable at the intercept (that is, the y value at $x = 0$) and the slope is calculated as

$$Slope = \frac{\log_{10} y_2 - \log_{10} y_1}{x_2 - x_1} \tag{22.2}$$

where $(x_1, \log_{10} y_1)$ and $(x_2, \log_{10} y_2)$ are two points from the straight line.

In order to transform this equation back to a form that is appropriate for non-logarithmic scales, we first must subtract $\log_{10} y_0$ from both sides of Eq. (22.1) to give

$$\log_{10} y - \log_{10} y_0 = (slope)x$$

or

$$\log_{10}(y/y_0) = (\text{slope})x$$

The antilogarithm of this equation is

$$\frac{y}{y_0} = 10^{(\text{slope})x}$$

or

$$y = y_0 10^{(\text{slope})x} \tag{22.3}$$

Thus this equation can be used to compute values for Fig. 22.19a. As seen in the next example, it can prove useful in making predictions.

EXAMPLE 22.2   *Deriving a Mathematical Model from a Semilog Plot*

PROBLEM STATEMENT: The population (p) of a small community on the outskirts of a city grows rapidly over a 20-year period:

t	0	5	10	15	20
p	100	212	448	949	2009

This data is plotted in Fig. 22.19a. As an engineer working for a utility company, you must forecast the population 5 years into the future in order to anticipate the demand for power. Employ a semilog plot to make this prediction.

SOLUTION: A semilog plot of the data can be developed by taking the logarithm of the population:

t	0	5	10	15	20
$\log_{10} p$	2	2.326	2.651	2.977	3.303

These values can be graphed (Fig. 22.19b) to verify that the data follow a straight line relationship. Then the graph can be used to determine an intercept of 2 and a slope of

$$\text{Slope} = \frac{3.041 - 2.260}{16 - 4} = 0.0651$$

Therefore, according to Eq. (22.3), the model for Fig. 29.19a is

$$p = 100(10)^{0.0651t} \tag{E22.2.1}$$

Substituting $t = 25$ into this equation allows us to predict

$$p = 100(10)^{0.0651(25)} = 4241$$

Therefore in 5 years the population will more than double if present trends continue.

The utility of semilogarithmic paper is that it can be used to produce plots such as Fig. 22.19b without actually taking the logarithms of the data. Figure 22.20 shows how the data from Example 22.2 can be directly plotted on such paper. If, as is the case in Fig. 22.20, a straight line results, then the analysis from Example 22.2 can be performed to derive the predictive equation. The only need for logarithms is in determining the slope with Eq. (22.2).

Note that the foregoing analysis could have also been accomplished with natural logarithms. If so, the equation for the transformed line [comparable to Eq. (22.1)] is

$$\ln y = \ln y_0 + (\text{slope}')x \tag{22.4}$$

The slope would be [comparable to Eq. (22.2)]

$$\text{Slope}' = \frac{\ln y_2 - \ln y_1}{x_2 - x_1} \tag{22.5}$$

and the final predictive equation is [comparable to Eq. (22.3)]

$$y = y_0 e^{(\text{slope}')x} \tag{22.6}$$

FIGURE 22.20
Plot of the data from Example 22.2 on semilog graph paper.

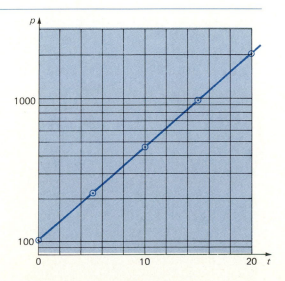

Either Eqs. (22.1) through (22.3) or Eqs. (22.4) through (22.6) can be fit to the same data. For example, applying the natural log version to the data in Example 22.2 yields the following predictive equation:

$$p = 100e^{0.15t}$$

This equation yields results similar to those given by Eq. (E22.2.1). For example,

$$p = 100e^{0.15(25)} = 4250$$

The slope' is marked with a prime in Eqs. (22.5) and (22.6) to distinguish it from the slope in Eqs. (22.2) and (22.3). The two slopes are related by the formula

$$\text{Slope} = \frac{\text{slope}'}{2.3025}$$

which derives from the fact that

$$\log_{10} x = \frac{\ln x}{2.3025} \tag{22.7}$$

Note that the use of semilog paper does not hinge on whether you employ common (base 10) or natural (base e) logarithms. This is due to the fact that the two types of logarithms are linearly related by Eq. (22.7). However, it is very important to determine the slope properly. If the base 10 model [Eq. (22.3)] is used, the slope must be determined with Eq. (22.2), whereas the base e model [Eq. (22.6)] requires use of Eq. (22.5).

Log-log plots.   Another way that an equation can be fit to curved data is to employ a log-log plot (Fig. 22.18b). Again, the intent is to see whether the data "straightens out," or is linearized, in the process. If so, the following equation holds:

$$\log_{10} y = \log_{10} y_0 + (\text{slope}) \log_{10} x \tag{22.8}$$

where y_0 is equal to the value of the dependent variable at the intercept. For a log-log plot the intercept corresponds to the value of $\log_{10} y$ at $\log_{10} x = 0$ (that is, at $x = 1$). The slope is computed by [compare with Eq. (22.2)]

$$\text{Slope} = \frac{\log_{10} y_2 - \log_{10} y_1}{\log_{10} x_2 - \log_{10} x_1} \tag{22.9}$$

Just as with the semilog model, Eq. (22.8) can be reexpressed by subtracting $\log_{10} y_0$ from both sides, taking the antilog, and rearranging the result to yield

$$y = y_0 x^{\text{slope}} \tag{22.10}$$

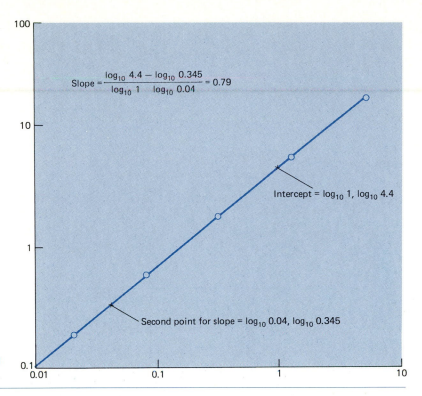

$$\text{Slope} = \frac{\log_{10} 4.4 - \log_{10} 0.345}{\log_{10} 1 - \log_{10} 0.04} = 0.79$$

Intercept = $\log_{10} 1, \log_{10} 4.4$

Second point for slope = $\log_{10} 0.04, \log_{10} 0.345$

FIGURE 22.21
A plot of the data from Fig. 22.18
on log-log graph paper.

This model can then be used to make predictions of y as a function of x.

For this case, special log-log graph paper is available to facilitate making the initial plot without having to actually take the logarithm of the data. Then if you ascertain that the data traces out a straight line, the slope and the intercept can be determined. Both the plot and the evaluation of the slope and intercept are shown in Fig. 22.21. These values can be substituted into Eq. (22.10) to give the model equation,

$$y = 4.4x^{0.79}$$

This equation can then be used to make predictions. Note that again it does not matter whether common or natural logs are employed as long as you are consistent. Equation (22.9) yields the same slope, regardless of the type of logarithm employed.

22.3.2 Polar Graphs, Nomographs, and Charts

Polar graphs. Polar coordinates are employed when the magnitude of a variable is to be displayed with regard to angular position. One application would be to display the intensity of light or sound at distances around a source such as a lamp or speaker (Fig. 22.22). Other applications range from the study of rotating objects to the analysis of mathematical functions.

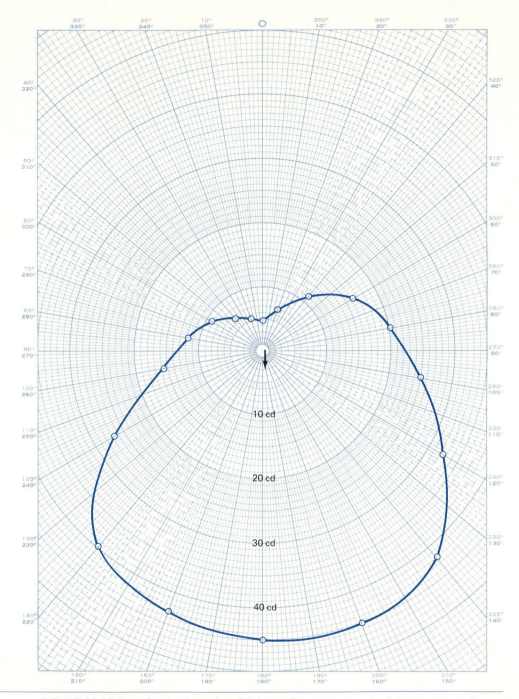

FIGURE 22.22 Data for the intensity of light at distances around a light source, as displayed on polar coordinates. The arrow indicates the direction in which the source is oriented.

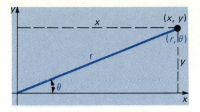

FIGURE 22.23
The relationship of cartesian and polar coordinates.

Sometimes it is useful to make transformations between cartesian and polar coordinates. The following equations are used for this purpose (see Fig. 22.23). To convert from polar to cartesian,

$$x = r \sin \theta \qquad (22.11)$$

and

$$y = r \cos \theta \qquad (22.12)$$

To convert from cartesian to polar,

$$r = \sqrt{x^2 + y^2} \qquad (22.13)$$

and [note, if $x < 0$, then add π radians to Eq. (22.14)]

$$\theta = \arctan (y/x) \qquad (22.14)$$

Nomographs. A nomograph is a graphical display of an established relationship between variables. Figure 22.24 shows a very simple example for the conversion of temperature between the Fahrenheit and the Celsius scales. This nomograph is made by simply placing the scales side by side vertically so that the corresponding values line up. Thus, if you know the Celsius value, you merely have to read the corresponding Fahrenheit value from the adjacent scale. More complex nomographs include several vertical scales and intricate schemes for predicting values. However, the fundamental idea exemplified by Fig. 22.24—that is, the visual alignment of the relationships—holds.

FIGURE 22.24
A simple nomograph relating the Celsius and Fahrenheit scales for temperature.

Bar and pie charts. Although more commonly applied in the economic and social sciences, bar and pie charts can prove useful in engineering. Figure 22.25 shows some

FIGURE 22.25 Some engineering examples of (a) pie charts and (b) bar diagrams.

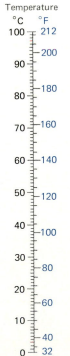

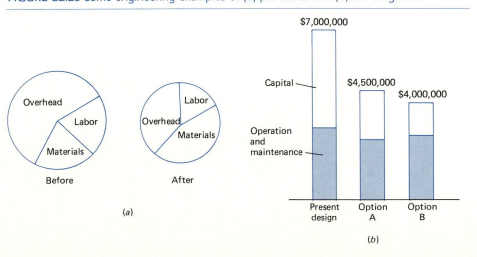

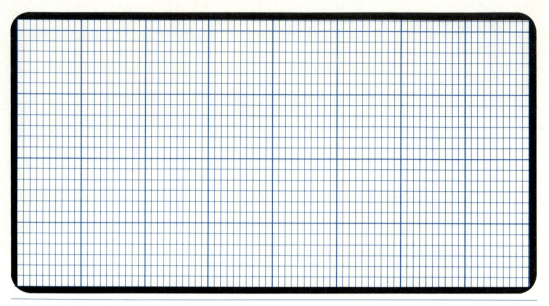

FIGURE 22.26 The screen of a typical computer monitor. When in text mode, the screen consists of 25 rows and 80 columns. (Note that the 25th row is sometimes dedicated to function key prompts.) It should be recognized that some computers have different numbers of rows and columns. However, the basic idea of the monitor screen as a two-dimensional space holds for all computers.

examples. We will not discuss rules or guidelines for their construction here. In addition, computer software packages are available to generate such charts accurately and efficiently. These displays are particularly useful when communicating technical information to clients and the general public.

22.4 COMPUTER PRINTER PLOTS

As with other aspects of engineering problem solving, the construction of graphs can be facilitated by the computer. In a later section of the book (Chap. 39), we will delve more deeply into advanced preprogrammed graphics packages and their impact on engineering. In the present chapter we will introduce the basic rationale underlying computer graphics and present some subroutines for drawing line plots.

Computer graphics is based on a simple idea that relates back to the very definition of plotting itself. As should be clear to you by now, plotting consists of locating points on a two-dimensional space. Computer output media, such as printer paper or monitor screens, also represent two-dimensional spaces. Figure 22.26 shows the screen of a typical computer monitor. Such a screen is said to be in the *text mode*. As an example, this figure shows a text mode that consists of 25 rows and 80 columns, defining a total of 2000 rectangles, or cells.*

*Note that this version of text mode is employed on the latest models of the IBM PC (with the KEY OFF command in effect for BASIC). Other computers sometimes utilize text modes with different dimensions.

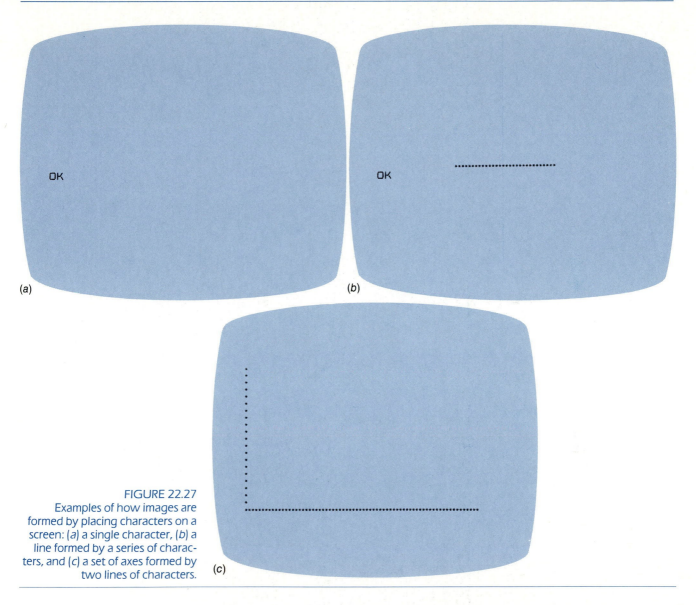

FIGURE 22.27
Examples of how images are formed by placing characters on a screen: (a) a single character, (b) a line formed by a series of characters, and (c) a set of axes formed by two lines of characters.

Computer graphics in text mode are built up from the fundamental operation of placing a character into one of these cells. For example, as seen in Fig. 22.27a, a character such as an asterisk may be placed in one of the cells. If several asterisks are placed in a row of cells, a line can be formed (Fig. 22.27b). Similarly, one vertical column and one horizontal row may be positioned to form a set of axes (Fig. 22.27c). Figure 22.28 lists the subroutines used to produce the three screens shown in Fig. 22.27.

In a similar fashion, all manner of images may be constructed on the screen. Of course, to attain more refined patterns, additional commands beyond the simple input/

```
100 REM ***** SUBROUTINE TO PRODUCE AN ASTERISK *****
110 REM            AT ANY LOCATION ON THE SCREEN
120 REM
130 REM        VARIABLES
140 REM
150 REM        I = ROW (ANY NUMBER FROM 1 TO 24)
160 REM        J = COLUMN (ANY NUMBER FROM 1 TO 80)
170 REM
180 REM              Input row and column
190 REM
200 INPUT "ROW = ";I
210 INPUT "COLUMN = ";J
220 REM
230 REM     Clear Screen & Skip Down to I-TH Row
240 REM
250 CLS
260 FOR I = 1 TO I - 1
270     PRINT
280 NEXT I
290 REM
300 REM      Print out Asterisk in J-TH Column
310 REM
320 PRINT TAB(J);"*"
330 END
```

```
C     ***** SUBROUTINE TO PRODUCE AN ASTERISK *****
C                 AT ANY LOCATION ON THE PAGE
C
C         VARIABLES
C
C         I = ROW (ANY NUMBER FROM 1 TO 55)
C         J = COLUMN (ANY NUMBER FROM 1 TO 80)
C
C               Input row and column
C
      READ (5,*) I, J
C
C           Skip Down to I-TH Row
C
      DO 10 K=1, I-1
        WRITE(6,*) ' '
  10  CONTINUE
C
C         Print out Asterisk in J-TH Column
C
      WRITE(6,*) ((' ',K=1,J),'*')
      STOP
      END
```

(a)

```
100 REM ***** SUBROUTINE TO PRODUCE A LINE OF *****
110 REM         30 ASTERISKS ALONG THE I-TH ROW
120 REM                  ON THE SCREEN
130 REM
140 REM     DEFINITION OF VARIABLES
150 REM
160 REM     I = ROW (ANY NUMBER FROM 1 TO 24)
170 REM
180 REM            Input Row and Column
190 REM
200 INPUT "ROW = ";I
210 REM
220 REM     Clear Screen & Skip Down to I-TH Row
230 REM
240 CLS
250 FOR I = 1 TO I - 1
260     PRINT
270 NEXT I
280 REM
290 REM          Print out 30 Asterisks
300 REM
310 PRINT TAB(25);
320 FOR I = 1 TO 30
330     PRINT "*";
340 NEXT I
350 END
```

```
C     ***** SUBROUTINE TO PRODUCE A LINE OF *****
C             30 ASTERISKS ALONG THE I-TH ROW
C                      ON THE SCREEN
C
C         DEFINITION OF VARIABLES
C
C         I = ROW (ANY NUMBER FROM 1 TO 24)
C
C               Input Row and Column
C
      READ(5,*) I
C
C         Clear Screen & Skip Down to I-TH Row
C
      DO 10 K = 1, I - 1
        WRITE(6,*) ' '
  10  CONTINUE
C
C               Print out 30 Asterisks
C
      WRITE(6,100) ('*',J=1,30)
 100  FORMAT(T25,30(A1))
      STOP
      END
```

(b)

```
100 REM ******  SUBROUTINE TO PRODUCE AXES  *******
110 REM
120 REM             Loop to Produce Y Axes
130 REM
140 CLS
150 FOR I = 1 TO 20
160     PRINT "*"
170 NEXT I
180 REM          Loop to Produce X Axes
190 FOR J = 1 TO 70
200     PRINT "*";
210 NEXT J
220 RETURN
```

```
C     ******  SUBROUTINE TO PRODUCE AXES  *******
C
C             Loop to Produce Y Axes
C
      DO 10 I=1,20
        WRITE(6,*)  '*'
  10  CONTINUE
C             Loop to Produce X Axes
      WRITE(6,100) ('*',J = 1,70)
 100  FORMAT(70(A1))
      STOP
      END
```

(c)

FIGURE 22.28 BASIC and FORTRAN subroutines to form the images from Fig. 22.27:
(a) point, (b) line, and (c) axes.

output statements in Fig. 22.28 are needed. For example, in some forms of BASIC, a single command is available to directly position a character anywhere on the screen in text mode. Thus the awkward skipping of spaces exhibited by the simple programs in Fig. 22.28 is avoided. In addition, on some personal computers color can be added to enhance the image.

Beyond the text mode, higher-resolution graphic modes that employ finer grids are also available to attain more refined patterns. In addition, canned software has been developed to facilitate graphics. Discussion of high-resolution graphics and canned software will be deferred to Chap. 39. For the time being, we will use simple input/output commands to develop line graphs in the text mode. The graphs we will develop are called *computer-printer plots* because they can be implemented on a standard line printer without recourse to special graphics commands or software. Although these are among the most primitive examples of computer graphics and are not considered of professional quality, they serve to introduce you to the fundamental concepts underlying computer graphics and provide you with the working capability to produce a line graph on your computer. In addition, they have the advantage that they can be implemented on both personal and mainframe computers employing either BASIC or FORTRAN.

22.4.1 Subroutine to Generate a Sideways Line Graph

Figure 22.29 lists BASIC and FORTRAN programs containing a subroutine to generate a line graph of a function. Figure 22.30 is an example of the type of plot that can be developed with the program for $y = \cos x$. As can be seen from both figures, the program first draws the y axis as a series of pluses horizontally across the screen, and an asterisk is placed at the location of the y intercept. Then the program advances to the next x value (which corresponds to the next horizontal line on the screen). At this location, a plus is printed to designate the x axis, and an asterisk is printed at the proper location of the y value. This process is repeated until the end of the x axis is reached.

In order to place the asterisks at the proper position, a linear transformation is necessary. This is done by the following simple equation:*

$$Y = INT((Y - YMIN)/YINC + 0.5) \tag{22.15}$$

where YMIN is the minimum value of the dependent variable and YINC is the value of Y corresponding to a single horizontal scale unit. This equation takes the given value of Y on the right-hand side of Eq. (22.15) and transforms it to the comparable number of horizontal scale units. The INT function is used so that the result is expressed as a whole number, and the 0.5 is added so that the result is rounded properly.

Notice how the computer graph in Fig. 22.30 comes out sideways. This result is attributable to the fact that we have positioned the origin at the upper left corner of the plot. This was done to be consistent with the way the computer's line printer

*Note that variables in the BASIC version are preceded by the letters SP to make them local as discussed in Sec. 13.3.

```
100 REM      ***************************************************
110 REM      *              SIDEWAYS PRINTER PLOT              *
120 REM      *                 (BASIC version)                *
130 REM      *                 by  S.C. Chapra                *
140 REM      *                                                *
150 REM      * This program plots Y as a function of X. The   *
160 REM      * axes will be rotated clockwise 90 degrees.     *
170 REM      ***************************************************
180 REM
190 REM      .-------------------------------------------.
200 REM      |            DEFINITION OF VARIABLES         |
210 REM      |                                           |
220 REM      | XMIN   = MINIMUM INDEPENDENT VABIABLE      |
230 REM      | XMAX   = MAXIMUM INDEPENDENT VABIABLE      |
240 REM      | YMIN   = MINIMUM DEPENDENT VABIABLE        |
250 REM      | YMAX   = MAXIMUM DEPENDENT VABIABLE        |
260 REM      `-------------------------------------------'
270 REM
280 DEF FNF(X) = COS(X)
290 REM
300 REM ****************** MAIN PROGRAM  ******************
310 REM
320 GOSUB 1000
330 GOSUB 2000
340 END
350 REM
1000 REM ************ SUBROUTINE TO ENTER DATA  ***********
1010 REM
1020 INPUT "MINIMUM VALUE FOR X = ";XMIN
1030 INPUT "MAXIMUM VALUE FOR X = ";XMAX
1040 INPUT "MINIMUM VALUE FOR Y = ";YMIN
1050 INPUT "MAXIMUM VALUE FOR Y = ";YMAX
1060 RETURN
1070 REM
2000 REM *************** SUBROUTINE S - PLOT  *************
2010 REM
2020 REM      DRAWS AXES ROTATED 90 DEGREES, CALCULATES
2030 REM      VALUES FOR Y vs. X, AND PLOTS ASTERIKS AT
2040 REM      THE PROPER LOCATIONS.
2050 REM
2060 REM      DATA REQUIRED:    XMIN, XMAX, YMIN, YMAX
2070 REM      RESULTS RETURNED: NONE
2080 REM
2090 REM      LOCAL VARIABLES:
2100 REM
2110 REM      SPX      = INDEPENDENT VARIABLE
2120 REM      SPY      = DEPENDENT VARIABLE
2130 REM      SPXINC   = INCREMENT FOR INDEPENDENT VARIABLE
2140 REM      SPXLNGTH = LENGTH OF X-AXIS
2150 REM      SPYINC   = INCREMENT FOR DEPENDENT VARIABLE
2160 REM      SPYLNGTH = LENGTH OF Y-AXIS
2170 REM
2180 REM --------------- BODY OF S - PLOT ----------------
2190 SPXLNGTH = 20
2200 SPXINC = (XMAX-XMIN)/SPXLNGTH
2210 SPYLNGTH = 55
2220 SPYINC = (YMAX-YMIN)/SPYLNGTH
2230 PRINT:PRINT
2240 PRINT TAB(17);
2250 PRINT YMIN;
2260 FOR I = 0 TO SPYLNGTH - 5
2270     PRINT " ";
2280 NEXT I
2290 PRINT YMAX
2300 PRINT TAB(10);XMIN;
2310 FOR SPX = XMIN TO XMAX STEP SPXINC
2320     SPY = FNF(SPX)
2330     SPY=INT((SPY-YMIN)/SPYINC+.5)
2340     PRINT TAB(20);
2350     FOR J = 0 TO SPYLNGTH
2360         VALUE = 0
2370         IF SPY = J THEN VALUE = VALUE + 1
```

```
C       ***************************************************
C       *              SIDEWAYS PRINTER PLOT              *
C       *                (FORTRAN version)                *
C       *                 by  S.C. Chapra                 *
C       *                                                 *
C       * This program plots Y as a function of X. The    *
C       * axes will be rotated clockwise 90 degrees.      *
C       ***************************************************
C
C       .-------------------------------------------.
C       |            DEFINITION OF VARIABLES         |
C       |                                           |
C       | XMIN   = MINIMUM INDEPENDENT VABIABLE      |
C       | XMAX   = MAXIMUM INDEPENDENT VABIABLE      |
C       | YMIN   = MINIMUM DEPENDENT VABIABLE        |
C       | YMAX   = MAXIMUM DEPENDENT VABIABLE        |
C       `-------------------------------------------'
C
C
****************** MAIN PROGRAM  ******************
C
        REAL XMIN, XMAX, YMIN, YMAX
        CALL ENTER(XMIN, XMAX, YMIN, YMAX)
        CALL SPLOT(XMIN, XMAX, YMIN, YMAX)
        STOP
        END
C
C
************ SUBROUTINE TO ENTER DATA  ***********
C
        SUBROUTINE ENTER(XMIN, XMAX, YMIN, YMAX)
        READ(5,*) XMIN, XMAX, YMIN, YMAX
        RETURN
        END
C
C
*************** SUBROUTINE S - PLOT  *************
C
        SUBROUTINE SPLOT(XMIN, XMAX, YMIN, YMAX)
C       DRAWS AXES ROTATED 90 DEGREES, CALCULATES
C       VALUES FOR Y vs. X, AND PLOTS ASTERIKS AT
C       THE PROPER LOCATIONS.
C
C       DATA REQUIRED:    XMIN, XMAX, YMIN, YMAX
C       RESULTS RETURNED: NONE
C
C       LOCAL VARIABLES:
C
C       X      = INDEPENDENT VARIABLE
C       Y      = DEPENDENT VARIABLE
C       XINC   = INCREMENT FOR INDEPENDENT VARIABLE
C       XLNGTH = LENGTH OF X-AXIS
C       YINC   = INCREMENT FOR DEPENDENT VARIABLE
C       YLNGTH = LENGTH OF Y-AXIS
C
C       --------------- BODY OF S - PLOT ----------------
        CHARACTER LINE*132
        XLNGTH = 50
        XINC = (XMAX-XMIN)/XLNGTH
        YLNGTH = 55
        YINC = (YMAX-YMIN)/YLNGTH
        WRITE(6,100) YMIN, YMAX
100     FORMAT(' ',T10,F5.1,T72,F5.1)
        WRITE(6,200) XMIN
200     FORMAT(' ',T10,F5.1)
        DO 20 X = XMIN, XMAX, XINC
            Y = FNF(X)
            Y = INT((Y-YMIN)/YINC+.5)
            LINE = ' '
            DO 10 J = 1, YLNGTH - 1
                VALUE = 0
                IF (Y .EQ. J) THEN
                    VALUE = VALUE+1
                ELSE
                ENDIF
```

```
2380        REM CASE                                        C
2390           IF VALUE > 0 THEN 2400 ELSE 2420             C            CASE
2400              PRINT "*";                                             IF (VALUE .GT. 0) THEN
2410              GOTO 2470                                                 LINE (J:J) = '*'
2420           IF J = 0 OR SPX = XMIN THEN 2430 ELSE 2450                ELSE IF ((J .EQ. 1) .OR. (X .EQ. XMIN)) THEN
2430              PRINT "+";                                                LINE (J:J) = '+'
2440              GOTO 2470                                              ELSE
2450           REM ELSECASE                                                LINE (J:J) = ' '
2460              PRINT " ";                                            ENDIF
2470        REM ENDCASE                                     C            ENDCASE
2480     NEXT J                                             C
2490 NEXT SPX                                                   10    CONTINUE
2500 PRINT:PRINT TAB(10);XMAX                                         WRITE(6,300),LINE
2510 RETURN                                                    300    FORMAT(' ',T20,A55)
                                                               20    CONTINUE
                                                                     WRITE(6,400) XMAX
                                                              400    FORMAT('+',T10,F5.1)
                                                                     RETURN
                                                                     END
                                                        C
                                                        C       FUNCTION FOR EQUATION
                                                        C
                                                                     REAL FUNCTION FNF(X)
                                                                     FNF = COS(X)
                                                                     RETURN
                                                                     END
```

FIGURE 22.29 BASIC and FORTRAN programs to generate a sideways printer plot of a function.

operates—that is, from left to right and from top to bottom. By positioning the origin at the upper left corner, we avoided some significant complications, involved in plotting the graph with an upright orientation—that is, with the origin at the conventional location in the lower left corner. These involve further scale transformations, which are covered in the next section.

FIGURE 22.30
A text-mode computer-generated plot of the cosine of x. The plot was generated using the programs from Fig. 22.29.

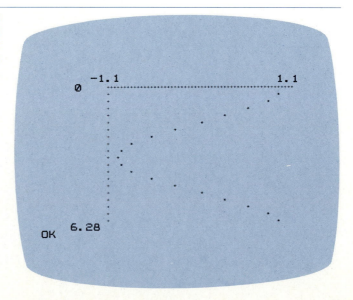

22.4.2 Subroutine to Generate an Upright Line Graph

In the previous section, we developed a simple printer plot of a line graph. One shortcoming of this plot was that it was positioned sideways on the page. The fact that the graph was not printed upright relates to fundamental differences between the coordinates of a typical computer screen and a conventional line graph. In essence, as depicted in Fig. 22.31, one major difference is that the former has its origin at the upper left corner, whereas the latter has its lowest values at the lower left corner. Additionally, the scaling of the axes differs. In the text mode, the horizontal coordinates of the screen refer (usually) to 80 discrete columns and the vertical (usually) to 25 discrete rows. In contrast, the horizontal axis of the line graph represents a continuous range of an independent variable and the vertical a continuous range of a dependent variable.

In order to take a given set of x, y coordinates and display them on a computer screen, we must express the x's and the y's so that they correspond to the proper column and row integer coordinates of the screen. In other words, we must transform one set of coordinates into the other and, in effect, flip the plot around mathematically.

FIGURE 22.31
The steps in taking a conventional x, y plot (*b*) and positioning it on the monitor screen (*a*).

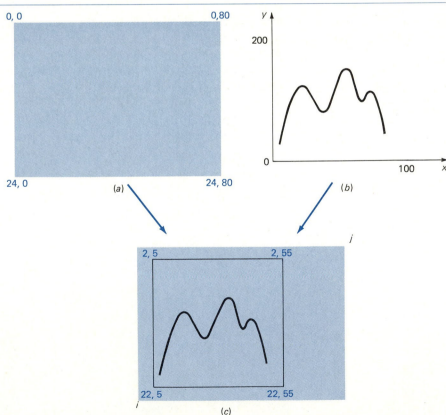

Because the scales are linear, these transformations are actually very easy to determine. All that is required is to derive a linear equation that establishes the one-to-one correspondence between the coordinate systems. The following example illustrates how this is done.

EXAMPLE 22.3 Transformations to Construct a Printer Plot

PROBLEM STATEMENT: Develop the transformation equations needed to establish the one-to-one correspondence between the upright x, y coordinates (Fig. 22.31b) and the i, j coordinates of a computer monitor (Fig. 22.31a).

SOLUTION: We need to develop equations to compute j, given x, and i, given y. First let's concentrate on developing the equation to compute j as a function of x. From Fig. 22.31c it can be seen that at the origin of the plot $x = 0$ corresponds to $j = 5$. In addition, the range of the abscissa, $x = 0$ to 100, corresponds to $j = 5$ to 55. Therefore a simple linear relationship between the two scales is

$$j = 5 + \frac{55 - 5}{100 - 0} x = 5 + 0.5x$$

You can check this result by substituting $x = 0$ to obtain 5 and $x = 100$ to obtain 55. Thus the transformation correctly computes these two end points. Because the relationship is linear, the transformation will hold for all the other points.

In a similar manner, we can derive the equation to predict i as a function of y. For this case, the origin $y = 0$ corresponds to $i = 22$, and the range of the ordinate from $y = 0$ to 200 corresponds to $i = 22$ to 2. Thus the linear transformation for the vertical axis is

$$i = 22 + \frac{2 - 22}{200 - 0} y = 22 - 0.1y$$

The transformation developed in the foregoing example can be written in more general form. For the horizontal axis,

$$j = \frac{x(j_{max} - j_{min}) - x_{min}j_{max} + x_{max}j_{min}}{x_{max} - x_{min}} \tag{22.16}$$

where j_{min} is the value of j at the origin and j_{max} and x_{max} are the values of j and x at the right end of the horizontal axis. For the vertical coordinates,

$$i = \frac{y(i_{max} - i_{min}) - y_{max}i_{max} + y_{min}i_{min}}{y_{min} - y_{max}} \tag{22.17}$$

where i_{max} is the value of i at the origin and i_{min} and y_{max} are the values of i and y at

```
100 REM      ***************************************************
110 REM      *         UPRIGHT PRINTER PLOT USING ARRAY        *
120 REM      *                 (BASIC version)                 *
130 REM      *                 by  S.C. Chapra                 *
140 REM      *                                                 *
150 REM      *   This program plots Y as a function of X.      *
160 REM      ***************************************************
170 REM
180 REM          .---------------------------------------.
190 REM          |          DEFINITION OF VARIABLES       |
200 REM          |                                        |
210 REM          |    XMIN = MINIMUM INDEPENDENT VABIABLE  |
220 REM          |    XMAX = MAXIMUM INDEPENDENT VABIABLE  |
230 REM          |    YMIN = MINIMUM DEPENDENT VABIABLE    |
240 REM          |    YMAX = MAXIMUM DEPENDENT VABIABLE    |
250 REM          '---------------------------------------'
260 REM
270 DIM A$(24,80),X(15),Y(15)
280 REM ----------------------------------------------------
290 REM       Functions to Transform Real Values to
300 REM                 Array Subscripts
310 REM
320 DEF FNI(UPY)=(UPY*(UPIMAX-UPIMIN)-YMAX*UPIMAX+YMIN*UPIMIN)
                                                /(YMIN-YMAX)
330 DEF FNJ(UPX)=(UPX*(UPJMAX-UPJMIN)-XMIN*UPJMAX+XMAX*UPJMIN)
                                                /(XMAX-XMIN)
340 REM
350 REM            Function to Round to Nearest Integer
360 REM
370 DEF FNR(K) = INT(K+.5)
380 REM
390 REM         Function to Compute Y Values for Curve.
400 REM
410 DEF FNF(X) = COS(X)
420 REM
430 REM ****************** MAIN PROGRAM ********************
440 REM
450 GOSUB 1000
460 GOSUB 2000
470 END
480 REM
1000 REM ************ SUBROUTINE TO ENTER DATA  ***********
1010 REM
1020 INPUT "MINIMUM VALUE FOR X = ";XMIN
1030 INPUT "MAXIMUM VALUE FOR X = ";XMAX
1040 INPUT "MINIMUM VALUE FOR Y = ";YMIN
1050 INPUT "MAXIMUM VALUE FOR Y = ";YMAX
1060 RETURN
1070 REM
2000 REM *************** SUBROUTINE U - PLOT  *************
2010 REM
2020 REM        CALCULATES VALUES FOR Y vs. X, DRAWS AXES,
2030 REM        AND PLOTS ASTERIKS AT THE PROPER LOCATIONS.
2040 REM
2050 REM        DATA REQUIRED:   XMIN, XMAX, YMIN, YMAX
2060 REM        RESULTS RETURNED: NONE
2070 REM
2080 REM        LOCAL VARIABLES:
2090 REM
2100 REM        UPX     = INDEPENDENT VARIABLE
2110 REM        UPY     = DEPENDENT VARIABLE
2120 REM        UPIMIN  = MINIMUM ROW SUBSCRIPT OF ARRAY
2130 REM        UPIMAX  = MAXIMUM ROW SUBSCRIPT OF ARRAY
2140 REM        UPJMIN  = MINIMUM COLUMN SUBSCRIPT OF ARRAY
2150 REM        UPJMAX  = MAXIMUM COLUMN SUBSCRIPT OF ARRAY
2160 REM        UPIDUM  = DUMMY VARIABLE
2170 REM        UPJDUM  = DUMMY VARIABLE
2180 REM
2190 REM ------------------ BODY OF U - PLOT ---------------
2200 CLS
2210 UPIMIN = 1
2220 UPIMAX = 20
2230 UPJMIN = 1
2240 UPJMAX = 70
2250 PRINT "PLOT BEING CONSTRUCTED. PLEASE WAIT."
2260 REM
2270 REM                     Clear Plot
2280 REM
2290 FOR I = UPIMIN TO UPIMAX
2300    FOR J = UPJMIN TO UPJMAX
```

```fortran
C        ***************************************************
C        *         UPRIGHT PRINTER PLOT USING ARRAY        *
C        *                (FORTRAN version)                *
C        *                by  S.C. Chapra                  *
C        *                                                 *
C        *   This program plots Y as a function of X.      *
C        ***************************************************
C
C            .---------------------------------------.
C            |          DEFINITION OF VARIABLES       |
C            |                                        |
C            |    XMIN = MINIMUM INDEPENDENT VABIABLE  |
C            |    XMAX = MAXIMUM INDEPENDENT VABIABLE  |
C            |    YMIN = MINIMUM DEPENDENT VABIABLE    |
C            |    YMAX = MAXIMUM DEPENDENT VABIABLE    |
C            '---------------------------------------'
C
      REAL X, Y, XMAX, XMIN, YMAX, YMIN
      CHARACTER A(60,80)*1
C     ----------------------------------------------------
C
C     ***************** MAIN PROGRAM  *****************
C
      CALL ENTER(XMIN, XMAX, YMIN, YMAX)
      CALL UPLOT(XMIN, XMAX, YMIN, YMAX)
      STOP
      END
C
C     ************ SUBROUTINE TO ENTER DATA  **********
C
      SUBROUTINE ENTER(XMIN, XMAX, YMIN, YMAX)
      READ(5,*) XMIN, XMAX, YMIN, YMAX
      RETURN
      END
C
C     ************** SUBROUTINE U - PLOT  *************
C
      SUBROUTINE UPLOT(XMIN, XMAX, YMIN, YMAX)
C           CALCULATES VALUES FOR Y vs. X, DRAWS AXES,
C           AND PLOTS ASTERIKS AT THE PROPER LOCATIONS.
C
C           DATA REQUIRED:   XMIN, XMAX, YMIN, YMAX
C           RESULTS RETURNED: NONE
C
C           LOCAL VARIABLES:
C
C           X      = INDEPENDENT VARIABLE
C           Y      = DEPENDENT VARIABLE
C           IMIN   = MINIMUM ROW SUBSCRIPT OF ARRAY
C           IMAX   = MAXIMUM ROW SUBSCRIPT OF ARRAY
C           JMIN   = MINIMUM COLUMN SUBSCRIPT OF ARRAY
C           JMAX   = MAXIMUM COLUMN SUBSCRIPT OF ARRAY
C           IDUMMY = DUMMY VARIABLE
C           JDUMMY = DUMMY VARIABLE
C
C     ----------------- BODY OF U - PLOT --------------
      REAL XMIN, XMAX, YMIN, YMAX, X, Y
      INTEGER FNR
      CHARACTER A(60,80)*1
      IMIN = 1
      IMAX = 50
      JMIN = 1
      JMAX = 70
C
C                   Clear Plot
C
      DO 20 I = IMIN, IMAX
         DO 10 J = JMIN, JMAX
            A(I,J) = ' '
10       CONTINUE
20    CONTINUE
C
C                   Set Up Axes
C
      DO 30 J = JMIN, JMAX
         A(IMAX,J) = '+'
30    CONTINUE
      DO 40 I = IMIN, IMAX
         A(I,1) = '+'
40    CONTINUE
```

```
2310      A$(I,J) = " "
2320    NEXT J
2330 NEXT I
2340 REM
2350 REM                     Set Up Axes
2360 REM
2370 FOR J = UPJMIN TO UPJMAX
2380    A$(UPIMAX,J) = "+"
2390 NEXT J
2400 FOR I = UPIMIN TO UPIMAX
2410    A$(I,1) = "+"
2420 NEXT I
2430 REM
2440 REM      Transform X,Y Coordinates of Points to I,J
2450 REM          and Enter Curve into Array.
2460 REM
2470 XCOUNT = UPJMAX - UPJMIN
2480 XINC = (XMAX - XMIN)/XCOUNT
2490 UPX = XMIN
2500 FOR J = UPJMIN TO UPJMAX
2510    UPY = FNF(UPX)
2520    UPIDUM = FNI(UPY)
2530    UPJDUM = FNJ(UPX)
2540    UPIDUM = FNR(UPIDUM)
2550    UPJDUM = FNR(UPJDUM)
2560    A$(UPIDUM,UPJDUM) = "*"
2570    UPX = UPX + XINC
2580 NEXT J
2590 CLS
2600 REM                   Print Out Plot
2610 REM
2620 FOR I = UPIMIN TO UPIMAX
2630    FOR J = UPJMIN TO UPJMAX-1
2640       PRINT A$(I,J);
2650    NEXT J
2660    PRINT A$(I,J)
2670 NEXT I
2680 RETURN
```

```
C
C          Transform X,Y Coordinates of Points to J,I
C              and Enter Curve into Array.
C
      XCOUNT = JMAX - JMIN
      XINC = (XMAX - XMIN)/XCOUNT
      X = XMIN
      DO 50 J = JMIN, JMAX
         Y = FNF(X)
         IDUMMY = FNI(Y,IMIN,IMAX,YMIN,YMAX)
         JDUMMY = FNJ(X,JMIN,JMAX,XMIN,XMAX)
         IDUMMY = FNR(IDUMMY)
         JDUMMY = FNR(JDUMMY)
         A(IDUMMY,JDUMMY) = '*'
         X = X + XINC
50    CONTINUE
C
C                   Print Out Plot
C
      DO 60 I = IMIN, IMAX
         WRITE(6,*) (A(I,J),J = JMIN,JMAX -1)
60    CONTINUE
      RETURN
      END
C    --------------------------------------------------------
C          Functions to Transform Real Values to
C                  Array Subscripts
C
      FUNCTION FNI(Y,IMIN,IMAX,YMIN,YMAX)
      REAL YMIN,YMAX,Y
      FNI = (Y*(IMAX-IMIN)-YMAX*IMAX+YMIN*IMIN)/(YMIN-YMAX)
      RETURN
      END
C
      REAL FUNCTION FNJ(X,JMIN,JMAX,XMIN,XMAX)
      REAL XMIN,XMAX,X
      FNJ = (X*(JMAX-JMIN)-XMIN*JMAX+XMAX*JMIN)/(XMAX-XMIN)
      RETURN
      END
C    --------------------------------------------------------
C
C          Function to Round to Nearest Integer
C
      INTEGER FUNCTION FNR(K)
      FNR = INT(K+.5)
      RETURN
      END
C    --------------------------------------------------------
C
C          Function to Compute Y Values for Curve.
C
      REAL FUNCTION FNF(X)
      FNF = COS(X)
      RETURN
      END
```

FIGURE 22.32 BASIC and FORTRAN programs to generate a conventional line plot of a function.

the top end of the vertical axis. In Example 22.3, $x_{min} = 0$ and $x_{max} = 100$ which correspond to $j_{min} = 5$ and $j_{max} = 55$. Similarly, $y_{min} = 0$ and $y_{max} = 200$ correspond to $i_{min} = 2$ and $i_{max} = 22$. These values can be substituted into Eqs. (22.16) and (22.17). When these two equations are rounded to integers, they provide the means to transform any x or y value into the appropriate j or i coordinate on the screen.

Figure 22.32 lists BASIC and FORTRAN programs that employ these transformations to produce an upright plot of a function. Figure 22.33 is an example of the type of plot that can be generated with these programs.

Aside from their use of the transformations, the programs from Fig. 22.32 differ from those in Fig. 22.29 in another fundamental way. That is, the programs in Fig. 22.32 use an array to construct and to output the plot. The reason that the array is employed relates back to the physical limitations of the printer and the types of BASIC and FORTRAN commands that are available to output data on these devices. On the conventional line printer, all movement is from left to right and from top to bottom. It is not possible to move back up the page from bottom to top or to backspace along a line from right to left. The programs in Fig. 22.29 are compatible with and, in fact, capitalize on these constraints.

By transforming the data and setting the plot upright, we lose this compatibility. Although there are ways to use standard input and output statements to program around this dilemma, most are fairly complicated. A simple and more direct approach hinges on the availability of a more powerful and flexible type of statement—that is, one that would allow us to position a character at will at any location on the screen. Because some versions of BASIC and many of FORTRAN do not include such a statement, we will defer discussion of it to the next section. For the time being, the program in Fig. 22.32 utilizes an array to accomplish the same objective—that is, to position a character anywhere on a two-dimensional space. The subscripts of the array provide a means for doing this.

As seen in the BASIC program of Fig. 22.32, we define an array with the same dimensions as the screen—that is, 24 by 80. Initially this array is filled with blanks. Next, the axes are defined by entering plus signs into a left-hand column and a bottom row of the array. Then we can use the transformation equations to determine the appropriate subscripts to enter asterisks at the locations in the array that are to contain characters. Finally, this array can be output to give the plot shown in Fig. 22.33.

FIGURE 22.33
A computer-generated plot of the cosine of x versus x. The plot was generated with the programs from Fig. 22.32.

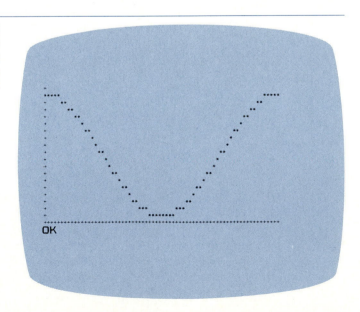

The foregoing approach has two primary disadvantages. First, it is slow. This is because we have to fill out every element of the array with either blanks, plus signs, or asterisks and also because we have to print out the entire array to obtain the plot. Second, it requires the storage of 1920 (that is, 24 × 80) characters. On a small personal computer, this might represent a significant sacrifice of memory. However, the program has a number of advantages. It is conceptually easy to understand, and it can be implemented on any type of computer, using either BASIC or FORTRAN. Additionally, it has served to highlight the need for a single statement to position a character anywhere on a two-dimensional space. We will discuss such a statement in the next section.

22.4.3 Upright Plots Using the LOCATE Statement of BASIC

Certain dialects of BASIC include a single command that can be used to position the cursor anywhere on the screen. For the IBM PC, the general form of this statement is

ln LOCATE *rn, cn*

where *rn* and *cn* are the row number (1 to 25*) and the column number (1 to 80), respectively, that designate the cell where the cursor is to be positioned. For example, it could be positioned in the upper left corner of the screen by

20 LOCATE 1,1

or in the lower right corner by

30 LOCATE 25,80

For computers such as the Apple II, the comparable operation is accomplished with the commands HTAB and VTAB, which have the general form

ln HTAB *cn*

and

ln VTAB *rn*

Once the cursor is in position, any permissible character may be printed. Figure 22.34 shows a computer program that uses the LOCATE statement to construct a plot of a line graph of the same sort developed with an array in Fig. 22.32. The resulting plot is identical to that in Fig. 22.33. However, it will be generated much more quickly because of the increased efficiency of the LOCATE command.

---

*Assuming that KEY OFF is used to suppress the function key prompts.

```
100 REM    *****************************************************
110 REM    *          UPRIGHT PRINTER PLOT USING LOCATE        *
120 REM    *                 (BASIC version)                   *
130 REM    *                by  S.C. Chapra                    *
140 REM    *                                                   *
150 REM    *     This program uses the LOCATE command to       *
160 REM    *     plot Y as a function of X.                    *
170 REM    *****************************************************
180 REM
190 REM        .-------------------------------------------.
200 REM        |          DEFINITION OF VARIABLES          |
210 REM        |                                           |
220 REM        |   XMIN = MINIMUM INDEPENDENT VABIABLE      |
230 REM        |   XMAX = MAXIMUM INDEPENDENT VABIABLE      |
240 REM        |   YMIN = MINIMUM DEPENDENT VABIABLE        |
250 REM        |   YMAX = MAXIMUM DEPENDENT VABIABLE        |
260 REM        |                                           |
270 REM        '-------------------------------------------'
280 DIM X(15),Y(15)
290 REM  --------------------------------------------------
300 REM         Functions to Transform Real Values
310 REM               to Array Subscripts
320 REM
330 DEF FNI(LCY)=(LCY*(LCIMAX-LCIMIN)-YMAX*LCIMAX+YMIN*LCIMIN)
                                                 /(YMIN-YMAX)
340 DEF FNJ(LCX)=(LCX*(LCJMAX-LCJMIN)-XMIN*LCJMAX+XMAX*LCJMIN)
                                                 /(XMAX-XMIN)
350 REM
360 REM        Function to Round to Nearest Integer
370 REM
380 DEF FNR(K) = INT(K+.5)
390 REM
400 REM        Function to Compute Values for Curve
410 REM
420 DEF FNF(X) = COS(X)
430 REM
440 REM ****************** MAIN PROGRAM  ******************
450 REM
460 GOSUB 1000
470 GOSUB 2000
480 END
1000 REM *********** SUBROUTINE TO ENTER DATA *************
1010 REM
1020 INPUT "MINIMUM VALUE FOR X = ";XMIN
1030 INPUT "MAXIMUM VALUE FOR X = ";XMAX
1040 INPUT "MINIMUM VALUE FOR Y = ";YMIN
1050 INPUT "MAXIMUM VALUE FOR Y = ";YMAX
1060 RETURN
1070 REM
```

```
2000 REM ************* SUBROUTINE L - PLOT  *************
2010 REM
2020 REM        CALCULATES VALUES FOR Y vs. X, DRAWS AXES,
2030 REM        AND PLOTS ASTERIKS USING THE LOCATE COMMAND.
2040 REM
2050 REM        DATA REQUIRED:   XMIN, XMAX, YMIN, YMAX
2060 REM        RESULTS RETURNED: NONE
2070 REM
2080 REM        LOCAL VARIABLES:
2090 REM
2100 REM        LCX    = INDEPENDENT VARIABLE
2110 REM        LCY    = DEPENDENT VARIABLE
2120 REM        LCIMIN = MINIMUM ROW SUBSCRIPT OF ARRAY
2130 REM        LCIMAX = MAXIMUM ROW SUBSCRIPT OF ARRAY
2140 REM        LCJMIN = MINIMUM COLUMN SUBSCRIPT OF ARRAY
2150 REM        LCJMAX = MAXIMUM COLUMN SUBSCRIPT OF ARRAY
2160 REM        LCIDUM = DUMMY VARIABLE
2170 REM        LCJDUM = DUMMY VARIABLE
2180 REM
2190 REM ----------------- BODY OF L - PLOT -----------------
2200 REM
2210 REM          Set Limits of Array Used for Plot
2220 CLS
2230 LCIMIN = 1
2240 LCIMAX = 20
2250 LCJMIN = 1
2260 LCJMAX = 70
2270 REM                     Set up Axes
2280 REM
2290 FOR I = LCIMIN TO LCIMAX - 1
2300    PRINT "|"
2310 NEXT I
2320 FOR J = LCJMIN TO LCJMAX
2330    PRINT "-";
2340 NEXT J
2350 REM     Transform X,Y Coordinates of Points to I,J
2360 REM             and Print Out Points
2370 REM
2380 XCOUNT = LCJMAX - LCJMIN
2390 XINC = (XMAX - XMIN)/XCOUNT
2400 LCX = XMIN
2410 FOR J = LCJMIN TO LCJMAX
2420    LCY = FNF(LCX)
2430    LCIDUM = FNI(LCY)
2440    LCJDUM = FNJ(LCX)
2450    LCIDUM = FNR(LCIDUM)
2460    LCJDUM = FNR(LCJDUM)
2470    LOCATE LCIDUM,LCJDUM
2480    PRINT "*"
2490    LCX = LCX + XINC
2500 NEXT J
2510 LOCATE LCIMAX + 1, LCJMIN
2520 RETURN
```

FIGURE 22.34 A program in BASIC to generate a conventional plot using the LOCATE command from the IBM PC.

PROBLEMS

22.1 For a graph in rectangular cartesian coordinates, on which axes are the dependent and the independent variables plotted?

22.2 Determine the polar coordinates for a point with the rectangular coordinates: $(5, 2)$, $(8, -1)$, $(-5, -6)$.

22.3 Name the six steps for constructing a graph.

22.4 Why is it desirable to scale a graph so that the graduations are a power of 10 times 1, 2, or 5?

22.5 Give some reasons for employing semilog and log-log plots.

22.6 Will the equation $y = 5e^{-x}$ plot as a straight line on log-log paper? If not, is there a type of paper on which it will plot straight?

22.7 The following data was gathered in an experiment to determine how the flow rate through a valve (Q) varied with the valve opening (V):

Q, ft³/s	V (% valve opened)	Q, ft³/s	V (% valve opened)
0.000	0	0.770	50
0.021	5	0.980	60
0.096	10	1.100	70
0.200	20	1.200	80
0.350	30	1.300	90
0.560	40	1.400	100

Plot the data on a graph with appropriate labels and titles and determine visually the best-fit curve through the points.

22.8 Fill in the steps omitted in the text to derive Eq. (22.10). Why is the intercept of a line on a log-log plot taken to be at the value of $x = 1$?

22.9 The following data was taken during an experiment to determine the strength of a new concrete mixture as a function of the time after it has been poured:

Time, d	3	4	5	6	7	10	14	21
Compression strength, ksi	0.7	1.9	3.2	4.1	5.4	7.6	8.3	9.1

Plot this data on rectilinear and semilog graph paper and extrapolate from these plots the compression strength after 28 days. Interpret your results.

22.10 A variable y is said to be inversely proportional to a variable x if the relationship between the two can be expressed as

$$y = \frac{1}{x}$$

On what graph paper would this equation plot as a straight line?

22.11 On what graph paper would

$$y = \frac{1}{x^3}$$

plot as a straight line?

22.12 Determine all the types of graph paper on which the following functions would plot as straight lines:

(a) $y = -5 + x$ (d) $y = 23e^x$

(b) $y = \dfrac{30}{x}$ (e) $y = 23e^{x^2}$

(c) $y = \dfrac{10^x}{x}$ (f) $y = \dfrac{x}{33 + x}$

Employ transformations if necessary.

22.13 Develop your own printer-plot program based on the material in this chapter. Express the program in subroutine form so that in later chapters you can use it to add a plotting capability to other programs. Have the following data serve as input to the subroutine: one-dimensional arrays, X and Y, holding the x, y coordinates of the points; N, the total number of points; and the minimum and maximum limits of the X and Y axes—XMIN, XMAX, YMIN, YMAX. Use a structured approach for the program and develop both external and internal documentation. Try to incorporate as many features of a well-designed, hand-drawn graph as possible.

22.14 Test the program developed in Prob. 22.13 by using it to plot the points from Probs. 22.7 and 22.9.

22.15 Develop a program as described in Prob. 22.13 but design it to plot a function.

22.16 Test the program developed in Prob. 22.15 by using it to plot the functions in Probs. 22.10 and 22.11 from $x = 0.5$ to 5 and from $x = -3$ to 3.

22.17 Develop a program as described in Prob. 22.13 but design it to plot both a function and discrete data points on the same graph.

CHAPTER 23
SORTING

A fundamental programming operation that has a variety of engineering applications is the sorting of a group of items in some specified order. Aside from its relevance to engineering problem solving, sorting also serves to demonstrate the utility of arrays.

23.1 SORTING NUMBERS BY SELECTION

When dealing with numbers, it is often necessary to sort them in ascending order. There are many methods for accomplishing this objective. A simple approach, called a *selection sort*, is probably very close to the commonsense approach you might use to solve the problem. First you would scan the set of numbers and pick the smallest. Next you would bring this value to the head of the list. Then you would repeat the procedure on the remaining numbers and bring the next smallest value to the second place on the list. This procedure would be continued until the numbers were ordered. A simple program to perform the selection sort is displayed in Fig. 23.1. The following example illustrates how it works.

EXAMPLE 23.1   Selection Sort

PROBLEM STATEMENT: Use a selection sort to arrange the first five flows from Table 14.1 (or Table 21.1) in ascending order. The flows are

125   102   147   76   95

```
1000 REM ************    SUBROUTINE SELECTION  ************
1010 REM
1020 REM      PERFORMS A SELECTION SORT
1030 REM
1040 REM      DATA REQUIRED: NONE
1050 REM      RESULTS RETURNED: NONE
1060 REM
1070 REM      LOCAL VARIABLES:
1080 REM
1090 REM      SEVAR() = ARRAY TO BE SORTED
1100 REM      SENUM = NUMBER OF ELEMENTS IN
1110 REM                  SEVAR()
1120 REM --------------- BODY OF SELECT ---------------
1130 DIM SEVAR(20)
1140 FOR I = 1 TO SENUM - 1
1150   REM
1160   REM         Set Initial Element as Smallest
1170   SMALL = SEVAR(I)
1180   K = I
1190   REM             Search for Smallest Element
1200   FOR J = I+1 TO SENUM
1210     IF SEVAR(J) < SMALL THEN 1220 ELSE 1240
1220       SMALL = SEVAR(J)
1230       K = J
1240     REM ENDIF
1250   NEXT J
1260   REM SWITCH
1270   SEVAR(K) = SEVAR(I)
1280   SEVAR(I) = SMALL
1290 NEXT I
1300 RETURN
```

```
C
C     ************    SUBROUTINE SELECTION  ************
C
C          PERFORMS A SELECTION SORT
C
C          DATA REQUIRED: NONE
C          RESULTS RETURNED: NONE
C
C          LOCAL VARIABLES:
C
C          VAR() = ARRAY TO BE SORTED
C          NUM = NUMBER OF ELEMENTS IN
C                    VAR()
C     --------------- BODY OF SELECT ---------------
C
      REAL VAR(20)
      READ(5,*) (VAR(J),J=1,10)
      NUM = 10
      DO 20 I=1, NUM - 1
C
C              Set Initial Element as Smallest
      SMALL = VAR(I)
      K = I
C              Search for Smallest Element
      DO 10 J=I+1, NUM
        IF (VAR(J) .LT. SMALL) THEN
            SMALL = VAR(J)
            K = J
        ENDIF
10    CONTINUE
C
C          SWITCH
      VAR(K) = VAR(I)
      VAR(I) = SMALL
20    CONTINUE
      DO 30 K=1, 10
        WRITE(6,*) VAR(K)
30    CONTINUE
      STOP
      END
```

FIGURE 23.1 BASIC and FORTRAN programs to perform a selection sort of an array of numbers.

SOLUTION: First we must find the lowest number, which we designate as SMALL. We start by setting SMALL equal to the first number on the list, VAR(1), and noting that its subscript, which we designate as K, is equal to 1. Then we compare SMALL with the remaining numbers on the list. If we find a smaller number, we set SMALL equal to this number and store the subscript as K. When this search is completed, SMALL = 76 and K = 4. Then we switch VAR(1) = 125 with VAR(4) = 76 in order to bring the smallest number to the head of the list. This switch is also required to retain the number (125) that was originally first. After this is completed the list looks like

$$76 \quad 102 \quad 147 \quad 125 \quad 95$$

Now we repeat the process but with SMALL = VAR(2) = 102. On this pass, we find that VAR(5) is the smallest remaining number, and so we switch VAR(5) and VAR(2) to give

$$76 \quad 95 \quad 147 \quad 125 \quad 102$$

Next we start at SMALL = VAR(3) = 147 and switch VAR(3) with VAR(5) to yield

$$76 \quad 95 \quad 102 \quad 125 \quad 147$$

At this point, the numbers are ordered. However, the computer cannot perceive this, and so it must take the final step of SMALL = VAR(4) = 125, even though this does not result in a switch.

23.2   SORTING STRING DATA

Recall that in Chap. 4 we mentioned that the letters of the alphabet in string variables have numerical values that follow the natural ordering of the alphabet. For example, A is less than B. Thus we can compare and sort string variables using algorithms similar to the ones employed for numbers.

EXAMPLE 23.2   Sorting String Data

PROBLEM STATEMENT: Given the following student names and grades for a course,

Elam Jane	93
Rychlik Ed	79
Clark Karen	85
Taylor Susan	85
Arella Adam	97
Dollar Bill	71

use the selection sort to print out the names and grades in alphabetical order.

SOLUTION: The program in Fig. 23.1 has to be modified in several ways to allow it to sort the alphabetic and numeric data for this example. First the names and grades have to be input as separate subscripted string and numeric variables (Fig. 23.2). Then the sort is performed on the basis of the string names. For example, if the names are stored (for the BASIC version) in N$(I), line 1210 in Fig. 23.1 has to be rewritten as

```
1210 IF N$(I) < S$ THEN 1220 ELSE 1240
```

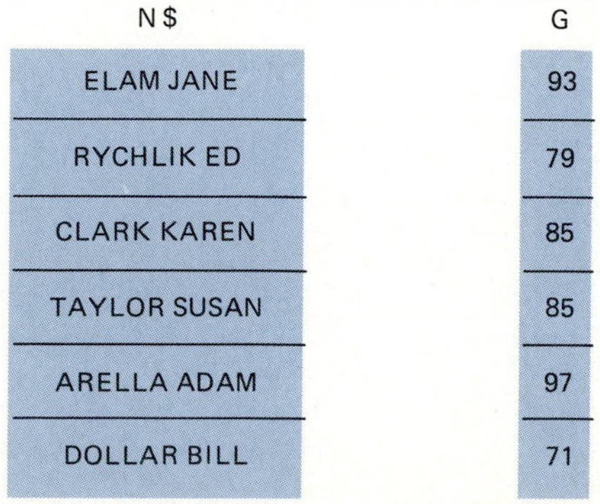

FIGURE 23.2
Parallel arrays N$ and G used to store the names and the grades for Example 23.2.

N$	G
ELAM JANE	93
RYCHLIK ED	79
CLARK KAREN	85
TAYLOR SUSAN	85
ARELLA ADAM	97
DOLLAR BILL	71

where S$ would be the string variable where the current smallest (that is, the first in alphabetical order) name would be stored. Although the sort is based on the name, the grades have to be saved and switched along with the names so that the correct grades are associated with the correct students. The result is

Arella Adam	97
Clark Karen	85
Dollar Bill	71
Elam Jane	93
Rychlik Ed	79
Taylor Susan	85

23.3   BUBBLE AND SHELL SORTS

Although the selection sort works well enough, there are other algorithms that perform more efficiently (Knuth, 1973). One such method, the Shell sort, is relatively simple to understand and executes approximately twice as fast as the selection sort for arrays of about 200 members. For larger arrays, its relative advantage becomes even more pronounced.

Before describing this technique, we will first briefly present another method that is very slow but will help us to understand the Shell sort. Called the *bubble sort*, this method consists of starting at the beginning of the array and comparing adjacent values. If the first value is smaller than the second, nothing is done. However, if, as seen in Fig. 23.3*a*, the first is larger than the second, the two values are switched. Next the second and third values are compared, and if they are not in the correct order (that is, smallest first) they are also switched. This process is repeated until the end of the array is reached. This entire sequence is called a *pass*. Figure 23.3*a* through *d* shows the switches that occur on the first pass through an array of five numbers.

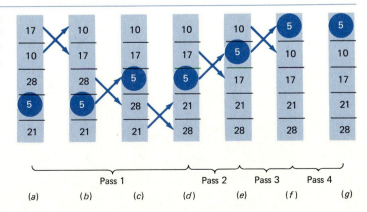

FIGURE 23.3
The bubble sort. Notice how the smallest number rises like a bubble as the sort proceeds.

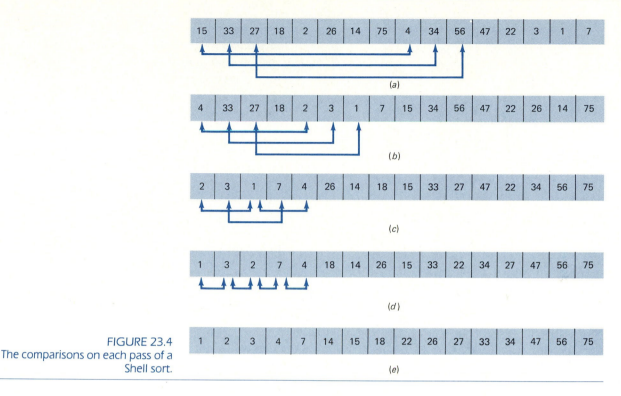

FIGURE 23.4
The comparisons on each pass of a Shell sort.

After the first pass the process returns to the top of the list and another pass is implemented. The procedure is repeated until a pass with no switches occurs (Fig. 23.3*g*). At this point, the sort is complete. The name "bubble sort" comes from the fact that, as depicted in Fig. 23.3, the smallest value tends to rise to the top of the list like a bubble rising through a liquid.

The *Shell sort* (named for its originator, Donald Shell) is similar to the bubble sort in that values in an array are compared and switched. However, in contrast to the bubble sort, nonadjacent elements are swapped. On the first pass the interval over which the comparison is made is one-half the entire length of the array. As seen in Fig. 23.4*a*, in the first pass through a 16-element array, the first and ninth items are compared, then the second and tenth, etc. On the second pass, the comparison interval is halved. As seen in Fig. 23.4*b*, the first is compared with the fifth, the second with the sixth, etc. For each pass these comparisons are repeated until no switches occur. (Note that, if switches occur during a pass, the pass is repeated.) Then the next pass with a halved comparison interval is implemented. Thus, as shown in Fig. 23.4*d*, the last pass amounts to a bubble sort.

Although the Shell sort is slightly more complicated than either the bubble or the selection sort, its associated computer algorithm is still quite simple (Fig. 23.5). Its increased efficiency more than justifies the slight increase in complexity.

```
1000 REM **************** SUBROUTINE SHELL ***************
1010 REM
1020 REM      PERFORMS A SHELL SORT
1030 REM
1040 REM      DATA REQUIRED:    N  --) NUMBER OF ELEMENTS IN X()
1045 REM                        X() --) UNSORTED
1050 REM      RESULTS RETURNED: X() --) SORTED
1060 REM
1070 REM      LOCAL VARIABLES:
1080 REM
1090 REM      SHGAP   = GAP SIZE FOR ELEMENT COMPARISON
1100 REM                DURING SORT
1110 REM      SHSWCH  = INDICATES THAT ELEMENTS HAVE
1120 REM                BEEN SWITCHED
1130 REM      N       = NUMBER OF ELEMENTS IN X()
1140 REM      SHDUM   = TEMPORARY DUMMY VARIABLE
1150 REM
1160 REM ---------------- BODY OF SHELL ------------------
1170 REM
1180 REM              Subroutine for Shell Sort
1190 REM
1200 SHGAP = N
1210 SHGAP = INT(SHGAP/2)
1220 REM PERFORM LOOP WHILE INTERVAL )= 1
1230 WHILE SHGAP )= 1
1240     REM
1250     REM          Perform Loop While Swithces Occur
1260     SHSWCH = 1
1270     WHILE SHSWCH = 1
1280         REM        Initialize Switch Indicator
1290         SHSWCH = 0
1300         REM Loop to Compare and, if Necessary, Switch
1310         FOR J = 1 TO N-SHGAP
1320             REM
1330             REM If Former Greater than Latter
1340             IF X(J) ) X(J+SHGAP) THEN 1360 ELSE 1410
1350                 REM
1360                 REM Then Switch and Set Switch Indicator to 1
1370                 SHDUM = X(J)
1380                 X(J) = X(J + SHGAP)
1390                 X(J + SHGAP) = SHDUM
1400                 SHSWCH = 1
1410             REM ENDIF
1420         NEXT J
1430     WEND
1440     SHGAP = INT(SHGAP/2)
1450 WEND
1460 RETURN
```

```
C
C
      **************** SUBROUTINE SHELL ***************
C
      SUBROUTINE SHELL(X,N)
C
C         PERFORMS A SHELL SORT
C
C         DATA REQUIRED:    N  --) NUMBER OF ELEMENTS IN X()
C                           X() --) UNSORTED
C         RESULTS RETURNED: X() --) SORTED
C
C         LOCAL VARIABLES:
C
C         GAP     =   GAP SIZE FOR ELEMENT COMPARISON
C                     DURING SORT
C         SWCH    =   INDICATES THAT ELEMENTS HAVE
C                     BEEN SWITCHED
C         N       =   NUMBER OF ELEMENTS IN X()
C         DUMMY   =   TEMPORARY DUMMY VARIABLE
C
      REAL X(20),DUMMY
      INTEGER SWCH,GAP,N
C     ---------------- BODY OF SHELL ------------------
C
C               Subroutine for Shell Sort
C
      GAP = N
      GAP = INT(GAP/2)
C
C     DO WHILE GAP ) = 1
1000  IF (GAP .GE. 1) THEN
          SWCH = 1
C
C         DO WHILE SWCH = 1
3000      IF (SWCH .EQ. 1) THEN
C
C             Initialize Switch Indicator
              SWCH = 0
C
C             Loop to Compare and, if Necessary, Switch
              DO 10 J=1, N - GAP
C
C                 If Former Greater than Latter
                  IF (X(J) .GT. X(J+GAP)) THEN
C
C                     Then Switch and Set Switch Indicator to 1
                      DUMMY = X(J)
                      X(J) = X(J + GAP)
                      X(J + GAP) = DUMMY
                      SWCH = 1
                  ENDIF
10            CONTINUE
              GOTO 3000
          ENDIF
C         END WHILE
C
          GAP = INT(GAP/2)
          GOTO 1000
      ENDIF
C     END WHILE
C
      RETURN
      END
```

FIGURE 23.5 BASIC and FORTRAN subroutines for the Shell sort.

23.4 ADVANCED SORTING ALGORITHMS

Beyond the Shell sort, there are a variety of other algorithms for putting information in order. Barron (1983) presents a nice review of some of the methods, along with BASIC code and flowcharts. Knuth (1973) offers a comprehensive treatment of the subject.

PROBLEMS

23.1 In the course of a flood analysis, it is often necessary to analyze past rainfall records to determine the maximum rainfall intensities. As part of this analysis, it is your assignment to develop a program to sort such data and then print it out. The data is given as:

Amount	Duration	Date
2.4	1.5	4-12-50
3.5	1.75	5-27-52
7.6	5.2	1-1-55
5.2	2.4	6-26-62
3.5	3.1	5-21-65
10.2	26.2	6-20-68
8.4	2.1	3-29-73
3.9	2.7	4-29-76
2.5	0.4	7-5-78
7.5	3.1	6-28-82

Use this data to compute storm intensity (amount/duration) for each reading. Then sort the data in ascending order according to storm intensity. Print out the intensities in ascending order.

23.2 Repeat Prob 23.1 but print out the date, amount, and duration, along with the intensity, in a tabular format. *Note:* The easiest way to handle the dates are with a character-variable array.

23.3 Develop a computer program to sort a group of names in alphabetical order. Employ the Shell sort as the algorithm for ordering the names.

23.4 In engineering it is sometimes necessary to acquire data from many sources and then consolidate and sort the data. For example, suppose a chemical manufacturing firm wants to assess the efficiency of each of its five manufacturing plants. Efficiency tests are conducted at each of the factories on 5 successive days. Due to unforeseen circumstances, plants 2 and 4 were able to successfully conduct only 3 and 4 of the tests, respectively. Therefore the data is returned as:

Day	Plant 1	Plant 2	Plant 3	Plant 4	Plant 5
1	70	—	53	72	44
2	55	—	48	70	48
3	62	52	52	65	52
4	73	37	61	75	51
5	57	50	56	—	47

Enter this data into your program and print it out as shown. Determine the average efficiency for each plant, sort these averages, and then print out the plant number, the number of successful tests, and the average efficiency in order of performance. Use a Shell sort as the algorithm to order the data.

23.5 Repeat Prob. 23.4 but replace the Shell with a selection sort. Note that if good modular programming technique has been employed, the modification will only involve changing the sort subroutine.

23.6 Develop a subroutine for the Shell sort. Use structured programming techniques and modular design so that the subroutine can be integrated into other programs throughout the remainder of the book. Document the subroutine both internally and externally.

23.7 Repeat Prob. 23.6 but for the bubble sort.

23.8 Repeat Prob. 23.6 but for the selection sort.

23.9 Perform Probs. 23.6 through 23.8. Use a random-number generator to generate one-dimensional arrays containing 10, 100, and 1000 numbers. Store these arrays on files. Use each of the sorting algorithms to order each of the files. Time each of them to assess their efficiency. Plot the results on log-log paper (log of sort time versus log of number being sorted). Interpret your results.

23.10 Aside from sorting, another important operation performed on data is searching—that is, locating an individual item in a group of data. One common method for accomplishing this objective is the *binary search*. Suppose that you are given an array $A(I)$ of length N and must determine whether the value V is included in this array. After sorting the array in ascending order with a technique like the Shell sort, you employ the following algorithm to conduct a binary search:

Compare V with the middle term $A(M)$. (*a*) If $V = A(M)$, the search is successful; (*b*) if $V < A(M)$, V is in the first half; and (*c*) if $V > A(M)$, V is in the second half. In the case of (*b*) or (*c*), the search is repeated in the half in which V is known to lie. Then V is again compared with $A(M)$, where M is now redefined as the middle value of the smaller list. The process is repeated until V is located or the array is exhausted.

Develop a subroutine to implement a binary search. Use structured programming techniques and a modular design so that the subroutine can be easily integrated into other programs. Document the subroutine both internally and externally.

CHAPTER 24
ERROR ANALYSIS AND STATISTICS

Engineers must continually deal with data that contains errors. This statement at first might seem contrary to what one normally conceives of as sound engineering. Students and practicing engineers constantly strive to limit errors in their work. When taking examinations or doing homework problems, you are penalized, not rewarded, for your errors. In professional practice, errors can be costly and sometimes catastrophic. If a structure or device fails, lives can be lost.

Although perfection is a laudable goal, it is rarely, if ever, attained. For example, it is unlikely that an engineering survey crew would be able to measure the exact length of an automobile test track. There would always be some discrepancy or error between the measurement and the true value. If the discrepancy is large, the measurement is unacceptable. However, if it is sufficiently small, the error is considered negligible and the measurement deemed adequate. In every case, the key equation is "How much error is tolerable?"

The present chapter deals with this question by showing how errors can be quantified. In the initial sections we will discuss some general issues and focus on the error of a single measurement. Then in later sections of the chapter we will cover some of the statistical techniques used when repeated measurements are performed to obtain an estimate.

24.1 SIGNIFICANT FIGURES

Almost all of engineering deals with the manipulation of numbers. Consequently, before discussing errors, it may be useful to review basic concepts related to the approximate representation of numbers themselves.

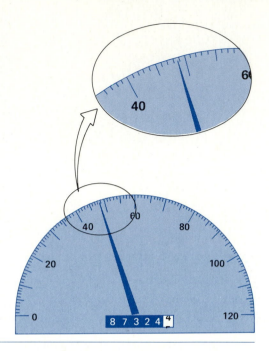

FIGURE 24.1
An automobile speedometer and
odometer.

Whenever we employ a number in a computation, we must have assurance that it can be used with confidence. For example, Fig. 24.1 depicts a speedometer and an odometer from an automobile. Visual inspection of the speedometer indicates that the car is traveling between 48 and 49 km/h. Because the indicator is higher than the midpoint between the markers on the gauge, we can say with assurance that the car is traveling at approximately 49 km/h. We have confidence in this result because two or more reasonable individuals reading this gauge would arrive at the same conclusion. However, let's say that we insist that the speed be estimated to one decimal place. For this case, one person might read 48.7, whereas another might read 48.8 km/h. Therefore, because of the limits of this instrument, only the first two digits can be considered certain. Estimates of the third digit (or higher) must be viewed as suspect. It would be ludicrous to claim, on the basis of the speedometer, that the automobile is traveling at 48.7642138 km/h. In contrast, the odometer provides up to six certain digits. From Fig. 24.1, we can conclude that the car has traveled slightly less than 87,324.5 km during its lifetime. In this case, the seventh digit (and higher) is uncertain.

The concept of a significant figure, or digit, has been developed to formally designate the reliability of a numerical value. The *significant digits* of a number are those that can be used with confidence. They correspond to the certain digits plus one estimated digit. For example, the speedometer and the odometer in Fig. 24.1 yield readings of three and seven significant figures, respectively. For the speedometer, the two certain digits are 48. It is conventional to set the estimated digit at one-half of the smallest scale division on the measurement device. Thus the speedometer reading would consist of the three significant figures—48.5. In a similar fashion, the odometer would yield a seven-significant-figure reading of 87324.45.

Significant figures must be considered in all engineering work involving numbers. Whether we are reading an instrument or performing calculations, we should report only the significant figures of a result so that anyone who uses the numbers can do so with confidence.

Although it is usually a straightforward procedure to ascertain the significant figures of a number, some cases can lead to confusion. For example, zeros are not always significant figures because they may be necessary just to locate a decimal point. The numbers 0.00001845, 0.0001845, and, 0.001845 all have four significant figures. Similarly, when trailing zeros are used in large numbers, it is not clear how many, if any, of the zeros are significant. For example, at face value the number 45,300 may have three, four, or five significant digits, depending on whether the zeros are known with confidence. Such uncertainty can be resolved by using scientific notation, where 4.53×10^4, 4.530×10^4, and 4.5300×10^4 designate that the number is known to three, four, and five significant figures, respectively.

24.2 ACCURACY AND PRECISION

The errors associated with engineering calculations and measurements can be characterized with regard to their precision and accuracy. *Accuracy* refers to how closely a computed or measured value agrees with the true value. *Precision* refers to how closely individual computed or measured values agree with each other.

These concepts can be illustrated graphically using an analogy from marksmanship (Fig. 24.2). *Inaccuracy* (also called *bias*) is defined as systematic deviation from

FIGURE 24.2
Examples from marksmanship illustrating the concepts of accuracy and precision: (*a*) inaccurate and imprecise; (*b*) accurate and imprecise; (*c*) inaccurate and precise; (*d*) accurate and precise.

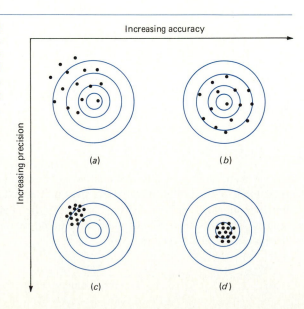

the truth. Thus, although the shots in Fig. 24.2c are more tightly grouped than in Fig. 24.2a, the two cases are equally biased because they are both centered on the upper left quadrant of the target. *Imprecision* (also called *uncertainty*), on the other hand, refers to the magnitude of the scatter. Therefore, although Fig. 24.2b and d are equally accurate (that is, centered on the bull's-eye), the latter is more precise because the shots are tightly grouped.

Significant figures can be used to signify both the accuracy and the precision of calculations and measurements. In order to do this, we have to first introduce some quantitative definitions of error.

24.3   ERROR

Now that we have introduced the qualitative notations of accuracy and precision, we can show how they are expressed quantitatively. Engineering calculations and measurements should be sufficiently accurate, or unbiased, to meet the requirements of a particular problem. They should also be precise enough for adequate engineering design. In this book we use the collective term *error* to represent both the inaccuracy and the imprecision of our calculations and measurements. As to the specific components of the error, we employ the term *bias* to represent inaccuracy and *uncertainty* to represent imprecision.

It should be noted that other terms can be used to signify these quantities. For example, some engineers refer to inaccuracy as *systematic error* and imprecision as *random error* (Ang and Tang, 1975). Whatever terminology is employed, it is critical to recognize the fundamental difference between the two types of error and how it relates to the reliability of your work.

24.3.1   Bias

Bias is the discrepancy between an exact, or true, value and an estimate. The relationship between the true value and the estimate can be formulated as

$$\text{True value} = \text{estimate} + \text{bias} \tag{24.1}$$

By rearranging Eq. (24.1), we find that bias is equal to the discrepancy between the truth and the estimate, as in

$$\text{Bias} = \text{true value} - \text{estimate} \tag{24.2}$$

A shortcoming of this quantitative definition is that it takes no account of the magnitude of the value under examination. For example, a bias of a millimeter is much more significant if we are measuring the diameter of a wire rather than its length. One way to account for the magnitude of the quantity being evaluated is to normalize the bias to the true value, as in

$$\text{Fractional relative bias} = \frac{\text{bias}}{\text{true value}} \tag{24.3}$$

where, as specified by Eq. (24.2), bias = true value − estimate. The fractional relative bias can also be multiplied by 100 percent and expressed as

$$\text{Percent relative bias} = \frac{\text{bias}}{\text{true value}} \times 100\%$$

(24.4)

An important advantage of relative bias is that it is independent of units. Whereas a change in units can have a marked effect on Eq. (24.2), it has no effect on Eqs. (24.3) and (24.4) because both the numerator and denominator of their right-hand sides would be multiplied by the same conversion factors.

The signs of Eqs. (24.2) through (24.4) may be either positive or negative. If the estimate is greater than the true value, the bias will be negative; if the estimate is less than the true value, the bias will be positive. The denominator for Eqs. (24.3) and (24.4) may be less than zero, which can also lead to a negative relative bias. Often we may not be concerned with the sign of the bias but are interested only in its magnitude. Therefore it is often useful to employ the absolute value of Eqs. (24.2) through (24.4).

EXAMPLE 24.1   Calculating Bias

PROBLEM STATEMENT: Suppose that you have the task of measuring the lengths of a bridge and a rivet, and you come up with 9999 and 11 cm, respectively. If the true values are 10,000 and 10 cm, respectively, compute (*a*) the bias and (*b*) the absolute value of the percent relative bias for each case.

SOLUTION
(*a*) The bias for measuring the bridge is [Eq. (24.2)]

$$\text{Bias} = 10{,}000 - 9999 = 1 \text{ cm}$$

and for the rivet is

$$\text{Bias} = 10 - 11 = -1 \text{ cm}$$

Thus both biases have the same magnitude but different signs because one measurement is an underestimate and the other an overestimate.
(*b*) The absolute value of the percent relative bias for the bridge is

$$|\text{Percent relative bias}| = \left| \frac{1}{10{,}000} \right| 100\% = 0.01\%$$

and for the rivet is

$$|\text{Percent relative bias}| = \left| \frac{-1}{10} \right| 100\% = 10\%$$

Although both measurements represent discrepancies of 1 cm, the absolute value for the percent relative bias for the rivet is much greater. You would, therefore, conclude that you have done an adequate job of measuring the bridge, whereas your estimate for the rivet leaves something to be desired.

When performing actual measurements, the true value is rarely known. If you knew the true value, there would be little point in performing the measurement in the first place. Therefore, you might wonder whether Eqs. (24.1) through (24.4) have more than merely theoretical significance.

One practical application occurs when an equipment manufacturer or user checks out the accuracy of an instrument. For example, an automobile manufacturer could place a car on rollers that can be set at a prespecified velocity. Then the speedometer can be checked to ensure that it provides an accurate reading. If not—that is, if the measurement instrument is biased—it would have to be adjusted to conform to the true value. This adjustment process is called *calibration*. Some instruments are adjusted at the factory, whereas others must be adjusted by the user. In either case, the calibration involves comparison of a measurement with a true value in order to evaluate the bias.

24.3.2 Uncertainty

When using an unbiased instrument, the key question in the measurement process reduces to one of precision. A formulation to represent such cases is

$$\text{True value} = \text{estimate} \pm \text{uncertainty} \tag{24.5}$$

In contrast to Eq. (24.1), the true value is not defined exactly but rather lies somewhere within the range specified by the uncertainty. The estimate is usually expressed to the number of significant figures corresponding to the precision of the measuring device. For example, for the speedometer in Fig. 24.1, we could conclude that the velocity was 48.5 with an error of 0.5, or, using Eq. (24.5),

$$\text{True value} = 48.5 \pm 0.5$$

In other words, we can say with confidence that the true value lies somewhere between 48 and 49. It must be stressed, however, that our confidence in this result is predicated on the assumption that the graduations on the speedometer are a valid expression of the instrument's precision. Although this is often true, there are certain instances where the precision is poorer than implied by the instrument's gauge. In such cases, Eq. (24.5) would merely be modified to reflect the increase or decrease in uncertainty. For example, if we knew that the speedometer had a precision of 0.7 km/h, the reading would be expressed as

$$\text{True value} = 48.5 \pm 0.7$$

This signifies that the result is between 47.8 and 49.2.

Many equipment manufacturers specify the precision of their instruments. In other cases, you may have to estimate the uncertainty yourself. In Sec. 24.8, we present methods for accomplishing this objective.

Just as we developed a relative fractional bias to remove scale effects from our estimate of accuracy, we can develop a fractional relative uncertainty, as in

$$\text{Fractional relative uncertainty} = \frac{\text{uncertainty}}{\text{estimate}}$$

and a percent relative uncertainty, as in

$$\text{Percent relative uncertainty} = \frac{\text{uncertainty}}{\text{estimate}} \, 100\%$$

Note that in both cases the uncertainty is normalized to the estimate rather than to the true value as is the case for the relative biases. Also note that absolute values are usually employed in these formulas because the magnitude rather than the sign is typically of interest.

EXAMPLE 24.2   Calculating Uncertainty

PROBLEM STATEMENT: An automobile manufacturer places an automobile on rollers that maintain the car's velocity at a steady level of 48.5 km/h. During this test, the car is subjected to various environmental factors such as different temperatures and simulated road surfaces. At various times, the speedometer is read and the results recorded as in Table 24.1. Use this information to determine (a) the uncertainty and (b) the percent relative uncertainty for the speedometer.

SOLUTION
(a) Although the average value for the data in Table 24.1 is 48.5 km/h (which indicates no bias), recorded values range from 46.5 to 50.5. If this range is considered a valid estimate of the instrument's precision, the uncertainty can be calculated as

$$\text{Uncertainty} = \frac{50.5 - 46.5}{2} = 2 \text{ km/h}$$

TABLE 24.1   Speedometer Readings under Test Conditions with True Value of Velocity Held at 48.5 km/h

48.5	47.5	49.5	48.5
48.5	48.5	47.5	49.5
49.5	50.5	48.5	
47.5	48.5	46.5	

Therefore, if a value of 48.5 is read, we can state that

True value = 48.5 ± 2 km/h

That is, we can state that the true velocity falls somewhere between 46.5 and 50.5 km/h.

(*b*) The percent relative uncertainty is

$$\text{Percent relative uncertainty} = \frac{2}{48.5} \, 100\% = 4\%$$

24.3.3 Total Error

For certain cases, a measurement will be both inaccurate and imprecise. In these situations the total error consists of both bias and uncertainty, as in

Total error = bias ± uncertainty

Therefore the true value is defined as

True value = estimate + bias ± uncertainty

The above formula provides a comprehensive framework for considering measurement error. We will now take a closer look at the uncertainty term in this formula. As shown in the following sections, the field of statistics has been developed to quantify random errors of this type.

24.4   DESCRIPTIVE STATISTICS

In the previous section we introduced the notion of the uncertainty, or random error, as a measure of the imprecision of a measurement. As a first example, we quantified the imprecision of an automobile speedometer. To do this, we placed the car on rollers that maintained a constant velocity of 48.5 km/h. Then we took a reading at various times from the speedometer and compiled the results in Table 24.1. The 14 results indicated an average of 48.5 km/h and a range from 46.5 to 50.5. This range was used to quantify the uncertainty, or spread of the readings, as ± 2 km/h.

We could continue this experiment to obtain additional readings. Table 24.2 contains the original 14 readings along with 10 additional readings. As with the previous case, most of the readings are close to 48.5. However, notice that the range now stretches from 45.5 to 52.5. Thus the range has grown from a level of 4 km/h for the 14 readings to a level of 7 km/h for the 24 readings. The fact that the range grows as the number of readings increases makes it a poor measure of uncertainty. For this reason, an alternative measure of spread must be devised.

TABLE 24.2 Speedometer Readings from Table 24.1 plus 10 Additional Readings (Shaded)

48.5	47.5	49.5	48.5
48.5	48.5	47.5	49.5
49.5	50.5	48.5	52.5
47.5	48.5	46.5	47.5
45.5	46.5	51.5	48.5
46.5	50.5	50.5	49.5

The field of statistics has been developed to address such problems. *Statistics* can be defined as the science of systematic collection and organization of quantifiable data. It also provides methods for using the data to draw conclusions.

One of the most fundamental objectives of statistics (and the origin of its name) is the summarization of data in one or more well-chosen ''statistics'' that convey as much information as possible about specific characteristics of the data set. These *descriptive statistics* are most often selected to represent (1) the location of the center of the distribution of the data and (2) the degree of spread of the data set. We already took a first step in this direction in the previous section when we examined the average value and the range of the speedometer readings. The following sections are devoted to a closer look at these and other descriptive statistics. First, however, we will discuss some general concepts related to the statistical characterization of data.

24.5 POPULATIONS AND SAMPLES

A common exercise in engineering is to collect information in order to describe and draw conclusions regarding an entire *population* of individuals, objects, or characteristics. A problem with such an exercise is that it is often impossible or impractical to study each and every individual item of a large population. In such cases, the engineer must attempt to make the characterization on the basis of a limited *sample*.

One example is the industrial engineer who must assess the quality of a population of a particular manufactured product, say an automobile. It would clearly be impractical to test each automobile individually. Consequently, the engineer could randomly select a number of the cars and, on the basis of the sample, attempt to characterize the quality of the entire population.

Another example of a sample of a population was our effort to characterize the uncertainty or imprecision of the speedometer in the previous chapter. For this case, the population is the total number of speedometer readings that could conceivably be taken to measure the car's velocity. This type of population is theoretically infinite in the sense that, assuming the car never malfunctioned, the readings could be taken indefinitely. Thus the total population could *never* be completely characterized. Our only option is to take a limited sample of readings and to use this sample to characterize the population.

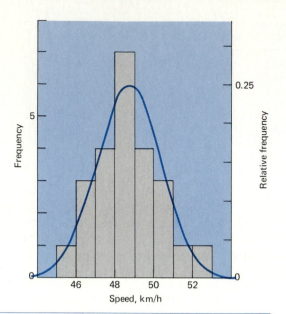

FIGURE 24.3
A histogram for the speedometer
data. An ideal bell-shaped or normal distribution is superimposed.

In certain cases you may have the opportunity to analyze an entire population. A good example is a census. However, more frequently your task will be to characterize an "infinite" population on the basis of a limited sample. With this as background, we can proceed to the first method of descriptive statistics—the histogram.

24.6   THE DISTRIBUTION OF DATA AND HISTOGRAMS

Aside from determining their range, another way to characterize a set of random measurements is to determine their distribution—that is, the shape with which the data is spread out. A histogram is a simple visual way to do this.

A histogram is constructed by sorting the measurements into intervals. The total number that falls within each interval is called the *frequency*. These frequencies are plotted in the form of a bar diagram. The units of measurement are plotted on the horizontal axis and the frequency of occurrence of each interval is plotted on the vertical axis.

Figure 24.3 is a histogram that was constructed for the speedometer data from Table 24.2. In addition to the frequency, this plot also shows the relative frequency as a right-hand-side ordinate. The *relative frequency* is the frequency divided by the total number of measurements. Thus, the height of each bar is the fraction of the total that falls within a particular interval.

Notice how most of the measurements group around the center of the distribution and fall off in an almost symmetrical fashion to each side. Many distributions have this characteristic shape. If we have a very large set of data and employ narrow intervals, the histogram will often be transformed to a smooth shape. The symmetric, bell-shaped curve superimposed on Fig. 24.3 is one such characteristic shape—the

TABLE 24.3   Measurements* of the Coefficient of Thermal Expansion
of a Structural Steel

6.555	6.625	6.465	6.655	6.455	6.545
6.675	6.445	6.725	6.485	6.665	6.715
6.405	6.645	6.785	6.645	6.535	6.495
6.495	6.685	6.525	6.585	6.755	6.705
6.745	6.515	6.695	6.685	6.555	6.695
6.505	6.535	6.565	6.705	6.675	6.545

*10^{-6} in/in/°F.

normal distribution. Given enough addition measurements, the histogram for this particular case could eventually approach the normal distribution. Because of its importance in statistical analysis, we will return to the normal distribution at a later point in this chapter.

It is a relatively easy task to construct Fig. 24.3 because of the manner in which the speedometer readings were taken. Recall from Sec. 24.1 that the recording of the speedometer readings constituted a sorting process. Whenever the indicator fell between two of the speedometer graduations, we recorded the velocity as the midpoint. Thus all the readings were automatically placed within intervals of 1 km/h. Consequently, one reading fell between 45 and 46, three fell between 46 and 47, and so on. Therefore, for this particular case, the construction of the histogram was easy. Other cases are not so automatic and require judgment on the part of the individual constructing the histogram.

An example is the data for the coefficient of thermal expansion of a structural steel, as reported in Table 24.3. The instrument used to make these measurements has a gauge that allows readings to four significant figures. Because of this, each measurement falls within an interval that is 0.01 unit wide. If each of these is taken as the width of the intervals for the histogram, Fig. 24.4*a* results. Because of the range of the measurements and the amount of the data, it is difficult to perceive the shape of the data distribution on the basis of this histogram. To obtain a more coherent and smoother-looking histogram we must use broader intervals to characterize the data. As seen in Fig. 24.4*b*, grouping the data into 0.04-unit intervals reveals that the data seems to be concentrated in two groups, that is, a *bimodal distribution*. Note that if the intervals are made too broad, as shown Fig. 24.4*c*, the true shape of the data will also be obscured.

A few general guidelines can help to create meaningful histograms. It is good practice to use about 10 intervals when making a first attempt to construct the histogram. In any event, try to avoid using less than 6 or more than 18. Of course, the proper choice will depend on the amount of data. If you only have 10 to 20 items of data, 10 intervals will probably be too many because none of the intervals would have frequencies large enough to make the distribution evident. Conversely, for a very large quantity of data, more than 20 intervals might be used effectively.

Another rule is to ensure that each measurement falls into a single interval. That is, make sure that none of the measurements falls exactly on the upper or lower bound

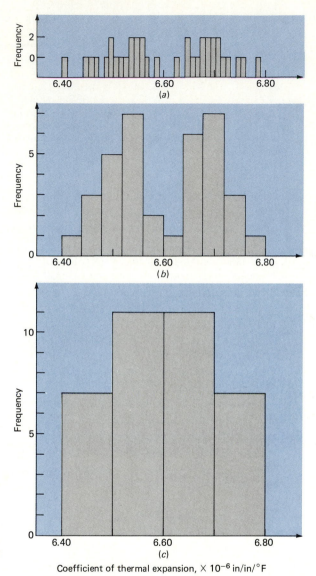

FIGURE 24.4
Three histograms constructed with
the same data but using different
segment widths.

Coefficient of thermal expansion, $\times 10^{-6}$ in/in/°F

of an interval. If this occurs, it will be unclear into which interval the measurement
should be placed.

Finally, and most important, be flexible. Try a number of possibilities in order
to ensure that the final histogram represents the true shape of the data distribution.
As we will see in the next section, the ability to try a number of different cases is
greatly facilitated by computer-generated histograms.

Comparisons of Fig. 24.3 and 24.4 should also make it evident that histograms

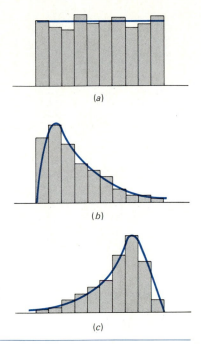

FIGURE 24.5
Some characteristic shapes of histograms; (a) uniform, (b) skewed right, and (c) skewed left. Idealized shapes for such distributions are superimposed.

come in a variety of characteristic shapes. Figure 24.5 illustrates three other types. Figure 24.5a is called a *uniform distribution* because the number of instances within a range occur with equal likelihood. Figure 24.5b is similar to the normal distribution in that the data is concentrated around one particular location. However, the distribution is not symmetrical for this case but is skewed to one side. The particular distribution in Fig. 24.5b is said to be "skewed to the right" because the right tail is drawn out. Figure 24.5c shows a distribution that is skewed to the left.

24.6.1 Computer Programs for the Histogram

Figure 24.6 shows BASIC and FORTRAN programs to determine and print out a histogram. After the data is entered, the programs implement a Shell sort to place the numbers in ascending order (recall Sec. 23.3). This provides the lower and upper bounds for the histogram [X(1) and X(N)]. In addition, sorting the data makes it easier to organize the numbers into the segments of the histogram.

The histogram itself consists of 10 segments ranging from the lower to the upper bound. A loop is used to determine each segment's frequency—that is, to count the number of values falling within each segment. Finally, the resulting histogram, along with the lower and upper limits for each segment, is printed out. Notice how in Fig. 24.7, the histogram is printed sideways. This is the most direct way to print a bar diagram because of the manner in which the computer's printer outputs results—that is, from left to right and from top to bottom of a printed page.

```
100 REM          **************************************************
110 REM          *              HISTOGRAM PROGRAM                  *
120 REM          *                (BASIC version)                  *
130 REM          *                 by  S.C. Chapra                 *
140 REM          *                                                 *
150 REM          *  This program plots a frequency histogram.      *
160 REM          **************************************************
170 REM
180 REM            .-------------------------------------------.
190 REM            |          DEFINITION OF VARIABLES           |
200 REM            |                                            |
210 REM            |  X() = VALUES OF A RANDOM VARIABLE          |
220 REM            |  N   = TOTAL NUMBER OF VALUES OF X          |
230 REM            '--------------------------------------------'
240 REM
250 DIM X(100),PLFREQ(10)
260 REM ---------------------------------------------------------
270 REM           Function to Round to 3 Decimal Places
280 REM
290 DEF FNR(X) = INT(X*10^3+.5)/10^3
300 REM
310 REM ******************  MAIN PROGRAM  ******************
320 REM
330 GOSUB 800
340 GOSUB 1000
350 GOSUB 2000
360 END
800 REM ************* SUBROUTINE TO ENTER DATA ***********
810 REM
820 CLS
830 FOR I = 1 TO 100
840    INPUT "VALUE = (-999 TO STOP)?   ",X(I)
850    IF X(I) = -999 THEN 860 ELSE 880
860       N = I - 1
870       I = 100
880    REM ENDIF
890 NEXT I
900 RETURN
1000 REM ******************  SUBROUTINE SHELL  **************
1010 REM
1020 REM      PERFORMS A SHELL SORT
1030 REM
1040 REM      DATA REQUIRED:    N, X() --) UNSORTED
1050 REM      RESULTS RETURNED: X() --) SORTED
1060 REM
1070 REM      LOCAL VARIABLES:
1080 REM
1090 REM      SHGAP   = GAP SIZE FOR ELEMENT COMPARISON
1100 REM                DURING SORT
1110 REM      SHSWCH  = INDICATES THAT ELEMENTS HAVE
1120 REM                BEEN SWITCHED
1130 REM      SHDUM   = TEMPORARY DUMMY VARIABLE
1140 REM
1150 REM ------------------ BODY OF SHELL ------------------
1160 REM
1170 REM              Subroutine for Shell Sort
1180 REM
1190 SHGAP = N
1200 SHGAP = INT(SHGAP/2)
1210 REM PERFORM LOOP WHILE INTERVAL )= 1
1220 WHILE SHGAP )= 1
1230    REM
1240    REM       Perform Loop While Swithces Occur
1250    SHSWCH = 1
1260    WHILE SHSWCH = 1
1270       REM       Initialize Switch Indicator
1280       SHSWCH = 0
1290       REM Loop to Compare and, if Necessary, Switch
1300       FOR J = 1 TO N-SHGAP
1310          REM
1320          REM If Former Greater than Latter
1330          IF X(J) ) X(J+SHGAP) THEN 1350 ELSE 1400
1340             REM
1350             REM Then Switch and Set Switch Indicator to 1
1360             SHDUM = X(J)
1370             X(J) = X(J + SHGAP)
1380             X(J + SHGAP) = SHDUM
1390             SHSWCH = 1
1400          REM ENDIF
1410       NEXT J
1420    WEND
1430    SHGAP = INT(SHGAP/2)
1440 WEND
1450 RETURN
2000 REM ************* SUBROUTINE PLOT HISTOGRAM ************
2010 REM
2020 REM      DIVIDES FREQUENCY DATA INTO CLASSES
2030 REM      AND PLOTS A HISTOGRAPH.
2040 REM
2050 REM      DATA REQUIRED:    X(), N
2060 REM      RESULTS RETURNED: NONE
2070 REM
2080 REM      LOCAL VARIABLES:
2090 REM
2100 REM      PLLOBND  = SMALLEST VALUE INPUT BY USER
2110 REM      PLUPBND  = LARGEST VALUE INPUT BY USER
2120 REM      PLWDTH   = WIDTH OF EACH CLASS
2130 REM      PLUPLIM  = UPPER LIMIT OF A CLASS
2140 REM      PLSEGMNT = CLASS NUMBER DURING ITERATION
2150 REM      PLFREQ() = FREQUENCY OF OCCURANCES WITHIN A CLASS
2160 REM      PLLOWER  = LOWER LIMIT OF AN ITERATION CLASS
2170 REM      PLUPPER  = UPPER LIMIT OF AN ITERATION CLASS
2180 REM
2190 REM ---------------- BODY OF HISTOGRAM ----------------
2200 REM
2210 REM         Determine Frequency for Each Segment
2220 PLLOBND = X(1)
2230 PLUPBND = X(N) + X(N)/1000000!
2240 PLWDTH = (PLUPBND - PLLOBND)/10
2250 PLUPLIM = PLLOBND + PLWDTH
2260 PLSEGMNT = 1
2270 FOR I = 1 TO N
2280    WHILE X(I) ) PLUPLIM
2290       PLUPLIM = PLUPLIM + PLWDTH
2300       PLSEGMNT = PLSEGMNT + 1
2310    WEND
2320    PLFREQ(PLSEGMNT) = PLFREQ(PLSEGMNT) + 1
2330 NEXT I
2340 REM
2350 REM              Print Out Histogram
2360 REM
2370 CLS
2380 PLLOWER = PLLOBND
2390 FOR I = 1 TO 10
2400    PLUPPER = PLLOWER + PLWDTH
2410    PRINT USING "####.##   ";PLLOWER,PLUPPER;
2420    IF PLFREQ(I) () 0 THEN 2430 ELSE 2460
2430       FOR J = 1 TO PLFREQ(I)
2440          PRINT "*";
2450       NEXT J
2460    REM ENDIF
2470    PRINT
2480    PLLOWER = PLUPPER
2490 NEXT I
2500 RETURN
```

FIGURE 24.6 BASIC and FORTRAN programs to generate a histogram.

```
C
C          ***********************************************
C          *               HISTOGRAM PROGRAM             *
C          *               (FORTRAN version)             *
C          *               by  S.C. Chapra               *
C          *                                             *
C          *  This program plots a frequency histogram.  *
C          ***********************************************
C
C                .---------------------------------.
C                |         DEFINITION OF VARIABLES  |
C                |                                  |
C                | X() = VALUES OF A RANDOM VARIABLE|
C                | N   = TOTAL NUMBER OF VALUES OF X |
C                |                                  |
C                `---------------------------------'
C
           REAL X(100), FREQ(10)
C
C          ****************** MAIN PROGRAM *******************
C
           CALL ENTER(X,N)
           CALL SHELL(X,N)
           CALL HISTOG(X,N)
           STOP
           END
C
C          ************* SUBROUTINE TO ENTER DATA ***********
C
           SUBROUTINE ENTER(X,N)
           REAL X(100)
           INTEGER N
           DO 10 I=1,100
             READ(5,*) X(I)
             IF (X(I).EQ.-999) THEN
                GO TO 20
             ENDIF
     10    CONTINUE
     20    N = I - 1
           RETURN
           END
C          ***************** SUBROUTINE SHELL **************
C
           SUBROUTINE SHELL(X,N)
C
C          PERFORMS A SHELL SORT
C
C          DATA REQUIRED:    N   --) NUMBER OF ELEMENTS IN X()
C                            X() --) UNSORTED
C          RESULTS RETURNED: X() --) SORTED
C
C          LOCAL VARIABLES:
C
C          GAP   =    GAP SIZE FOR ELEMENT COMPARISON
C                     DURING SORT
C          SWCH  =    INDICATES THAT ELEMENTS HAVE
C                     BEEN SWITCHED
C          N     =    NUMBER OF ELEMENTS IN X()
C          DUMMY =    TEMPORARY DUMMY VARIABLE
C
           REAL X(20),DUMMY
           INTEGER SWCH,GAP,N
C          ------------------ BODY OF SHELL ------------------
C
C                    Subroutine for Shell Sort
C
           GAP = N
           GAP = INT(GAP/2)
C
C          DO WHILE GAP ) = 1
   1000    IF (GAP .GE. 1) THEN
             SWCH = 1
C
C            DO WHILE SWCH = 1
   3000      IF (SWCH .EQ. 1) THEN
C
C              Initialize Switch Indicator
               SWCH = 0
C
C              Loop to Compare and, if Necessary, Switch
               DO 10 J=1, N - GAP
C
C                If Former Greater than Latter
                 IF (X(J) .GT. X(J+GAP)) THEN
C
C                    Then Switch and Set Switch Indicator to 1
                     DUMMY = X(J)
                     X(J) = X(J + GAP)
                     X(J + GAP) = DUMMY
                     SWCH = 1
                 ENDIF
     10        CONTINUE
               GOTO 3000
             ENDIF
C          END WHILE
C
             GAP = INT(GAP/2)
             GOTO 1000
           ENDIF
C          END WHILE
C
           RETURN
           END
C          ************* SUBROUTINE PLOT HISTOGRAM ***********
C
           SUBROUTINE HISTOG(X,N)
C
C          DIVIDES FREQUENCY DATA INTO CLASSES
C          AND PLOTS A HISTOGRAPH.
C
C          DATA REQUIRED:    X(), N
C          RESULTS RETURNED: NONE
C
C          LOCAL VARIABLES:
C
C          LOBND  = SMALLEST VALUE INPUT BY USER
C          UPBND  = LARGEST VALUE INPUT BY USER
C          WDTH   = WIDTH OF EACH CLASS
C          UPLIM  = UPPER LIMIT OF A CLASS
C          SEGMNT = CLASS NUMBER DURING ITERATION
C          FREQ() = FREQUENCY OF OCCURANCES WITHIN A CLASS
C          LOWER  = LOWER LIMIT OF AN ITERATION CLASS
C          UPPER  = UPPER LIMIT OF AN ITERATION CLASS
C
           REAL    X(100), FREQ(10), UPLIM, WDTH,
          +        LOBND, UPBND, LOWER, UPPER
           INTEGER N, SEGMNT
C          ---------------- BODY OF HISTOGRAM ----------------
C
C                 Determine Frequency for Each Segment
           LOBND = X(1)
           UPBND = X(N) + X(N)/1000000.
           WDTH = (UPBND - LOBND)/10
           UPLIM = LOBND + WDTH
           SEGMNT = 1
           DO 10 I=1, N
C
C            DO WHILE X(I) ) UPLIM
   1000      IF (X(I) .GT. UPLIM) THEN
                 GOTO 2000
             ELSE
                 GOTO 3000
   2000          UPLIM = UPLIM + WDTH
                 SEGMNT = SEGMNT + 1
                 GOTO 1000
   3000      ENDIF
C            END WHILE
C
             FREQ(SEGMNT) = FREQ(SEGMNT) + 1
     10    CONTINUE
C
C                     Print Out Histogram
C
           LOWER = LOBND
           DO 20 I=1,10
             UPPER = LOWER + WDTH
             IF (FREQ(I) .EQ. 0) THEN
               WRITE(6,*) LOWER,UPPER
             ELSE
               WRITE (6,*) (LOWER,UPPER,('*',J=1,FREQ(I)))
             ENDIF
             LOWER = UPPER
     20    CONTINUE
           RETURN
           END
C          ----------------------------------------------------
C                 Function to Round to 3 Decimal Places
C
           INTEGER FUNCTION FNR(X)
           INTEGER X
           FNR = INT(X*10**3+.5)/10**3
           RETURN
           END
```

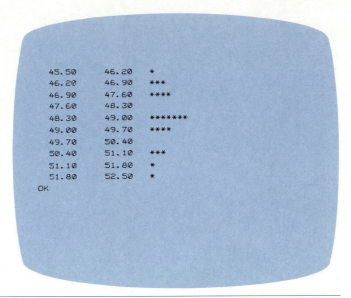

FIGURE 24.7
A histogram produced with the
programs from Fig. 24.6 for the
data from Table 24.2.

EXAMPLE 24.3   Generating Histograms with the Computer

PROBLEM STATEMENT: A user-friendly computer program to create histograms is contained in the ENGINCOMP software associated with the text. We can use the software to generate a histogram for the data in Tables 24.2 and 24.3.

FIGURE 24.8 Computer screens displaying the ENGINCOMP software to create a histogram for the data from Table 24.2: (a) the way in which the parameters of the histogram are entered and (b) the resulting histogram.

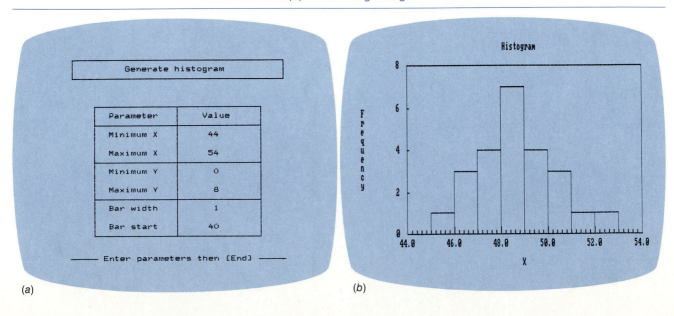

SOLUTION: Figure 24.8*a* shows the screen used to enter the parameters for the histogram. Notice that the parameters allow you to scale the height of the histogram, the limits of the abscissa, and the width of the bars. This allows you to quickly try different values until you obtain an acceptable histogram. The results for the data from Table 24.2 are shown in Fig. 24.8*b*.

An example of three histograms developed for the data from Table 24.3 is shown in Fig. 24.9. Notice how the choice of the bar width results in different shapes. The version in Fig. 24.9*b* is the most satisfactory in that it captures the bimodal nature of this distribution. Because software allows you to construct the histograms in seconds,

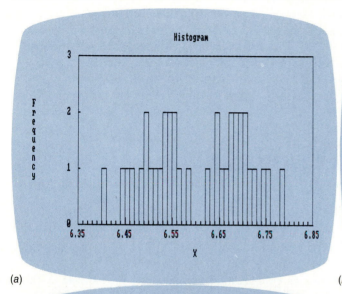

(a)

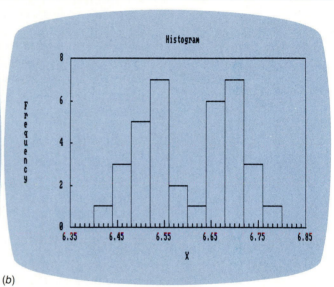

(b)

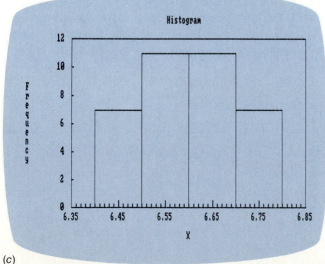

(c)

FIGURE 24.9 Three computer-generated histograms for the data from Table 24.3 using bar widths of (a) 0.01, (b) 0.04, and (c) 0.1. Computerized software such as ENGINCOMP allows you to generate different versions in an efficient manner. This sort of exploratory data analysis is invaluable for building insight regarding data distributions.

you have the capability of trying many versions. Such exploratory data analysis is one of the strengths of computerized statistical software.

24.7   MEASURES OF LOCATION

The histogram is more descriptive than a simple listing of data because histograms are presented in an organized manner and convey information that may be apparent only after a careful study of a table of data. There are many times, however, when we may want to compare or summarize sets of data. For example, we might want to determine whether, in a collective sense, the values in one distribution are greater or less than in another. For such cases, histograms are cumbersome and statistics will be better suited to our needs. As described in the next section, the most fundamental statistics for this purpose provide measures of the central location of the data set.

One of the first steps in analyzing data of the sort found in Table 24.2 is to determine the average value. However, as described in this section, a simple average or mean is but one of several ways to quantify the location of a data set.

24.7.1   The Mean

The most common location statistic is the arithmetic mean. The *arithmetic mean* $\bar{y}$ of a sample is defined as the sum of the individual data points y_i divided by the number of points n, or

$$\bar{y} = \frac{\Sigma y_i}{n} \tag{24.6}$$

where the summation is from $i = 1$ through n.* For example, for the data in Table 24.2, the summation is 1168.0 and n is 24. Therefore the mean is 1168.0/24, or 48.7.

It should be noted that the symbol $\bar{y}$ is used to designate the arithmetic mean of a *sample* of a population. For cases where we are dealing with the mean of the entire *population*, Eq. (24.6) is used for the computation, but the mean is designated by the symbol μ.

EXAMPLE 24.4   Computing the Mean of Samples and Populations

PROBLEM STATEMENT: The final grades for all the students enrolled in a particular course are

Juniors	67	99	79	71	62
Juniors	96	84	68	77	81
Sophomores	88	71	73	52	86

*Unless noted otherwise, all summations in the remainder of this chapter are also from i through n.

Determine the arithmetic mean for the entire population. Also, compute the arithmetic mean for a sample consisting of the juniors—that is, the first two rows.

SOLUTION: The summation of all the grades is 1154 and $n = 15$. Therefore the mean is

$$\mu = \frac{1154}{15} = 76.93$$

where μ is used for this particular case because we are characterizing the mean of the entire population.

For the first two rows, the summation is 784 and $n = 10$ and the mean is

$$\bar{y} = \frac{784}{10} = 78.40$$

where $\bar{y}$ is used because we are taking a sample from the population to characterize the mean. The distinction between the sample and the population is mainly of theoretical interest. However, when we turn to measures of spread in Sec. 24.8, it will actually have a bearing on the proper choice of formulas.

Although the mean is usually the preferred statistic for defining central location, it can sometimes be misleading in this regard. In particular, the mean can be very sensitive to a few extreme measurements when the sample size is small. Consequently, alternative measures are sometimes employed.

One alternative, a *trimmed mean*, is computed after excluding a fixed percentage of the smallest and largest numbers. For example, a 5 percent trimmed mean is calculated after excluding the smallest and largest 5 percent of the numbers.

EXAMPLE 24.5   Computing Trimmed Means

PROBLEM STATEMENT: The following data represents the concentration of a chemical feeding into a reactor:

1.755	1.835	1.785	1.775
1.945	0.115	1.815	
1.555	1.615	1.755	

where all the concentrations are in moles per liter. Construct a histogram and compute the mean and the 10 percent trimmed mean for the readings.

SOLUTION: Because of the small amount of data, we will use intervals of 0.1 mol/L in width. The resulting histogram (Fig. 24.10) indicates that most of the data is grouped around 1.75, while a single extreme point falls in the interval from 0.1 to 0.2. The mean is computed as

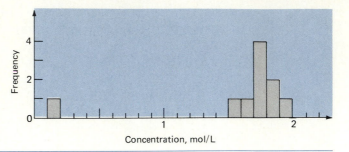

FIGURE 24.10
A histogram for concentrations of
a chemical feeding into a reactor
(from Example 24.5). The concentration off to the side is sometimes
called an "outlier."

$$\bar{y} = \frac{15.950}{10} = 1.595$$

Thus, although the points are clearly clustered around 1.75, the extreme point has a pronounced effect on the mean.

For the trimmed mean, 10 percent of the values (that is, one value) are trimmed from the bottom and from the top of the data set prior to determining the mean. Thus the values of 0.115 and 1.945 are excluded. The 10 percent trimmed mean is calculated as

$$\bar{y}_{t,10} = \frac{13.890}{8} = 1.736$$

Therefore the result is closer to the visual depiction of the distribution's center evident in Fig. 24.10.

Sometimes the trimmed portion of a distribution will not be an integer. For example, a 10 percent trimmed mean for 24 data points involves trimming 2.4 points from each end. Since this is clearly impossible, the usual strategy is to compute two trimmed means—one excluding two and the other excluding three extreme points. Then linear interpolation (to be discussed in Chap. 27) can be used to determine the 10 percent trimmed mean.

Although trimmed means have their advantages, they should be implemented with caution. They are particularly appropriate for cases when extreme values, or *outliers*, are, in fact, erroneous. For example, the lab technician making the readings for Example 24.5 could have made a blunder when recording the measurement of 0.115. On the other hand, the reading could be true, in which case it is a valid piece of information that should be considered when evaluating the data.

Aside from trimmed means, other location statistics are available that are insensitive to outliers. One of the most common is the median.

24.7.2 Medians and Percentiles

The *median* is the midpoint of a group of data. It is calculated by first putting the data in ascending order. If the number of measurements is odd, the median is the middle value of the measurements. If the number is even, the median is the arithmetic mean of the two middle values.

EXAMPLE 24.6   Computing the Median

PROBLEM STATEMENT: Determine the median of the data from Table 24.2 and from Example 24.5.

SOLUTION: First the data from Table 24.2 is put in ascending order:

45.5	47.5	48.5	(48.5)	49.5	50.5
46.5	47.5	48.5	48.5	49.9	50.5
46.5	47.5	48.5	48.5	49.5	51.5
46.5	47.5	(48.5)	48.5	50.5	52.5

The two middle values are 48.5; therefore the median is 48.5.

For Example 24.5, the two middle values are 1.755 and 1.775, for which the arithmetic mean is $(1.755 + 1.775)/2$, or 1.765. Thus, in contrast to the arithmetic mean calculated in Example 24.5 (1.595), this result is insensitive to the outlier value of 0.115 and is much closer to the center of the distribution indicated by Fig. 24.10.

The median is even less sensitive to extreme values than the trimmed mean. The median is, in fact, the 50 percent trimmed mean. This is because the process of ordering the data and picking the middle value is identical to trimming the values one at a time from each end until you arrive at the middle.

The median is just one of a family of statistics, called *percentiles*, that divide an ordered set of data into portions. For example, the three *quartiles* divide the data at the 25, 50, and 75 percent levels. Of course, the second quartile is the median. There are also percentiles that divide the data up to a specified percent. For example, the 95 percentile is the value that separates the lower 95 from the upper 5 percent of the data.

Aside from some advanced applications, the median is not as useful as the arithmetic mean for general mathematical statistics. However, for cases with extreme values or asymmetric distributions (that is, strongly skewed), it can be a useful descriptor of central location.

24.7.3   The Mode

The *mode* is the value that occurs most frequently. For example, the value of 48.5 occurs most often in Table 24.2 and is, therefore, the mode. Often a single value for a mode has little meaning. For example, Fig. 24.4a indicates that nine values occur two times. In such cases the *modal interval* is more meaningful. This refers to the interval with the highest frequency from the histogram that is used to describe the distribution. In Fig. 24.4b, there are two modal intervals from 6.52 to 6.56 and from 6.64 to 6.72. These two peaks are what led us to call this distribution "bimodal." Distributions with more than one peak are called *multimodal*.

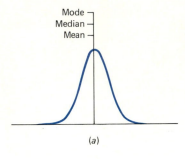

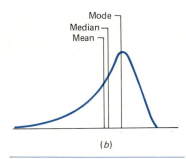

FIGURE 24.11
The relative position of measures of central tendency for (a) a symmetric and (b) an asymmetric distribution.

24.7.4 Comparison of Central Location Statistics

The arithmetic mean is generally the preferred measure of central location because it has a number of advantages. Among other things, it is much easier to work with mathematically. On the other hand, the median and mode are less sensitive to extreme values. The trimmed mean offers a compromise that somewhat remedies this defect.

For data that is unimodal and symmetric, the mean, median, and mode are identical (Fig. 24.11*a*). For a unimodal, asymmetric distribution, such as the skewed distributions in Fig. 24.5, the mean will be closest to the drawn out tail, then the median and then the mode. An easy way to remember this is that they fall in alphabetical order away from the long tail (Fig. 24.11*b*).

24.7.5 The Geometric Mean

Aside from the arithmetic mean, the median, and the mode, there are other measures of central location. Many of these involve transformations. For log-transformed data, a *geometric mean GM* can be computed as

$$GM = \text{antilog}\left(\frac{1}{n} \Sigma \log_{10} y_i\right)$$

Because the addition of logarithms is equivalent to the multiplication of their antilogarithms, the geometric mean can be represented alternatively as

$$GM = \sqrt[n]{y_1 y_2 y_3, \ldots, y_n} = \sqrt[n]{\prod_{i=1}^{n} y_i}$$

where Π means "the product of." Note that y_i must be > 0.

Many beginners in statistics have difficulty accepting the fact that measures of central tendency other than the arithmetic mean are valid. However, there are cases in engineering where measures such as the geometric mean are even preferable. Additional information on descriptive statistics can be found in the many fine books on the subject—for example, Wonnacott and Wonnacott (1972), Snedecor and Cochran (1967), and Box, Hunter, and Hunter (1978).

24.8 MEASURES OF SPREAD

The three distributions in Fig. 24.12 all have approximately the same mean. However, there are obvious differences in how they are spread out around the central value. Thus additional statistics, beyond those for central location, are required to characterize the spread, or dispersion, of data.

24.8.1 Range and Average Deviation

To this point we have used the range as our first estimate of dispersion. However, as noted previously, the range is inadequate because it is strongly affected by outliers.

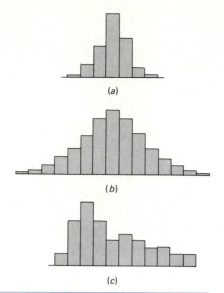

FIGURE 24.12
Three distributions with the same arithmetic mean but different spreads.

An alternative measure that minimizes the effect of extreme values would be to compute an average deviation from the mean, as in

$$\frac{\Sigma(\bar{y} - y_i)}{n}$$

This statistic is faulty because it is always equal to zero, a result of the fact that the mean marks the spot at which positive and negative deviations are balanced. Consequently, the positive and negative deviations cancel when they are summed.

A possible alternative would be to average the absolute values of the deviations, as in

$$\frac{\Sigma|\bar{y} - y_i|}{n}$$

Although this statistic can be used under some circumstances, experience as well as theoretical difficulties (that are beyond the scope of this text) have shown it to be generally inadequate. As discussed in the next section, the average of the squared deviations is the preferred statistic for measuring dispersion.

24.8.2 Variance and Standard Deviation

The primary statistic to characterize the dispersion of the data is the *variance*, which is defined as the average of the squared deviations from the mean. The squared deviations for a population are represented by

$$S_t = \Sigma(\mu - y_i)^2 \tag{24.7}$$

where S_t is called the *total sum of the squares of the residuals*. The formula for the variance is

$$\sigma^2 = \frac{\Sigma(\mu - y_i)^2}{n}$$

where σ^2 is the symbol for the variance. Squaring the deviations overcomes the fact that the sum of the deviations is zero.

The units for the variance are the square of the unit of y—for example, square feet, kilograms squared, or kilometers squared per hour squared. In order to have a statistic that has the same units as the quantity being measured, another statistic called the *standard deviation* (σ) is defined as the square root of the variance, or

$$\sigma = \sqrt{\frac{\Sigma(\mu - y_i)^2}{n}}$$

EXAMPLE 24.7   Computing the Variance and Standard Deviation of a Population

PROBLEM STATEMENT: Determine the variance and the standard deviation of the data from Example 24.4.

SOLUTION: The mean for the data from Example 24.4 is 76.93, and therefore the sum of the squared deviations can be calculated as

y_i	$(\mu - y_i)^2$
67	99
96	364
88	123
99	487
84	50
71	35
79	4
68	80
73	15
71	35
77	0
52	622
62	223
81	17
86	82
	$\Sigma(\mu - y_i)^2 = 2236$

Therefore the variance is ($n = 15$)

$$\sigma^2 = \frac{2236}{15} = 149$$

and the standard deviation is

$$\sigma = \sqrt{149} = 12.2$$

When dealing with samples rather than with the entire population, slightly different formulas are involved. For the total sum of the squared deviations,

$$S_t = \Sigma(\bar{y} - y_i)^2 \tag{24.8}$$

For the sample variance,

$$s_y^2 = \frac{\Sigma(\bar{y} - y_i)^2}{n - 1} \tag{24.9}$$

and for the sample standard deviation,

$$s_y = \sqrt{\frac{\Sigma(\bar{y} - y_i)^2}{n - 1}} \tag{24.10}$$

Notice that for these sample statistics, we divide by $n - 1$ rather than by n.

EXAMPLE 24.8   Computing the Sample Variance and Standard Deviation

PROBLEM STATEMENT: Determine the variance and the standard deviation for the first two rows of data in Example 24.4.

SOLUTION: The summation of the squared deviations for the first 10 values in Example 24.4 can be calculated as

y_i	$(\bar{y} - y_i)^2$
67	130
96	310
99	424
84	31
79	0
68	108
71	55
77	2
62	269
81	7
	$\Sigma(\bar{y} - y_i)^2 = 1336$

Therefore the variance is

$$s_y^2 = \frac{1336}{10 - 1} = 148.4$$

and the standard deviation is

$$s_y = \sqrt{148.4} = 12.2$$

We divide by $n - 1$ in Eq. (24.9) so that the sample variance s_y^2 is an unbiased estimator of the population variance σ^2. Mathematical analysts have proven that if n is used as the divisor, then the resulting value of s_y^2 will tend to underestimate σ^2. This is because the y_i's will tend to be closer to $\bar{y}$ than to the true mean μ of the population. Dividing by $n - 1$ corrects for this effect. Of course, when n is large, the resulting correction will be negligible as in Examples 24.7 and 24.8.

The quantity $n - 1$ is referred to as the *degrees of freedom*. Hence s_y^2 is said to be based on $n - 1$ degrees of freedom. This nomenclature derives from the fact that the sum of the quantities upon which s_y^2 is based (that is, $\bar{y} - y_1$, $\bar{y} - y_2$, . . ., $\bar{y} - y_n$) add up to zero. Consequently, if any $n - 1$ of the values is specified, the remaining value is fixed. Thus only $n - 1$ of the values are said to be freely determined. Another justification for dividing by $n - 1$ is the fact that there is no such thing as the spread of a single data point. For the case where $n = 1$, Eq. (24.9) yields a meaningless result of infinity.

It should be noted that we almost always deal with statistics for samples rather than for the entire population. Consequently, the standard deviation and variance calculated on a pocket calculator or in most software packages are invariably determined by Eqs. (24.9) and (24.10). Practically speaking, the only time that the distinction really matters is when dealing with very small values of n. Otherwise, the difference between the $n - 1$ in the sample variance and the n in the population variance is negligible.

24.8.3   The Coefficient of Variation

Recall that in Sec. 24.3 we introduced relative errors to remove scale effects from error estimates. In the same fashion, we can normalize the standard deviation to the mean, as in (for a sample of a population)

$$CV = \frac{s_y}{\bar{y}} 100\%$$

where CV is referred to as the *coefficient of variation*. For a population, CV would be computed as $(\sigma/\mu) \times 100\%$.

EXAMPLE 24.9   *Computing the Coefficient of Variation*

PROBLEM STATEMENT: Use the results of Examples 24.4 and 24.8 to determine the coefficient of variation for the first two rows of grades from Example 24.4.

SOLUTION: From Example 24.4, the mean of the first two rows is $\bar{y} = 78.4$ and from Example 24.8, the standard deviation is $s_y = 12.2$. Therefore the coefficient of variation is

$$CV = \frac{12.2}{78.4} \, 100\% = 15.6\%$$

Recall that in Sec. 24.3.2 we used the range to quantify the uncertainty of a group of measurements. For instance, in Example 24.2, we determined that the true value of the speedometer readings in Table 24.1 could be expressed as 48.5 ± 2 km/h. We also indicated that the 2 km/h uncertainty could be expressed in a percent relative uncertainty of 4 percent. The standard deviation and coefficient of variation provide alternative and superior statistics for the same purpose.

EXAMPLE 24.10   *Using the Standard Deviation and Coefficient of Variation as Measures of Uncertainty*

PROBLEM STATEMENT: Repeat Example 24.2 but use the standard deviation and the coefficient of variation to quantify the uncertainty.

SOLUTION: The mean and standard deviation of the data in Table 24.1 can be determined to be $\bar{y} = 48.5$ and $s_y = 1.04$. Therefore the uncertainty of the readings can be expressed as 48.5 ± 1.04 km/h. The coefficient of variation is

$$CV = \frac{1.04}{48.5} \, 100\% = 2.1\%$$

Notice that, in the foregoing example, the values for the standard deviation ($s_y = 1.04$ km/h) and the coefficient of variation (CV = 2.1 percent) are smaller than the uncertainty (2 km/h) and the percent relative uncertainty (4 percent) based on the range. This is as expected because the former values are averaged deviations rather than the extreme deviation represented by the range. As such, the standard deviation is less sensitive to outlying values and does not tend to grow as the number of measurements increases. Rather, it converges on a constant value—the population standard deviation—as *n* increases. This makes it preferable to the range. In addition, there are mathematical advantages, discussion of which is beyond the scope of this text, that make it preferable.

```
100 REM      ****************************************************
110 REM      *              DESCRIPTIVE STATISTICS              *
120 REM      *                 (BASIC version)                  *
130 REM      *                  by  S.C. Chapra                 *
140 REM      *                                                  *
150 REM      *    This program determines the mean, median,     *
160 REM      *    variance, coefficient of variation, and       *
170 REM      *    standard deviation for a set of numbers.      *
180 REM      ****************************************************
190 'REM
200 REM               .----------------------------------.
210 REM               |        DEFINITION OF VARIABLES    |
220 REM               |                                   |
230 REM               | X() = VALUES OF A RANDOM VARIABLE |
240 REM               | N   = TOTAL NUMBER OF VALUES OF X |
250 REM               `----------------------------------'
260 REM
270 DIM X(100)
280 REM
290 REM ****************** MAIN PROGRAM ******************
300 REM
310 GOSUB 800
320 GOSUB 1000
330 GOSUB 2000
340 END
350 REM
800 REM ************* SUBROUTINE TO ENTER DATA ***********
810 REM
820 CLS
830 FOR I = 1 TO 100
840    INPUT "VALUE = (-999 TO STOP)?   ", X(I)
850    IF X(I) = -999 THEN 860 ELSE 880
860        N = I - 1
870        I = 100
880    REM ENDIF
890 NEXT I
900 RETURN
1000 REM *************** SUBROUTINE SHELL ***************
1010 REM
1020 REM      PERFORMS A SHELL SORT
1030 REM
1040 REM      DATA REQUIRED:    N, X() --) UNSORTED
1050 REM      RESULTS RETURNED: X() --) SORTED
1060 REM
1070 REM      LOCAL VARIABLES:
1080 REM
1090 REM      SHGAP   = GAP SIZE FOR ELEMENT COMPARISON
1100 REM                DURING SORT
1110 REM      SHSWCH  = INDICATES THAT ELEMENTS HAVE
1120 REM                BEEN SWITCHED
1130 REM      SHDUM   = TEMPORARY DUMMY VARIABLE
1140 REM
1150 REM ------------------- BODY OF SHELL -----------------
1160 REM
1170 REM                Subroutine for Shell Sort
1180 REM
1190 SHGAP = N
1200 SHGAP = INT(SHGAP/2)
1210 REM PERFORM LOOP WHILE INTERVAL )= 1
1220 WHILE SHGAP )= 1
1230    REM
1240    REM        Perform Loop While Swithces Occur
1250    SHSWCH = 1
1260    WHILE SHSWCH = 1
1270       REM       Initialize Switch Indicator
1280       SHSWCH = 0
1290       REM Loop to Compare and, if Necessary, Switch
1300       FOR J = 1 TO N-SHGAP
1310          REM
1320          REM If Former Greater than Latter
1330          IF X(J) ) X(J+SHGAP) THEN 1350 ELSE 1400
1340             REM
1350             REM Then Switch and Set Switch Indicator to 1
1360             SHDUM = X(J)
1370             X(J) = X(J + SHGAP)
1380             X(J + SHGAP) = SHDUM
1390             SHSWCH = 1
1400          REM ENDIF
1410       NEXT J
1420    WEND
1430    SHGAP = INT(SHGAP/2)
1440 WEND
1450 RETURN
1460 REM
2000 REM ************** SUBROUTINE STATISTICS  *************
2010 REM
2020 REM        COMPUTES AND PRINTS OUT BASIC
2030 REM        STATISTICAL INFORMATION.
2040 REM
2050 REM        DATA REQUIRED:    N, X()
2060 REM        RESULTS RETURNED: NONE
2070 REM
2080 REM        LOCAL VARIABLES:
2090 REM
2100 REM        STSUM    = TOTAL OF ALL NUMBERS INPUT
2110 REM        STMEAN   = ARITHMETIC MEAN OF DATA
2120 REM        STMEDIAN = MEDIAN OF DATA
2130 REM        STVAR    = VARIANCE OF DAT
2140 REM        STDDV    = STANDARD DEVIATION OF DATA
2150 REM        STCOEVAR = COEFFICIENT OF VARIATION OF DATA
2160 REM
2170 REM ---------------- BODY OF STATISTICS --------------
2180 REM
2190 REM             Determine Arithmetic Mean
2200 STSUM = 0
2210 FOR I = 1 TO N
2220    STSUM = STSUM + X(I)
2230 NEXT I
2240 STMEAN = STSUM/N
2250 REM
2260 REM                  Determine Median
2270 REM
2280 IF N/2 - INT(N/2) = 0 THEN 2290 ELSE 2310
2290    STMEDIAN = X(N/2)
2300    GOTO 2330
2310    REM ELSE
2320    STMEDIAN = (X(N/2-.5)+X(N/2+.5))/2
2330 REM ENDIF
2340 REM                  Determine Variance
2350 REM
2360 STSUM = 0
2370 FOR I = 1 TO N
2380    STSUM = STSUM + (X(I) - STMEAN)^2
2390 NEXT I
2400 STVAR = STSUM/(N-1)
2410 REM
2420 REM              Determine Standard Deviation
2430 REM
2440 STDDEV = SQR(STVAR)
2450 REM
2460 REM         Determine Coefficient of Variation (%)
2470 STCOEVAR = (STDDEV/STMEAN)*100
2480 REM
2490 REM                 Print out Results
2500 PRINT "NUMBER OF VALUES = ";N
2510 PRINT "MEAN = ";STMEAN
2520 PRINT "MEDIAN = ";STMEDIAN
2530 PRINT "LOWEST VALUE = ";X(1)
2540 PRINT "HIGHEST VALUE = ";X(N)
2550 PRINT "VARIANCE = ";STVAR
2560 PRINT "STANDARD DEVIATION = ";STDDEV
2570 PRINT "COEFFICIENT OF VARIATION(%) = ";STCOEVAR
2580 RETURN
```

FIGURE 24.13 FORTRAN and BASIC programs to determine descriptive statistics.

```
C
C
C
C

1000

C
C
3000
C
C

C
C

C
```

```
***************************************************
*             DESCRIPTIVE STATISTICS              *
*               (FORTRAN version)                 *
*                 by  S.C. Chapra                 *
*                                                 *
*     This program determines the mean, median,   *
*     variance, coefficient of variation, and     *
*     standard deviation for a set of numbers.    *
***************************************************

       .-------------------------------------------.
       |         DEFINITION OF VARIABLES           |
       |                                           |
       | X() = VALUES OF A RANDOM VARIABLE         |
       | N   = TOTAL NUMBER OF VALUES OF X         |
       '-------------------------------------------'

       REAL X(100)

       ***************** MAIN PROGRAM *****************

       CALL ENTER(X,N)
       CALL SHELL(X,N)
       CALL STAT(X,N)
       STOP
       END


       ************* SUBROUTINE TO ENTER DATA **********

       SUBROUTINE ENTER(X,N)
       REAL X(100)
       INTEGER N
       DO 10 I=1,100
          READ(5,*) X(I)
          IF (X(I).EQ.-999) THEN
             GO TO 20
          ENDIF
10     CONTINUE
20     N = I - 1
       RETURN
       END
       ***************** SUBROUTINE SHELL **************
       SUBROUTINE SHELL(X,N)

          PERFORMS A SHELL SORT

          DATA REQUIRED:    N  --> NUMBER OF ELEMENTS IN X()
                            X() --> UNSORTED
          RESULTS RETURNED: X() --> SORTED

          LOCAL VARIABLES:

          GAP   =   GAP SIZE FOR ELEMENT COMPARISON
                    DURING SORT
          SWCH  =   INDICATES THAT ELEMENTS HAVE
                    BEEN SWITCHED
          N     =   NUMBER OF ELEMENTS IN X()
          DUMMY =   TEMPORARY DUMMY VARIABLE

       REAL X(20),DUMMY
       INTEGER SWCH,GAP,N
       ------------------ BODY OF SHELL ------------------

                 Subroutine for Shell Sort

       GAP = N
       GAP = INT(GAP/2)

       DO WHILE GAP >= 1
       IF (GAP .GE. 1) THEN
          SWCH = 1

          DO WHILE SWCH = 1
          IF (SWCH .EQ. 1) THEN

             Initialize Switch Indicator
             SWCH = 0

             Loop to Compare and, if Necessary, Switch
             DO 10 J=1, N - GAP

                If Former Greater than Latter
                IF (X(J) .GT. X(J+GAP)) THEN
```

```
C
C                If Former Greater than Latter
                 IF (X(J) .GT. X(J+GAP)) THEN
C
C                   Then Switch and Set Switch Indicator to 1
                    DUMMY = X(J)
                    X(J) = X(J + GAP)
                    X(J + GAP) = DUMMY
                    SWCH = 1
                 ENDIF
10           CONTINUE
             GOTO 3000
          ENDIF
C         END WHILE
C
          GAP = INT(GAP/2)
          GOTO 1000
       ENDIF
C      END WHILE
C
       RETURN
       END

       ************** SUBROUTINE STATISTICS ************

       SUBROUTINE STAT(X,N)
C
C         COMPUTES AND PRINTS OUT BASIC
C         STATISTICAL INFORMATION.
C
C         DATA REQUIRED:    N, X()
C         RESULTS RETURNED: NONE
C
C         LOCAL VARIABLES:
C
C         SUM    = TOTAL OF ALL NUMBERS INPUT
C         MEAN   = ARITHMETIC MEAN OF DATA
C         MEDIAN = MEDIAN OF DATA
C         VAR    = VARIANCE OF DAT
C         STDDV  = STANDARD DEVIATION OF DATA
C         COEVAR = COEFFICIENT OF VARIATION OF DATA

       REAL X(100),SUM,MEAN,MEDIAN,VAR,STDDV,COEVAR
       INTEGER N
C      ---------------- BODY OF STATISTICS --------------
C                   Determine Arithmetic Mean
       SUM = 0
       DO 10 I=1, N
          SUM = SUM + X(I)
10     CONTINUE
       MEAN = SUM/N
C
C                      Determine Median
C
       IF ((N/2 - INT(N/2)) .EQ. 0) THEN
          MEDIAN = X(N/2)
       ELSE
          MEDIAN = (X(N/2-.5)+X(N/2+.5))/2
       ENDIF
C                     Determine Variance
C
       SUM = 0
       DO 20 I=1, N
          SUM = SUM + (X(I) - MEAN)**2
20     CONTINUE
       VAR = SUM/(N-1)
C
C                 Determine Standard Deviation
C
       STDDEV = SQRT(VAR)
C
C          Determine Coefficient of Variation (%)
       COEVAR = (STDDEV/MEAN)*100
C
C                       Print out Results
       WRITE(6,*) 'NUMBER OF VALUES = ',N
       WRITE(6,*) 'MEAN = ',MEAN
       WRITE(6,*) 'MEDIAN = ',MEDIAN
       WRITE(6,*) 'LOWEST VALUE = ',X(1)
       WRITE(6,*) 'HIGHEST VALUE = ',X(N)
       WRITE(6,*) 'VARIANCE = ',VAR
       WRITE(6,*) 'STANDARD DEVIATION = ',STDDEV
       WRITE(6,*) 'COEFFICIENT OF VARIATION(%) = ',COEVAR
       RETURN
       END
```

```
NUMBER OF VALUES =    24
MEAN = 48.66667
MEDIAN = 48.5
LOWEST VALUE =    45.5
HIGHEST VALUE =    42.5
VARIANCE =  2.84058
STANDARD DEVIATION =   1.685402
COEFFICIENT OF VARIATION(%) =   3.463155
OK
```

Although the standard deviation is preferable, to this point its actual meaning is probably less clear than is that of the range. We know that it represents a measure of dispersion, but it may not be evident just how much of the dispersion it measures. Even though the range has its deficiencies, we at least know it represents the entire uncertainty of the data. Thus we can say with confidence that 100 percent of our measurements fall within the range. To this point, a similar statement cannot be made regarding the standard deviation. In Sec. 24.9, we will make the meaning of this statistic more tangible. However, before doing this we will first present some programs for determining descriptive statistics on the computer.

24.8.4 Computer Programs for Descriptive Statistics

The computer is an ideal tool for determining descriptive statistics. Figure 24.13 shows BASIC and FORTRAN programs that have been developed for this purpose. After the data is entered, the program implements a Shell sort (recall Sec. 23.3) to place the numbers in ascending order. This serves two purposes. First, it defines the range of the data. After sorting, the first and the last elements, $X(1)$ and $X(N)$, will contain the lower and upper bounds for the data. Second, it allows us to determine the median in a simple manner.

The program calculates and prints out two measures of central tendency—the arithmetic mean and the median—and four measures of spread—the range, variance, standard deviation, and coefficient of variation. Figure 24.14 shows the output for the speedometer data from Table 24.2.

24.9   THE NORMAL DISTRIBUTION

As mentioned in Sec. 24.6, the most common distribution in statistics is the normal distribution. It is common because (1) it closely approximates many data sets and (2) it has advantages in mathematical statistics. It can be represented formally by the normal probability density function

$$f(y) = \frac{1}{\sigma\sqrt{2\pi}} e^{-(y-\mu)^2/2\sigma^2}$$

(24.11)

where $f(y)$ is the *density* of a particular value of y. The density is similar to the relative frequency shown in Fig. 24.3. Recall from our discussion of histograms that the relative frequency is the frequency of occurrence divided by the total number of occurrences. Thus the summation of all the relative frequencies should add up to 1. The *density* is a similar quantity that is suited for a *continuous* function (in contrast to the *discrete* histogram) such as the normal distribution. The area under the normal distribution curve between two values of y is equal to the probability that readings will fall between these values. Similarly, the area from $-\infty$ to ∞ is equal to 1, connoting the obvious fact that there will be a 100 percent probability that a reading will fall between these extreme limits.

If various values of y are employed, Eq. (24.11) can be used to calculate a bell-shaped curve. Notice that the population mean μ and the population variance σ appear as parameters in this equation. Consequently, there is not just one normal distribution but an infinite number of such curves, depending on the values of μ and σ that are used in Eq. (24.11).

EXAMPLE 24.11   Calculating the Normal Probability Density Function

PROBLEM STATEMENT: Use Eq. (24.11) to compute values of the normal probability density function for (*a*) $\mu = 5$, $\sigma = 2.2$; (*b*) $\mu = 10$, $\sigma = 2.2$; and (*c*) $\mu = 5$, $\sigma = 1.0$. Compute values from $y = 0$ to 20 and plot the results.

SOLUTION: For $\mu = 5$ and $\sigma = 2.2$, Eq. (24.11) can be used to compute

$$f(y) = \frac{1}{2.2\sqrt{2\pi}} e^{-(y-5)^2/2(2.2)^2}$$

Values of $f(y)$ can be determined by substituting different y's into this equation. For example, if $y = 8$,

$$f(8) = \frac{1}{2.2\sqrt{2\pi}} e^{-(8-5)^2/2(2.2)^2} = 0.072$$

The results for other values of y as well as for cases (b) and (c) are

y	(a)	(b)	(c)
1	0.035	0.000	0.000
2	0.072	0.000	0.004
3	0.120	0.001	0.054
4	0.164	0.004	0.242
5	0.181	0.014	0.399
6	0.164	0.035	0.242
7	0.120	0.072	0.054
8	0.072	0.120	0.004
9	0.035	0.164	0.000
10	0.014	0.181	0.000
11	0.004	0.164	0.000
12	0.001	0.120	0.000

These values are plotted in Fig. 24.15. Notice that because they have the same variance, the curves for (a) and (b) have the same spread. However, because they have different means, they are centered on different values of y. On the other hand, because curves (a) and (c) have the same means, they are both centered on $y = 5$. However, because the variance of (c) is smaller, its spread is narrower than that of (a).

As stated previously, the area under a normal probability density function between two values of y is equal to the probability that readings will fall between those values. For example, in Fig. 24.16, the area between 0.5 and 2.0 is equal to 0.2857. This means that 28.57 percent of the values will fall between these limits. Obviously, the area under the entire function—that is, from $y = -\infty$ to ∞—is equal to 1.0, meaning that 100 percent of the values fall within this interval.

A useful property of the normal distribution is that the intervals defined by

FIGURE 24.15
Three bell-shaped curves with different means and standard deviations.

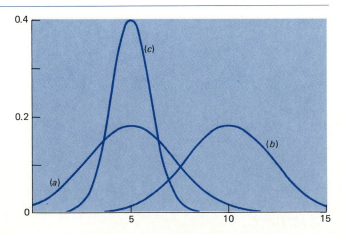

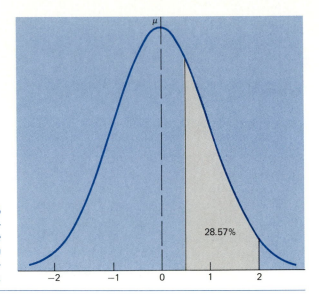

FIGURE 24.16
A normal probability curve showing how the area under the curve represents the probability that a random event will occur at locations between the abscissa values.

multiples of the standard deviation contain a fixed percentage of the items. For example, $\mu \pm \sigma$ contains 68.3 percent of the items, $\mu \pm 2\sigma$ contains 95.5 percent of the items, etc. (Fig. 24.17). These percentages hold for any and all normal distribu-

FIGURE 24.17
The percent probabilities for the normal distribution encompassed by various multiples of the standard deviation.

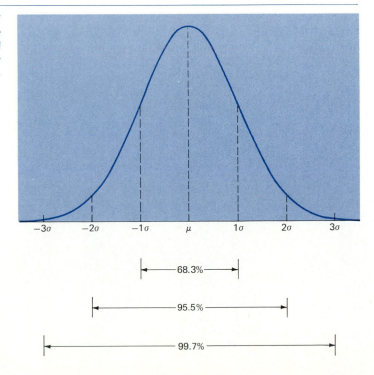

tions. Consequently, if the speedometer readings from Table 24.1 are truly normally distributed and if $\bar{y}$ and s_y are adequate measures of μ and σ, then the result of Example 24.10, 48.5 ± 1.04 km/h, implies that 68.3 percent of the values fall between 47.46 and 49.54. Similarly for two standard deviations, 48.5 ± 2.08 km/h implies that 95.5 percent of the values fall between 46.42 and 50.58.

24.10 PREPACKAGED STATISTICS SOFTWARE

The material in this chapter provides you with the capability of determining descriptive statistics with the computer. In addition to your own programs, prepackaged, or canned, software is also available for statistical analyses. As with other forms of commercially available software, these programs are designed to make your analytical work as effortless as possible.

The ENGINCOMP software that accompanies this text includes a statistics program that is representative of such software. As shown in the following example, it represents a potent tool for analyzing engineering data.

EXAMPLE 24.12   Using Canned Software to Determine Descriptive Statistics

FIGURE 24.18
Screens generated by the statistics package from ENGINCOMP. The package is used to analyze the data from Example 24.5. (a) shows the analysis for all the points and (b) shows the results when the outlier value of 0.115 is excluded.

PROBLEM STATEMENT: Employ the statistics program from ENGINCOMP to analyze the data from Example 24.5.

SOLUTION: A user-friendly computer program to determine descriptive statistics is contained on the ENGINCOMP software associated with the text. The results of applying this program to the data from Example 24.5 are shown in Fig. 24.18. The

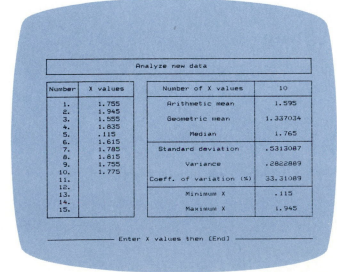

Analyze new data			
Number	X values	Number of X values	10
1.	1.755	Arithmetic mean	1.595
2.	1.945		
3.	1.555	Geometric mean	1.337034
4.	1.835		
5.	.115	Median	1.765
6.	1.615		
7.	1.785	Standard deviation	.5313087
8.	1.815		
9.	1.755	Variance	.2822889
10.	1.775		
11.		Coeff. of variation (%)	33.31089
12.			
13.		Minimum X	.115
14.			
15.		Maximum X	1.945

Enter X values then [End]

(a)

Analyze new data			
Number	X values	Number of X values	9
1.	1.755	Arithmetic mean	1.759444
2.	1.945		
3.	1.555	Geometric mean	1.756002
4.	1.835		
5.	1.615	Median	1.775
6.			
7.	1.785	Standard deviation	.1155542
8.	1.815		
9.	1.755	Variance	1.335278E-02
10.	1.775		
11.		Coeff. of variation (%)	6.567655
12.			
13.		Minimum X	1.555
14.			
15.		Maximum X	1.945

Enter X values then [End]

(b)

data to be analyzed is entered in a tabular format that is easy to edit. As each point is entered, the statistics are automatically updated. This feature is included because of its utility for determining the impact of outliers—that is, atypical values that lie far outside the normal range of points.

For example, Fig. 24.18a shows the result when all the data is analyzed. Because the software is designed to easily add or delete data, the effect of removing the outlying value of 0.115 can be readily determined.

The result, as shown in Fig. 24.18b, indicates that the outlier has a pronounced effect on the statistics. Thus you might use this result as the basis for reexamining this point to determine whether it was erroneous.

PROBLEMS

24.1 Give examples of the errors in common, everyday measuring devices. Define what might introduce bias and uncertainty into the readings obtained from these devices and how you might be able to assess the magnitude of the error.

24.2 Name at least four measures of central tendency.

24.3 Plot the normal distribution given by Eq. (24.11) and indicate the symmetrical range on the plot where 90 percent of the values would fall. Employ a mean and variance of 1 for this plot.

24.4 Three survey crews are sent out to measure the distance between two immovable landmarks. From prior measurements, it is known that the distance is 74.37 ft, with the last digit uncertain. The three crews measure the distance three times each, with the results:

Crew 1	Crew 2	Crew 3
75.01	74.23	74.30
74.75	74.37	74.75
74.80	74.30	74.29

Identify the bias and uncertainty of each crew.

24.5 The following weights were determined for a group of students in a class;

90	200	150	135	138
175	188	177	142	112
137	.115	192	172	160
128	122	134	176	181
140	151	155	166	175
194	220	140	142	126

Develop a histogram for this data. How would you classify the distribution? What reason can you give to explain the shape of the distribution?

24.6 Determine the mean, median, modal interval, and geometric mean of the weights given in Prob. 24.5.

24.7 Determine the range, standard deviation, and variance of the data given in Prob. 24.5.

24.8 Determine the mean, median, modal interval, range, variance, and standard deviation for the data from Table 14.1.

24.9 Construct a histogram for the data from Table 14.1.

24.10 Why is the standard deviation for a sample calculated by a different formula from that used for the entire population? It is not necessary to show this mathematically; use qualitative reasons.

24.11 Based on Fig. 24.6, develop a computer program to determine the histogram of a group of data. Test it by analyzing the data given in Table 24.3.

24.12 Employ the program developed in Prob. 24.11 to solve Probs. 24.5 and 24.9.

24.13 Based on Fig. 24.13, develop a computer program to determine descriptive statistics. Test it by analyzing the data from Table 24.3.

24.14 Employ the program developed in Prob. 24.13 to solve Probs. 24.6 to 24.8.

24.15 Combine the programs developed in Probs. 24.11 and 24.13 into a package that is capable of generating both histograms and descriptive statistics. Use a modular approach and employ structured programming techniques. Document the program both internally and externally and test it with the data from Table 24.3.

24.16 Employ the program developed in Prob. 24.15 to analyze the data from Table 14.1.

24.17 Employ the ENGINCOMP statistics package to perform Probs. 24.5 through 25.9.

24.18 Develop a program to determine the mean, the median, and the percent trimmed means (from 0 to 100 percent in increments of 10). Apply this program to the data from Table 24.3. Plot the values of the percent trimmed mean versus percent trimmed and indicate the location of the mean and the median on this plot.

24.19 Perform the same analysis as in Prob. 24.18 but apply it to the data from Example 24.5. Interpret the resulting plot and how it might prove useful as an aid in determining the best measure of central tendency.

24.20 You measure the inflow concentration of two reactors and come up with the following estimates (the true values are also shown):

	Estimate	True Value
Reactor A	15	17
Reactor B	278	272

Compute the bias and the percent relative bias for each reactor. On the basis of your results, which measurement is better? Why?

CHAPTER 25
SIMULATION

At the beginning of Chap. 1, we set forth a vignette wherein an engineer used computer-aided design and computer-aided manufacturing (CAD/CAM) to develop and produce a new automobile. An important aspect of the process was "to evaluate its . . . behavior under simulated road conditions." One way to do this would be to build a prototype and investigate its performance at a test track. Although this is usually done during the later stages of the design, the engineer might desire some preliminary indications of performance prior to actually building a prototype. Oftentimes the discovery of design flaws during the early stages can save considerable expense and effort further down the line.

The computer can be used to perform this preliminary testing. Just as a prototype is subjected to simulated road conditions at the test track, the computer can be used to mathematically simulate its behavior. In Chaps. 9 and 16 we have already shown how the random-number generator can be used to simulate a very simple physical process—the toss of a coin. To refresh your memory, the next example deals with a similar but slightly more complicated example—the toss of a die. For this case, random integers between 1 and 6 must be generated. The following example employs both the RND and INT functions to perform this task.

EXAMPLE 25.1   *Simulating the Random Toss of a Die with a Computer*

PROBLEM STATEMENT: Use the RND and INT functions to simulate the random toss of a die.

SOLUTION: The BASIC and FORTRAN programs in Fig. 25.1 are set up to generate integer values between a lower and an upper limit. Thus, as seen in Fig. 25.1, if the

```
100 REM ****** SUBROUTINE TO REPRESENT DIE TOSSES ******
110 REM
120 REM       DEFINITION OF VARIABLES
130 REM
140 REM       LOWER  = LOWEST POSSIBLE DIE ROLL
150 REM       UPPER  = HIGHEST POSSIBLE DIE ROLL
160 REM       NUMTOS = NUMBER OF DIE ROLLS
170 REM       TOSS   = THE NUMBER THAT CAME UP ON THE ROLL
180 REM       N()    = ARRAY THAT KEEPS COUNT HOW MANY
190 REM                TIMES EACH NUMBER HAS BEEN ROLLED
200 REM
210 REM         Define function to simulate die roll
220 REM
230 DEF FNRAN(I) = INT((UPPER - LOWER + 1) * RND + LOWER)
240 REM
250 REM             Input number of Die Rolls
260 REM
270 CLS: INPUT "NUMBER OF RANDOM TOSSES = ";NUMTOS
280 LOWER = 1
290 UPPER = 6
300 REM
310 REM             Generate and Print Die Rolls
320 REM
330 PRINT:PRINT "RANDOM TOSS":PRINT
340 FOR I = 1 TO NUMTOS
350     TOSS = FNRAN(I)
360     N(TOSS) = N(TOSS) + 1
370 NEXT I
380 FOR I = LOWER TO UPPER
390     PRINT "# of ";I;"'s = ";N(I)
400 NEXT I
410 END
```

```
C    ****** ALGORITHM TO REPRESENT DIE TOSSES ******
C
C          DEFINITION OF VARIABLES
C
C          LOWER  = LOWEST POSSIBLE DIE ROLL
C          UPPER  = HIGHEST POSSIBLE DIE ROLL
C          NUMTOS = NUMBER OF DIE ROLLS
C          TOSS   = THE NUMBER THAT CAME UP ON THE ROLL
C          N()    = ARRAY THAT KEEPS COUNT HOW MANY
C                   TIMES EACH NUMBER HAS BEEN ROLLED
C
      INTEGER N(1000),UPPER,LOWER,TOSS,FNRAN
C
C              Input number of Die Rolls
C
      READ(5,*) NUMTOS
      LOWER = 1
      UPPER = 6
C
C              Generate and Print Die Rolls
C
      WRITE(6,*) 'RANDOM TOSS
      DO 10 I=1, NUMTOS
         TOSS = FNRAN(UPPER,LOWER)
         N(TOSS) = N(TOSS) + 1
10    CONTINUE
      DO 20 I=LOWER, UPPER
         WRITE(6,*) '# of ', I ,'s = ', N(I)
20    CONTINUE
      STOP
      END
C -------------------------------------------------
C          Define function to simulate die roll
C
      INTEGER FUNCTION FNRAN(UPPER,LOWER)
      INTEGER UPPER,LOWER
      FNRAN = INT((UPPER - LOWER + 1) * RNDNUM(1000)  + LOWER)
      RETURN
      END
C -------------------------------------------------
C          Define function to generate random numbers
C
      FUNCTION RNDNUM(SEED)
      INTEGER A,X,SEED
      DATA A,M,I/1027,1048576,1/
      IF (I.EQ.1) THEN
         X = SEED
         I = 0
         XM = M
      ENDIF
      X = MOD(A*X,M)
      XX = X
      RNDNUM = XX/XM
      RETURN
      END
```

FIGURE 25.1 A program to simulate the random toss of a die.

limits are set at 1 and 6, the output simulates the toss of a die. The key to the program is the function FNRAN which has the value

```
INT ((UPPER - LOWER + 1) * RND + LOWER)
```

For the case where LOWER = 1 and UPPER = 6, this relationship amounts to

```
INT (6 * RND + 1)
```

The random number generator will generate numbers between (but not including) 0 and 1. Therefore the lowest number generated might be 0.0000000001. For this case, the value of FNRAN would be

TABLE 25.1   Results of Computer Simulations of the Toss of a Die. (Note that the percentages may not add up to 100 due to roundoff error.)

	12 Tosses		60 Tosses		600 Tosses		6000 Tosses	
	Frequency	%	Frequency	%	Frequency	%	Frequency	%
1	3	25.0	6	10.0	83	13.8	971	16.2
2	4	33.3	15	25.0	90	15.0	999	16.7
3	1	8.3	7	11.7	108	18.0	1007	16.8
4	0	0	16	26.7	116	19.3	1029	17.2
5	3	25.0	8	13.3	101	16.8	968	16.1
6	1	8.3	8	13.3	102	17.0	1026	17.1

```
INT(1.0000000006)
```

The INT then truncates the 1.0000000006 to yield a value of 1.

In a similar fashion, the highest number generated by the random number generator might be 0.9999999999, which gives

```
INT(6.999999994)
```

which INT would truncate to yield a value of 6. The random numbers between the two limits would result in integer values between 1 and 6.

Table 25.1 shows the results for 12 tosses. The expected values for this case would be two of each integer. However, as with a real die, the randomness of the toss means that for a limited number of tosses, it is possible for certain numbers to show up with more or less frequency than expected. As shown in Table 25.1, there are four 2s, whereas there are no 4s.

Table 25.1 and Fig. 25.2 shows these results, along with results of some additional runs where we generate greater numbers of tosses. The percent of each value is also shown to give a feel for the uniformity of the distribution of the tosses. Notice how, as the tosses are increased, the frequencies exhibit less variability. For example, at 600 tosses, there is a 5.5 percent difference between the lowest frequency and the highest frequency. At 6000 tosses, the difference is reduced to 1.1 percent. Thus, as more tosses are simulated, the distribution becomes more uniform.

The previous example provides a way to generate a uniform distribution of numbers between two limits (Fig. 25.2). If the INT function had not been used, the program would have provided continuous random numbers between the lower and the upper limits. As shown in the next example, such uniformly distributed random numbers can be used to perform a simulation analysis on an engineering system.

EXAMPLE 25.2   Simulating the Falling Parachutist Problem

PROBLEM STATEMENT: In Chap. 22, the following equation was presented to compute the velocity of a falling parachutist [Eq. (E22.1.1)]:

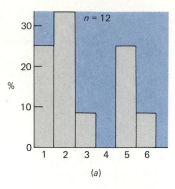

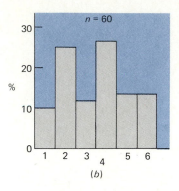

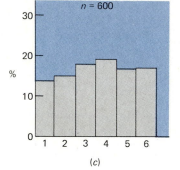

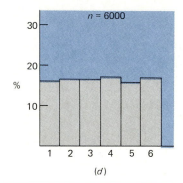

FIGURE 25.2
Histograms for the four cases from Table 25.1. As the number of tosses increases, the results more and more closely approach a uniform distribution.

$$v = \frac{gm}{c}(1 - e^{-(c/m)t}) \qquad\qquad (E25.2.1)$$

where v = velocity, m/s
g = acceleration due to gravity, m/s^2
m = mass, kg
c = drag coefficient, kg/s
t = time, s

This equation can be used to compute v as a function of time. If $g = 9.8$ m/s^2, $m = 68.1$ kg, and $c = 12$ kg/s, Eq. (E25.2.1) can be used to compute the velocity at $t = 3$ s as

$$v = \frac{9.8(68.1)}{12}(1 - e^{-(12/68.1)3}) = 22.835 \text{ m/s}$$

Now it is a known fact that drag coefficients are difficult to measure. In separate experiments, it has been established that the value of 12 kg/s is only known to a precision of ± 20 percent. Thus the value could range anywhere from 9.6 to 14.4 kg/s. Although you suspect that the drag coefficients are probably normally distributed, there is no evidence of this. Therefore as a conservative estimate you assume that it varies uniformly over the range. Use the general equation developed in Example 25.1 to randomly generate a number of drag coefficients from 9.6 to 14.4. Use each of these values in conjunction with Eq. (E25.2.1) to compute a value of v. Then use a histogram to investigate how the variability of the drag coefficient affects the variability of the predicted velocity.

SOLUTION: A computer program to implement this simulation analysis is shown in Fig. 25.3. The uniformly distributed random numbers are generated with

```
DRAG = (UPPER-LOWER) * RND + LOWER
```

FIGURE 25.3 BASIC and FORTRAN programs to simulate the velocity of a falling parachutist subject to drag coefficients ranging from 9.6 to 14.4 kg/s.

```
100 REM    ******************************************
110 REM    *          FALLING PARACHUTE MODEL        *
120 REM    *             (BASIC version)             *
130 REM    *              by  S.C. Chapra            *
140 REM    *                                         *
150 REM    *   This program simulates the velocity of *
160 REM    *   a falling parachutist for random values of *
170 REM    *   the drag coefficient.                  *
180 REM    ******************************************
190 REM
200 REM         .----------------------------------.
210 REM         |        DEFINITION OF VARIABLES    |
220 REM         |                                  |
230 REM         | X() = VALUES OF A RANDOM VARIABLE |
240 REM         | N   = TOTAL NUMBER OF VALUES OF X |
250 REM         '----------------------------------'
260 REM
270 REM ***************** MAIN PROGRAM *****************
280 REM
290 GOSUB 1000
300 GOSUB 2000
310 PRINT : PRINT "SORTED VEL."
320 FOR I = 1 TO N
330    PRINT X(I)
340 NEXT I
350 GOSUB 3000
360 END
1000 REM **************** SUBROUTINE DRAG ***************
1010 REM
1020 REM      GENERATES RANDOM VELOCITIES
1030 REM
1040 REM      REQUIRED DATA:    NONE
1050 REM      RESULTS RETURNED: N,X()
1060 REM
1070 REM ---------------- BODY OF DRAG -----------------
1080 REM
1090 CLS
1100 INPUT "NUMBER OF SIMULATIONS = ";N
1110 PRINT:PRINT "   DRAG        VELOCITY"
1120 LOWER = 9.600001
1130 UPPER = 14.4
1140 DIM X(N)
1150 FOR I = 1 TO N
1160    DRAG = (UPPER - LOWER) * RND + LOWER
1170    X(I) = 9.8*68.1/DRAG*(1 - EXP(-(DRAG/68.1)*3))
1180 PRINT DRAG,X(I)
1190 NEXT I
1200 RETURN
2000 REM **************** SUBROUTINE SHELL **************
2010 REM
2020 REM      PERFORMS A SHELL SORT
2030 REM
2040 REM      DATA REQUIRED:    N  --> # OF ELEMENTS IN X()
2050 REM                        X() --> UNSORTED
2060 REM      RESULTS RETURNED: X() --> SORTED
2070 REM
2080 REM      LOCAL VARIABLES:
2090 REM
2100 REM      SHGAP   = GAP SIZE FOR ELEMENT COMPARISON
2110 REM                DURING SORT
2120 REM      SHSWCH  = INDICATES THAT ELEMENTS HAVE
2130 REM                BEEN SWITCHED
2140 REM      N       = NUMBER OF ELEMENTS IN X()
2150 REM      SHDUM   = TEMPORAY DUMMY VARIABLE
2160 REM
2170 REM ---------------- BODY OF SHELL ----------------
2180 REM
2190 REM           Subroutine for Shell Sort
2200 REM
2210 SHGAP = N
2220 SHGAP = INT(SHGAP/2)
2230 REM PERFORM LOOP WHILE INTERVAL )= 1
```

```
C      ******************************************
C      *          FALLING PARACHUTE MODEL        *
C      *            (FORTRAN version)            *
C      *              by  S.C. Chapra            *
C      *                                         *
C      *   This program simulates the velocity of a *
C      *   falling parachutist for random values of *
C      *   the drag coefficient.                  *
C      ******************************************
C
C         .----------------------------------.
C         |        DEFINITION OF VARIABLES    |
C         |                                  |
C         | X() = VALUES OF A RANDOM VARIABLE |
C         | N   = TOTAL NUMBER OF VALUES OF X |
C         '----------------------------------'
       REAL X(100), LOWER, UPPER
       INTEGER N
C
C      ***************** MAIN PROGRAM *****************
C
       CALL DRAG(N,X)
       CALL SHELL(N,X)
       CALL HISTOG(N,X)
       STOP
       END
C      **************** SUBROUTINE DRAG ***************
C
       SUBROUTINE DRAG(N,X)
C
C           GENERATES RANDOM VELOCITIES
C
C           REQUIRED DATA:    NONE
C           RESULTS RETURNED: N,X()
C
       REAL X(100)
       INTEGER N
C      ---------------- BODY OF DRAG -----------------
C
       READ(5,*) N
       LOWER = 9.600001
       UPPER = 14.4
       DO 10 I=1, N
         DRG = (UPPER - LOWER) * RNDNUM(1) + LOWER
         X(I) = 9.8*68.1/DRG*(1 - EXP(-(DRG/68.1)*3))
10     CONTINUE
       RETURN
       END
C      **************** SUBROUTINE SHELL **************
C
       SUBROUTINE SHELL(N,X)
C
C           PERFORMS A SHELL SORT
C
C           DATA REQUIRED:    N  --> NUMBER OF ELEMENTS IN X()
C                             X() --> UNSORTED
C           RESULTS RETURNED: X() --> SORTED
C
C           LOCAL VARIABLES:
C
C           GAP     = GAP SIZE FOR ELEMENT COMPARISON
C                     DURING SORT
C           SWCH    = INDICATES THAT ELEMENTS HAVE
C                     BEEN SWITCHED
C           N       = NUMBER OF ELEMENTS IN X()
C           DUMMY   = TEMPORARY DUMMY VARIABLE
C
       REAL X(20),DUMMY
       INTEGER SWCH, GAP, N
C      ---------------- BODY OF SHELL ----------------
C
C           Subroutine for Shell Sort
```

```
2240 WHILE SHGAP >= 1
2250    REM
2260    REM         Perform Loop While Swithces Occur
2270    SHSWCH = 1
2280    WHILE SHSWCH = 1
2290       REM        Initialize Switch Indicator
2300       SHSWCH = 0
2310       REM Loop to Compare and, if Necessary, Switch
2320       FOR J = 1 TO N-SHGAP
2330          REM
2340          REM If Former Greater than Latter
2350          IF X(J) > X(J+SHGAP) THEN 2370 ELSE 2420
2360             REM
2370             REM  Switch and Set Switch Indicator to 1
2380             SHDUM = X(J)
2390             X(J) = X(J + SHGAP)
2400             X(J + SHGAP) = SHDUM
2410             SHSWCH = 1
2420          REM ENDIF
2430       NEXT J
2440    WEND
2450    SHGAP = INT(SHGAP/2)
2460 WEND
2470 RETURN
3000 REM ************** SUBROUTINE PLOT HISTOGRAM ************
3010 REM
3020 REM      DIVIDES FREQUENCY DATA INTO CLASSES
3030 REM      AND PLOTS A HISTOGRAM.
3040 REM
3050 REM      DATA REQUIRED:    X(), N
3060 REM      RESULTS RETURNED: NONE
3070 REM
3080 REM      LOCAL VARIABLES:
3090 REM
3100 REM      PLLOBND  = SMALLEST VALUE INPUT BY USER
3110 REM      PLUPBND  = LARGEST VALUE INPUT BY USER
3120 REM      PLWDTH   = WIDTH OF EACH CLASS
3130 REM      PLUPLIM  = UPPER LIMIT OF A CLASS
3140 REM      PLSEGMNT = CLASS NUMBER DURING ITERATION
3150 REM      PLFREQ() = FREQUENCY OF OCCURANCES WITHIN A CLASS
3160 REM      PLLOWER  = LOWER LIMIT OF AN ITERATION CLASS
3170 REM      PLUPPER  = UPPER LIMIT OF AN ITERATION CLASS
3180 REM
3190 REM --------------- BODY OF HISTOGRAM ---------------
3200 REM
3210 REM             Determine Frequency for Each Segment
3220 PLLOBND = X(1)
3230 PLUPBND = X(N) + X(N)/1000000!
3240 PLWDTH = (PLUPBND - PLLOBND)/10
3250 PLUPLIM = PLLOBND + PLWDTH
3260 PLSEGMNT = 1
3270 FOR I = 1 TO N
3280    WHILE X(I) > PLUPLIM
3290       PLUPLIM = PLUPLIM + PLWDTH
3300       PLSEGMNT = PLSEGMNT + 1
3310    WEND
3320    PLFREQ(PLSEGMNT) = PLFREQ(PLSEGMNT) + 1
3330 NEXT I
3340 REM
3350 REM             Print Out Histogram
3360 REM
3370 CLS
3380 PLLOWER = PLLOBND
3390 FOR I = 1 TO 10
3400    PLUPPER = PLLOWER + PLWDTH
3410    PRINT USING "####.##    ";PLLOWER,PLUPPER;
3420    IF PLFREQ(I) <> 0 THEN 3430 ELSE 3460
3430       FOR J = 1 TO PLFREQ(I)
3440          PRINT "*";
3450       NEXT J
3460    REM ENDIF
3470    PRINT
3480    PLLOWER = PLUPPER
3490 NEXT I
3500 RETURN
```

```
C
      GAP = N
      GAP = INT(GAP/2)
C
C     DO WHILE GAP > = 1
1000  IF (GAP .GE. 1) THEN
         SWCH = 1
C
C        DO WHILE SWCH = 1
         IF (SWCH .EQ. 1) THEN
C
C           Initialize Switch Indicator
            SWCH = 0
C
C           Loop to Compare and, if Necessary, Switch
            DO 10 J=1, N - GAP
C
C              If Former Greater than Latter
               IF (X(J) .GT. X(J+GAP)) THEN
C
C                 Then Switch and Set Switch Indicator to 1
                  DUMMY = X(J)
                  X(J) = X(J + GAP)
                  X(J + GAP) = DUMMY
                  SWCH = 1
               ENDIF
10          CONTINUE
            GOTO 3000
         ENDIF
C        END WHILE
C
         GAP = INT(GAP/2)
         GOTO 1000
      ENDIF
C     END WHILE
C
      RETURN
      END
C     ************** SUBROUTINE PLOT HISTOGRAM ************
C
      SUBROUTINE HISTOG(N, X)
C
C           DIVIDES FREQUENCY DATA INTO CLASSES
C           AND PLOTS A HISTOGRAM.
C
C           DATA REQUIRED:    X(), N
C           RESULTS RETURNED: NONE
C
C           LOCAL VARIABLES:
C
C           LOBND  = SMALLEST VALUE INPUT BY USER
C           UPBND  = LARGEST VALUE INPUT BY USER
C           WDTH   = WIDTH OF EACH CLASS
C           UPLIM  = UPPER LIMIT OF A CLASS
C           SEGMNT = CLASS NUMBER DURING ITERATION
C           FREQ() = FREQUENCY OF OCCURANCES WITHIN A CLASS
C           LOWER  = LOWER LIMIT OF AN ITERATION CLASS
C           UPPER  = UPPER LIMIT OF AN ITERATION CLASS
C
      REAL    X(100), FREQ(10), UPLIM, LOBND, WDTH,
     +        LOWER, UPPER
      INTEGER N, SEGMNT
C     --------------- BODY OF HISTOGRAM ---------------
C
C           Determine Frequency for Each Segment
      LOBND = X(1)
      UPBND = X(N) + X(N)/1000000.
      WDTH = (UPBND - LOBND)/10
      UPLIM = LOBND + WDTH
      SEGMNT = 1
      DO 10 I=1, N
C
C        DO WHILE X(I) > UPLIM
1000     IF (X(I) .GT. UPLIM) THEN
            GOTO 2000
         ELSE
            GOTO 3000
2000     UPLIM = UPLIM + WDTH
         SEGMNT = SEGMNT + 1
         GOTO 1000
3000     ENDIF
C        END WHILE
C
         FREQ(SEGMNT) = FREQ(SEGMNT) + 1
10    CONTINUE
C
```

```
C                                     Print Out Histogram
C
      LOWER = LOBND
      DO 20 I=1,10
         UPPER = LOWER + WDTH
         IF (FREQ(I) .EQ. 0) THEN
            WRITE(6,*) LOWER,UPPER
         ELSE
            WRITE (6,*) (LOWER,UPPER,('*',J=1,FREQ(I)))
         ENDIF
         LOWER = UPPER
   20 CONTINUE
      RETURN
      END
C     ----------------------------------------------------------------
C                    Function to Round to 3 Decimal Places
C
      INTEGER FUNCTION FNR(X)
      INTEGER X
      FNR = INT(X*10**3+.5)/10**3
      RETURN
      END
C     ----------------------------------------------------------------
C                    Define function to generate random numbers
C
      FUNCTION RNDNUM(SEED)
      INTEGER A,X,SEED
      DATA A,M/1027,1048576/
      IF (I.EQ.0) THEN
         X = SEED
      ENDIF
      X = MOD(A*X,M)
      RNDNUM = REAL(X)/REAL(M)
      I = 1
      RETURN
      END
```

FIGURE 25.3 *Continued.*

where LOWER = 9.6 and UPPER = 14.4. This parameter is used to compute a value of velocity. Then these velocities are developed into a histogram. The results are shown in Fig. 25.4 and in Table 25.2 for 10, 100, and 1000 simulations. As might be expected, the results are also uniformly distributed, with a range that is very close to those which would occur using the minimum and maximum values of drag. As more simulations are performed, the distribution more closely approximates the uniform distribution. This is a direct result of the assumption of uniform distribution for the drag coefficient. In addition, note that the average values are close to the value of 22.835 computed with the value of $c = 12$ kg/s at the midpoint of the range.

FIGURE 25.4 Histograms resulting from 10, 100, and 1000 simulations of the velocity of the falling parachutist.

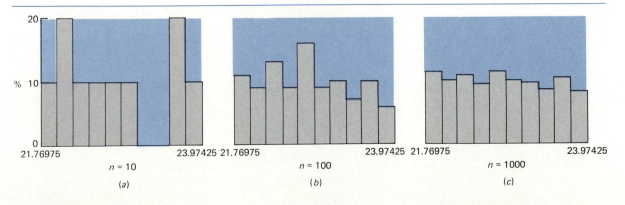

TABLE 25.2   Results of Simulations of Velocity of Falling Parachutist for
Random Values of Drag

Simulation	Range	Average Value
10	21.87212–23.80148	22.79522
100	21.79023–23.95303	22.78995
1000	21.77182–23.96855	22.83121

The foregoing example is formally referred to as a *Monte Carlo simulation*. Although it yields intuitive results for this simple problem, there are instances where such computer simulations yield surprising outcomes and provide insights that would otherwise be impossible to determine. The approach is only feasible because of the computer's ability to implement tedious, repetitive computations in an efficient manner.

PROBLEMS

25.1   Write a program to simulate the toss of two dice; run this program for 20, 100, and 500 tosses. Plot the distributions resulting from the analysis and comment on their shapes.

25.2   Write a program to simulate a hand of cards for black jack. Consider an ace of spades to have a value of 1 and the jack, queen, and king of spades to have values of 11, 12, and 13, respectively. Number the other suits with a similar scheme, starting with the ace of diamonds as 14, the ace of hearts as 27, and the ace of clubs as 40. Therefore, the fifty-second card will be the king of clubs. When developing this program make sure that you incorporate the fact that, as each card is dealt, the deck has been depleted by one, and the program should not be capable of dealing this card again.

25.3   Often in the simulation of engineering processes, the occurrence or magnitude of an event is not a purely random process but is dependent on the previous event along with a random component. Put mathematically, the magnitude of the occurrence of an event can be given as

$$E_n = f(E_{n-1}) + er$$

where E_n and E_{n-1} are the magnitude of events—the current (n) and the prior ($n - 1$) events, respectively—e is a weighting factor, and r is a random number between 0 and 1. For example, suppose it was determined that the average yearly flow of water in a channel can be represented by

$$F_n = \sin\left(\frac{\pi F_{n-1}}{2\overline{F}}\right) F_{n-1} + 0.3r$$

where $\overline{F}$ is the average flow for a 30-year sampling period and F_{n-1} is the previous year's flow. If the average flow is taken to be 300 ft³/s and the first year's flow is taken to be the average, simulate the yearly average flow for the next 20 years. Plot the values versus time and interpret your results as to whether they appear realistic. Base your interpretation in part on some statistical measure.

25.4 Perform the same computation as in Prob. 9.20 (or 16.20) but consider the initial velocity to be a random variable that varies $\pm$ 10 percent around the average value. Have the computer perform 100 and 1000 simulations and develop histograms (preferably generated by the computer) of the resulting ranges. Interpret your results.

25.5 Repeat Example 25.2 but treat both c and m as random variables. Allow the drag to vary ± 20 percent and the mass to vary ± 10 percent. Generate 50, 100, and 200 simulations in order to characterize the distribution of the resulting velocity.

25.6 The "drunkard's walk" is a term used to describe the type of random motion found in many diffusion processes. (The name "drunkard's walk" stems from the similarity between this type of motion and the random stumbling around that a drunkard might exhibit.) In one-dimensional space, the random motion can only be to the right or to the left. A random-number generator can be employed to simulate the process by generating random numbers in the same fashion we used for the coin in Examples 9.6 and 16.7. For this case, however, rather than having the two states as heads or tails, we consider a 0 to 0.5 to represent a step left and a 0.5 to 1 a step right. Write a program that will simulate a 100-step random walk and run this routine 50 times to obtain 50 different results for the final location of the drunkard (that is, the number of steps away from the starting position, where steps to the right are positive and steps to the left are negative). Plot your results as a histogram and comment on the result.

25.7 Repeat Prob. 25.6 but this time let the number of steps taken by the drunkard vary uniformly between 25 and 75. Again, obtain 50 different results and comment on the distribution of the resulting histogram.

25.8 It is known that in an industrial process a pump is likely to fail once every 300 h. This is not critical unless the valve on the pump line also fails to operate. If both failures occur, the results are costly. An experiment was performed and it was determined that the valve fails to operate once every 200 h. Determine the frequency at which the two fail simultaneously. Note that the correct solution will involve some scaling.

PART SEVEN

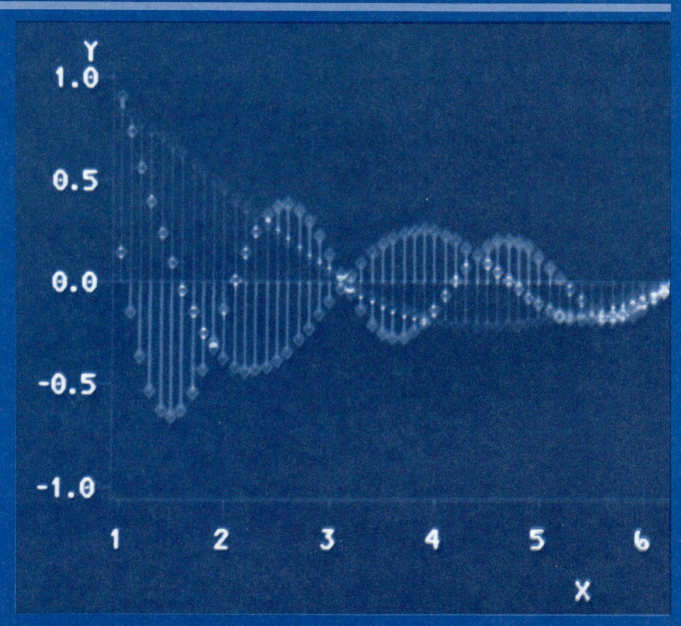

COMPUTER MATHEMATICS

Now that you have been introduced to some aspects of data analysis, the next step in the problem-solving process is to investigate how the computer is used to manipulate these numbers and solve mathematical problems. At the core of computer mathematics is the area of study known as numerical methods.

Numerical methods are techniques whereby mathematical problems are formulated so that they can be solved by arithmetic operations. Because computers are very good at performing arithmetic, they are ideal vehicles for implementing numerical methods. Because of this, the field of numerical methods is sometimes loosely defined as "computer mathematics."

The present part of the book constitutes a brief introduction to some of the elementary ideas and techniques in the field of numerical methods. It is intended to provide you with a feeling for how computers can be effectively utilized to solve mathematical problems. The material is divided into three major groups of techniques. *Chapters 26* and *27* are devoted to methods for fitting curves to data. *Chapters 28* and *29* deal with solving equations for unknowns. Finally, *Chaps. 30, 31,* and *32* introduce techniques for solving calculus problems with the computer.

CHAPTER26
CURVE FITTING: REGRESSION

A great deal of engineering work is based on cause-effect interactions. One of our routine tasks is to establish the relationship between two variables. A graph provides one means to accomplish this objective. For cartesian coordinate systems (recall Fig. 22.5), it is conventional to plot the *dependent variable* on the vertical axis, or *ordinate*, and the *independent variable* on the horizontal axis, or *abscissa*. The dependent variable is then said to be plotted "versus" or "as a function of" the dependent variable.

By displaying the data on a two-dimensional space, patterns emerge that might not be evident from the tabulated information alone. For example, close inspection of the table in Fig. 26.1*a* suggests that higher values of *y* seem to be associated with higher values of *x*. Although this positive relationship between the variables can be inferred from the table, the graph in Fig. 26.1*b* provides the additional insight that the relationship curves upward, or accelerates, as *x* increases. Such insights are one of the great strengths of graphical approaches.

Although graphs have utility for establishing the qualitative relationship between variables, an even more powerful approach is to fit a curve through the points. As depicted in Fig. 26.2, such a curve can then be used to predict values of *y* as a function of *x*. The simplest method for deriving such curves is to merely "eyeball" the data and sketch the curve. Although this is certainly valid for "quick and dirty" estimates, it is fundamentally flawed because it is subjective. That is, for most cases, each individual would come up with a slightly different version of a "best" curve.

The following two chapters are devoted to some mathematical techniques that have been developed to remove this subjectivity from the curve-fitting process. These techniques are objective in that they yield consistent results regardless of the individual performing the analysis.

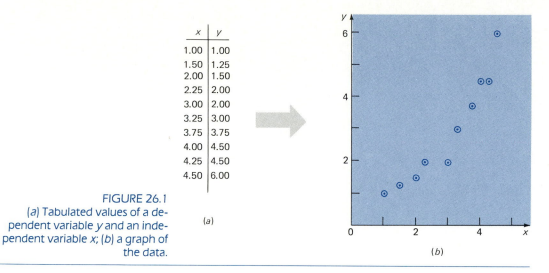

x	y
1.00	1.00
1.50	1.25
2.00	1.50
2.25	2.00
3.00	2.00
3.25	3.00
3.75	3.75
4.00	4.50
4.25	4.50
4.50	6.00

(a)

(b)

FIGURE 26.1
(a) Tabulated values of a dependent variable y and an independent variable x; (b) a graph of the data.

The approaches can be divided into two general categories depending on the amount of error that is associated with the data. First, where the data exhibits a significant degree of error, or uncertainty, the strategy is to derive a curve that represents the general trend of the data and minimizes some measure of error. Because any individual point may be incorrect, no effort is made to intersect every point. Rather, the curve is derived to follow the pattern of the points taken as a group (Fig. 26.3a). This approach, called *least-squares regression*, is the subject of the present chapter.

FIGURE 26.2
A curve fit of the data from Fig. 26.1a. This fit was based on an "eyeball" sketch. Such curves can be used to predict values of the dependent variable as a function of the independent variable. Here it is used to predict that at x = 2.75, y is approximately equal to 2.25.

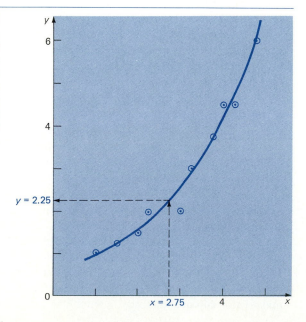

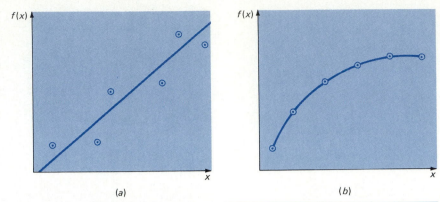

FIGURE 26.3
Two fundamentally different ways
to fit curves to data: (a) regression
and (b) interpolation.

Second, where the data is known to be very precise, the basic approach is to fit a curve or a series of curves that pass directly through each of the points. Common sources of such data are the tables found in such references as engineering and scientific handbooks. Examples range from engineering economics tables to tables of physical properties. The estimation of values between well-known discrete points is called *interpolation* (Fig. 26.3b). It will be discussed in Chap. 27.

26.1 LINEAR REGRESSION

Because of their interest in cause-effect interactions, engineers have frequent occasion to establish the relationship between a set of paired variables: (x_1, y_1), (x_2, y_2), . . . , (x_n, y_n). Although examples abound throughout engineering, this task is especially common in experimental studies. In this context, the independent variable x is usually a characteristic that is under the control of the experimenter (that is, x is the cause), whereas the dependent variable y is the resultant behavior that is being measured (that is, y is the effect).

For a materials engineer, x might be the force per unit area (that is, the stress) imposed on a steel rod, whereas y is the resulting deformation per unit length (that is, the strain). In electrical engineering, x might be the current imposed on a resistor, whereas y is the resulting voltage drop. Numerous other examples exist in all fields of engineering. In any case, because x is controlled and y is measured, the former is usually considered to be error-free, while the latter can have error associated with it.

Table 26.1 contains a set of six paired observations of the type compiled during a typical experimental study. In this hypothetical example, we are interested in characterizing factors influencing the downward velocity of a free-falling parachutist. From physics, it is known that a falling object is subject to two vertical forces: the downward force of gravity and the upward, retarding force of air resistance. Preliminary observations suggest that air resistance is positively correlated with fall velocity. That is, the higher the vertical velocity, the greater the air resistance. In order to characterize this relationship quantitatively, a wind-tunnel experiment is conducted to measure the

TABLE 26.1 Six Values of Air Resistance (y_i) versus Velocity (x_i) Collected during a Wind-Tunnel Experiment

x_i, m/s	y_i, N
10	110
15	230
20	210
25	350
30	330
35	460

air-resistance force as a function of various velocities (see Table 26.1). When this data is plotted, it suggests the linear or straight-line pattern seen in Fig. 26.4.

Linear regression is the technique for determining the "best" straight line through data such as that contained in Table 26.1. The equation for the straight line can be expressed as

$$\hat{y} = a_0 + a_1 x \tag{26.1}$$

where $\hat{y}$ (read y hat or y caret) is the predicted value of the dependent variable and a_0 and a_1 are the intercept and the slope, respectively, of the straight line. The object of linear regression is to determine the straight line (as defined by a_0 and a_1) that passes as close to as many points as possible. Because most of the error is associated with the y values, "closeness" is defined in terms of the vertical distance between the points and the line. This distance is referred to as the *residual* (Fig. 26.5) and can be computed as

$$\text{Residual}_i = y_i - \hat{y}_i$$

where $\hat{y}_i$ is the y value of the straight line at x_i as calculated by Eq. (26.1).

FIGURE 26.4
A plot of the data from Table 26.1 along with the best-fit line computed in Example 26.1.

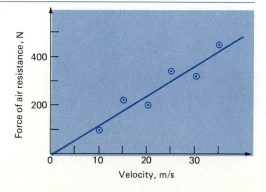

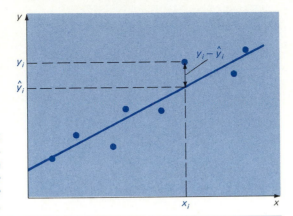

FIGURE 26.5
Plot of dependent versus independent variables along with regression line. The regression line chosen minimizes the sum of the squares of the residuals $y_i - \hat{y}_i$.

For a variety of theoretical reasons that are beyond the scope of this text, the "best" straight line is defined as the one that minimizes the *sum of the squares of the residuals* between the points and the line. Thus the object of linear regression is to determine values of a_0 and a_1 that result in a minimum value of

$$S_r = \sum_{i=1}^{n} (y_i - \hat{y}_i)^2 \tag{26.2}$$

where S_r is the *sum of the squares of the residuals*. Using a derivation that employs calculus (see Chapra and Canale, 1985, for details), the values of a_1 and a_0 that minimize Eq. (26.2) can be calculated using the formulas

$$a_1 = \frac{n \sum x_i y_i - \sum x_i \sum y_i}{n \sum x_i^2 - (\sum x_i)^2} \tag{26.3}$$

and

$$a_0 = \frac{\sum y_i - a_1 \sum x_i}{n} \tag{26.4}$$

where all the summations are from $i = 1$ to n.

EXAMPLE 26.1   Linear Regression

PROBLEM STATEMENT: Use linear regression to fit a straight line to the data from Table 26.1.

SOLUTION: The summations needed for Eqs. (26.3) and (26.4) can be computed as

x_i	y_i	$x_i y_i$	x_i^2
10	110	1100	100
15	230	3450	225
20	210	4200	400
25	350	8750	625
30	330	9900	900
35	460	16,100	1225
135	1690	43,500	3475

These values can be substituted into Eq. (26.3) to calculate the slope:

$$a_1 = \frac{6(43,500) - 135(1690)}{6(3475) - (135)^2} = 12.5143$$

And Eq. (26.4) can be used to determine the intercept:

$$a_0 = \frac{1690 - 12.5143(135)}{6} = 0.095$$

Therefore, the linear least-squares fit is

$$\hat{y} = 0.095 + 12.5143x$$

The line, along with the data, is shown in Fig. 26.4.

26.2   QUANTIFYING THE "GOODNESS" OF THE LEAST-SQUARES FIT

Any line other than the one determined in Example 26.1 results in a larger sum of the squares of the residuals. Thus the line is unique and in terms of our chosen criterion is a "best" line through the points. A number of additional properties of this fit can be elucidated by examining more closely the way in which the residuals were computed. Recall that the sum of the squares is defined as [Eq. (26.2)]

$$S_r = \sum_{i=1}^{n} (\hat{y}_i - y_i)^2 \tag{26.5}$$

Notice the similarity between Eq. (26.5) and the total sum of the squared residuals from Eq. (24.8). In the case of Eq. (24.8), the residuals represented the discrepancy between a measurement and an estimate of central tendency—the mean. For Eq. (26.5), the residuals represent the discrepancy between a measurement and another estimate of central tendency—the straight line.

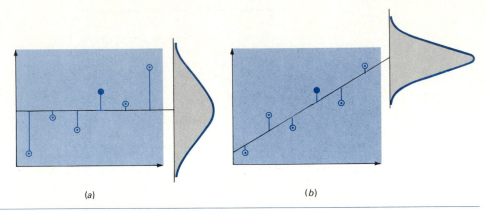

FIGURE 26.6 Regression data showing (a) the spread of the data around the mean of the dependent variable and (b) the spread of the data around the best-fit line. The reduction in the spread in going from (a) to (b), as indicated by the bell-shaped curves at the right, represents the improvement due to linear regression.

Just as the standard deviation was used to quantify the spread of data around the mean in Eq. (24.10), a similar estimate of spread around the regression line can be determined with

$$s_{y/x} = \sqrt{\frac{S_r}{n - 2}}$$

(26.6)

where $s_{y/x}$ is called the *standard error of the estimate*. We divide by $n - 2$ because two data-derived estimates—a_0 and a_1—were used to compute S_r; thus we have lost two degrees of freedom. Note that, as with our discussion of the standard deviation in Sec. 24.8.2, another justification for dividing by $n - 2$ is that there is no such thing as the "spread of data" around a straight line connecting two points. Thus, for the case where $n = 2$, Eq. (26.6) yields a meaningless result of infinity.

Just as was the case with the standard deviation, the standard error of the estimate quantifies the spread of the data. However, $s_{y/x}$ quantifies the spread *around the regression line* as shown in Fig. 26.6b in contrast to the original standard deviation s_y that quantified the spread *around the mean* (Fig. 26.6a).

The above concepts can be used to quantify the "goodness" of our fit. This is particularly useful for comparison of several regressions (see Fig. 26.7). To do this, we return to the original data and determine the *total sum of the squares* around the mean for the dependent variable (in our case, y). As was the case for Eq. (24.8), this quantity is designated S_t. This is the uncertainty associated with the dependent variable prior to regression. After performing the regression, we can compute S_r, the sum of the squares of the residuals around the regression line. This represents the uncertainty that remains after the regression. It is, therefore, sometimes called the *unexplained sum of the squares*. The difference between the two quantities, $S_t - S_r$, quantifies the improvement or error reduction due to describing the data in terms of a straight

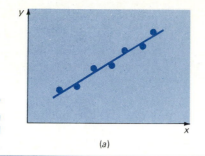

FIGURE 26.7
Examples of linear regression with
(a) small and (b) large residual
errors.

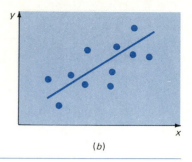

(a)   (b)

line rather than as an average value. Because the magnitude of this quantity is scale-dependent, the difference is normalized to the total error to yield

$$r^2 = \frac{S_t - S_r}{S_t} \qquad (26.7)$$

where r^2 is called the *coefficient of determination* and r is the *correlation coefficient* $(= \sqrt{r^2})$. For a perfect fit, $S_r = 0$ and $r = r^2 = 1$, signifying that the line explains 100 percent of the variability of the data. For $r = r^2 = 0$, $S_r = S_t$ and the fit represents no improvement. An alternative formulation for r that is more convenient for computer implementation is

$$r = \frac{n \sum x_i y_i - (\sum x_i)(\sum y_i)}{\sqrt{n \sum x_i^2 - (\sum x_i)^2} \sqrt{n \sum y_i^2 - (\sum y_i)^2}} \qquad (26.8)$$

EXAMPLE 26.2   "Goodness" of a Linear Least-Squares Fit

PROBLEM STATEMENT: Compute the total standard deviation, the standard error of the estimate, and the correlation coefficient for the data in Example 26.1.

SOLUTION: The mean of the y values from Example 26.1 is $\bar{y} = 1690/6 = 281.67$. Therefore the additional summations needed to solve this example can be computed as

x_i	y_i	y_i^2	$(y_i - \bar{y})^2$	$\hat{y}_i$	$(y_i - \hat{y}_i)^2$
10	110	12,100	29,469	125	225
15	230	52,900	2669	188	1764
20	210	44,100	5136	250	1600
25	350	122,500	4669	313	1369
30	330	108,900	2336	376	2116
35	460	211,600	31,803	438	484
		$S_t = 552,100$	$S_r = 76,082$		7558

These, along with the summations determined previously in Example 26.1, can be used to compute the total standard deviation [Eq. (24.10)]:

$$s_y = \sqrt{\frac{76,082}{6-1}} = 123$$

And the standard error of the estimate is [Eq. (26.6)]

$$s_{y/x} = \sqrt{\frac{7558}{6-2}} = 43.5$$

Thus, because $s_{y/x} < s_y$, we can see that the linear regression has decreased our uncertainty regarding the measurements. The extent of the improvement can be quantified by the coefficient of determination [Eq. (26.7)]:

$$r^2 = \frac{76,082 - 7558}{76,082} = 0.901$$

or by the correlation coefficient

$$r = \sqrt{0.901} = 0.949$$

An alternative computation of r can be made using Eq. (26.8):

$$r = \frac{6(43,500) - 135(1690)}{\sqrt{6(3475) - (135)^2} \sqrt{6(552,100) - (1690)^2}} = 0.949$$

Before we proceed to the computer program for linear regression, a word of caution is in order. Although the correlation coefficient provides a handy measure of goodness of fit, you should be careful not to ascribe more meaning to it than is warranted. Just because r is "close" to 1 does not mean that the fit is necessarily "good." For example, it is possible to obtain a relatively high value of r when the underlying relationship between y and x is not even linear! Draper and Smith (1981) provide guidance and additional material regarding assessment of results for linear regression. In addition, you should always, at the minimum, inspect a plot of the data along with your best-fit line whenever you employ regression.

26.3　COMPUTER PROGRAMS FOR LINEAR REGRESSION

As seen in Fig. 26.8, it is a relatively trivial matter to develop a program for linear regression. Notice that the programs in Fig. 26.8 do not compute the standard error of the estimate, the coefficient of determination, or the correlation coefficient. In Prob. 26.1 you will have the task of including these capabilities in your version of the program.

```
100 REM     ********************************************    C     ********************************************
110 REM     *              LINEAR REGRESSION          *    C     *              LINEAR REGRESSION          *
120 REM     *               (BASIC version)           *    C     *               (FORTRAN version)         *
130 REM     *                by  S.C. Chapra          *    C     *                by  S.C. Chapra          *
140 REM     *                                         *    C     *                                         *
150 REM     *   This program applies linear regression to   C     *   This program applies linear regression to  *
160 REM     *   fit a straight line to the data input *    C     *   to fit a straight line to the data input  *
170 REM     *   by the user.                          *    C     *   by the user.                          *
180 REM     ********************************************    C     ********************************************
190 REM     .----------------------------------------.    C     .----------------------------------------.
200 REM     |         DEFINITION OF VARIABLES       |    C     |         DEFINITION OF VARIABLES       |
210 REM     |                                        |    C     |                                        |
220 REM     | N     = NUMBER OF DATA POINTS          |    C     | N     = NUMBER OF DATA POINTS          |
230 REM     | X     = INDEPENDENT VARIABLE           |    C     | X     = INDEPENDENT VARIABLE           |
240 REM     | Y     = DEPENDENT VARIABLE             |    C     | Y     = DEPENDENT VARIABLE             |
250 REM     | SUMX  = SUM OF X'S                     |    C     | SUMX  = SUM OF X'S                     |
260 REM     | SUMY  = SUM OF Y'S                     |    C     | SUMY  = SUM OF Y'S                     |
270 REM     | SUMX2 = SUM OF SQUARE OF X'S           |    C     | SUMX2 = SUM OF SQUARE OF X'S           |
280 REM     | SUMXY = SUM OF PRODUCT OF X AND Y      |    C     | SUMXY = SUM OF PRODUCT OF X AND Y      |
290 REM     | XMEAN = MEAN OF X'S                    |    C     | XMEAN = MEAN OF X'S                    |
300 REM     | YMEAN = MEAN OF Y'S                    |    C     | YMEAN = MEAN OF Y'S                    |
310 REM     | SLOPE = SLOPE OF BEST FIT LINE         |    C     | SLOPE = SLOPE OF BEST FIT LINE         |
320 REM     | YINT  = Y INTERCEPT OF BEST FIT LINE   |    C     | YINT  = Y INTERCEPT OF BEST FIT LINE   |
330 REM     '----------------------------------------'    C     '----------------------------------------'
340 REM                                                   C
350 REM ****************** MAIN PROGRAM *****************
360 REM                                                         REAL X,Y,SUMX,SUMY,SUMX2,SUMXY,XMEAN,YMEAN,SLOPE,YINT
370 REM               Input Data and Determine Sums            INTEGER N
380 REM                                                   C
390 CLS                                                   C     ****************** MAIN PROGRAM *****************
400 INPUT "Number of X,Y pairs to input? ", N             C
410 PRINT: PRINT                                          C                Input Data and Determine Sums
420 FOR I = 1 TO N                                              READ(5,*) N
430     INPUT "X,Y = ? ",X,Y                                    DO 10 I=1, N
440     SUMX = SUMX + X                                            READ(5,*) X,Y
450     SUMY = SUMY + Y                                            SUMX = SUMX + X
460     SUMX2 = SUMX2 + X*X                                        SUMY = SUMY + Y
470     SUMXY = SUMXY + X*Y                                        SUMX2 = SUMX2 + X*X
480 NEXT I                                                         SUMXY = SUMXY + X*Y
490 REM                                               10    CONTINUE
500 REM                  Compute Means                    C
510 REM                                                   C                   Compute Means
520 XMEAN = SUMX/N                                         C
530 YMEAN = SUMY/N                                              XMEAN = SUMX/N
540 REM                                                         YMEAN = SUMY/N
550 REM            Compute Slope and Intercept            C
560 REM                                                   C            Compute Slope and Intercept
570 SLOPE = (N*SUMXY - SUMX*SUMY)/(N*SUMX2-SUMX^2)        C
580 YINT = YMEAN - SLOPE*XMEAN                                  SLOPE = (N*SUMXY - SUMX*SUMY)/(N*SUMX2-SUMX**2)
590 REM                                                         YINT = YMEAN - SLOPE*XMEAN
600 REM                 Print out Results                 C
610 PRINT:PRINT                                           C                Print out Results
620 PRINT "Slope = "; SLOPE                                    WRITE(6,*) 'Slope = ', SLOPE
630 PRINT "Y intercept = "; YINT                               WRITE(6,*) 'Y intercept = ', YINT
640 END                                                        STOP
                                                               END
```

FIGURE 26.8 BASIC and FORTRAN programs to implement linear regression.

Additionally, because the graphical capabilities of computers are so varied, we have not included a plotting option in Fig. 26.8. However, as mentioned at the end of the previous section, such an option is critical for the effective use and interpretation of regression results. If your computer system has plotting capabilities, we recommend that you expand your program to include a plot of y versus x, showing both the data and the regression line. The printer plots described at the end of Chap. 22 can be modified for this purpose. In any event, the inclusion of this capability will greatly enhance the program's utility for problem solving.

Figure 26.9 shows some screens generated using a commercial software package that has a linear-regression algorithm very similar to that given in Fig. 26.8. The package is used to solve the problem from Examples 26.1 and 26.2. The regression

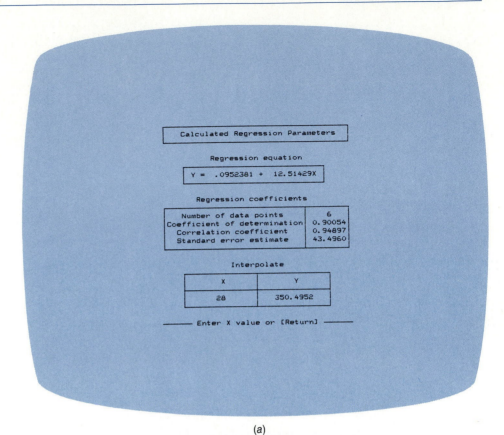

(a)

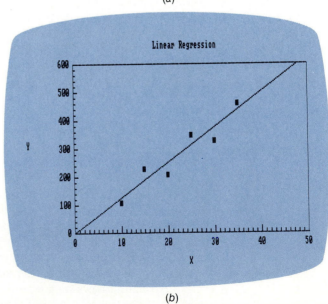

FIGURE 26.9
Computer screens generated by a linear regression program from a software package (NUMERICOMP by Canale and Chapra, 1985). The package is used to solve the problem previously analyzed in Examples 26.1 and 26.2. The regression results are shown in a; a plot of the data along with the best-fit line is shown in b.

(b)

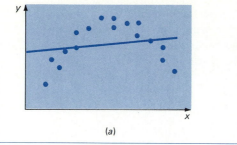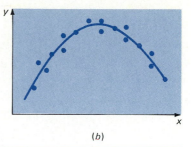

(a)   (b)

FIGURE 26.10
(a) Data that is ill-suited for linear least-squares regression; (b) indication that a parabola is preferable.

results are presented in Fig. 26.9a, and the best-fit line along with the data is plotted in Fig. 26.9b. These screens illustrate how output can be designed in a clear and accessible fashion. You can employ them as models for the software you develop from Fig. 26.8.

26.4 ALTERNATIVE AND ADVANCED METHODS

Linear regression provides a powerful technique for fitting a "best" straight line to data. However, it is predicated on the fact that the relationship between the dependent and independent variables is linear. This is not always the case, and the first step in any regression analysis should be to plot and visually inspect the data to ascertain whether a linear model applies.

Figure 26.10 shows some data that is obviously curved. A number of options are available for fitting a "best" curve to such data. First, the data can sometimes be "linearized" by transforming it logarithmically prior to regression. This is the same strategy as was used for the semi-log and log-log plots in Chap. 22. If transformations do not straighten the data out, two additional alternatives are polynomial and nonlinear regression. As the name implies, *polynomial regression* is used to fit a least-squares polynomial through data. Thus a best quadratic or a best cubic equation could be developed with this technique. In fact, linear regression is the simplest form of polynomial regression because a straight line is a first-order polynomial.

Nonlinear regression is employed to fit any equation to data. An introduction to some additional regression methods can be found in Chapra and Canale (1985). Draper and Smith (1981) provide a detailed description of the subject.

PROBLEMS

26.1 Utilize Fig. 26.8 as a starting point for developing your own program for linear regression. Employ structured programming techniques, modular design, and internal and external documentation in your development. Incorporate goodness-of-fit statistics into the program, and include the capability of plotting both the data and the best-fit line on the same line graph.

26.2 Develop the program from Prob. 26.1, but incorporate an algorithm that classifies each of the data points in terms of how far away it is from the line. This can be accomplished, by using a quantity such as the residual normalized to the standard error of the estimate:

$$\frac{|y_i - \hat{y}_i|}{s_{y/x}}$$

This quantity provides a relative measure of the distance of each point away from the line, and it can help you identify which points are farthest from the line. It should be noted that although the above quantity is adequate for the present context, there are better and more sophisticated measures and methods for flagging outliers.

26.3 Define the following terms. What do they indicate?
(a) Standard error of the estimate
(b) Standard deviation
(c) Correlation coefficient

26.4 Curve fitting to data points can generally be divided into two categories. Name them and contrast the differences between the two approaches. For the following situations select a method to fit the data to a curve and explain the reasons for your choice of method.
(a) A chemical engineer needs to find a very precise property of steam in a table. Unfortunately, the temperature for which the engineer wants this property is not given exactly in the table.
(b) An electrical engineer takes many measurements of a network's intrinsic capacitance for different inputs of current. The engineer wants to fit a curve through this data to be able to predict the effect on capacitance for all values of current.
(c) The average temperature of the ocean at a certain location has been measured very accurately for the past several years. This temperature has been dropping slightly, and an atmospheric scientist wants to fit a curve through this data to be able to predict the date of the next ice age.

26.5 An experiment is performed to determine the percent elongation of a material as a function of temperature. The resulting data is

Temperature, °F	400	500	600	700	800	900	1000	1100
Elongation, %	11	13	13	15	17	19	20	23

Predict the percent elongation for a temperature of 780°F.

26.6 The shear strengths, in kips per square foot (ksf), of nine specimens taken at various depths in a clay stratum are

Depth, m	1.9	3.1	4.2	5.1	5.8	6.9	8.1	9.3	10.0
Strength, ksf	0.3	0.6	0.5	0.8	0.7	1.1	1.5	1.3	1.6

Estimate the shear stress at a depth of 7.5 m.

26.7 The distance required to stop an automobile is a function of its speed. The following experimental data was collected to quantify this relationship:

Speed, mi/h	15	20	25	30	40	50	60	70
Stopping distance, ft	15	21	45	46	65	90	111	98

Estimate the stopping distance for a car traveling at 50 mi/h?

26.8 You are provided with the following stress/strain data for an aluminum alloy:

Stress	1	2	3	4	5	6	7	8	9	10	11	12
Strain	2	4	6	6	6	7	8	7.5	7	7.5	8	7.5

Use linear regression to determine the strain corresponding to a stress of 7.4.

26.9 A transportation engineering study was conducted to determine the proper design of bike lanes. Data was gathered on bike-lane widths and average distance between bikes and passing cars. The data from 11 streets is

Lane width x, ft	5	10	7	7.5	7	6	10	9	5	5.5	8
Distance y, ft	3	8	5	8	6	6	10	10	4	5	7

(a) Plot the data.
(b) Fit a straight line to the data with linear regression. Add this line to the plot.
(c) If the minimum, safe average distance between bikes and passing cars is considered to be 6 ft, determine the corresponding minimum lane width.

26.10 An experiment is performed to define the relationship between applied stress and the time to fracture for a stainless steel. Eight different values of stress are applied, and the resulting data is

Applied stress x, kg/mm^2	5	10	15	20	25	30	35	40
Fracture time y, h	40	30	25	40	18	20	22	15

(a) Plot the data.
(b) Fit a straight line to the data with linear regression. Superimpose this line on your plot.
(c) Use the best-fit equation to predict the fracture time for an applied stress of 33 kg/mm^2.

26.11 In water-resources engineering the sizing of reservoirs depends on accurate estimates of water flow in the river that is being impounded. For some rivers, long-term historical records of such flow data are difficult to obtain. In contrast, meteorological data on precipitation is often available for many years past. Therefore it is often useful to determine a relationship between flow and precipitation. This relationship can then be used to estimate flows for years when only precipitation measurements were made. The following data is available for a river that is to be dammed:

Annual precipitation x, in	35	40	41	55	52	37	46	48	39	45
Annual water flow y, ft^3/s	4050	6075	5400	9500	7290	5700	6210	8440	4590	8000

(a) Plot the data.
(b) Fit a straight line to the data with linear regression. Superimpose this line on your plot.
(c) Use the best-fit line to predict the annual water flow if the precipitation is 50 in.

26.12 The concentration of total phosphorus (p in mg/m^3) and chlorophyll a (c in mg/m^3) for each of the Great Lakes is

	p	c
Lake Superior	4.5	0.8
Lake Michigan	8.0	2.0
Lake Huron	5.5	1.2
Lake Erie: west basin	39.0	11.0
central basin	19.5	4.4
east basin	17.5	3.3
Lake Ontario	21.0	5.5

Chlorophyll a is a parameter that indicates how much plant life is suspended in the water. As such, it indicates how unclear and unsightly the water appears. Use the above data to determine a relationship to predict c as a function of p. Use this equation to predict the level of chlorophyll that can be expected if waste treatment is used to lower the phosphorus concentration of western Lake Erie to 10 mg/m^3.

26.13 It is known that the tensile strength of a plastic increases as a function of the time it is heat-treated. The following data is collected:

Time	10	15	20	30	40	50	55	60	75
Tensile strength	4	20	18	50	33	48	80	105	78

Fit a straight line to this data and use the equation to determine the tensile strength at a time of 70 min.

26.14 The following data was gathered to determine the relationship between pressure and temperature of a fixed volume of 1 kg of nitrogen. The volume is 10 m^3.

T, °C	-20	0	20	40	50	70	100	120
p, N/m³	7500	8104	8700	9300	9620	10,200	10,500	11,700

Employ the ideal gas law $pV = nRT$ to determine R on the basis of this data. Note that for the law T must be expressed in kelvins.

26.15 The following data was taken from an experiment that measured the current in a wire for various imposed voltages:

Voltage, V	0	2	3	4	5	7	10
Current, A	0	5.2	7.8	10.7	13	19.3	26.5

On the basis of a linear regression of this data, determine current for a voltage of 5 V. Plot the line and the data and evaluate the fit.

26.16 It is known that the voltage drop across an inductor follows Faraday's law:

$$V_L = L \frac{di}{dt}$$

where V_L is the voltage drop (in volts), L is inductance (in henrys; 1 H = 1 Vs/A) and i is current (in amperes). Employ the following data to estimate L:

di/dt, A/s	1	2	4	6	8	10
V_L, V	5	11	19	31	39	50

What is the meaning, if any, of the intercept of the regression equation derived from this data?

26.17 In Sec. 22.3.1, we discussed semi-log plots for which the following equation holds [Eq. (22.3)]:

$$y = y_0 \, 10^{(\text{slope})x}$$

Employ linear regression to fit the data from Example 22.2 and predict population at $t = 25$. Compare the prediction with the one obtained in the example.

26.18 In Sec. 22.3.1 we discussed log-log plots for which the following equation holds [Eq. (22.10)]:

$$y = y_0 x^{\text{slope}} \qquad\qquad \text{(P26.18)}$$

Employ this model and linear regression to fit the following data for viscosity (v) as a function of temperature (T):

T, °F	40	50	60	70	80
v, 10^{-5} ft²/s	1.7	1.4	1.2	1.05	0.95

Determine the best-fit equation of the form of Eq. (P26.18) and use this relationship to predict v at $T = 63$ °F.

CHAPTER 27

CURVE FITTING:
INTERPOLATION

As described in the previous chapter, there are many situations where engineers must use regression to fit a curve to uncertain data. Just as often, however, we must estimate intermediate values between precise data. A common source of such information is the reference tables found in engineering and scientific handbooks. Examples range from tables of well-known physical and chemical properties to tables giving the values of mathematical functions.

Table 27.1 is an example from engineering economics. Because it is impossible to tabulate every possible bit of information, such tables usually include data spaced at uniform intervals. For example, in Table 27.1, the economic data corresponds to various interest rates from 10 to 25 percent. If we require information for an interest rate of 11 percent, the proper value of 8.0623 can be read directly from the table. However, if we require an intermediate value, say at 11.5 percent, interpolation is required.

The most common method used for this purpose is called *linear interpolation*. The rationale and shortcomings of this approach can be best illustrated by reexpressing the data from Table 27.1 in graphical form. Just as with any set of paired values, the economic data can be plotted on cartesian coordinates. As shown in Fig. 27.1, the interest rate is the independent variable and F/P, the ratio of future to present worth, is the dependent variable. In linear interpolation, intermediate values are estimated to lie on the straight line connecting two adjacent points. As will be shown in this chapter, the straight line can be expressed mathematically as

$$f(x) = a_0 + a_1 x \tag{27.1}$$

where a_0 is the intercept, a_1 is the slope, and $f(x)$ and x are the dependent and the

TABLE 27.1 Data from an Engineering Economics Table. The independent variable is interest rate *i*, and the dependent variable is the ratio of future to present worth, *F/P*, for a 20-year investment period.

i, %	F/P
10	6.7275
11	8.0623
12	9.6463
13	11.5231
14	13.7435
15	16.3665
20	38.3376
25	86.7362

independent variables, respectively. Thus Eq. (27.1) is used to calculate intermediate values between any two points.

The strengths and shortcomings of this approach are both evident from Fig. 27.1. Because of the close spacing of the points between 11 and 12 percent, this approach yields an excellent result when predicting a value, say, at 11.5 percent. However, suppose that we desired a prediction at 17.5 percent. For this case, the wide spacing makes the curvature of the data more prominent and, as is evident from Fig. 27.1, a substantial error results.

FIGURE 27.1
A plot of *F/P* versus *i* for the data from Table 27.1. Notice how linear interpolation would yield an almost exact result in the interval between *i* = 11 and 12, whereas a significant error results in the wider interval from *i* = 15 to 20. If you do not think the error looks significant, try convincing your bank to give you the higher *F/P* computed with linear interpolation for your savings account!

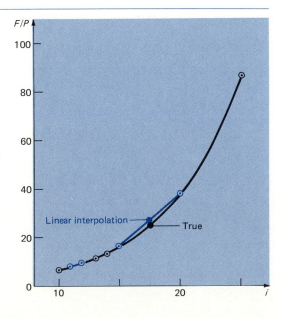

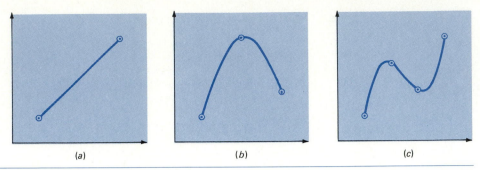

FIGURE 27.2 Examples of interpolating polynomials: (a) first-order (linear) connecting two points; (b) second-order (parabolic or quadratic) connecting three points; and (c) third-order (cubic) connecting four points.

One remedy for such situations is based on a simple extension of the reasoning underlying linear interpolation. As seen in Fig. 27.2a, linear interpolation is based on the premise that there is one and only one straight line connecting two points. Similarly, there is also a unique parabola,

$$f(x) = a_0 + a_1x + a_2x^2$$

that connects any three points (Fig. 27.2b). An interpolating parabola is capable of capturing some of the curvature of data and, hence, usually yields a prediction that is superior to linear interpolation.

Carrying this reasoning one step farther, we see that there is one and only one cubic equation,

$$f(x) = a_0 + a_1x + a_2x^2 + a_3x^3$$

that fits four points (Fig. 27.2c). This leads us to the general conclusion that there is a unique nth-order polynomial,

$$f(x) = a_0 + a_1x + a_2x^2 + \cdots + a_nx^n$$

that passes exactly through $n + 1$ data points. (This relationship results from the fundamental theorem of algebra.) As described in this chapter, *polynomial interpolation* is the technique used to derive the nth-order polynomial equation to fit $n + 1$ points. This polynomial can then be used to predict intermediate values between the points.

Although there is one and only one nth-order polynomial that fits $n + 1$ points, there are a variety of mathematical formats in which the polynomial can be expressed. In the present chapter, we will describe one version that is well suited for implementation on computers—the *Lagrange interpolating polynomial*. Before presenting the general equation, we will introduce the linear and parabolic versions because of their simple visual interpretation.

27.1   LINEAR INTERPOLATION

The simplest form of interpolation is to connect two data points with a straight line. This technique, called *linear interpolation*, is depicted graphically in Fig. 27.3. Using similar triangles,

$$\frac{f(x) - f(x_0)}{x - x_0} = \frac{f(x_1) - f(x_0)}{x_1 - x_0}$$

which can be rearranged to give

$$f(x) = f(x_0) + \frac{f(x_1) - f(x_0)}{x_1 - x_0}(x - x_0) \tag{27.2}$$

This particular form is called the *first-order Newton interpolating polynomial*. Notice that the equation is in the form of a straight line:

Prediction = constant + (slope)(distance)

The Lagrange interpolating polynomial can be derived directly from this formulation. In order to derive the Lagrange version, we reformulate the slope as

$$\frac{f(x_1) - f(x_0)}{x_1 - x_0} = \frac{f(x_1)}{x_1 - x_0} + \frac{f(x_0)}{x_0 - x_1} \tag{27.3}$$

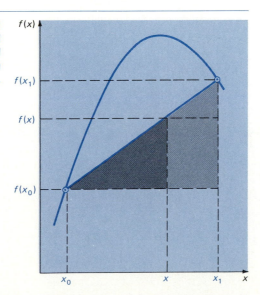

FIGURE 27.3
Linear interpolation: The shaded areas are the similar triangles used to derive the linear interpolation formula.

which is referred to as the *symmetric form*. Substituting Eq. (27.3) into Eq. (27.2) and collecting terms yields

$$f_1(x) = \frac{x - x_1}{x_0 - x_1} f(x_0) + \frac{x - x_0}{x_1 - x_0} f(x_1) \tag{27.4}$$

which is the *first-order Lagrange interpolating polynomial*. The prediction, $f_1(x)$, has been subscripted with a 1 to designate that this is the first-order, or linear, version.

EXAMPLE 27.1   The First-Order Lagrange Interpolating Polynomial

PROBLEM STATEMENT: Use linear interpolation to evaluate F/P at $i = 17.5$ percent on the basis of the data from Table 27.1. Note that the exact value computed from economic theory is 25.1627.

SOLUTION: From Table 27.1, the following data can be used to interpolate between $i = 15$ and 20 percent:

$$x_0 = 15 \qquad f(x_0) = 16.3665$$

$$x_1 = 20 \qquad f(x_1) = 38.3376$$

Therefore the estimate at $i = 17.5$ percent is [Eq. (27.4)]

$$f_1(17.5) = \frac{17.5 - 20}{15 - 20} 16.3665 + \frac{17.5 - 15}{20 - 15} 38.3376 = 27.352$$

This result represents a percent relative error of

$$\frac{25.1627 - 27.352}{25.1627} 100\% = -8.7\%$$

27.2   PARABOLIC INTERPOLATION

As depicted in Fig. 27.1, the error in Example 27.1 is due to the fact that we used a straight line to approximate a curve. Consequently, a strategy for improving the estimate is to introduce some curvature into the line connecting the points. If three data points are available, this can be accomplished with a second-order polynomial or parabola. The Lagrange polynomial for this purpose is

$$f_2(x) = \frac{(x - x_1)(x - x_2)}{(x_0 - x_1)(x_0 - x_2)} f(x_0) + \frac{(x - x_0)(x - x_2)}{(x_1 - x_0)(x_1 - x_2)} f(x_1)$$

$$+ \frac{(x - x_0)(x - x_1)}{(x_2 - x_0)(x_2 - x_1)} f(x_2) \tag{27.5}$$

EXAMPLE 27.2:   The Second-Order Lagrange Interpolating Polynomial

PROBLEM STATEMENT: Estimate F/P at $i = 17.5$ percent from the data in Table 27.1 by fitting a parabola to the points at $i = 15$, 20 and 25 percent.

SOLUTION: From Table 27.1, the following data can be used for the parabolic interpolation:

$$x_0 = 15 \qquad f(x_0) = 16.3665$$
$$x_1 = 20 \qquad f(x_1) = 38.3376$$
$$x_2 = 25 \qquad f(x_2) = 86.7362$$

Equation (27.5) can be used to compute the estimate at $i = 17.5$ percent:

$$f_2(17.5) = \frac{(17.5 - 20)(17.5 - 25)}{(15 - 20)(15 - 25)} 16.3665$$

$$+ \frac{(17.5 - 15)(17.5 - 25)}{(20 - 15)(20 - 25)} 38.3376$$

$$+ \frac{(17.5 - 15)(17.5 - 20)}{(25 - 15)(25 - 20)} 86.7362 = 24.049$$

which represents a percent relative error of 4.43 percent. Thus the error is about half that obtained in Example 27.1 with linear interpolation.

Figure 27.4 shows the parabolic interpolation. In addition, the entire parabola

FIGURE 27.4
Parabolic interpolation.

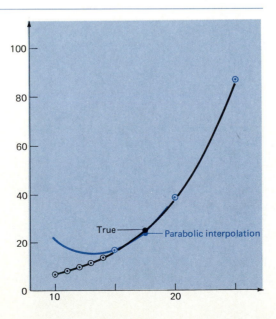

[as computed by substituting other values of x into Eq. (27.5)] is also plotted. Notice that by intersecting the three points at $i = 15$, 20 and 25 percent, the parabola captures some of the curvature of the data.

27.3   THE GENERAL FORM OF THE LAGRANGE POLYNOMIAL

Just as we have developed first- and second-order Lagrange polynomials, a third-order version can be derived as

$$
f_3(x) = \frac{(x - x_1)(x - x_2)(x - x_3)}{(x_0 - x_1)(x_0 - x_2)(x_0 - x_3)} f(x_0) + \frac{(x - x_0)(x - x_2)(x - x_3)}{(x_1 - x_0)(x_1 - x_2)(x_1 - x_3)} f(x_1)
$$
$$
+ \frac{(x - x_0)(x - x_1)(x - x_3)}{(x_2 - x_0)(x_2 - x_1)(x_2 - x_3)} f(x_2) + \frac{(x - x_0)(x - x_1)(x - x_2)}{(x_3 - x_0)(x_3 - x_1)(x_3 - x_2)} f(x_3)
$$
(27.6)

Inspection of Eqs. (27.4) through (27.6) leads to the following general representation of the nth order Lagrange polynomial:

$$
f_n(x) = \sum_{i=0}^{n} L_i(x) \, f(x_i)
$$
(27.7)

where

$$
L_i(x) = \prod_{\substack{j=0 \\ (j \neq 1)}}^{n} \frac{x - x_j}{x_i - x_j}
$$
(27.8)

where the symbol Π stands for "the product of." Equation (27.8) is evaluated for $j = 0$ to n, with the exception of $j = i$.

Although Eqs. (27.7) and (27.8) might seem complicated, they are really quite simple. This can be appreciated by noticing that each term, $L_i(x)$, will be 1 at $x = x_i$ and 0 at all the other data points. Thus each product, $L_i(x) \, f(x_i)$, will be exactly equal to $f(x_i)$ at x_i and will be zero at all the x values for the other data points (Fig. 27.5). Consequently, the summation of all the products designated by Eq. (27.7) represents the unique nth-order polynomial that passes exactly through all $n + 1$ data points.

An important feature of the Lagrange polynomial is that the points do not have to be equispaced nor be in any particular order. Such is the case for the following example.

EXAMPLE 27.3   *The Third-Order Lagrange Interpolating Polynomial*

PROBLEM STATEMENT: Estimate F/P at $i = 17.5$ percent from the data in Table 27.1 by fitting a third-order polynomial to the points at $i = 15$, 20, 25, and 14 percent.

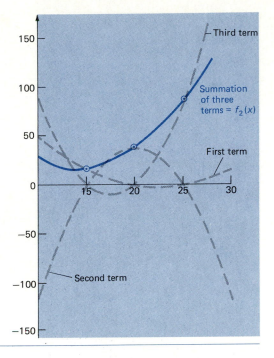

FIGURE 27.5
A visual depiction of the rationale behind the Lagrange polynomial. This figure shows the second-order case from Example 27.2. Each of the three terms in Eq. (27.5) passes through one of the data points and is zero at the other two. The summation of the three terms must, therefore, be the unique second-order polynomial $f_2(x)$ that passes exactly through the three points.

SOLUTION: The following data from Table 27.1 can be used for the interpolation:

$$x_0 = 15 \qquad f(x_0) = 16.3665$$

$$x_1 = 20 \qquad f(x_1) = 38.3376$$

$$x_2 = 25 \qquad f(x_2) = 86.7362$$

$$x_3 = 14 \qquad f(x_3) = 13.7435$$

Equation (27.6) can be used to compute the estimate at 17.5 percent:

$$f_3(17.5) = \frac{(17.5 - 20)(17.5 - 25)(17.5 - 14)}{(15 - 20)(15 - 25)(15 - 14)} \, 16.3665$$

$$+ \frac{(17.5 - 15)(17.5 - 25)(17.5 - 14)}{(20 - 15)(20 - 25)(20 - 14)} \, 38.3376$$

$$+ \frac{(17.5 - 15)(17.5 - 20)(17.5 - 14)}{(25 - 15)(25 - 20)(25 - 14)} \, 86.7362$$

$$+ \frac{(17.5 - 15)(17.5 - 20)(17.5 - 25)}{(14 - 15)(14 - 20)(14 - 25)} \, 13.7435 = 25.043$$

which represents an error of 0.48%. Thus, for this case, cubic interpolation represents a great improvement over the linear and parabolic versions from Examples 27.1 and 27.2.

```
100 REM    ********************************************        C    ********************************************
110 REM    *         LAGRANGE INTERPOLATION       *            C    *         LAGRANGE INTERPOLATION       *
120 REM    *            (BASIC version)            *            C    *           (FORTRAN version)           *
130 REM    *            by  S.C. Chapra            *            C    *            by  S.C. Chapra            *
140 REM    *                                       *            C    *                                       *
150 REM    *  This program applies Lagrange interpolation *     C    *  This program applies Lagrange interpolation *
160 REM    *  to the data input by the user.       *            C    *  to the data input by the user.       *
170 REM    ********************************************        C    ********************************************
180 REM                                                        C
190 REM    .------------------------------------------.        C    .------------------------------------------.
200 REM    |        DEFINITION OF VARIABLES           |        C    |        DEFINITION OF VARIABLES           |
210 REM    |                                          |        C    |                                          |
220 REM    |   N     = NUMBER OF POINTS               |        C    |   N     = NUMBER OF POINTS               |
230 REM    |   X     = VECTOR OF X COORDINATES        |        C    |   X     = VECTOR OF X COORDINATES        |
240 REM    |   Y     = VECTOR OF Y COORDINATES        |        C    |   Y     = VECTOR OF Y COORDINATES        |
250 REM    |   XINT  = X COORDINATE AT WHICH INTERPOLATION |    C    |   XINT  = X COORDINATE AT WHICH INTERPOLATION |
260 REM    |           IS DESIRED                     |        C    |           IS DESIRED                     |
270 REM    |   PRDCT = RUNNING PRODUCT FROM LAGRANGE  |        C    |   PRDCT = RUNNING PRODUCT FROM LAGRANGE  |
280 REM    |           FORMULA                        |        C    |           FORMULA                        |
290 REM    |   K     = ORDER OF INTERPOLATING POLYNOMIAL |     C    |   K     = ORDER OF INTERPOLATING POLYNOMIAL |
300 REM    |   YINT  = INTERPOLATED VALUE OF Y        |        C    |   YINT  = INTERPOLATED VALUE OF Y        |
310 REM    |   NUMINT = N - 1 = HIGHEST POSSIBLE ORDER OF |     C    |   NUMINT = N - 1 = HIGHEST POSSIBLE ORDER OF |
320 REM    |           INTERPOLATING POLYNOMIAL       |        C    |           INTERPOLATING POLYNOMIAL       |
330 REM    '------------------------------------------'        C    '------------------------------------------'
340 REM                                                        C
350 DIM X(10), Y(10)                                           REAL X(10), Y(10), XINT, PRDCT, K, YINT, NUMINT
360 REM                                                        INTEGER N
370 REM *****************  MAIN PROGRAM  *****************      C
380 REM                                                        C    *****************  MAIN PROGRAM  *****************
390 REM                   Input Data                           C
400 REM                                                        C                    Input Vector Data
410 CLS                                                        READ(5,*) N
420 INPUT "Number of X,Y pairs? ", N                           DO 10 I=1, N
430 PRINT:PRINT                                                   READ(5,*) X(I), Y(I)
440 FOR I = 1 TO N                                           10 CONTINUE
450    INPUT "X,Y = ? ",  X(I),Y(I)                            C
460 NEXT I                                                     C    Input Value of X at Which Interpolation is Desired
470 REM                                                        C
480 REM   Input Value of X at Which Interpolation is Desired   READ(5,*) XINT
490 REM                                                        WRITE(6,*) ' ORDER           INTERPOLATION'
500 PRINT                                                      C
510 INPUT "Value of X at interpolation point = ? ", XINT       C    The following nested loops compute values
520 PRINT: PRINT                                               C    of Y for Lagrange interpolating Polynomials
530 PRINT " ORDER  INTERPOLATION"                              C    of order 1 through N - 1.
540 REM                                                        C
550 REM    The following nested loops compute values           NUMINT = N - 1
560 REM    of Y for Lagrange interpolating Polynomials          DO 40 K=1, NUMINT
570 REM    of order 1 through N - 1.                              YINT = 0
580 REM                                                           KP = K + 1
590 NUMINT = N - 1                                                DO 30 I=1, KP
600 FOR K = 1 TO NUMINT                                             PRDCT = Y(I)
610    YINT = 0                                                     DO 20 J=1, KP
620    KP = K + 1                                                     IF (I .NE. J) THEN
630    FOR I = 1 TO KP                                                  PRDCT = PRDCT*(XINT - X(J))/(X(I) - X(J))
640       PRDCT = Y(I)                                                ENDIF
650       FOR J = 1 TO KP                                      20     CONTINUE
660          IF I <> J THEN 670 ELSE 680                            YINT = YINT + PRDCT
670          PRDCT = PRDCT*(XINT - X(J))/(X(I) - X(J))          30   CONTINUE
680       NEXT J                                                    WRITE(6,*) K,YINT
690       YINT = YINT + PRDCT                                  40 CONTINUE
700    NEXT I                                                   STOP
710    PRINT K,YINT                                             END
720 NEXT K
730 END
```

FIGURE 27.6 BASIC and FORTRAN computer programs for Lagrange interpolation.

27.4   COMPUTER PROGRAMS FOR LAGRANGE INTERPOLATION

As seen in Fig. 27.6, it is a relatively simple matter to develop programs for Lagrange interpolation. Notice how nested loops are used to implement the summations and products from Eqs. (27.7) and (27.8).

The programs in Fig. 27.6 do not have user-friendly input or output. In Prob.

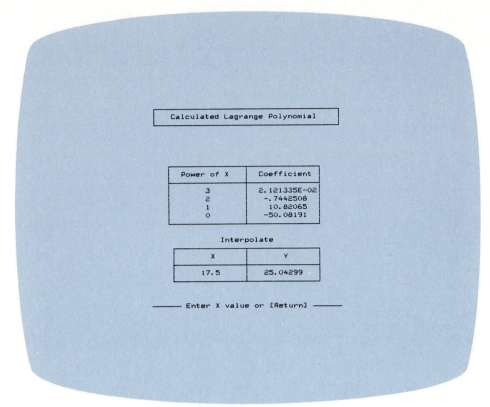

Power of X	Coefficient
3	2.121335E-02
2	-.7442508
1	10.82065
0	-50.08191

Interpolate

X	Y
17.5	25.04299

(a)

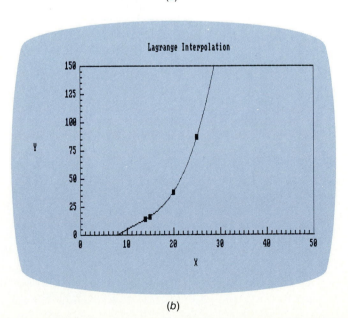

(b)

FIGURE 27.7
Computer screens generated by an interpolation software package (NUMERICOMP by Canale and Chapra, 1985). The package is used to solve the problem previously analyzed in Example 27.3.

27.1 you will have the task of incorporating such features into your version of the program.

Additionally, because the graphical capabilities of computers are so varied, we have not included a plotting option in Fig. 27.6. However, such an option is helpful for the effective use and interpretation of interpolation results. If your computer system has plotting capabilities, we recommend that you expand your program to include a plot of y versus x, showing both the data and the interpolating polynomial. The printer plots described at the end of Chap. 22 can be modified for this purpose. In any event, the inclusion of this capability will greatly enhance the program's utility for problem solving.

Figure 27.7 shows some screens generated using a commercial software package that has a Lagrange interpolation algorithm very similar to the one used in Fig. 27.6. The package is used to solve Example 27.3. The solution is shown in Fig. 27.7*a* and a plot of the interpolating polynomial and the data are shown in Fig. 27.7*b*. These

FIGURE 27.8

Fit of a fifth-order Lagrange interpolating polynomial to experimentally derived stress-strain data. Although the curve passes directly through each of the points, the error in the data induces wild oscillations between points.

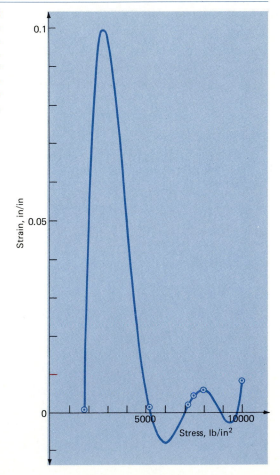

screens illustrate how output can be designed in a clear and accessible fashion. You can employ them as models for the software you develop from Fig. 27.6.

27.5   PITFALLS AND ADVANCED METHODS

Although there is one and only one nth order polynomial that fits $n + 1$ points, there are a variety of mathematical formats in which this polynomial can be expressed. Aside from the Lagrange equation, an alternative formulation is *Newton's interpolating polynomial*. We have already seen the linear form of the Newton polynomial in Eq. (27.2). Higher-order versions are also available. Newton's polynomial is usually the preferred method when the appropriate order is not known beforehand. This is because Newton's method allows for convenient error analysis that can provide insight into the proper order.

There are situations where both the Lagrange and Newton formats are inappropriate. Of course, one example is for data that has error associated with it. As you learned in the previous chapter, regression is a more suitable alternative for fitting curves to such data. If you try to fit a higher-order polynomial to uncertain data, the results are often unacceptable (see Fig. 27.8).

FIGURE 27.9 A visual representation of a situation where splines are superior to higher-order interpolating polynomials. The function to be fit undergoes an abrupt increase at $x = 0$. Parts (a) through (c) indicate that the abrupt change induces oscillations in interpolating polynomials. In contrast, because it is limited to third-order curves with smooth transitions, the cubic spline (d) provides a much more acceptable approximation.

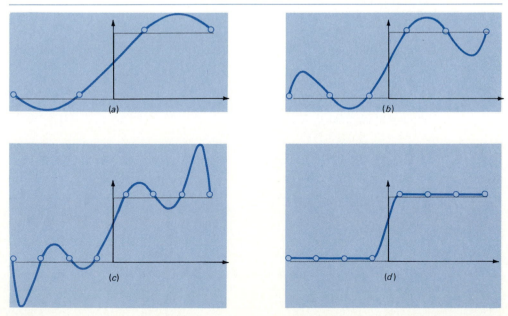

Another case for which interpolating polynomials do not work well is for data that is generally smooth but which has abrupt local changes (see Fig. 27.9). *Cubic splines* are interpolating polynomials that are better suited for such cases. In this approach, a separate cubic equation is fit between each pair of adjacent data points. The cubics are formulated so that the connections between the separate cubic equations are smooth. By limiting the polynomials to cubics, the wild oscillations of the higher-order versions do not occur. Thus, as seen in Fig. 27.9d, the resulting fit is much more acceptable for data with abrupt, local changes.

Beyond interpolating polynomials and splines, there are a variety of other techniques available for determining intermediate values between precisely-known points. Chapra and Canale (1985) provide an introduction to the subject. A more detailed treatment can be found in other references (for example, Ralston and Rabinowitz, 1975, or Gerald and Wheatley, 1984).

PROBLEMS

27.1 Utilize Fig. 27.6 as a starting point for developing your own program for Lagrange interpolation. Employ structured programming techniques, modular design, and internal and external documentation in your development. Include the capability to plot both the data and the interpolating polynomial on the same line graph.

27.2 Show that

$$L_i(x) = \prod_{\substack{j=0 \\ (j \neq i)}}^{n} \frac{x - x_j}{x_i - x_j} = 1$$

at $x = x_i$ and is equal to zero at all other data points. Also, show that

$$\sum_{i=0}^{n} L_i(x) = 1$$

for all the data points.

27.3 An experiment is used to determine the ultimate moment capacity of a concrete beam as a function of cross-sectional area. The experiment, which is performed with great precision, yields the following data:

Capacity, in · kips	932.3	1785.2	2558.6	3252.7	3867.4	4402.6	4858.4	5234.8
Area, in^2	1	2	3	4	5	6	7	8

Determine the ultimate moment capacity (to one decimal place) for an area of 5.4 in.

27.4 The acceleration due to gravity at an altitude y above the surface of the earth is given by

y, m	0	20,000	40,000	60,000	80,000
g, m/s²	9.8100	9.7487	9.6879	9.6278	9.5682

Compute g at $y = 55,000$ m to four decimal places of accuracy.

27.5 The ratio of the purchase price to annual payments (P/A) can be found in economic tables as a function of interest rate. For a 20-year loan, these values are

i	0.10	0.12	0.14	0.16	0.18	0.20	0.25	0.30
P/A	8.514	7.469	6.623	5.929	5.353	4.870	3.954	3.316

Compute P/A for an interest rate of 0.195.

27.6 The specific volume of a superheated steam is listed in steam tables for various temperatures. For example, at a pressure of 2950 lb/in², absolute:

T, °F	700	720	740	760	780
v	0.1058	0.1280	0.1462	0.1603	0.1703

Determine v at $T = 728$°F.

27.7 The vertical stress σ_z under the corner of a rectangular area subjected to a uniform load of intensity q is given by the solution of Boussinesq's equation:

$$\sigma_z = \frac{q}{4\pi}\left[\frac{2mn\sqrt{m^2 + n^2 + 1}}{m^2 + n^2 + 1 + m^2n^2}\frac{m^2 + n^2 + 2}{m^2 + n^2 + 1} + \sin^{-1}\left(\frac{2mn\sqrt{m^2 + n^2 + 1}}{m^2 + n^2 + 1 + m^2n^2}\right)\right]$$

Because this equation is inconvenient to solve manually, it has been reformulated as

$$\sigma_z = qf_{z(m,n)}$$

where $f_{z(m,n)}$ is called the *influence value* and m and n are dimensionless ratios, with $m = a/z$ and $n = b/z$ and a and b as defined in Fig. P27.7. The influence value is then listed in a table, a portion of which is given here:

FIGURE P27.7

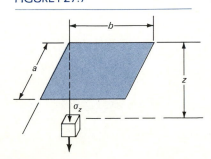

m	n = 1.2	n = 1.4	n = 1.6
0.1	0.02926	0.03007	0.03058
0.2	0.05733	0.05894	0.05994
0.3	0.08323	0.08561	0.08709
0.4	0.10631	0.10941	0.11135
0.5	0.12626	0.13003	0.13241
0.6	0.14309	0.14749	0.15027
0.7	0.15703	0.16199	0.16515
0.8	0.16843	0.17389	0.17739

If $a = 5.6$ and $b = 14$, compute σ_z at a depth 10 m below the corner of a rectangular footing that is subject to a total load of 100 t (metric tons). Express your answer in tonnes per square meter. Note that q is equal to the load per area.

27.8 Bessel functions often arise in advanced engineering analysis. These functions are usually not amenable to straightforward evaluation and, therefore, are often compiled in standard mathematical tables. For example,

x	$J_0(x)$
1.8	0.3400
2.0	0.2239
2.2	0.1104
2.4	0.0025
2.6	0.0968

Estimate $J_0(2.1)$. Note that the true value is 0.1666.

27.9 The current in a wire is measured with great precision as a function of time:

t	0	0.1250	0.2500	0.3750	0.5000
i	0	6.2402	7.7880	4.8599	0.0000

Determine i at $t = 0.32$.

CHAPTER 28

EQUATION SOLVING: ROOTS OF AN EQUATION

Years ago you learned to use the *quadratic formula*

$$x = \frac{-b \pm \sqrt{b^2 - 4ac}}{2a} \tag{28.1}$$

to solve

$$f(x) = ax^2 + bx + c \tag{28.2}$$

The values calculated with Eq. (28.1) are called the "roots" of Eq. (28.2). They represent the values of x that make Eq. (28.2) equal to zero. Thus we can define the root of an equation as the value of x that makes $f(x) = 0$. For this reason, roots are sometimes called the *zeros* of the equation.

Although the quadratic formula is handy for solving Eq. (28.2), there are many other functions for which the root cannot be determined so easily. For example, the following equation can be used to predict the vertical velocity of an object during free-fall:

$$v = \frac{gm}{c} (1 - e^{-(c/m)t}) \tag{28.3}$$

where velocity v is the dependent variable; time t is the independent variable, and the gravitational constant g, the drag coefficient c, and mass m are parameters. If the parameters are known, Eq. (28.3) can be used to predict the falling object's velocity as a function of time. Such computations can be performed because *v* is expressed *explicitly* as a function of time. That is, it is isolated on one side of the equal sign.

However, suppose that we had to determine the drag coefficient for a parachutist of a given mass to attain a prescribed velocity in a set time period. Although Eq. (28.3) provides a mathematical representation of the interrelationship among the model variables and parameters, it cannot be solved explicitly for the drag coefficient. Try it; there is no way to rearrange the equation so that *c* is isolated on one side of the equal sign. In such cases, *c* is said to be *implicit*.

This represents a real dilemma because many engineering design problems involve determining the properties or composition of a system (as represented by its parameters) in order to ensure that it performs in a desired manner (as represented by its variables). Thus these problems often require the determination of implicit parameters.

The solution to the dilemma is provided by numerical methods for roots of equations. To solve the problem with numerical methods, it is customary to express Eq. (28.3) in an alternative form. This is done by subtracting the dependent variable *v* from both sides of the equation in order to express the equation in a form similar to Eq. (28.2):

$$f(c) = \frac{gm}{c} (1 - e^{-(c/m)t}) - v \qquad\qquad (28.4)$$

The value of *c* that makes $f(c) = 0$ is, therefore, the root of the equation. This value also represents the drag coefficient that solves the design problem.

Before the advent of digital computers, such equations were usually solved by trial and error. This "technique" consists of guessing a value of *c* and evaluating whether $f(c)$ is zero. If not (as is almost always the case), another guess is made and $f(c)$ is again evaluated to determine whether the new value provides a better estimate of the root. The process is repeated until a guess is obtained that results in an $f(c)$ that is close to zero.

Such haphazard methods are obviously inefficient and inadequate for the requirements of engineering practice. A numerical method of the sort described in the present chapter represents an alternative that also starts with guesses but then employs a systematic strategy to home in on the true root. In addition, it is ideally suited for implementation on personal computers. As elaborated in the following pages, the combination of such systematic methods and computers makes the solution of most applied-roots-of-equations problems a simple and efficient task. However, before discussing this technique, we will briefly introduce a graphical method for depicting functions and their roots.

28.1   THE GRAPHICAL METHOD

A simple approach for obtaining an estimate of the root of an equation, $f(x) = 0$, is to make a plot of the function and observe where it crosses the *x* axis. This point, which represents the x value for which $f(x) = 0$, provides a rough approximation of the root.

EXAMPLE 28.1   The Graphical Approach

PROBLEM STATEMENT: Use the graphical approach to determine the drag coefficient c needed for a parachutist of mass $m = 68.1$ kg to have a velocity of 40 m/s after free-falling for time $t = 10$ s. *Note:* The acceleration due to gravity is 9.8 m/s².

SOLUTION: This problem can be solved by determining the root of Eq. (28.4) using the parameters $t = 10$, $g = 9.8$, $v = 40$, and $m = 68.1$:

$$f(c) = \frac{9.8(68.1)}{c}(1 - e^{-(c/68.1)10}) - 40$$

or

$$f(c) = \frac{667.38}{c}(1 - e^{-0.146843c}) - 40 \qquad\qquad (E28.1.1)$$

Various values of c can be substituted into this equation to compute

c	$f(c)$
4	34.115
8	17.653
12	6.067
16	-2.269
20	-8.401

These points are plotted in Fig. 28.1. The resulting curve crosses the c axis between 12 and 16. Visual inspection of the plot provides a rough estimate of the root of 14.75. The validity of the graphical estimate can be checked by substituting it into Eq. (E28.1.1) to yield

$$f(14.75) = \frac{667.38}{14.75}(1 - e^{-0.146843(14.75)}) - 40$$

$$= 0.059$$

which is close to zero. It can also be checked by substituting it into Eq. (28.3) along with the parameter values from this example to give

$$v = \frac{9.8(68.1)}{14.75}(1 - e^{-(14.75/68.1)10}) = 40.059$$

which is very close to the desired fall velocity of 40 m/s.

FIGURE 28.1
The graphical approach for determining the roots of an equation.

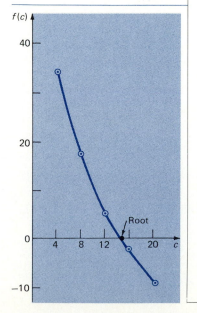

28.2   THE BISECTION METHOD

When applying the graphical method in Example 28.1, you have observed (Fig. 28.1) that the function changed sign on opposite sides of the root. In general, if a function $f(x)$ is real and continuous in an interval from x_l to x_u and $f(x_l)$ and $f(x_u)$ have opposite signs—that is, if

$$f(x_l)f(x_u) < 0$$

then there is at least one real root between x_l (the lower bound of the interval) and x_u (the upper bound).

Bisection, which is alternatively called *binary chopping, interval halving*, or *Bolzano's method*, capitalizes on this observation by locating an interval where the function changes sign. Then the location of the sign change (and consequently, the root) is identified more precisely by dividing the interval in half—that is by "bisecting" it into two equal subintervals. Each of these subintervals is searched to locate the sign change. The process is repeated to obtain refined estimates. An algorithm for bisection is listed in Fig. 28.2, and a graphical depiction is provided in Fig. 28.3. The following example goes through the actual computations involved in the method.

EXAMPLE 28.2   Bisection

PROBLEM STATEMENT: Use bisection to solve the same problem approached graphically in Example 28.1.

SOLUTION: The first step in bisection is to guess two values of the unknown (in the present problem, c) that give values for $f(c)$ with different signs. From Fig. 28.1,

---

FIGURE 28.2
An algorithm for bisection. This procedure is continued until the root estimate is accurate enough to meet your requirements.

Step 1:   Choose lower x_l and upper x_u, guesses for the root, so that the function changes sign over the interval. This can be checked by ensuring that $f(x_l)f(x_u) < 0$.

Step 2:   An estimate of the root x_r is determined by

$$x_r = \frac{x_l + x_u}{2}$$

Step 3:   Make the following evaluations to determine in which subinterval the root lies:

(a) If $f(x_l)f(x_r) < 0$, the root lies in the first subinterval. Therefore set $x_u = x_r$ and return to step 2.

(b) If $f(x_l)$ and $f(x_r) > 0$, the root lies in the second subinterval. Therefore set $x_l = x_r$ and return to step 2.

(c) If $f(x_l)f(x_r) = 0$, the root equals x_r; terminate the computation.

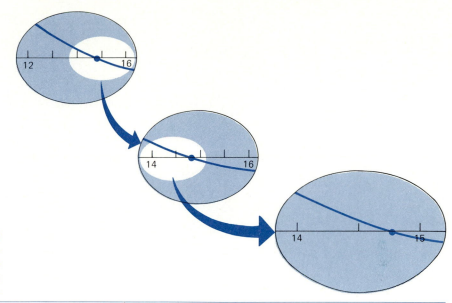

FIGURE 28.3
A graphical depiction of the bisection method. This plot conforms to the first three iterations from Example 28.2.

we can see that the function changes sign between values of 12 and 16. Therefore the initial estimate of the root lies at the midpoint of the interval:

$$x_r = \frac{12 + 16}{2} = 14$$

Next we compute the product of the function value at the lower bound and at the midpoint:

$$f(12)\,f(14) = 6.067(1.569) = 9.517$$

which is greater than zero, and hence no sign change occurs between the lower bound and the midpoint. Consequently, the root must be located between 14 and 16. Therefore we bisect the interval by redefining the lower bound as 14, and a revised root estimate is calculated as

$$x_r = \frac{14 + 16}{2} = 15$$

The process can be repeated to obtain refined estimates. For example,

$$f(14)\,f(15) = 1.569\,(-0.425) = -0.666$$

Therefore the root is between 14 and 15. The upper bound is redefined as 15, and the root estimate for the third iteration is calculated as

$$x_r = \frac{14 + 15}{2} = 14.5$$

The method can be repeated until the result is accurate enough to satisfy your needs.

We ended Example 28.2 with the statement that the method could be continued until the result was "accurate enough" to meet your needs. We must now develop an objective criterion for deciding when enough is enough—that is, when to terminate the method. An initial suggestion might be to end the calculation when the error in the root falls below some prescribed level. This strategy is flawed because, as described in Sec. 24.3, determination of the error depends on knowledge of the true value of the root. Such would not be the case in an actual situation because there would be no point in using the method if we already knew the root.

Therefore we require an error estimate that needs no foreknowledge of the root. Fortunately, there is a very neat way to accomplish this using bisection. Each time an approximate root is located using bisection as $x_r = (x_l + x_u)/2$, we know that the true root lies somewhere within an interval of $(x_u - x_l)/2 = \Delta x/2$. Therefore the root must lie within $\pm \Delta x/2$ of our estimate (see Fig. 28.4). For instance, when Example

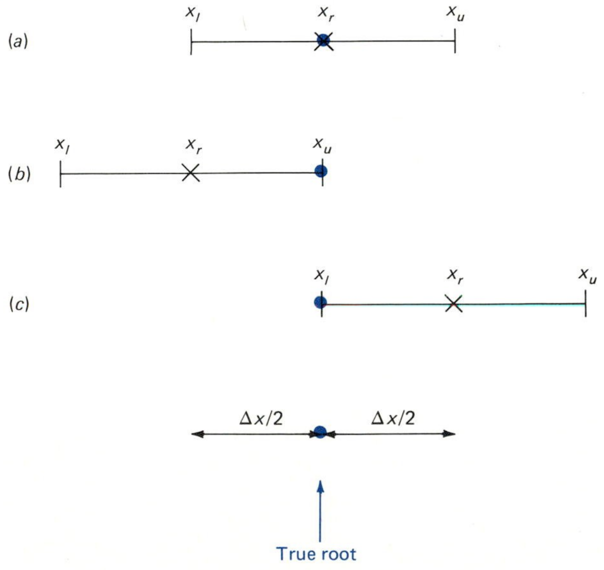

FIGURE 28.4
Three ways in which the interval may bracket the root: In (*a*) the true value lies at the center of the interval, whereas in (*b*) and (*c*) the true value lies near the extreme. Notice that the discrepancy between the true value and the midpoint of the interval never exceeds half the interval length $\Delta x/2$.

28.2 was terminated, we could make the definitive statement that

$$x_r = 14.5 \pm 0.5$$

For this bound to be exceeded, the true root would have to fall outside the interval, which, by definition, can never occur with the bisection method.

The bisection error estimate can be neatly incorporated into the computer algorithm (Fig. 28.2) as follows. Before starting the method, the user can specify an acceptable error level. Because it takes account of scale effects, the best approach is to specify a percent relative error [recall Eq. (24.4) and accompanying discussion]. This acceptable percent relative error can be designated as ϵ_s. Then after each iteration an approximate percent relative error ϵ_a can be calculated, as in

$$\epsilon_a = \left| \frac{x_u - x_l}{2x_r} \right| 100\% \tag{28.5}$$

which is the absolute value of the error interval $\Delta x/2$ multiplied by 100 percent and divided by the root estimate x_r. Note that the absolute value is used because we are not concerned with the sign of the error but only with its magnitude. When ϵ_a falls below ϵ_s, the computation can be terminated with assurance that the root is known to *at least* the level specified by ϵ_s. Because Eq. (28.5) represents a conservative estimate, the root is usually known better.

EXAMPLE 28.3   Error Estimates for Bisection

PROBLEM STATEMENT: Continue Example 28.2 until the approximate error ϵ_a falls below $\epsilon_s = 0.5$ percent. After each iteration, compute, along with the approximate error, a true error ϵ_t, in order to confirm that ϵ_t is always less than ϵ_a. In order to calculate ϵ_t, we will provide you with the true value of the root—14.7802, as determined with a computer program.

SOLUTION: For the first iteration, $x_l = 12$, $x_u = 16$, and $x_r = 14$. Therefore Eq. (28.5) can be used to compute

$$\epsilon_a = \left| \frac{16 - 12}{2(14)} \right| 100\% = 14.286\%$$

and the absolute value of the true error is

$$\epsilon_t = \left| \frac{14.7802 - 14}{14.7802} \right| 100\% = 5.279\%$$

Because ϵ_a is greater than $\epsilon_s = 0.5$ percent, the bisection method would be continued.

```
100 REM        *********************************************         C        ***********************************************
110 REM        *              BISECTION              *              C        *              BISECTION              *
120 REM        *          (BASIC version)            *              C        *          (FORTRAN version)          *
130 REM        *           by  S.C. Chapra           *              C        *           by  S.C. Chapra           *
140 REM        *                                     *              C        *                                     *
150 REM        *  This program uses bisection to find the  *        C        *  This program uses bisection to find the  *
160 REM        *  roots of an equation.              *              C        *  roots of an equation.              *
170 REM        *********************************************         C        ***********************************************
180 REM                                                             C
190 REM               .---------------------------------.          C
200 REM               |     DEFINITION OF VARIABLES     |          C               .---------------------------------.
210 REM               |                                 |          C               |     DEFINITION OF VARIABLES     |
220 REM               |   LOWER = LOWER GUESS           |          C               |                                 |
230 REM               |   UPPER = UPPER GUESS           |          C               |   LOWER = LOWER GUESS           |
240 REM               |   ES    = ACCEPTABLE ERROR (%)  |          C               |   UPPER = UPPER GUESS           |
250 REM               |   MAXIT = MAXIMUM ITERATIONS    |          C               |   ES    = ACCEPTABLE ERROR (%)  |
260 REM               |   XR    = ROOT ESTIMATE         |          C               |   MAXIT = MAXIMUM ITERATIONS    |
270 REM               |   EA    = ESTIMATED ERROR (%)   |          C               |   XR    = ROOT ESTIMATE         |
280 REM               |   NI    = NUMBER OF ITERATIONS  |          C               |   EA    = ESTIMATED ERROR (%)   |
290 REM               '---------------------------------'          C               |   NI    = NUMBER OF ITERATIONS  |
300 REM                                                             C               '---------------------------------'
310 REM                                                             C
320 REM           Function for which root is to be found.          C        REAL LOWER
330 REM                                                             C
340 DEF FNF(X) = 9.8*68.1/X * (1-EXP(-X/68.1*10)) - 40              C        ****************  MAIN PROGRAM  *****************
350 REM                                                             C
360 REM ***************** MAIN PROGRAM  *****************           C           Loop to Input Acceptable Guesses That Bracket
370 REM                                                             C
380 REM     Loop to Input Acceptable Guesses That Bracket          C        READ(5,*) LOWER, UPPER, ES, MAXIT
390 REM                                                                      IF ((FNF(LOWER) * FNF(UPPER)) .GE. 0) THEN
400 CLS                                                                         WRITE(6,*) 'GUESSES DO NOT BRACKET ROOT'
410 PRINT "LOW.GUESS, UP.GUESS, ERROR (%), MAX.ITER.";                         WRITE(6,*) 'PLEASE TRY NEW GUESSES'
420 INPUT LOWER,UPPER,ES,MAXIT                                              ELSE
430 WHILE FNF(LOWER)*FNF(UPPER) >= 0                           C               EA = 1.1 * ES
440    PRINT "GUESSES DO NOT BRACKET ROOT"                                     WHILE LOOP
450    INPUT LOWER,UPPER,ES,MAXIT                                 10         IF ((EA .GT. ES) .AND. (NI .LT. MAXIT)) THEN
460 WEND                                                                        XR = (LOWER + UPPER)/2
470 EA = ES *1.1                                                                NI = NI + 1
480 WHILE EA > ES AND NI < MAXIT                                                IF (XR .NE. 0) THEN
490    XR = (LOWER + UPPER)/2                                                      EA = ABS((UPPER - LOWER)/XR/2)*100
500    NI = NI + 1                                                              ENDIF
510    IF XR <> 0 THEN 520 ELSE 530                            C               CASE
520       EA = ABS((UPPER - LOWER)/XR/2)*100                                       AA = FNF(LOWER)*FNF(XR)
530    REM endif                                                                   IF (AA .LT. 0) THEN
540    REM CASE                                                                       UPPER = XR
550       AA = FNF(LOWER)*FNF(XR)                                               ELSE IF (AA .GT. 0) THEN
560       IF AA < 0 THEN 570 ELSE 590                                              LOWER = XR
570          UPPER = XR                                                         ELSE
580          GOTO 640                                                              EA = 0.
590       IF AA > 0 THEN 600 ELSE 620                                           ENDIF
600          LOWER = XR                                        C               END CASE
610          GOTO 640                                                          GO TO 10
620       REM ELSE CASE                                                     ENDIF
630          EA = 0                                            C
640    REM ENDCASE                                             C                       Print out Results
650 WEND                                                                       WRITE(6,*)  XR,EA,NI
660 REM                                                                     ENDIF
670 REM                     Print out Results                              STOP
680 PRINT XR,EA,NI                                                         END
690 END                                                       C        ------------------------------------------------
                                                              C                 Function for which root is to be found.
                                                              C
                                                                       REAL FUNCTION FNF(X)
                                                                       REAL X
                                                                       FNF = 9.8*68.1/X * (1-EXP(-X/68.1*10)) - 40
                                                                       RETURN
                                                                       END
```

FIGURE 28.5 BASIC and FORTRAN computer programs for bisection.

The results for all the iterations are

Iteration	x_l	x_u	x_r	ϵ_a, %	ϵ_t, %
1	12	16	14	14.286	5.279
2	14	16	15	6.667	1.487
3	14	15	14.5	3.448	1.896
4	14.5	15	14.75	1.695	0.204
5	14.75	15	14.875	0.840	0.641
6	14.75	14.875	14.8125	0.422	0.219

Thus after six iterations ϵ_a finally falls below $\epsilon_s = 0.5$ percent, and the computation can be terminated. As anticipated, ϵ_t is less than ϵ_a for every iteration. Therefore we can be confident that our final result of 14.8125 is within *at least* 0.5 percent of the true value. As indicated, the true error at this point is actually only 0.219 percent.

28.3 COMPUTER PROGRAMS FOR BISECTION

As seen in Fig. 28.5, it is a relatively simple matter to develop programs for bisection. Notice that the programs in Fig. 28.5 do not have user-friendly input or output. In Prob. 28.1 you will have the task of including such features in your version of the program.

Additionally, because the graphical capabilities of computers are so varied, we have not included a plotting option in Fig. 28.5. However, as mentioned at the beginning of this chapter, such an option is critical for making effective initial guesses for bisection. If your computer system has plotting capabilities, we recommend that you expand your program to include a plot of y versus x. The printer plots described at the end of Chap. 22 can be modified for this purpose. In any event, the inclusion of this capability will greatly enhance the program's utility for problem solving.

Figure 28.6 shows some screens generated using a commercial software package that has a bisection algorithm very similar to that shown in Fig. 28.5. The package is used to solve Examples 28.2 and 28.3. A plot of the function is shown in Fig. 28.6*a*, and the solution is shown in Fig. 28.6*b*. These screens illustrate how output can be designed in a clear and accessible fashion. You can employ them as models for the software you develop from Fig. 28.5.

28.4 PITFALLS AND ADVANCE METHODS

Pitfalls. Aside from providing rough estimates of the root, graphical interpretations are important tools for understanding the properties of functions and anticipating possible pitfalls of root-location methods such as bisection. For example, Fig. 28.7 shows a number of ways in which roots can occur in an interval prescribed by a lower

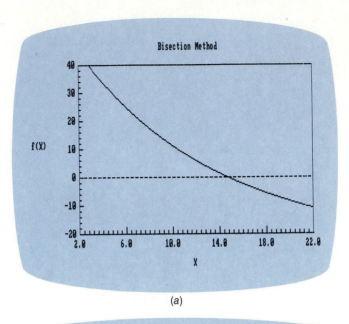

(a)

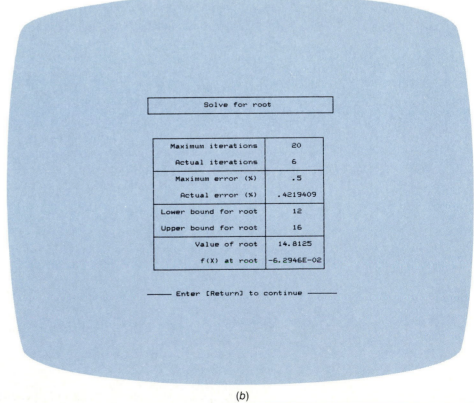

FIGURE 28.6
Computer screens generated by a bisection software package (NUMERICOMP by Canale and Chapra, 1985). The package is used to solve the problem previously analyzed in Examples 28.2 and 28.3. The screen in (a) shows a plot that can be used to make initial guesses. The screen in (b) shows the results of bisection.

(b)

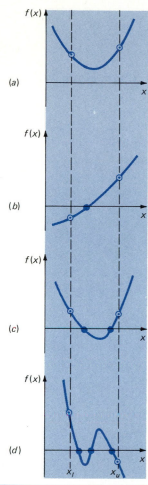

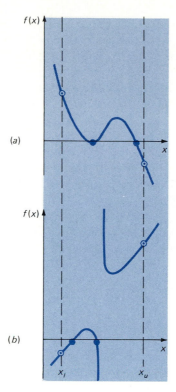

FIGURE 28.8
Illustration of some exceptions to the general cases depicted in Fig. 28.7. (a) Multiple root that occurs when the function is tangential to the x axis. For this case, although the end points are of opposite signs, there is an even number of real roots for the interval. (b) Discontinuous functions where end points of opposite signs also bracket an even number of real roots. Special strategies are required for determining the roots for these cases.

FIGURE 28.7
Illustration of a number of general ways that a real root may occur in an interval prescribed by a lower bound x_1 and an upper bound x_u. Parts (a) and (c) indicate that if both $f(x_1)$ and $f(x_u)$ have the same sign, either there will be no real roots or there will be an even number of real roots within the interval. Parts (b) and (d) indicate that if the function has different signs at the end points, there will be an odd number of real roots in the interval.

bound x_l and an upper bound x_u. Figure 28.7b depicts the case where a single root is bracketed by negative and positive values of $f(x)$. However, Fig. 28.7d, where $f(x_l)$ and $f(x_u)$ are also on opposite sides of the x axis, shows three roots occurring within the interval. In general, if $f(x_l)$ and $f(x_u)$ have opposite signs, there is an odd number of roots in the interval. As indicated by Fig. 28.7a and c, if $f(x_l)$ and $f(x_u)$ have the same sign, there are no roots or an even number of roots between the values.

Although these generalizations are usually true, there are cases where they do not hold. For example, *multiple roots*—that is, functions that are tangential to the x axis at the root (Fig. 28.8a)—and discontinuous functions (Fig. 28.8b) can violate these principles. An example of a function that has a multiple root is the cubic equation $f(x) = (x - 2)(x - 2)(x - 4)$. Notice that $x = 2$ makes two terms in this polynomial equal to zero. Hence $x = 2$ is called a *double root*. In addition to these cases, complex roots can occur.

The existence of functions of the type depicted in Fig. 28.8 makes it difficult to develop general computer algorithms guaranteed to locate all the roots in an interval. However, when used in conjunction with graphical approaches, numerical methods are extremely useful for solving many roots-of-equations problems confronted routinely by engineers and applied mathematicians.

Advanced methods. Beyond bisection, there are a variety of other techniques that are available to determine the roots of equations. These can be divided into bracketing and open methods (Fig. 28.9). *Bracketing methods* require two initial guesses that are located on either side of (that is, they "bracket") the root. Bisection is one bracketing method. Others, such as the *method of false position*, use more sophisticated and efficient strategies to systematically home in on the root. One strength of bracketing methods is that they always work.

Open methods also employ an iterative approach to locate the root. Some examples are the *Newton-Raphson* and the *secant methods*. The former requires only one initial guess, whereas the latter requires two. Open methods are usually much more efficient at locating roots than are bracketing methods. In addition, they can be used to locate multiple roots, whereas the bracketing methods, by definition, cannot. However, if the initial guess (or guesses) is too distant from the true root, the iterations sometimes *diverge*. That is, as the iterations proceed they actually move farther away from rather than homing in on the root. Detailed descriptions, along with computer algorithms for all these methods, can be found elsewhere (Chapra and Canale, 1985).

PROBLEMS

28.1 Utilize Fig. 28.5 as a starting point for developing your own program for bisection. Employ structured programming techniques, modular design, and internal and external documentation in your development. Include the capability to plot the function in order to facilitate making the initial guesses. Also, substitute the final result back into the function to verify whether the result is close to zero.

28.2 The number of iterations performed by the bisection method to reduce the error below a predefined value can be calculated before an analysis is undertaken based on the initial interval and the desired error of the final result. Derive a relationship to compute the number of iterations. *Hint*: The relation between the first, or initial, interval, $\Delta x_1 = x_u - x_l$, and the second interval, Δx_2, is $\Delta x_2 = \Delta x_1 / 2$. Additionally, the error after any iteration is $\Delta x/2$. (Recall Fig. 28.4). Remember to adjust your relationship so that it accounts for the fact that the number of iterations must be a whole number, or integer.

28.3 Why might multiple roots pose problems for a technique such as bisection?

28.4 If the bisection method is employed on an interval where the function changes sign, is it always going to locate a root? Why? Give examples to support your claim.

28.5 The upward velocity of a rocket can be computed by the following formula:

$$v = u \ln \frac{m_0}{m_0 - qt} - gt$$

where v is upward velocity, u is the velocity at which fuel is expelled relative to the rocket, m_0 is the initial mass of the rocket at time $t = 0$, q is the fuel consumption rate, and g is the downward acceleration of gravity (assumed constant = 9.8 m/s²). If u = 2200 m/s, m_0 = 160,000 kg, and q = 2680 kg/s, compute the time at which v = 1000 m/s. *Hint:* t is somewhere between 10 and 50 s. Determine your result so that it is within 1 percent of the true value. Check your answer.

28.6 The annual worth of an engineering project is characterized by the equation

$$A_t = \frac{-1900\,(1.18)^n}{1.18^n - 1} + \frac{45n}{1.18^n - 1} + 3000$$

where n is the number of years after the project begins operation. At what value of n will A_t = 0? That is, at what n will the project start to yield a profit? (*Hint:* It is less than 10 years.) Solve so that n is within 5 percent of the true value.

28.7 In environmental engineering the following equation can be used to compute the oxygen level in a river downstream from a sewage discharge:

$$c = 10 - 15(e^{-0.1x} - e^{-0.5x})$$

where x is distance downstream in miles. Determine the distance downstream where the oxygen level first falls to a reading of 4. (*Hint:* It is within 5 mi of the discharge.) Determine your answer to a 1 percent error.

28.8 Use bisection to determine the root of

$$f(x) = 1.9 - 0.56x - 0.8x^2 + 0.24x^3$$

Perform two iterations with initial guesses of x_l = 1 and x_u = 2. Compute the approximate error for each iteration.

28.9 Figure P28.9a shows a uniform beam subject to a linearly increasing distributed load. The equation for the resulting elastic curve is (see Fig. P28.9b)

$$y = \frac{w_0}{120EIL}(-x^5 + 2L^2x^3 - L^4x) \tag{P28.9}$$

If the derivative of the elastic curve is

$$\frac{dy}{dx} = \frac{w_0}{120EIL}(-5x^4 + 6L^2x^2 - L^4)$$

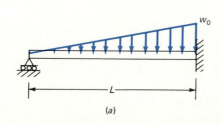

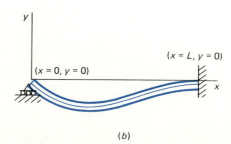

FIGURE P28.9   (a)   (b)

Use bisection to determine the point of maximum deflection (i.e., the value of x where $dy/dx = 0$). Then substitute this value into Eq. (P28.9) to determine the value of the maximum deflection. Use the following parameter values in your computation: $L = 180$ in, $E = 29 \times 10^6$ lb/in^2, $I = 723$ in^4, and $w_0 = 12$ kips/ft. Express your results in inches.

28.10 The height of a hanging cable is given by (recall Fig. 9.6)

$$y = \frac{H}{w} \cosh \left(\frac{w}{H} x \right) + y_0 - \frac{H}{w}$$

where H is the horizontal force at $x = a$, y_0 is the height at $x = a$, and w is the cable's weight per unit length. Determine the value of H, if $y = 12$ at $x = 50$ for $y_0 = 5$ and $w = 10$. (Recall the definition of the hyperbolic cosine from Sec 9.2.)

28.11 The secant formula defines the force per unit area, P/A, that causes a maximum stress σ_m in a column of given slenderness ratio, L_e/r:

$$\frac{P}{A} = \frac{\sigma_m}{1 + (ec/r^2) \sec [1/2(\sqrt{P/EA})(L_e/r)]}$$

If $E = 29 \times 10^3$ ksi, $ec/r^2 = 0.2$, and $\sigma_m = 36$ ksi, compute P/A for an $L_e/r = 100$. [Recall from Figs. 9.2 and 16.2 that sec $(x) = 1/\cos (x)$.]

28.12 The BASIC and FORTRAN intrinsic functions SQR and SQRT employ root-location methods in their routines. Because the roots must be located rapidly, more efficient methods than bisection are used. Nonetheless, it is instructive to formulate a subprogram that employs bisection to determine the square root of any number. Include appropriate checks to ensure accuracy and avoid any possible pitfalls.

28.13 Determine the roots of

$$f(x) = \frac{9 - 8x}{(1 - x)^2}$$

using bisection and initial guesses of $x_l = 0$ and $x_u = 3$. Perform the computation by hand to an error level of 1 percent. If any difficulties occur, explain them and state how they might be avoided in a computer program.

28.14 The van der Waals equation provides the relationship between the pressure and volume for a nonideal gas as

$$\left(p + \frac{a}{v^2} \right)(v - b) = RT$$

where p is pressure, a and b are parameters that relate to the type of gas, v is molal volume, R is the universal gas constant (0.082054 L · atm/mol · K), and T is absolute temperature. If $T = 350$ K and $p = 1.5$ atm, compute v for ethyl alcohol ($a = 12.02$ and $b = 0.08407$).

28.15 An oscillating current in an electric circuit is described by

$$I = 10e^{-t} \sin (2\pi t)$$

where t is in seconds. Determine all values of t such that $I = 2$.

28.16 The equation for a reflected standing wave in a harbor is given by

$$h = h_0 \left[\sin \left(\frac{2\pi x}{\lambda} \right) \cos \left(\frac{2\pi t v}{\lambda} \right) + e^{-x} \right]$$

Solve for x if $h = 0.5h_0$, $\lambda = 20$, $t = 10$, and $v = 50$.

CHAPTER 29

EQUATION SOLVING:
LINEAR ALGEBRAIC EQUATIONS

In Chap. 28 we determined the value x that satisfied a single equation, $f(x) = 0$. Now we deal with the case of determining the values x_1, x_2, . . . , x_n that simultaneously satisfy a set of equations:

$$f_1(x_1, x_2, \ldots, x_n) = 0$$
$$f_2(x_1, x_2, \ldots, x_n) = 0$$

.   .

.   .

.   .

$$f_n(x_1, x_2, \ldots, x_n) = 0$$

Such systems can be either linear or nonlinear. In the present chapter, we will deal exclusively with *linear algebraic equations* that are of the general form

$$a_{11}x_1 + a_{12}x_2 + \cdots + a_{1n}x_n = c_1$$
$$a_{21}x_1 + a_{22}x_2 + \cdots + a_{2n}x_n = c_2$$

.   .   .   .

.   .   .   .

.   .   .   .

$$a_{n1}x_1 + a_{n2}x_2 + \cdots + a_{nn}x_n = c_n$$

(29.1)

where the a's are constant coefficients, the c's are constants, and n is the number of equations. Note that the unknowns, x_1, x_2, . . . , x_n, are all raised to the first power and are multiplied by a constant. All other equations are nonlinear.

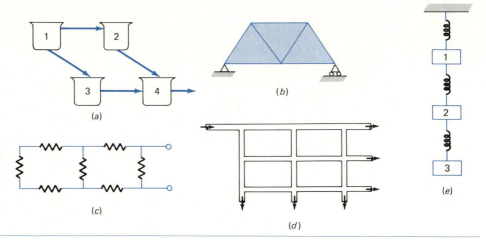

FIGURE 29.1
Some systems of interconnected elements that can be characterized as a system of equations: (*a*) reactors, (*b*) a structure, (*c*) an electric circuit, (*d*) a fluid pipe network, and (*e*) a series of blocks and springs.

Linear algebraic equations arise in a variety of problem contexts and in all fields of engineering. In particular, they originate in the mathematical modeling of large systems of interconnected elements such as structures, electric circuits, and fluid networks (see Fig. 29.1). In the previous chapter, you saw how single-component systems result in a single equation that can be solved using root-location techniques. Multicomponent systems of the type depicted in Fig. 29.1 result in a coupled set of mathematical equations that must be solved simultaneously. The equations are coupled because the individual parts of the system are influenced by other parts. For example, in Fig. 29.1*a*, reactor 4 receives chemical inputs from reactors 2 and 3. Consequently, its response is dependent on the quantity of chemical in these other reactors. When these dependencies are expressed mathematically, the resulting equations are often of the linear algebraic form of Eq. (29.1).

We will use a simple example to illustrate this in the present chapter. Suppose that a team of three parachutists is connected by a weightless cord while free-falling at a velocity of 5 m/s (Fig. 29.2). A simple problem might be to determine the tension in each cord and the acceleration of the team.

To study this problem, free-body diagrams can be developed specifying the various forces acting on each parachutist. As seen in Fig. 29.3, each parachutist is subjected to two major external forces, gravity and air resistance. If a downward force is assigned a positive sign, Newton's second law ($F = ma$) can be used to formulate the force due to gravity as

$$F_D = mg \tag{29.2}$$

where g is the gravitational constant, or the acceleration due to gravity, which is approximately equal to 9.8 m/s^2.

Air resistance can be formulated in a number of ways. A simple approach is to assume that it is linearly proportional to velocity, as in

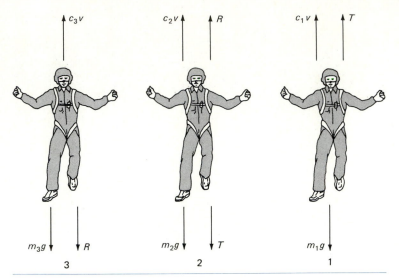

FIGURE 29.3 Free-body diagrams for each of the three falling parachutists.

$$F_U = -cv \qquad (29.3)$$

where c is a proportionality constant called the *drag coefficient* (in kilograms per second). Thus the greater the fall velocity, the greater the upward force due to air resistance.

In addition to these external forces, the cords connecting the parachutists also exert internal forces between the jumpers. The free-body diagrams in Fig. 29.3 specify all the forces for each parachutist. According to Newton's second law, the net force should equal mass times acceleration. Therefore for the first parachutist,

$$\underbrace{m_1g - T - c_1v}_{\text{Net force}} = m_1a$$

Similary, for the second and third parachutists,

$$m_2g + T - c_2v - R = m_2a$$

and

$$m_3g - c_3v + R = m_3a$$

These equations have three unknowns—a, T, and R. By bringing all the terms with

FIGURE 29.2
Three parachutists free-falling while connected by weightless cords.

unknowns to one side and all the knowns to the other, the following set of equations results:

$$m_1a + T \qquad = m_1g - c_1v \qquad (29.4a)$$

$$m_2a - T + R = m_2g - c_2v \qquad (29.4b)$$

$$m_3a \qquad - R = m_3g - c_3v \qquad (29.4c)$$

Suppose that the parachutists are free-falling at a velocity of 5 m/s; the following values are given for the masses and drag coefficients:

Parachutist	Mass, kg	Drag coefficient, kg/s
1	70	10
2	60	14
3	40	17

Substituting these values along with $g = 9.8$ m/s^2 into Eq. (29.4) yields

$$70a + T \qquad = 636 \qquad (29.5a)$$

$$60a - T + R = 518 \qquad (29.5b)$$

$$40a \qquad - R = 307 \qquad (29.5c)$$

Thus these are three linear algebraic equations with three unknowns that are of the exact form of Eq. (29.1). The method discussed in the present chapter, Gauss elimination, will be used to solve Eq. (29.5) for the unknowns.

It should be noted that a computer is really not required to solve such a small system of equations. Manual algebraic manipulations are quite satisfactory for solving two or three simultaneous equations. However, for four or more equations, manual solutions become arduous, and computers must be utilized. Because many engineering problems involving simultaneous equations can consist of hundreds and even thousands of equations, computers are an absolute necessity.

29.1 GAUSS ELIMINATION

In high school you probably learned to solve two simultaneous equations by eliminating unknowns. The basic strategy of this technique is to multiply one of the equations by a constant in order that one of the unknowns is eliminated when the equations are combined. This results in one equation with one unknown. This single equation can be solved for the unknown and the result substituted back into either of the original equations to determine the other unknown.

(a)
$$a_{11}\,x_1 \; + \; a_{12}\,x_2 \; + \; a_{13}\,x_3 \; = \; c_1$$
$$a_{21}\,x_1 \; + \; a_{22}\,x_2 \; + \; a_{23}\,x_3 \; = \; c_2$$
$$a_{31}\,x_1 \; + \; a_{32}\,x_2 \; + \; a_{33}\,x_3 \; = \; c_3$$

(b)
$$a_{11}\,x_1 \; + \; a_{12}\,x_2 \; + \; a_{13}\,x_3 \; = \; c_1$$
$$a_{22}'\,x_2 \; + \; a_{23}'\,x_3 \; = \; c_2'$$
$$a_{33}''\,x_3 \; = \; c_3''$$

FIGURE 29.4
The two phases of Gauss elimination: (b) forward-elimination and (c) back-substitution. The primes indicate the number of times that the coefficients and constants have been modified.

(c)
$$x_3 \; = \; c_3'' \, / \, a_{33}''$$
$$x_2 \; = \; (c_2' \; - \; a_{23}'\,x_3) \, / \, a_{22}'$$
$$x_1 \; = \; (c_1 \; - \; a_{12}\,x_2 \; - \; a_{13}\,x_3) \, / \, a_{11}$$

The basic approach to eliminating unknowns can be extended to large sets of equations by developing a systematic scheme to eliminate unknowns and then back-substitute. *Gauss elimination* is one of the simplest and most common of these schemes. As depicted in Fig. 29.4, the method consists of two phases. First, a *forward-elimination* is used to eliminate unknowns until the equations are in the form of Fig. 29.4b. Notice that the last equation has one unknown, the next to the last has two unknowns, and so on. Then a *back-substitution phase* is used to solve for the unknowns. The method is best illustrated by an example.

EXAMPLE 29.1   Gauss Elimination

PROBLEM STATEMENT: Solve Eq. (29.5) by Gauss elimination.

SOLUTION: The equations to be solved are

$$70a \; + \; T \qquad\quad = 636 \tag{E29.1.1a}$$

$$60a \; - \; T \; + \; R \; = 518 \tag{E29.1.1b}$$

$$40a \qquad\quad - \; R \; = 307 \tag{E29.1.1c}$$

First we will use forward-elimination to transform the equations into the form of Fig. 29.4*b*. The initial step will be to eliminate the first unknown, *a*, from the second and the third equations. To do this, multiply Eq. (E29.1.1*a*) by 60/70 to give

$$60a + 0.85714T = 545.14$$

Now this equation can be subtracted from Eq. (E29.1.1*b*) to eliminate the first term from the second equation. The result is

$$-1.8571T + R = -27.14$$

Similarly, the first equation can be multiplied by 40/70 and the result subtracted from the third equation to give

$$-0.57143T - R = -56.429$$

Therefore, the three equations are now

$$70a + \qquad T \qquad = 636 \qquad\qquad (E29.1.2a)$$
$$-1.8571 \;\; T + R = -27.14 \qquad\qquad (E29.1.2b)$$
$$-0.57143T - R = -56.429 \qquad\qquad (E29.1.2c)$$

Thus the first unknown is eliminated from the second and the third equations. During these manipulations, the first equation is referred to as the *pivot equation*, and the coefficient of the first unknown in the first equation (70) is called the *pivot element*.

The next step is to eliminate the second unknown from the third equation. For this step, the pivot equation is Eq. (E29.1.2*b*), and the pivot element is -1.8571. Equation (E29.1.2*b*) can be multiplied by $-0.57143/-1.8571$ and the result subtracted from Eq. (E29.1.2*c*) to give

$$-1.3077R = -48.078$$

With this manipulation, the forward-elimination phase is completed, and the three equations have been transformed to the desired form of Fig. 29.4*b*.

$$70a + \qquad T \qquad\qquad = 636 \qquad\qquad (E29.1.3a)$$
$$-1.8571T + \qquad R = -27.14 \qquad\qquad (E29.1.3b)$$
$$-1.3077\,R = -48.078 \qquad\qquad (E29.1.3c)$$

Now, back-substitution can be used to solve the third equation for

$$R = \frac{-48.078}{-1.3077} = 36.765$$

This result can be back-substituted into Eq. (E29.1.3b) to solve for

$$T = \frac{-27.14 - 36.765}{-1.8571} = 34.41$$

This result can be back-substituted into Eq. (E29.1.3a) to solve for

$$a = \frac{636 - 34.41}{70} = 8.59$$

Therefore, the final result is

$$a = 8.59 \text{ m/s}^2$$

$$T = 34.41 \text{ N}$$

$$R = 36.765 \text{ N}$$

We can check to see if this result is correct by substituting these values back into the original equations to see if they balance:

$$70(8.59) + 34.41 \qquad\qquad = 636$$

$$60(8.59) - 34.41 + 36.765 = 518$$

$$40(8.59) \qquad\quad - 36.765 = 307$$

Comparing the left- and right-hand sides shows that the equations balance:

$$635.71 \cong 636$$

$$517.755 \cong 518$$

$$306.835 \cong 307$$

The minor discrepancies are due to round-off error.

29.1.1 Pivoting

The technique described above is commonly referred to as *naive Gauss elimination*. It is called "naive" because during both the forward-elimination and the back-substitution phases, it is possible that a division by zero could occur. For example, if we used naive Gauss elimination to solve

$$2x_2 + 3x_3 = 8 \tag{29.6a}$$

$$2x_1 + 6x_2 + 7x_3 = -3 \tag{29.6b}$$

$$4x_1 + x_2 + 6x_3 = 5 \qquad\qquad (29.6c)$$

the initial operation consists of multiplying the first row by 2/0. Thus a division by zero occurs because the pivot element (a_{11}) is equal to zero. Problems may also arise when the pivot element is close to rather than exactly equal to zero, since, if the magnitude of the pivot element is very small compared with the other elements, round-off errors can be introduced.

Therefore before each column of coefficients is eliminated, it is advantageous to determine which of the coefficients in the column is largest. If this coefficient is larger than the pivot coefficient, the rows can be switched so that the largest element becomes the pivot element. For Eq. (29.6), the third equation has the largest coefficient (4) in the column that is to be eliminated. Therefore, Eq. (29.6a) and Eq. (29.6c) are switched to give

$$4x_1 + x_2 + 6x_3 = 5$$

$$2x_1 + 6x_2 + 7x_3 = -3$$

$$2x_2 + 3x_3 = 8$$

If this technique, which is called *partial pivoting*, is implemented prior to each elimination step, division by zero and other problems connected with very small pivot elements can be avoided. For this reason, as described in the next section, computer programs for Gauss elimination must include partial pivoting.

29.2 COMPUTER PROGRAMS FOR GAUSS ELIMINATION

As shown in Fig. 29.5, it is a relatively simple matter to develop programs for Gauss elimination. There are several features about these programs that are noteworthy. First, the coefficients are stored in a two-dimensional array, and the right-hand-side constants and the unknowns are stored as one-dimensional arrays. This allows efficient manipulation and storage of the constants and the results. Second, notice the extensive use of loops to perform the methodical steps of the algorithm. Third, a subroutine to perform partial pivoting is included. As mentioned in the previous section, this prevents division by zero from occurring in the event that a diagonal element is zero. Finally, a modular approach is taken. Thus each major step in the solution is represented as an autonomous subroutine.

The programs in Fig. 29.5 do not employ user-friendly input or output. In Prob. 29.1 you will have the task of including such features in your version of the program. Figure 29.6 shows some screens generated using a commercial software package that employs a Gauss elimination algorithm very similar to that shown in Fig. 29.5. The package is used to solve Example 29.1. The results are shown in Fig. 29.6a, and an error check is shown in Fig. 29.6b. The latter is important for detecting whether round-off error has occurred. These screens illustrate how output can be designed in a clear and accessible fashion. You can employ them as models for your own software.

```
100 REM    ********************************************
110 REM    *       GAUSS ELIMINATION WITH PIVOTING    *
120 REM    *             (BASIC version)              *
130 REM    *             by  S.C. Chapra              *
140 REM    *                                          *
150 REM    *     This program uses GAUSS elimination to *
160 REM    *     solve simultaneous equations.        *
170 REM    ********************************************
180 REM
190 REM    .------------------------------------------.
200 REM    |         DEFINITION VARIABLES             |
210 REM    |                                          |
220 REM    |  N  = NUMBER OF SIMULTANEOUS EQUATIONS   |
230 REM    |  A() = MATRIX OF COEFFICIENTS            |
240 REM    |  C() = VECTOR OF RIGHT-HAND-SIDE CONSTANTS |
250 REM    |  X() = UNKNOWNS                          |
260 REM    |                                          |
270 REM    `------------------------------------------'
280 REM DIM A(15,15), C(15), X(15)
290 REM
300 REM ******************* MAIN PROGRAM *******************
310 REM
320 GOSUB 1000
330 GOSUB 2000
340 GOSUB 3000
350 GOSUB 4000
360 END
1000 REM ************ SUBROUTINE TO ENTER DATA ***********
1010 REM
1020 CLS
1030 INPUT "NUMBER OF ROWS? ", N
1040 PRINT
1050 FOR I = 1 TO N
1060   FOR J = 1 TO N
1070     PRINT "ELEMENT(";I;",";J;") = ";: INPUT A(I,J)
1080   NEXT J
1090   PRINT "VECTOR(";I;") = ";: INPUT C(I)
1100   PRINT
1110 NEXT I
1120 RETURN
2000 REM ************ SUBROUTINE ELIMINATE ************
2010 REM
2020 REM      ALGORITHM FOR FORWARD ELIMINATION
2030 REM
2040 REM      DATA REQUIRED:   N, A()
2050 REM      RESULTS RETURNED: C(), A()
2060 REM
2070 REM      LOCAL VARIABLES: ELL, ELX
2080 REM
2090 REM ---------------- BODY OF ELIMINATE --------------
2100 ELL = N - 1
2110 FOR I1 = 1 TO ELL
2120   ELX = I1 + 1
2130   REM
2140   REM GOSUB Partial Pivoting
2150   GOSUB 5000
2160   FOR I2 = ELX TO N
2170     FOR J = ELX TO N
2180       A(I2,J) = A(I2,J) - A(I2,I1)/A(I1,I1)*A(I1,J)
2190     NEXT J
2200     C(I2) = C(I2) - A(I2,I1)/A(I1,I1)*C(I1)
2210   NEXT I2
2220 NEXT I1
2230 RETURN
2240 REM
3000 REM ************ SUBROUTINE SUBSTITUTE **************
3010 REM
3020 REM      PERFORMS BACK SUBSTITUTION
3030 REM
3040 REM      REQUIRED DATA:   C(),A(),N
3050 REM      RESULTS RETURNED: X()
3060 REM
3070 REM      LOCAL VARIABLES: SUL, SUI, SUJ, SUSUM
3080 REM
3090 REM ------------- BODY OF SUBSTITUTE ---------------
3100 SUL = N - 1
3110 X(N) = C(N)/A(N,N)
3120 FOR IX = 1 TO SUL
3130   SUSUM = 0
3140   SUI = N - IX
3150   SUJ = SUI + 1
3160   FOR JX = SUJ TO N
3170     SUSUM = SUSUM + A(SUI,JX) * X(JX)
3180   NEXT JX
3190   X(SUI) = (C(SUI) - SUSUM)/A(SUI,SUI)
3200 NEXT IX
3210 RETURN
```

```
4000 REM ******** SUBROUTINE TO PRINT OUT RESULTS *******
4010 REM
4020 PRINT:PRINT "SOLUTIONS TO THE EQUATIONS"
4030 FOR I = 1 TO N
4040   PRINT "X(" I ") ="; X(I)
4050 NEXT I
4060 RETURN
5000 REM *************** SUBROUTINE PIVOT ***************
5010 REM
5020 REM      PERFORMS PARTIAL PIVOTING
5030 REM
5040 REM      REQUIRED DATA:   A(), N
5050 REM      RESULTS RETURNED: A()
5060 REM
5070 REM      LOCAL VARIABLES: PIJ, PIBIG, PIDUM, PITEMP
5080 REM
5090 REM --------------- BODY OF PIVOT ----------------
5100 PIJ = I1
5110 PIBIG = ABS(A(I1,I1))
5120 FOR I = ELX TO N
5130   AM = ABS(A(I,I1))
5140   IF AM > PIBIG THEN 5160 ELSE 5172
5160     PIBIG = AM
5170     PIJ = I
5172   REM ENDIF
5180 NEXT I
5190 IF PIJ <> I1 THEN 5210 ELSE 5282
5210   FOR J = I1 TO N
5220     PIDUM = A(PIJ,J)
5230     A(PIJ,J) = A(I1,J)
5240     A(I1,J) = PIDUM
5250   NEXT J
5260   PITEMP = C(PIJ)
5270   C(PIJ) = C(I1)
5280   C(I1) = PITEMP
5282 REM ENDIF
5290 RETURN
```

```
C      ********************************************
C      *       GAUSS ELIMINATION WITH PIVOTING    *
C      *             (FORTRAN version)            *
C      *             by  S.C. Chapra              *
C      *                                          *
C      *     This program uses GAUSS elimination to *
C      *     solve simultaneous equations.        *
C      ********************************************
C
C      .------------------------------------------.
C      |         DEFINITION VARIABLES             |
C      |                                          |
C      |  N  = NUMBER OF SIMULTANEOUS EQUATIONS   |
C      |  A() = MATRIX OF COEFFICIENTS            |
C      |  C() = VECTOR OF RIGHT-HAND-SIDE CONSTANTS |
C      |  X() = UNKNOWNS                          |
C      `------------------------------------------'
C
       REAL A(15,15), C(15), X(15)
       INTEGER N
C
C      ***************** MAIN PROGRAM *****************
C
       CALL ENTER(N,A,C)
       CALL ELIM(N,A,C)
       CALL SUBST(N,A,C,X)
       CALL PRINT(N,X)
       STOP
       END
C      *********** SUBROUTINE TO ENTER DATA ***********
C
       SUBROUTINE ENTER(N,A,C)
       REAL A(15,15), C(15)
       INTEGER N
C
       READ(5,*) N
       DO 10 I=1, N
         READ(5,*) ((A(I,J),J=1,N), C(I))
10     CONTINUE
       RETURN
       END
C      *********** SUBROUTINE ELIMINATE ************
C
       SUBROUTINE ELIM(N,A,C)
C
C           ALGORITHM FOR FORWARD ELIMINATION
```

FIGURE 29.5 BASIC and FORTRAN programs for Gauss elimination with partial pivoting.

```fortran
C        DATA REQUIRED:    N, A(), C()
C        RESULTS RETURNED: C(), A()
C
C        LOCAL VARIABLES: L, IX, I1, I2, J
C
      REAL A(15,15), C(15)
      INTEGER L,N,X
C     --------------- BODY OF ELIMINATE ---------------
      L = N - 1
      DO 30 I1=1, L
         IX = I1 + 1
C
C        Use Partial Pivoting
         CALL PIVOT(N,A,C,IX,I1)
         DO 20 I2=IX, N
            DO 10 J=IX, N
               A(I2,J) = A(I2,J) - A(I2,I1)/A(I1,I1)*A(I1,J)
10          CONTINUE
            C(I2) = C(I2) - A(I2,I1)/A(I1,I1)*C(I1)
20       CONTINUE
30    CONTINUE
      RETURN
      END
C
C     ************* SUBROUTINE SUBSTITUTE **************
C
      SUBROUTINE SUBST(N,A,C,X)
C
C        PERFORMS BACK SUBSTITUTION
C
C        REQUIRED DATA:  C(),A(),N
C        RESULTS RETURNED: X()
C
C        LOCAL VARIABLES: L, I, J, SUM, IX, JX
C
      REAL A(15,15), C(15), X(15), SUM
      INTEGER L, N, IX, JX
C     --------------- BODY OF SUBSTITUTE ---------------
      L = N - 1
      X(N) = C(N)/A(N,N)
      DO 10 IX=1, L
         SUM = 0
         I = N - IX
         J = I + 1
         DO 20 JX=J, N
            SUM = SUM + A(I,JX) * X(JX)
20       CONTINUE
         X(I) = (C(I) - SUM)/A(I,I)
10    CONTINUE
      RETURN
      END
C     ******** SUBROUTINE TO PRINT OUT RESULTS *******
C
      SUBROUTINE PRINT(N,X)
      REAL X(15)
      INTEGER N
      WRITE(6,*) 'SOLUTIONS TO THE EQUATIONS'
      DO 10 I=1, N
         WRITE(6,*) 'X(',I,') =', X(I)
```

```fortran
10    CONTINUE
      RETURN
      END
C     *************** SUBROUTINE PIVOT ***************
C
      SUBROUTINE PIVOT(N,A,C,IX,I1)
C
C        PERFORMS PARTIAL PIVOTING
C
C        REQUIRED DATA:    N, A(), C(), IX, I1
C        RESULTS RETURNED: A(), C()
C
C        LOCAL VARIABLES: JJ, I, J, BIG, DUM, TEMP, AM
C
      REAL A(15,15), C(15), BIG, TEMP, DUM, AM
      INTEGER J,N
C     --------------- BODY OF PIVOT ---------------
      JJ = I1
      BIG = ABS(A(I1,I1))
      DO 10 I=IX, N
         AM = ABS(A(I,I1))
         IF (AM .GT. BIG) THEN
            BIG = AM
            JJ = I
         ENDIF
10    CONTINUE
      IF (JJ .NE. I1) THEN
         DO 20 J=I1, N
            DUM = A(JJ,J)
            A(JJ,J) = A(I1,J)
            A(I1,J) = DUM
20       CONTINUE
         TEMP = C(JJ)
         C(JJ) = C(I1)
         C(I1) = TEMP
      ENDIF
      RETURN
      END
```

29.3 PITFALLS AND ADVANCED METHODS

Pitfalls.   The adequacy of the solution of simultaneous equations depends on the condition of the system. Ill-conditioned systems are those where small changes in the coefficients can result in large changes in the solution. An alternative interpretation is that a wide range of answers can approximately satisfy the equations. Because round-off errors can induce small changes in the coefficients, these artificial changes can lead to large solution errors for ill-conditioned systems. One simple remedy is to use double-precision variables in your computer program. For a more detailed discussion of the problem see Chapra and Canale (1985). Fortunately, most linear algebraic equations derived from engineering-problem settings are naturally well-conditioned. Therefore, although you should be aware of the problem, it does not occur frequently in practice.

Another potential pitfall is the presence of round-off error for very large systems of equations. This is due to the fact that for a technique such as Gauss elimination,

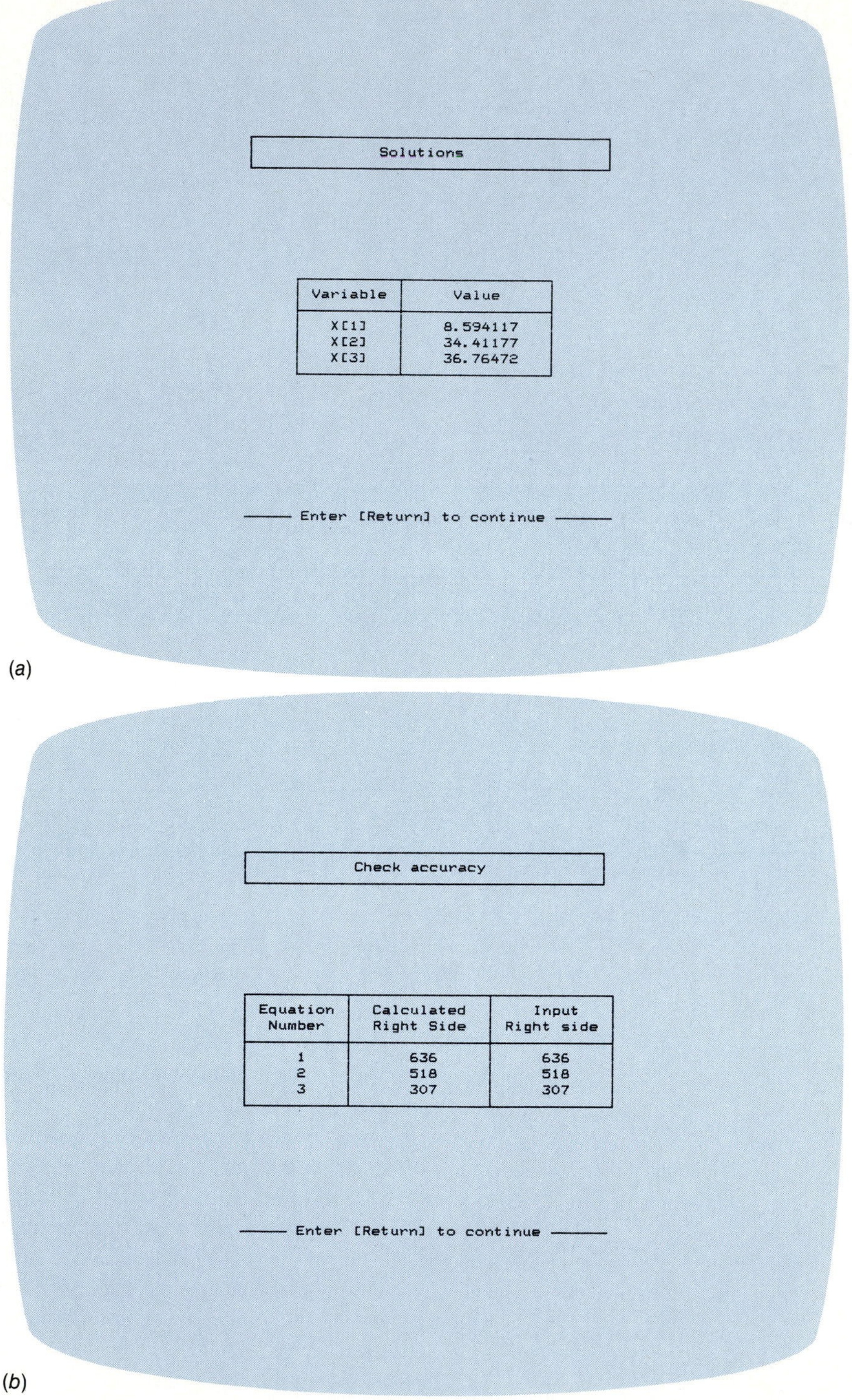

FIGURE 29.6
Computer screens generated by a Gauss elimination software package (NUMERICOMP by Canale and Chapra, 1985). The package is used to solve the problem previously analyzed in Example 29.1.

every result is dependent on previous results. Consequently, an error in the early steps will tend to propagate. In general, a rough rule of thumb is that round-off may be important when dealing with greater than 25 to 50 equations. For these cases, Gauss elimination sometimes performs poorly and, as discussed in the next section, other methods may be needed.

Advanced methods. A variety of methods is available for solving simultaneous equations. One approach that is particularly well suited for large systems is the *Gauss-Seidel method*. This technique is fundamentally different from Gauss elimination in that it is an approximate iterative method. That is, it employs initial guesses and then iterates to obtain refined estimates of the solution. The Gauss-Seidel method is particularly well-suited for large numbers of equations. In these cases, elimination methods can be subject to round-off errors. Because the error of the Gauss-Seidel method is controlled by the number of iterations, round-off error is not an issue of concern. However, there are certain instances where the Gauss-Seidel technique will not converge on the correct answer. These and other trade-offs between elimination and iterative methods are discussed elsewhere (Chapra and Canale, 1985; Ralston and Rabinowitz, 1978).

PROBLEMS

29.1 Utilize Fig. 29.5 as a starting point for developing your own program for Gauss elimination. Employ structured programming techniques, modular design, and internal and external documentation in your development. Include the capability of substituting the answers back into the original equations to substantiate that the results make the equations balance.

29.2 Three parachutists are connected by a weightless cord while free-falling. The team is accelerating at a rate of 5 m/s^2. Calculate the tension in each section of cord and the velocity, given the following:

Parachutist	Mass, kg	Drag coefficient, kg/s
1 (top)	50	12
2 (middle)	30	10
3 (bottom)	70	14

29.3 Three parachutists are connected by a weightless cord while free-falling. The upward force due to drag is represented by

$$F_U = -cv^2$$

where c is a drag coefficient (in kilograms per meter). If the team has a velocity of 10 m/s, calculate the tension in each rope and the acceleration, given the following:

Parachutist	Mass, kg	Drag coefficient, kg/m
1 (top)	50	0.6
2 (middle)	30	0.5
3 (bottom)	70	0.7

29.4 Solve by Gauss elimination:

$$x_1 + 2x_2 + 3x_3 = -3$$

$$x_1 - x_2 - 2x_3 = 1$$

$$4x_1 - 2x_2 + 5x_3 = -18$$

29.5 Idealized spring-mass systems have numerous applications throughout engineering. Figure P29.5 shows an arrangement of four springs in series being depressed with a force of 2000 lb. At equilibrium, force-balance equations can be developed defining the interrelationships between the springs:

$$k_2(x_2 - x_1) = k_1 x_1$$

$$k_3(x_3 - x_2) = k_2(x_2 - x_1)$$

$$k_4(x_4 - x_3) = k_3(x_3 - x_2)$$

$$F = k_4(x_4 - x_3)$$

where the k's are spring constants. If k_1 through k_4 are 100, 50, 75, and 200 lb/in, respectively, compute the x's.

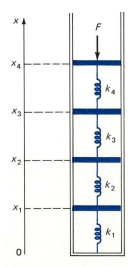

FIGURE P29.5

29.6 For Prob. 29.5, solve for the force F if x_3 is given as 60 in. All other parameters are as given in Prob. 29.5.

29.7 Three blocks are connected by a weightless cord and rest on an inclined plane (Fig. P29.7a). Employing a procedure similar to the one used in the analysis of the falling parachutists in Example 29.1 yields the following set of simultaneous equations (free-body diagrams are shown in Fig. P29.7b):

$$100a + T \qquad = 519.72$$

$$50a - T + R = 216.55$$

$$20a \qquad - R = \quad 86.62$$

Solve for acceleration a and the tensions T and R in the two ropes.

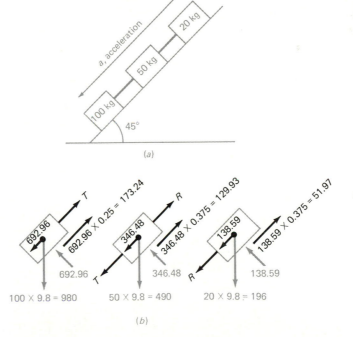

(a)

(b)

FIGURE P29.7

29.8 Perform a computation similar to that called for in Prob. 29.7 but for the system shown in Fig. P29.8.

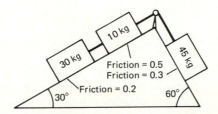

FIGURE P29.8

29.9 Given two equations, a graphical solution is possible:

$$3x_1 + 2x_2 = 18 \qquad -x_1 + 2x_2 = 2$$

Plot each of these equations on cartesian coordinates, with the ordinate as x_2 and the abscissa as x_1. What is the solution? Compare the graphical result to the analytical solution of the equations obtained by eliminating unknowns.

29.10 Obtain graphical solutions for the following equations (see Prob. 29.10):

(a) $\quad -\dfrac{1}{2}x_1 \ + x_2 \ = 1$

$\quad\quad -\dfrac{1}{2}x_1 \ + x_2 \ = \dfrac{1}{2}$

(b) $\quad -\dfrac{1}{2}x_1 \ + x_2 \ = 1$

$\quad\quad - x_1 \ + 2x_2 = 2$

(c) $\quad -\dfrac{2.3}{5}x_1 + x_2 \ = 1.1$

$\quad\quad -\dfrac{1}{2}x_1 \ + x_2 \ = 1$

Interpret your results for each case and speculate how these situations might pose problems for a computer algorithm. Such cases are said to be *ill-conditional*.

29.11 Although the Lagrange polynomial described in Chap. 27 provides a convenient means for determining intermediate values between points, it does not provide a convenient polynomial of the conventional form

$$f(x) = a_0 + a_1x + a_2x^2 + \cdots + a_nx^n$$

An interpolating polynomial of this form can be determined using simultaneous equations. For example, suppose that you desired to compute the coefficients of the parabola

$$f_2(x) = a_0 + a_1x + a_2x^2 \tag{P29.11}$$

Three points are required: $[x_0, f(x_0)]$; $[x_1, f(x_1)]$; and $[x_2, f(x_2)]$. Each can be substituted into Eq. (P29.11) to give

$$f(x_0) = a_0 + a_1x_0 + a_2x_0^2$$

$$f(x_1) = a_0 + a_1x_1 + a_2x_1^2$$

$$f(x_2) = a_0 + a_1x_2 + a_2x_2^2$$

Thus, for this case, the x's are the knowns and the a's are the unknowns. Because there are three equations with three unknowns, a technique such as Gauss elimination can be employed to solve for the a's. Test this approach by determining the parabola of the form of Eq. (P29.11) to interpolate between (1, 2), (3, 8), and (7, 6). Plot the equation and the points.

CHAPTER30

COMPUTER CALCULUS: DIFFERENTIATION

The computer's fundamental mathematical capability is limited to arithmetic operations and a handful of intrinsic functions. In the previous chapters, we have seen how this seemingly limited repertoire can be used to develop a variety of powerful problem-solving techniques. Now we will learn how these same simple capabilities can be used to solve calculus problems with the computer. The logical starting point is the mathematical operation that is at the heart of calculus—differentiation.

30.1 THE DERIVATIVE

Calculus is the mathematics of change. Because engineers must continuously deal with systems that move and grow, knowledge of calculus is an essential tool of our profession.

At the heart of calculus lies the mathematical concept of the derivative. It represents the rate of change of a dependent variable with respect to an independent variable. As depicted in Fig. 30.1a, the mathematical definition of the derivative begins with a difference approximation:

$$\frac{\Delta y}{\Delta t} = \frac{f(t_1 + \Delta t) - f(t_1)}{\Delta t} \tag{30.1}$$

where y and $f(t)$ are alternative representations of the dependent variable and t is the independent variable. If Δt is allowed to approach zero, as occurs in moving from Fig. 30.1a to 30.1c, the difference becomes a derivative:

$$\frac{dy}{dt} = \lim_{\Delta t \to 0} \frac{f(t_1 + \Delta t) - f(t_1)}{\Delta t}$$

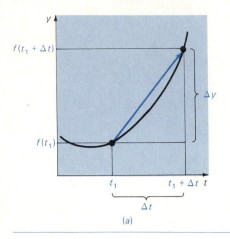

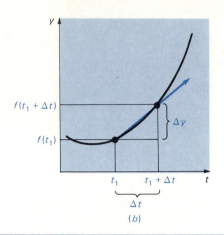

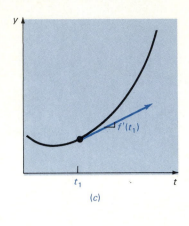

FIGURE 30.1
The graphical definition of a derivative: Going from (a) to (c), as Δt approaches zero and the difference approximation becomes a derivative.

where dy/dt [which can also be represented as y' or $f'(t_1)$] is called the *first derivative of y with respect to t evaluated at* t_1. As seen in the visual depiction of Fig. 30.1c, the derivative is the slope of the tangent to the curve at t_1.

Aside from providing the slope at a single point, it is possible to "take the derivative of," or, as it is called in calculus, to "differentiate" a function itself. In this way, a second function results that can be used to compute the derivative for different values of the independent variable. General rules are available for this purpose. In the case of polynomials, the following simple rule applies:

$$\frac{d}{dt} x^n = nx^{n-1} \tag{30.2}$$

FIGURE 30.2
Graphs of (a) distance, in meters; (b) velocity, in meters per second; and (c) acceleration, in meters per second squared versus time, in seconds, for the flight of a rocket.

For example, suppose that the position of a moving object is described by the function

$$y = 2t^2 \tag{30.3}$$

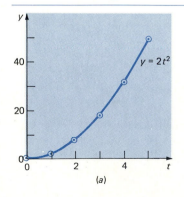

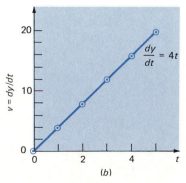

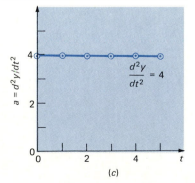

TABLE 30.1 Values of Distance (in Meters), Velocity (in Meters per Second), and Acceleration (in Meters per Second Squared) as a Function of Time (in Seconds) for a Rocket

Time t (a)	Distance y (b)	Velocity v (c)	Acceleration a (d)
0	0	0	4
1	2	4	4
2	8	8	4
3	18	12	4
4	32	16	4
5	50	20	4

In this case, y might represent the height of a rocket taking off from a launching pad. By substituting values of t into Eq. (30.3), we can calculate height as a function of time. As seen in Fig. 30.2a and column (b) of Table 30.1, the rocket seems to be accelerating. That is, it travels a progressively larger distance with each passing unit of time. Differentiation can be used to quantify this behavior. Equation (30.3) can be differentiated by applying Eq. (30.2) to yield

$$v = \frac{dy}{dt} = 4t \tag{30.4}$$

where v is velocity. Equation (30.4) provides the slope of Fig. 30.2a which is equivalent to the rocket's velocity. As plotted in Fig. 30.2b and listed in column (c) of Table 30.1, this result indicates that velocity is increasing linearly with time. That is, with every unit of time the velocity increases by 4.

Finally, Eq. (30.4) can also be differentiated to give the second derivative (that is, the derivative of the derivative) of y with respect to t:

$$a = \frac{d^2y}{dt^2} = 4$$

where d^2y/dt^2 is the second derivative. Thus the rocket has a constant acceleration of 4 [see Fig. 30.2c and column (d) of Table 30.1].

Differentiation of a function has many applications in engineering. However, computers are not capable of the type of analytical differentiation performed above. That is, we cannot input a function such as $y = 2t^2$ and have the computer output the derivative, $y' = 4t$. Nevertheless, the computer can determine slopes of functions at points. It is this sort of differentiation that is described in the present chapter.

30.2 NUMERICAL DIFFERENTIATION

The computer differentiates in a manner that is similar to the original difference representation (Fig. 30.1a) used in the definition of the derivative [Eq. (30.1)]:

$$\frac{dy}{dt} \cong \frac{\Delta y}{\Delta t} = \frac{f(t + \Delta t) - f(t)}{\Delta t} \tag{30.5}$$

Thus, if we know two points on a curve, the derivative can be approximated by a difference. In anticipation of applying Eq. (30.5) on the computer, we will reexpress it using slightly different nomenclature:

$$\frac{dy}{dt} \cong \frac{y_{i+1} - y_i}{h} \tag{30.6}$$

where y_i is the value of the dependent variable corresponding to the independent variable t_i at which the derivative is to be estimated; y_{i+1} is a value of the dependent variable at a later value of the independent variable t_{i+1}; and $h = t_{i+1} - t_i$ is called the *step size*. As depicted in Fig. 30.3a, Eq. (30.6) is referred to as a *forward difference*. It is called this because it utilizes data at the base point (that is, at i) and at a forward point (that is, at $i + 1$) to estimate the derivative.

Two similar approximations of the first derivative can also be made. As shown in Fig. 30.3b, a *backward difference* uses the base point and a previous point, as in

$$\frac{dy}{dt} \cong \frac{y_i - y_{i-1}}{h} \tag{30.7}$$

Finally, a centered or *central difference* uses the difference between the forward and the backward point to estimate the derivative at the base point (Fig. 30.3c):

$$\frac{dy}{dt} \cong \frac{y_{i+1} - y_{i-1}}{2h} \tag{30.8}$$

As is apparent from Fig. 30.3, the central difference is superior to the forward and backward forms. This superiority is also supported by the following example.

FIGURE 30.3 (a) Forward, (b) backward, and (c) central difference approximations of the derivative of y at t_i. Notice how the central difference provides a superior estimate of the derivative at t_i.

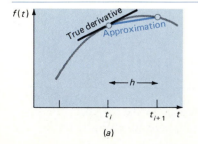
(a)

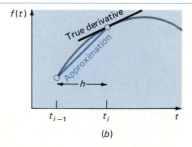
(b)

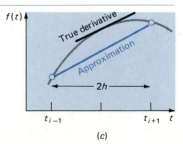
(c)

EXAMPLE 30.1   *Difference Approximations of the First Derivative*

PROBLEM STATEMENT: Use Eqs. (30.6) through (30.8) to estimate the velocity of the rocket from column (*c*) of Table 30.1 based on the times and distances from columns (*a*) and (*b*).

SOLUTION: The forward difference [Eq. (30.6)] must be used to estimate the derivative at $t = 0$ because no previous value is available to compute a central or backward difference. The result is

$$\frac{dy}{dt} \cong \frac{2 - 0}{1 - 0} = 2$$

In a similar fashion, the backward difference [Eq. (30.7)] can be employed to estimate the derivative at the end of the table at $t = 5$, as in

$$\frac{dy}{dt} \cong \frac{50 - 32}{5 - 4} = 18$$

Finally, the central difference [Eq. (30.8)] can be used to estimate the derivatives at each of the intermediate points. For example, at $t = 1$,

$$\frac{dy}{dt} \cong \frac{8 - 0}{2 - 0} = 4$$

The results for the entire computation, along with the true velocity (column *c* of Table 30.1) for comparison, are

Time	Distance	True Velocity	Difference Approximation
0	0	0	2
1	2	4	4
2	8	8	8
3	18	12	12
4	32	16	16
5	50	20	18

Notice that the forward and backward differences are somewhat erroneous at the end points but that the central difference approximations yield perfect estimates for the intermediate points. The perfect results are not typical and stem from the function used to characterize distance in this problem. However, it *is* generally true that the central difference will perform better than either the forward or backward difference.

30.3   COMPUTER PROGRAMS FOR NUMERICAL DIFFERENTIATION

As seen in Fig. 30.4, it is a relatively simple matter to develop programs for numerical differentiation. Notice that user-defined functions are employed to compute the derivatives. In addition, note that the programs in Fig. 30.4 do not have user-friendly input

```
100 REM     ****************************************************
110 REM     *     FIRST DERIVATIVES (FINITE DIFFERENCE)      *
120 REM     *              (BASIC version)                   *
130 REM     *              by  S.C. Chapra                   *
140 REM     *                                                *
150 REM     *  This program finds the first derivative of    *
160 REM     *  input data by using the finite difference     *
170 REM     *  method.                                       *
180 REM     ****************************************************
190 REM
200 REM     .------------------------------------------------.
210 REM     |             DEFINITION OF VARIABLES            |
220 REM     |                                               |
230 REM     | X   = INDEPENDENT VARIABLE                    |
240 REM     | X1  = FIRST X VALUE                           |
250 REM     | F   = DEPENDENT VARIABLE                      |
260 REM     | N   = NUMBER OF POINTS                        |
270 REM     | H   = STEP SIZE                               |
280 REM     | D   = DERIVATIVES                             |
290 REM     | FNF = FORWARD-DIFFERENCE FIRST DERIVATIVE     |
300 REM     | FNC = CENTRAL-DIFFERENCE FIRST DERIVATIVE     |
310 REM     | FNB = BACKWARD-DIFFERENCE FIRST DERIVATIVE    |
320 REM     '------------------------------------------------'
330 REM
340 DIM F(100),D(100)
350 REM
360 REM             Functions to Compute Derivatives
370 REM
380 DEF FNF(J) = (F(J+1) - F(J))/H
390 DEF FNC(J) = (F(J+1) - F(J-1))/(2*H)
400 DEF FNB(J) = (F(J) - F(J-1))/H
410 REM ****************  MAIN PROGRAM  ****************
420 REM
430 REM            Input Values to be Differentiated
440 REM
450 CLS
460 INPUT "NUMBER OF POINTS = ";N
470 INPUT "STEP SIZE = ";H
480 INPUT "FIRST VALUE OF X = ";X1
490 X = X1
500 FOR I = 1 TO N
510    PRINT "F(";X;") = ";
520    INPUT F(I)
530    X = X + H
540 NEXT I
550 REM
560 REM              Compute Derivatives
570 REM
580 D(1) = FNF(1)
590 FOR I = 2 TO N-1
600    D(I) = FNC(I)
610 NEXT I
620 D(N) = FNB(N)
630 REM
640 REM              Print out Results
650 REM
660 PRINT" X-VALUE     Y-VALUE       DERIVATIVE"
670 X = X1
680 FOR I = 1 TO N
690    PRINT X,F(I),D(I)
700    X = X + H
710 NEXT I
720 END
```

```
C     ****************************************************
C     *     FIRST DERIVATIVES (FINITE DIFFERENCE)      *
C     *              (FORTRAN version)                 *
C     *              by  S.C. Chapra                   *
C     *                                                *
C     *  This program finds the first derivative of    *
C     *  data by using the finite difference method.   *
C     ****************************************************
C
C
C     .------------------------------------------------.
C     |             DEFINITION OF VARIABLES            |
C     |                                               |
C     | X   = INDEPENDENT VARIABLE                    |
C     | X1  = FIRST X VALUE                           |
C     | F   = DEPENDENT VARIABLE                      |
C     | N   = NUMBER OF POINTS                        |
C     | H   = STEP SIZE                               |
C     | D   = DERIVATIVES                             |
C     | FNF = FORWARD-DIFFERENCE FIRST DERIVATIVE     |
C     | FNC = CENTRAL-DIFFERENCE FIRST DERIVATIVE     |
C     | FNB = BACKWARD-DIFFERENCE FIRST DERIVATIVE    |
C     '------------------------------------------------'
C
      REAL F(100), D(100), H, X, X1
C
C     ****************  MAIN PROGRAM  ****************
C
C           Input Values to be Differentiated
C
      READ(5,*) N, H, X1
      X = X1
      DO 10 I=1, N
         READ(5,*) F(I)
         X = X + H
10    CONTINUE
C
C              Compute Derivatives
C
      K = 1
      D(1) = FNF(K,H,F)
      DO 20 I=2, N-1
         D(I) = FNC(I,H,F)
20    CONTINUE
      D(N) = FNB(N,H,F)
C
C              Print out Results
C
      WRITE(6,*) ' X-VALUE       Y-VALUE        DERIVATIVE'
      X = X1
      DO 30 I = 1, N
         WRITE(6,*) X, F(I), D(I)
         X = X + H
30    CONTINUE
      STOP
      END
C     -------------------------------------------------------
C              Functions to Compute Derivatives
C
      REAL FUNCTION FNF(J,H,F)
      REAL F(100),H
      INTEGER J
      FNF = (F(J+1) - F(J))/H
      RETURN
      END
C
      REAL FUNCTION FNC(J,H,F)
      REAL F(100),H
      INTEGER J
      FNC = (F(J+1) - F(J-1))/(2*H)
      RETURN
      END
C
      REAL FUNCTION FNB(J,H,F)
      REAL F(100),H
      INTEGER J
      FNB = (F(J) - F(J-1))/H
      RETURN
      END
```

FIGURE 30.4 BASIC and FORTRAN computer programs for numerical differentiation.

```
NUMBER OF POINTS = ? 6
STEP SIZE = ? 1
FIRST VALUE OF X = ? 0
F( 0 ) = ? 0
F( 1 ) = ? 2
F( 2 ) = ? 8
F( 3 ) = ? 18
F( 4 ) = ? 32
F( 5 ) = ? 50
X-VALUE       Y-VALUE        DERIVATIVE
  0             0                2
  1             2                4
  2             8                8
  3            18               12
  4            32               16
  5            50               18
OK
```

FIGURE 30.5 A computer screen generated by Fig. 30.4. This application solves the problem previously analyzed in Example 30.1.

and output. In Prob. 30.1 you will have the task of including such features in your version of the programs.

Figure 30.5 shows an example of the output from a program that was developed from Fig. 30.4. This program is used to solve Example 30.1.

30.4 PITFALLS AND ADVANCED METHODS

Pitfalls. A shortcoming of numerical differentiation is that it tends to amplify errors in the data. Figure 30.6a and b shows the data and results for Example 30.1. Figure 30.6c uses the same data, but some points are raised and some are lowered slightly. This minor modification is barely apparent from Fig. 30.6c. However, the resulting effect in Fig. 30.6d is significant because the process of differentiation amplifies errors.

One remedy for this problem is to use polynomial or nonlinear regression to fit a lower-order curve to the data. As discussed in Chap. 26, regression does not attempt

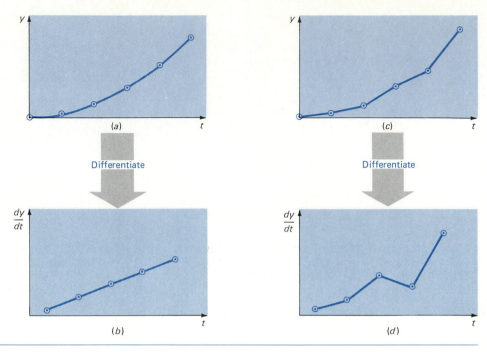

FIGURE 30.6
Illustration of how even small data errors introduce large variability into numerical estimates of derivatives: (a) data with no error; (b) the resulting numerical differentiation; (c) data modified slightly; (d) the resulting differentiation manifests significant variability.

to intersect each point but rather traces out the general trend. The regression equation can be differentiated analytically to determine derivative estimates. In addition, a technique such as cubic splines can also be used to fit the data, and the resulting cubic equation can be differentiated to estimate the derivatives. Although cubic splines intersect every point, the fact that they are limited to cubic equations usually results in smoother derivatives.

Advanced methods. Aside from the techniques described in the previous section, there are other advanced methods for numerical differentiation. For equispaced data, higher-order forward-, backward-, and central-difference formulas are available. These formulas are summarized elsewhere (Chapra and Canale, 1985).

Two major difficulties with Eqs. (30.6) through (30.8) is that they are only applicable for equispaced data and the forward and backward differences at the end points are less accurate than the central differences in the middle. One remedy is to fit a second-order Lagrange interpolating polynomial [recall Eq. (27.5)] to each set of three adjacent points. Remember that this polynomial does not require that the points be equispaced. The second-order polynomial can be differentiated analytically to give

$$
\frac{dy}{dt} \cong y_{i-1} \frac{2t - t_i - t_{i+1}}{(t_{i-1} - t_i)(t_{i-1} - t_{i+1})} + y_i \frac{2t - t_{i-1} - t_{i+1}}{(t_i - t_{i-1})(t_i - t_{i+1})}
$$
$$
+ y_{i+1} \frac{2t - t_{i-1} - t_i}{(t_{i+1} - t_{i-1})(t_{i+1} - t_i)}
$$

(30.9)

where t is the value at which you want to estimate the derivative. Although this equation is certainly more complex than Eqs. (30.6) to (30.8), it has some very important advantages. First, it can be used to estimate the derivatives anywhere within the range prescribed by the three points. Second, the points themselves do not have to be equally spaced. Third, the derivative estimate is of the same accuracy as the centered difference [(Eq. (30.8)]. In fact, for equispaced points, Eq. (30.9) evaluated at $t = t_i$ reduces to Eq. (30.8). Because of these properties, Eq. (30.9) is superior to Eqs. (30.6) through (30.8). In Prob. 30.1 you will have the task of incorporating Eq. (30.9) into the program in Fig. 30.4. This will provide you with an advanced capability for performing numerical differentiation.

PROBLEMS

30.1 Utilize Fig. 30.4 as a starting point for developing your own program for numerical differentiation. Employ structured programming techniques, modular design, and internal and external documentation in your development. Incorporate Eq. (30.9) in place of the finite divided differences in order to allow the use of unequally spaced data and to improve the accuracy of the derivative estimates.

30.2 What are the fundamental differences between the forward, backward, and central differences? Which is most accurate and why?

30.3 Compute the forward-, backward-, and central-difference approximations for $y = \sin x$ at $x = \pi/3$. Note that the derivative as determined by calculus is $dy/dx = \cos x$. Thus the value of the derivative at $\pi/3$ is 0.5. Compute the approximations for (a) $h = \pi/3$, (b) $h = \pi/6$, and (c) $h = \pi/12$. Calculate the percent relative error for each result. What conclusions can be drawn regarding the effect of halving the step size on the accuracy of each type of approximation?

30.4 Prove that for equispaced data points, Eq. (30.9) reduces to Eq. (30.8) at $t = t_i$.

30.5 Apply Eq. (30.9) to recompute the derivative estimates for Example 30.1. Interpret your results.

30.6 Compute the central difference for each of the following functions at the specified location and for the specified step size:

(a) $y = x^3 + 2x - 10$ at $x = 0$, $h = 0.5$
(b) $y = x^3 + 2x - 10$ at $x = 0$, $h = 0.1$
(c) $y = \tan x$ at $x = 4$, $h = 0.2$
(d) $y = \sin (0.5\sqrt{x})$ at $x = 1$, $h = 0.25$
(e) $y = e^x$ at $x = 0$, $h = 0.5$

30.7 A jet fighter's position on an aircraft carrier's runway was timed during landing:

t, s	0	0.51	1.03	1.74	2.36	3.24	3.82
x, m	154	186	209	250	262	272	274

where x is distance from the end of the carrier. Estimate (a) velocity (dx/dt) and (b) acceleration (dv/dt) using numerical differentiation.

30.8 The rate of cooling of a body (Fig. P30.8) can be expressed as

$$\frac{dT}{dt} = -k(T - T_a)$$

where T is the temperature of the body (in degrees Celsius), T_a is the temperature of the surrounding medium (in degrees Celsius), and k is a proportionality constant (per minute). Thus this equation (which is called *Newton's law of cooling*) specifies that the rate of cooling is proportional to the difference in the temperatures of the body and of the surrounding medium. If a metal ball heated to 90°C is dropped into water that is held constant at $T_a = 20$°C, the temperature of the ball changes as in

Time, min	0	5	10	15	20	25
Temperature, °C	90	62.5	45.8	35.6	29.5	25.8

Utilize numerical differentiation to determine dT/dt at each value of time. Plot dT/dt versus $T - T_a$ and employ linear regression to evaluate k.

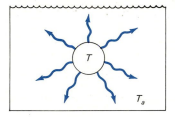

FIGURE P30.8

30.9 The following data was collected when a large oil tanker was loading:

t, min	0	15	30	45	60	90	120
V, 10^6 barrels	0.5	0.65	0.73	0.88	1.03	1.14	1.30

Calculate the flow rate Q(that is, dV/dt) for each time.

30.10 The *heat current H* is the quantity of heat flowing through a material per unit time. It can be computed with

$$H = -kA\frac{dT}{dx}$$

where H has units of joules (or kg · m^2/s^2) per second; k, a coefficient of thermal conductivity that parameterizes the heat-conducting properties of the material, has units of joules per second per meter per degree Celsius; A is the cross-sectional area perpendicular to the path of heat flow; T is temperature (in degrees Celsius); and x is distance (in centimeters) along the path of heat flow. If the temperature outside a house is $-20°C$ and inside is $21°C$, what is the rate of heat loss through a 6-in-thick, 1000-ft^2 wall made of

Material	k
Wood	0.08
Red brick	0.60
Concrete	0.80
Insulating brick	0.15

30.11 *Fick's first diffusion law* states that

$$\text{Mass flux} = -D\frac{dc}{dx} \qquad \text{(P30.11)}$$

where mass flux is the quantity of mass that passes across a unit area per unit time (in grams per square centimeter per second), D is a diffusion coefficient (in square centimeters per second), c is concentration, and x is distance (in centimeters). An environmental engineer measures the following concentration of a pollutant in the sediments underlying a lake ($x = 0$ at the sediment-water interface and increases downward):

x, cm	0	1	2.5
c, 10^{-6}g/cm^3	0.1	0.4	0.75

Use the best numerical differentiation technique available to estimate the derivative at $x = 0$. Employ this estimate in conjunction with Eq. (P30.11) to compute the mass flux of pollutant out of the sediments and into the overlying waters ($D = 10^{-6}$ cm/s^2). For a lake with 10^6 m^2 of sediments, how much pollutant would be transported into the lake over a year's time?

30.12 *Faraday's law* characterizes the voltage drop across an inductor as

$$V_L = L\frac{di}{dt}$$

where V_L is voltage drop (in volts), L is the inductance (in henrys; 1 H = 1 Vs/A), i is current (in amperes), and t is time (in seconds). Determine the voltage drop as a function of time from the following data:

t	0	0.1	0.2	0.4	0.6	0.8
i	0	0.1	0.2	0.6	1.3	2.7

The inductance is equal to 5 H.

CHAPTER 31
COMPUTER CALCULUS: INTEGRATION

In the previous chapter we learned how the computer is used to determine the derivative of data. Now we will turn to the opposite process in calculus—*integration*.

According to the dictionary definition, to *integrate* means "to bring together, as parts, into a whole; to unite; to indicate the total amount. . . ." Mathematically, integration is represented by

$$I = \int_a^b f(x) \, dx \tag{31.1}$$

which stands for the integral of the function $f(x)$ with respect to the independent variable x, evaluated between the limits $x = a$ to $x = b$.

As suggested by the dictionary definition, the "meaning" of Eq. (31.1) is the *total value*, or *summation*, of $f(x) \, dx$ over the range from $x = a$ to b. In fact, the symbol is actually a stylized capital S that is intended to signify the close connection between integration and summation.

Figure 31.1 represents a graphical manifestation of the concept. For functions lying above the x axis, the integral expressed by Eq. (31.1) corresponds to the area under the curve of $f(x)$ between $x = a$ and b.

We will have numerous occasions to refer back to this graphical conception as we develop computer-oriented methods for integration. In fact, some of the common precomputer methods for integration are based on graphs. For example, a simple intuitive approach is to plot the function on a grid (Fig. 31.2) and count the number of boxes that approximate the area. This number multiplied by the area of each box provides a rough estimate of the integral. The estimate can be refined, at the expense of additional effort, by using a finer grid.

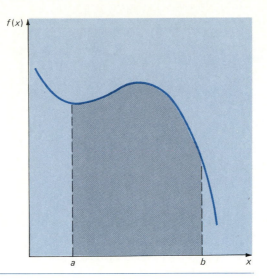

FIGURE 31.1
Graphical representation of the integral of $f(x)$ between the limits $x = a$ to b. The integral is equivalent to the area under the curve.

In the present chapter we will describe an alternative method for integration that is also based on a graphical approach and which is easier to use than the grid method. Also, it is ideal for implementation on personal computers.

Integration has numerous applications in engineering. A number of examples relate directly to the idea of the integral as the area under a curve. Figure 31.3 depicts a few cases where integration is used for this purpose.

Other common applications relate to the analogy between integration and summation. For example, the average or mean value of a continuous function between

FIGURE 31.2
The use of a grid to approximate an integral.

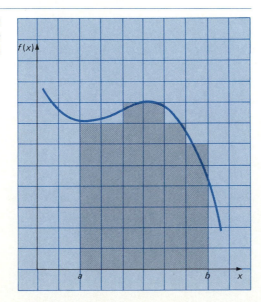

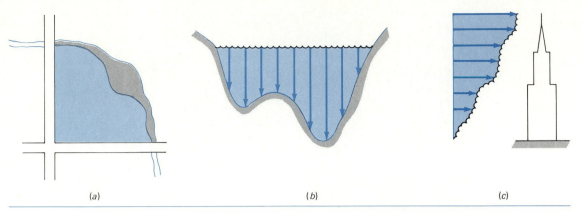

(a)   (b)   (c)

FIGURE 31.3 Examples of how integration is used to evaluate the areas of surfaces in engineering. (a) A surveyor might need to know the area of a field bounded by a meandering stream and two roads. (b) A water-resource engineer might need to know the cross-sectional area of a river. (c) A structural engineer might need to integrate a nonuniform force due to a wind blowing against the side of a skyscraper.

$x = a$ and b can be computed by

$$\text{Mean} = \frac{\int_a^b f(x)\ dx}{b - a} \tag{31.2}$$

Thus, the numerator represents the summation of $f(x)\ dx$. Then by dividing by the distance from a to b, we obtain the average, or mean, height of the function (Fig. 31.4). Equation (31.2) can be compared with Eq. (24.6) to contrast the calculation of the mean for continuous functions and discrete data.

Another application, which will serve as an example for this chapter, is to determine a quantity on the basis of its given rate of change. This example most clearly demonstrates the relationship and opposite natures of differentiation and integration. In the previous chapter we used differentiation to determine the velocity of an object, given displacement as a function of time. That is, as shown in Fig. 31.5a,

$$v(t) = \frac{d}{dt}\, y(t)$$

FIGURE 31.4
The integral can be used to evaluate the mean value of a function.

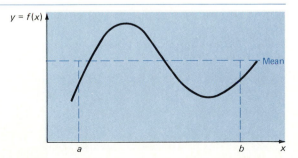

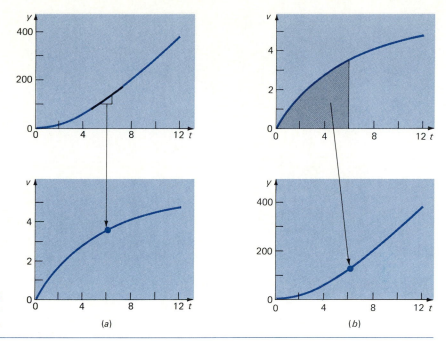

FIGURE 31.5
The contrast between differentiation and integration.

(a)   (b)

In the present chapter we are given velocity and use integration to determine displacement:

$$y(t) = \int_0^t v(t)\ dt \tag{31.3}$$

Thus by integrating or "summing" the product of velocity multiplied by time, we can determine how far the projectile travels over time t. As seen in Fig. 31.5b, the area under the velocity-time curve from 0 to t gives the distance traveled over that time. In the present chapter we will use the computer to evaluate this area.

31.1   THE TRAPEZOIDAL RULE

As depicted in Fig. 31.6, a simple approach for determining the integral of a function is as the area under the straight line connecting the function values at the ends of the integration interval. The formula to compute this area is

$$I \cong (b - a)\ \frac{f(a) + f(b)}{2} \tag{31.4}$$

Because the area under the line is a trapezoid, this formula is called the *trapezoidal rule*. Recall from geometry that the formula for computing the area of a trapezoid is

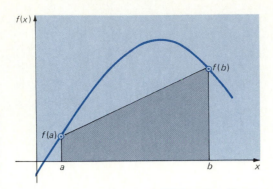

FIGURE 31.6
The trapezoidal rule consists of
taking the area under the straight
line connecting the function values
at the end points of the integration
interval.

the height times the average of the bases (Fig. 31.7*a*). In the present case, the concept is the same but the trapezoid is on its side (Fig. 31.7*b*). Therefore the integral estimate can be represented as

$$I \cong \text{width} \times \text{average height}$$

where for the trapezoidal rule, the average height is the average of the function values at the end points, or $[f(a) + f(b)]/2$.

EXAMPLE 31.1   The Trapezoidal Rule

PROBLEM STATEMENT: The velocity of a falling parachutist is given by the following function of time:

$$v(t) = \frac{gm}{c} (1 - e^{-(c/m)t}) \tag{E31.1.1}$$

where $v(t)$ is velocity in meters per second, g is the gravitational constant of 9.8 m/s^2, m is the mass of the parachutist equal to 68.1 g, and c is a drag coefficient

FIGURE 31.7
(*a*) The formula for computing the
area of a trapezoid: height times
the average of the bases. (*b*) For
the trapezoidal rule, the concept is
the same but the trapezoid is on its
side.

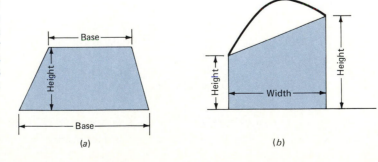

of 12.5 g/s. A plot of the function can be developed by substituting the parameters and various values of t into the equation to give

t, s	v, m/s
0	0.00
2	16.40
4	27.77
6	35.64
8	41.10
10	44.87
12	47.49
14	49.30
16	50.56

These points are plotted in Fig. 31.8. Use the trapezoidal rule to determine how far the parachutist has fallen after 12 s.

SOLUTION: According to Eq. (31.3), the distance can be determined by integrating the velocity, as in

$$y(t) = \int_0^t v(t) \, dt$$

or, substituting the function and the given values,

$$y(t) = \int_0^{12} 53.39(1 - e^{-0.18355t}) \, dt$$

Using calculus, this integral can be evaluated analytically to give an exact result of 381.958 m. This exact value will allow us to assess the accuracy of the trapezoidal-rule approximation.

FIGURE 31.8
A plot of the velocity of a falling parachutist versus time. The area under the curve from $t = 0$ to 12 is equal to the distance the parachutist travels over the first 12 s of free-fall. The use of a single application of the trapezoidal rule to approximate this area is shown. The area between the curve and the shaded portion represents the error of the trapezoidal rule approximation.

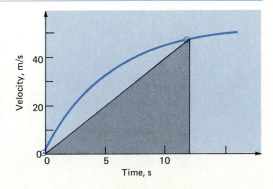

In order to implement the trapezoidal rule, the values at the end points can be substituted into Eq. (31.4) to give

$$I \cong (12 - 0)\frac{0 + 47.49}{2} = 284.94 \text{ m}$$

which represents a percent relative error of

$$\text{Error} = \frac{381.958 - 284.94}{381.958} 100\% = 25.4\%$$

The reason for this large error is evident from the graphical depiction in Fig. 31.8. Notice that the area under the straight line neglects a significant portion of the integral lying above the line.

One way to improve the accuracy of the trapezoidal rule is to divide the integration interval from a to b into a number of segments and apply the method to each segment. The areas of the individual segments can be added to yield the total integral. Figure 31.9 shows the general format and nomenclature we will employ for the multiple-segment trapezoidal rule. Notice that the integration interval is divided into n segments. If, as shown in Fig. 31.9, a and b are designated as x_0 and x_n, the total integral is

$$I = h_1\frac{f(x_0) + f(x_1)}{2} + h_2\frac{f(x_1) + f(x_2)}{2} + \cdots + h_n\frac{f(x_{n-1}) + f(x_n)}{2} \tag{31.5}$$

FIGURE 31.9
Nomenclature used for the multiple-segment trapezoidal rule.

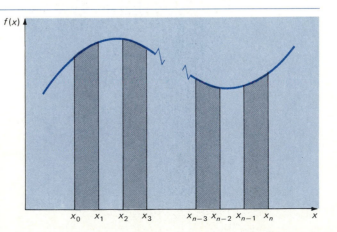

where h_i is the width of segment $i = x_i - x_{i-1}$. For the case where the segments are of equal width,

$$h = \frac{b - a}{n} \tag{31.6}$$

and the integral can be expressed more concisely by grouping terms, as in

$$I = \frac{h}{2} \left[f(x_0) + 2 \sum_{i=1}^{n-1} f(x_i) + f(x_n) \right]$$

or, substituting Eq. (31.6),

$$I = (b - a) \frac{f(x_0) + 2 \sum_{i=1}^{n-1} f(x_i) + f(x_n)}{2n} \tag{31.7}$$

$$\underbrace{}_{\text{Width}} \quad \underbrace{\phantom{\frac{f(x_0) + 2 \sum f(x_i) + f(x_n)}{2n}}}_{\text{Average height}}$$

Because the summation of the coefficients of $f(x)$ in the numerator divided by $2n$ is equal to 1, the average height is represented by a weighted average of the function values. According to Eq. (31.7), the interior points are given twice the weight of the two end points, $f(x_0)$ and $f(x_n)$.

EXAMPLE 31.2   The Multiple-Segment Trapezoidal Rule

PROBLEM STATEMENT: Solve the same problem given in Example 31.1 but use a three- and a six-segment trapezoidal rule to determine the integral.

SOLUTION: Because the data used to generate the plot in Example 31.1 is equi-spaced, Eq. (31.7) can be used to evaluate the integral. For the three-segment trap-ezoidal rule, the result is

$$I = (12 - 0) \frac{0 + 2(27.77 + 41.10) + 47.49}{6} = 370.46$$

which represents a percent relative error of 3 percent. Thus using the multiple-segment version represents a great improvement over the single-segment result of 25.4 percent obtained previously in Example 31.1. The improvement becomes even more pro-nounced for the six-segment version:

$$I = (12 - 0) \frac{0 + 2(16.40 + 27.77 + 35.64 + 41.10 + 44.87) + 47.49}{12}$$

$$= 379.05$$

which represents an error of 0.76 percent.

31.2 COMPUTER PROGRAMS FOR THE TRAPEZOIDAL RULE

As seen in Fig. 31.10, it is a relatively simple matter to develop programs for the trapezoidal rule. These programs are set up for the equal-sized, multisegment version of the method. Notice that a modular approach is taken and that a loop is employed

FIGURE 31.10 BASIC and FORTRAN computer programs for the multiple-segment trapezoidal rule.

```
100 REM     ***********************************************
110 REM     *     MULTIPLE SEGMENT TRAPEZOIDAL RULE      *
120 REM     *              (BASIC version)               *
130 REM     *              by  S.C. Chapra               *
140 REM     *                                            *
150 REM     *   This program uses the trapezoidal rule to *
160 REM     *   calculate the integral of a set of data. *
170 REM     ***********************************************
180 REM
190 REM          .------------------------------------------.
200 REM          |          DEFINITION OF VARIABLES         |
210 REM          |                                          |
220 REM          | N     = NUMBER OF POINTS                 |
230 REM          | SEGS  = NUMBER OF SEGMENTS               |
240 REM          | A     = LOWER LIMIT OF INTEGRATION       |
250 REM          | B     = UPPER LIMIT OF INTEGRATION       |
260 REM          | H     = SEGMENT WIDTH                    |
270 REM          | Y     = VECTOR CONTAINING VALUES OF      |
280 REM          |         DEPENDENT VARIABLE               |
290 REM          | AREA  = INTEGRAL                         |
300 REM          | HEIGHT = AVERAGE HEIGHT                  |
310 REM          '------------------------------------------'
320 DIM Y(20)
330 REM
340 REM ****************** MAIN PROGRAM ******************
350 REM
360 CLS
370 GOSUB 1000
380 GOSUB 2000
390 PRINT: PRINT "INTEGRAL = " AREA
400 END
1000 REM *********** SUBROUTINE TO ENTER DATA ***********
1010 REM
1020 INPUT "NUMBER OF POINTS = ? ",N
1030 SEGS = N - 1
1040 INPUT "LOWER AND UPPER LIMITS OF INTEGRATION? ", A, B
1050 H = (B - A)/SEGS
1060 FOR I = 1 TO N
1070   PRINT "Y(" I ") = ";
1080   INPUT Y(I)
1090 NEXT I
1100 RETURN
2000 REM ************* SUBROUTINE TRAPEZOIDAL ***********
2010 REM
2020 REM     MULTIPLE-SEGMENT TRAPEZOIDAL RULE ALGORITHM
2030 REM
2040 REM     DATA REQUIRED:   Y(), SEGS
2050 REM     RESULTS RETURNED: AREA
2060 REM
2070 REM     LOCAL VARIABLES:
2080 REM
2090 REM     TRSUM  = SUMMATION OF TWO TIMES EACH Y VALUE
2100 REM
2110 REM --------------- BODY OF TRAPEZOIDAL --------------
2120 TRSUM = Y(1)
2130 FOR I = 2 TO SEGS
2140   TRSUM = TRSUM + 2 * Y(I)
2150 NEXT I
2160 TRSUM = TRSUM + Y(N)
2170 HEIGHT = TRSUM/(2*SEGS)
2180 AREA = (B - A) * HEIGHT
2190 RETURN
```

```
C     ***********************************************
C     *     MULTIPLE SEGMENT TRAPEZOIDAL RULE      *
C     *              (FORTRAN version)             *
C     *              by  S.C. Chapra               *
C     *                                            *
C     *   This program uses the trapezoidal rule to *
C     *   calculate the integral of a set of data. *
C     ***********************************************
C
C          .------------------------------------------.
C          |          DEFINITION OF VARIABLES         |
C          |                                          |
C          | N     = NUMBER OF POINTS                 |
C          | SEGS  = NUMBER OF SEGMENTS               |
C          | A     = LOWER LIMIT OF INTEGRATION       |
C          | B     = UPPER LIMIT OF INTEGRATION       |
C          | H     = SEGMENT WIDTH                    |
C          | Y     = VECTOR CONTAINING VALUES OF      |
C          |         DEPENDENT VARIABLE               |
C          | AREA  = INTEGRAL                         |
C          | HEIGHT = AVERAGE HEIGHT                  |
C          '------------------------------------------'
      REAL A, B, Y(20), SEGS, AREA
C
C     ****************** MAIN PROGRAM ******************
C
      CALL ENTER(N,A,B,Y,SEGS)
      CALL TRAP(N,A,B,Y,AREA,SEGS)
      WRITE(6,*) 'INTEGRAL = ', AREA
      STOP
      END
C     *********** SUBROUTINE TO ENTER DATA ***********
      SUBROUTINE ENTER(N,A,B,Y,SEGS)
      REAL A,B,Y(20)
      READ(5,*) N
      SEGS = N - 1
      READ(5,*) A, B
      H = (B - A)/SEGS
      DO 10 I=1, N
        READ(5,*) Y(I)
10    CONTINUE
      RETURN
      END
C     ************* SUBROUTINE TRAPEZOIDAL ***********
      SUBROUTINE TRAP(N,A,B,Y,AREA,SEGS)
C
C         MULTIPLE-SEGMENT TRAPEZOIDAL RULE ALGORITHM
C
C         DATA REQUIRED:   Y(), SEGS
C         RESULTS RETURNED: AREA
C
C         LOCAL VARIABLES:
C
C         TRSUM  = SUMMATION OF TWO TIMES EACH Y VALUE
C
      REAL SUM,Y(20),SEGS,HEIGHT,AREA
C     --------------- BODY OF TRAPEZOIDAL --------------
      TRSUM = Y(1)
      DO 10 I=2, SEGS
        TRSUM = TRSUM + 2 * Y(I)
10    CONTINUE
      TRSUM = TRSUM + Y(N)
      HEIGHT = TRSUM/(2*SEGS)
      AREA = (B - A) * HEIGHT
      RETURN
      END
```

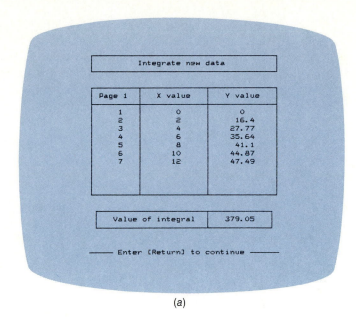

(a)

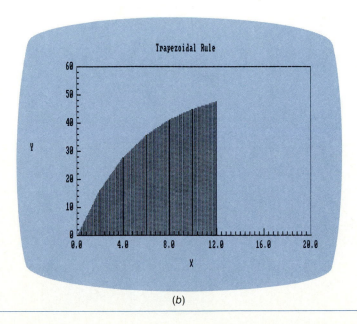

(b)

FIGURE 31.11
Computer screens generated by a trapezoidal-rule software package (NUMERICOMP by Canale and Chapra, 1985). The package is used to solve the problem previously analyzed in Example 31.2.

to perform the computation efficiently. In addition, note that the programs in Fig. 31.10 do not have the capability to integrate a function. In Prob. 31.1 you will have the task of including this feature in your version of the program.

Figure 31.11 shows some screens generated using a commercial software package that has a trapezoidal rule algorithm very similar to that shown in Fig. 31.10. The package is used to solve Example 31.2. The solution is shown in Fig. 31.11*a*, and a plot is shown in Fig. 31.11*b*. These screens illustrate how output can be designed in a clear and accessible fashion. You can employ them as models for your own software.

31.3   ADVANCED METHODS

In the present chapter we used a simple geometric derivation to develop the trapezoidal-rule formula (see Fig. 31.12*a*). An alternative derivation involves fitting a straight line between the function values at the end points with a Lagrange interpolating polynomial [Eq. (27.4)]. This equation can be integrated analytically, and the result is the same as Eq. (31.4).

Taking this reasoning a step farther, we develop a more accurate estimate by fitting a Lagrange parabola to three function values at the ends of the integration interval and in the middle (Fig. 31.12*b*). When the parabolic equation is integrated, the following formula results:

$$I = (b - a) \frac{f(x_0) + 4f(x_1) + f(x_2)}{6} \tag{31.8}$$

which is called *Simpson's 1/3 rule*. Just as was the case with the trapezoidal rule, a multisegment version can also be developed, as in

$$I = (b - a) \frac{f(x_0) + 4\sum_{i=1,3,5}^{n-1} f(x_i) + 2\sum_{j=2,4,6}^{n-2} f(x_j) + f(x_n)}{3n} \tag{31.9}$$

FIGURE 31.12
(*a*) The trapezoidal rule establishes the area under the straight line connecting the function values at the end points. (*b*) Simpson's 1/3 rule establishes the area under the parabola connecting values at each end and one in the middle.

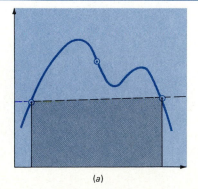

(a)

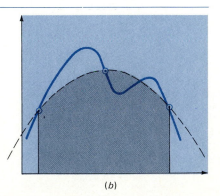

(b)

Equations (31.8) and (31.9) are much more accurate than the trapezoidal rule. In Prob. 31.3 you will have the task of developing a computer program to implement Simpson's rule.

In addition to Simpson's rule, other advanced integration techniques are available. Description of these methods is found elsewhere (Chapra and Canale, 1985).

PROBLEMS

31.1 Utilize Fig. 31.10 as a starting point for developing your own computer program for the multiple-segment trapezoidal rule. Employ structured programming techniques, modular design, and internal and external documentation in your development. Incorporate the capability to integrate a function as well as discrete data points by having the program employ the function to generate a table of values.

31.2 Utilize Fig. 31.10 as a starting point for developing your own computer program for a multiple-segment trapezoidal rule. Design the program so that it can evaluate unequally spaced tabular data [Eq. (31.5)]. Employ structured programming techniques, modular design, and internal and external documentation in your development.

31.3 Repeat Prob. 31.1 but for the multiple-segment Simpson's 1/3 rule.

31.4 If the velocity is known, then the distance that a parachutist falls over a time t is given by

$$d = \int_0^t v(t)\, dt$$

If the velocity is defined by

$$v(t) = 5000 \left(\frac{t}{3.75 + t} \right)$$

use numerical integration to determine how far the parachutist falls in 6 s.

31.5 The upward velocity of a rocket can be computed by the following formula:

$$v = u \ln \left(\frac{m_0}{m_0 - qt} \right) - gt$$

where v is upward velocity, u is the velocity at which fuel is expelled relative to the rocket, m_0 is the initial mass of the rocket at time $t = 0$, q is the fuel consumption rate, and g is the downward acceleration of gravity (assumed constant $= 9.8$ m/s^2). If $u = 2200$ m/s, $m_0 = 160{,}000$ kg, and $q = 2680$ kg/s, use a numerical method to determine how high the rocket will fly in 30 s.

31.6 In order to estimate the size of a new dam you have to determine the total volume of water (in gallons) that flows down a river in a year's time. You have available to you the following long-term average data for the river:

Date	mid-Jan.	mid-Feb.	mid-Mar.	mid-Apr.	mid-June	mid-Aug.	mid-Oct.	mid-Nov.	mid Dec.
Flow rate, ft³/s	1100	1300	2200	4200	3600	700	750	800	1000

Determine the volume. Be careful of units. Note that there are 7.481 gal/ft^3 and take care to make a proper estimate of flow at the end points.

31.7 You normally jog down a country road and want to determine how far you run. Although the odometer on your car is broken, you can use the speedometer to record the car's velocity at 1-min intervals as you follow your jogging route:

Time, min	0	1	2	3	4	5	6	6.33
Velocity, mi/h	0	35	40	48	50	55	32	36

Use this data to determine how many miles you jog. Be careful of units.

31.8 A transportation engineering study requires the calculation of the total number of cars that pass through an intersection over a 24-h period. An individual visits the intersection at various times over the course of the day and counts the number of cars that pass through the intersection in a minute. Utilize this data to estimate the total number of cars that pass through the intersection per day. (Be careful of units and use the most accurate numerical method wherever possible.)

Time	Rate, cars/min	Time	Rate, cars/min
12:00 midnight	2	12:30 P.M.	19
2:00 A.M.	2	2:00 P.M.	7
4:00 A.M.	0	4:00 P.M.	9
5:00 A.M.	2	5:00 P.M.	20
6:00 A.M.	5	6:00 P.M.	26
7:00 A.M.	18	7:00 P.M.	·12
8:00 A.M.	27	8:00 P.M.	9
9:00 A.M.	12	9:00 P.M.	15
10:30 A.M.	15	10:00 P.M.	8
11:30 A.M.	17	11:00 P.M.	5
		12:00 P.M.	3

31.9 A rod subject to an axial load (Fig. P31.9a) will be deformed as shown in the stress strain curve in Fig. P31.9b. The area under the curve from zero stress out to the point of rupture is called the *modulus of toughness* of the material. It provides a measure of the energy per unit volume required to cause the

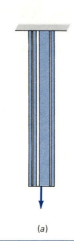

FIGURE P31.9
(a) A rod under axial loading and (b) the resulting stress-strain curve where stress is in kips per square inch (10^3 lb/in²) and strain is dimensionless.

s	e
0.02	40.0
0.05	37.5
0.10	43.0
0.15	52.0
0.20	60.0
0.25	55.0

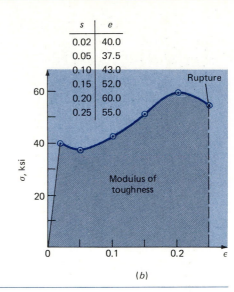

(a)

(b)

material to rupture. As such, it is representative of a structure's ability to withstand an impact load. Use numerical integration to compute the modulus of toughness for the stress-strain curve seen in Fig. P31.9b.

31.10 Employ the multiple-segment trapezoidal rule to evaluate the vertical distance traveled by a rocket if the vertical velocity is given by

$$v = 10t^2 \qquad\qquad 0 \le t \le 10$$
$$v = 1000 - 5t \qquad 10 < t \le 20$$
$$v = 45t + 2(t - 20)^2 \qquad 20 < t \le 30$$

31.11 The work done on an object is equal to the force times the distance moved in the direction of the force. The velocity of an object in the direction of a force is given by

$$v = 5t \qquad\qquad 0 \le t \le 7$$
$$v = 35 - (7 - t)5 \qquad 7 < t \le 14$$

Employ the multiple-segment trapezoidal rule to determine the work if a constant force of 200 lb is applied for all t.

31.12 If the velocity distribution of a fluid flowing through a pipe is known (Fig. P31.12), the flow rate Q (that is, the volume of water passing through the pipe per unit time) can be computed by $Q = \int v \, dA$, where v is the velocity and A is the pipe's cross-sectional area. (To grasp, the meaning of this relationship physically, recall the close connection between summation and integration).

FIGURE P31.12

For a circular pipe, $A = \pi r^2$ and $dA = 2\pi r\, dr$. Therefore

$$Q = \int_0^r v(2\pi r)\, dr$$

where r is the radial distance measured outward from the center of the pipe. If the velocity distribution is given by

$$v = 3.0 \left(1 - \frac{r}{r_0}\right)^{1/7}$$

where r_0 is the total radius (in this case, 3 in), compute Q using the multiple-segment trapezoidal rule.

31.13 If the velocity of a roller coaster in the horizontal direction is given by

$$v(t) = (60 - t)^2 + (60 - t) \sin (t^{1/2})$$

determine the horizontal distance traveled in 60 s using the multiple-segment trapezoidal rule.

31.14 The exponential integral

$$E = \int_a^b \frac{e^{-x}}{x}\, dx$$

arises frequently in engineering analysis. Evaluate this integral from $a = 1$ to $b = 5$ by the multiple-segment trapezoidal rule.

31.15 One method for improving integral estimates is to perform the integration twice—once with a certain number of segments (n) and again with twice as many segments ($2n$). It is then possible to combine these two values to obtain a third, improved estimate. This type of approach is known as *Richardson extrapolation*. [See Chapra and Canale (1985) for additional information.] For the trapezoidal rule, the formula to compute the improved estimate I_3 is

$$I_3 = \frac{4}{3} I_2 - \frac{1}{3} I_1$$

where I_2 is the integral value using $2n$ segments and I_1 is the value using n segments. Employ this approach to obtain the integral of

$$\int_{-3}^3 x^4 + 2x^2 + 4x - 4\, dx$$

by using the multiple-segment trapezoidal rule with $n = 6$ to compute I_1 and $n = 12$ to compute I_2. Determine the percent relative error for I_1, I_2, and I_3 (the exact value of the integral can be computed with calculus as 109.2) in order to substantiate that the Richardson extrapolation does, in fact, lead to superior results.

31.16 In Chap. 24 the following relationship was given to compute the height (that is, frequency) of a bell-shaped or normal distribution [Eq. (24.11)]:

$$f(y) = \frac{1}{\sigma \sqrt{2\pi}} e^{-(y-\mu)^2/2\sigma^2}$$ (P31.16)

We claimed that the area under the curve specified by this equation was equal to the probability that a random event would occur between two values of y (see Fig. P31.16). We further asserted that the area from $y = -\sigma$ to $y = \sigma$ was equal to a 68.3 percent probability. That is,

$$\int_{-\sigma}^{\sigma} \frac{1}{\sigma \sqrt{2\pi}} e^{-(y-\mu)^2/2\sigma^2} \, dy = 0.683$$

Employ the multiple-segment trapezoidal rule to verify that this is correct for $\mu = 0$ and $\sigma = 1$.

FIGURE P31.16

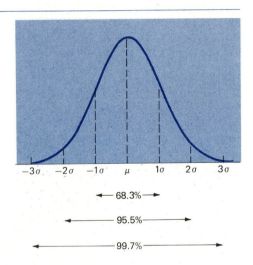

CHAPTER32

COMPUTER CALCULUS: RATE EQUATIONS

The subject of ordinary differential equations is traditionally not taught to freshman engineers. This is not unreasonable because a student usually needs several prerequisite college mathematics courses to fully grasp and appreciate the rich theory associated with the subject. Although it is generally true that ordinary differential equations are best taken after freshman year, it should be noted that most of the subject's complexity is associated with obtaining exact analytical solutions. In contrast, the simpler numerical solutions for solving ordinary differential equations with the computer are, in fact, quite easy to understand. For this reason, and because of their prominence in computer mathematics, we have chosen to devote the present chapter to solving simple ordinary differential equations. This material will provide you with an introduction to the subject, an introduction that will we hope, stimulate and motivate your future studies of the theory associated with differential equations. Before discussing how rate equations can be solved with a personal computer, we will first orient you by describing what they are mathematically and where they originate in engineering.

32.1 WHAT IS A RATE EQUATION?

In order to illustrate what we mean by a rate equation, let us start with a given function:

$$y = -0.5x^4 + 4x^3 - 10x^2 + 8.5x + 1 \qquad (32.1)$$

where y is the dependent variable and x is the independent variable. Equation (32.1) is a fourth-order polynomial. As shown in Fig. 32.1a, it can be used to compute

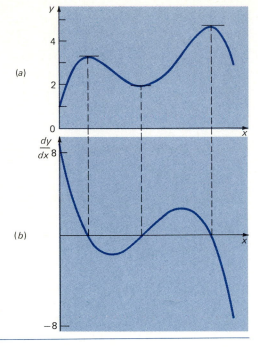

FIGURE 32.1
(a) Plot of y versus x and (b) plot of dy/dx versus x for the function $y = -0.5x^4 + 4x^3 - 10x^2 + 8.5x + 1$. Notice how the zero values of the derivatives correspond to the point at which the original function is flat—that is, has a zero slope. Also notice that the maximum absolute values of the derivatives are at the ends of the interval where the slopes of the function are greatest.

values of y for corresponding values of x. Thus the equation characterizes the relationship between y and x. Now if we apply the rule expressed previously in Eq. (30.2), we can differentiate Eq. (32.1) to give

$$\frac{dy}{dx} = -2x^3 + 12x^2 - 20x + 8.5 \tag{32.2}$$

This equation also characterizes the relationship between y and x but in a different manner from that of Eq. (32.1). Rather than explicitly representing the value of y for each value of x, Eq. (32.2) provides the rate of change of y with respect to x (that is, the slope) for each value of x. Figure 32.1 shows plots of both the function and its derivative.

Equations that contain derivatives, such as Eq. (32.2), are called *differential equations*. Although there are various types, the kind represented by Eq. (32.2) is called a *first-order ordinary differential equation*. The adjective "first-order" refers to the fact that the highest derivative in the equation is a first derivative. The adjective "ordinary" refers to the fact that the dependent variable y is a function of only one independent variable x. First-order ordinary differential equations can usually be written in the format of Eq. (32.2)—that is, with the derivative isolated on the left side of the equal sign. Thus the equation can be used to compute the rate of change of the dependent variable with respect to the independent variable at each value of the independent variable. For this reason, we give such equations a special name—*rate equations*.

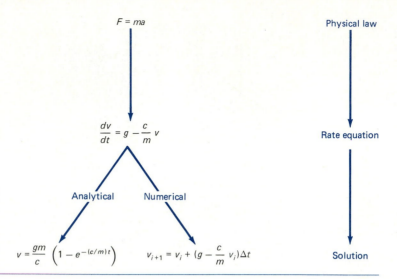

FIGURE 32.2
The sequence of events in the application of rate equations for engineering problem solving. The example shown is for the velocity of a falling parachutist.

Before moving on to the place of rate equations in engineering, we will briefly elaborate on what is meant by the "solution" of a rate equation. This is important because the solution is the primary result that we will find valuable in our engineering application of rate equations. As just demonstrated in going from Eq. (32.1) to (32.2), we can generate a differential equation given the function. In most engineering work, we are usually interested in the reverse process—that is, the object is usually to determine the function given the differential equation. The function then represents the solution.

As described in the next section, the differential equation is usually derived from a physical law. It typically characterizes the rate of change of a variable of interest with respect to an independent variable such as distance or time. The object of the exercise or the solution is, therefore, to come up with a nondifferential equation or function to allow us to compute the variable of interest itself as a function of the independent variable. The process is depicted schematically in Fig. 32.2.

32.2 WHERE DO RATE EQUATIONS COME FROM?

Although rate equations are the simplest type of differential equations, they have wide applicability in engineering. This is because many of the fundamental laws of physics, mechanics, electricity, and thermodynamics explain *variations* in physical properties and states of systems. Rather than describing the *state* of physical systems directly, the laws are usually couched in terms of spatial and temporal *changes*.

A simple example relates to the velocity of a falling parachutist. At several previous points in the text, we have used the following equation to characterize the velocity of a falling parachutist:

$$v = \frac{gm}{c}(1 - e^{-(c/m)t}) \tag{32.3}$$

FIGURE 32.3
Schematic diagram of the forces acting on a falling parachutist. F_D is the downward force due to gravity; F_U is the upward force due to air resistance.

where m is the parachutist's mass (in kilograms), c is a drag coefficient (in kilograms per second), and t is time (in seconds). We will now demonstrate how this relationship is actually the solution of a rate equation derived from Newton's second law of motion. Although this example is very simple, it is a good illustration of how rate equations are derived from physical laws in engineering.

As stated above, the derivation starts with Newton's second law:

$$F = ma \tag{32.4}$$

where F is the net force acting on a body (in newtons, or kilogram-meters per second squared) and a is its acceleration (in meters per second squared). This law can be formulated as a differential equation by expressing the acceleration as the time rate of change of the velocity (dv/dt) and dividing by m to yield

$$\frac{dv}{dt} = \frac{F}{m} \tag{32.5}$$

where v is velocity (in meters per second). Thus by isolating the derivative on the left side of the equal sign, we have expressed Newton's second law in the general format of a rate equation.

Next we must express the net force in terms of measurable variables and parameters. As shown in Fig. 32.3, the net force is composed of two individual forces, gravity and air resistance. We have already developed quantitative relationships for these forces in Chap. 29. We recall [Eq. (29.2)] that the downward force of gravity is represented by

$$F_D = mg$$

whereas, according to Eq. (29.3), the upward force of air resistance is represented by

$$F_U = -cv$$

Therefore the net force is obtained by adding these individual forces:

$$F = mg - cv$$

This result can then be substituted into Eq. (32.5) to give

$$\frac{dv}{dt} = g - \frac{c}{m} v \tag{32.6}$$

Equation (32.6) is a model that relates the acceleration of a falling object to the forces acting on it. It is also a rate equation that is similar in form to Eq. (32.2). That is, it characterizes the rate of change of a dependent variable v with respect to a single independent variable t. As mentioned previously, the "solution" for such an equation is another equation that characterizes the values of the dependent variable itself with respect to the independent variable. There are two fundamental approaches to obtaining such solutions, analytical and numerical.

An analytical or exact solution for Eq. (32.6) is obtained using the methods of calculus. For example, if the parachutist is originally at rest ($v = 0$ at $t = 0$), calculus can be used to solve Eq. (32.6) for

$$v = \frac{gm}{c}(1 - e^{-(c/m)t}) \tag{32.7}$$

Thus you can now see that this equation, which we have used on several previous occasions to characterize the parachutist's velocity, is actually the solution of a differential equation.

EXAMPLE 32.1   *Analytical Solution for the Falling Parachutist Problem*

PROBLEM STATEMENT: A parachutist with a mass of 68.1 kg jumps out of a stationary hot air balloon. Use Eq. (32.7) to compute velocity prior to opening the chute. The drag coefficient is equal to 12.5 kg/s.

SOLUTION: Inserting the parameters into Eq. (32.7) yields

$$v = \frac{9.8(68.1)}{12.5}(1 - e^{-(12.5/68.1)t})$$

$$= 53.39(1 - e^{-0.18355t})$$

which can be used to compute

t, s	v, m/s
0	0.00
2	16.40
4	27.77
6	35.64
8	41.10
10	44.87
12	47.49
∞	53.39

According to the model, the parachutist accelerates rapidly (Fig. 32.4). A velocity of 44.87 m/s (100.4 mi/h) is attained after 10 s. Note also that after a sufficiently long time, a constant velocity, called the *terminal velocity*, of 53.39 m/s (119.4 mi/h) is reached. This velocity is constant because, after a sufficient time, the force of gravity will be in balance with the air resistance. Thus the net force is zero and acceleration ceases.

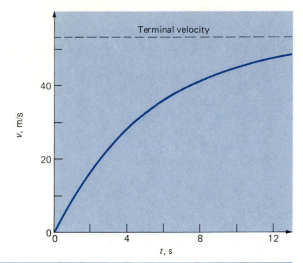

FIGURE 32.4
The analytical solution to the falling parachutist problem as computed in Example 32.1. Velocity increases with time and asymptotically approaches a terminal velocity.

Equation (32.7) is called an *analytical* or *exact solution* because it exactly satisfies the original differential equation. Unfortunately, there are many rate equations that cannot be solved exactly. In many of these cases, the only alternative is to develop a numerical solution that approximates the exact solution. This solution is not itself a continuous function. Rather, it is a table of numerical values that approximates $v(t)$ at various values of t. As described in the next section, the simplest technique available for this purpose is Euler's method.

32.3 EULER'S METHOD

One way to introduce Euler's method is to derive it to solve the falling parachutist problem. As should be obvious by now, the fundamental approach for solving a mathematical problem with the computer is to reformulate the problem so that it can be solved by arithmetic operations. Our one stumbling block for solving Eq. (32.6) in this way is the derivative term dv/dt. However, as we have already shown in Chap. 30, difference approximations can be used to express derivatives in arithmetic terms. For example, using a forward difference [Eq. (30.6)], the first derivative of v with respect to t can be approximated by

$$\frac{dv}{dt} \cong \frac{\Delta v}{\Delta t} = \frac{v_{i+1} - v_i}{h} \tag{32.8}$$

where v_i and v_{i+1} are velocities at a present and a future time, respectively, and h is the step size $\Delta t = t_{i+1} - t_i$. Substituting Eq. (32.8) into Eq. (32.6) yields

$$\frac{v_{i+1} - v_i}{h} = g - \frac{c}{m} v_i$$

which can be solved for

$$v_{i+1} = v_i + \left(g - \frac{c}{m} v_i \right) h \tag{32.9}$$

Notice that the term in parentheses is the rate equation itself [Eq. (32.6)]. That is, it provides a means to compute the rate of change or slope of v. Thus the differential equation has been transformed into an equation that can be used to determine the velocity algebraically at t_{i+1} using the slope and previous values of v. If you are given an initial value for velocity at some time t_i, you can easily compute velocity at a later time t_{i+1}. This new value of v at t_{i+1} can in turn be employed to extend the computation to v at t_{i+2} and so on. Thus at any time along the way,

New value = old value + slope × step size

EXAMPLE 32.2   *Numerical Solution to the Falling Parachutist Problem*

PROBLEM STATEMENT: Perform the same computation as in Example 32.1 but use Eq. (32.9) to compute velocity. Employ a step size of 2 s for the calculation.

SOLUTION: At the start of the computation ($t_i = 0$), the velocity of the parachutist is zero. Using this information and the parameter values from Example 32.1, Eq. (32.9) can be used to compute velocity at $t_{i+1} = 2$ s:

$$v = 0 + \left[9.8 - \frac{12.5}{68.1} (0) \right] 2 = 19.60 \text{ m/s}$$

For the next interval (from $t = 2$ to 4 s), the computation is repeated, with the result

$$v = 19.6 + \left[9.8 - \frac{12.5}{68.1} (19.6) \right] 2 = 32.00 \text{ m/s}$$

The calculation is continued in a similar fashion to obtain additional values:

t, s	v, m/s
0	0.00
2	19.60
4	32.00
6	39.85
8	44.82
10	47.97
12	49.96
∞	53.39

The results are plotted in Fig. 32.5 along with the exact solution. It can be seen that the numerical method accurately captures the major features of the exact solution.

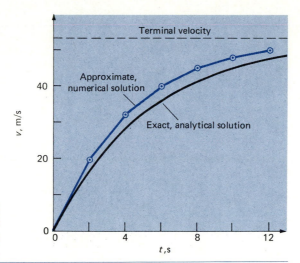

FIGURE 32.5
Comparison of the numerical and analytical solutions for the falling parachutist problem.

However, because we have employed straight-line segments to approximate a continuously curving function, there is some discrepancy between the two results. One way to minimize such discrepancies is to use a smaller step size. For example, applying Eq. (32.9) at 1-s intervals results in a smaller error, as the straight-line segments track closer to the true solution. Using hand calculations, the effort associated with using smaller and smaller step sizes would make such numerical solutions impractical. However, with the aid of the computer, large numbers of computations can be performed easily. Thus you can accurately model the velocity of the falling parachutist without having to solve the differential equation exactly.

According to Eq. (32.9), the slope estimate provided by the rate equation is used to extrapolate linearly from an old value to a new value over a step size h. This approach can be represented generally as

$$y_{i+1} = y_i + \left(\frac{dy}{dx}\right)_i h \qquad (32.10)$$

where $(dy/dx)_i$ represents the value of the rate equation evaluated at x_i. This formula is referred to as *Euler's* (or the *point-slope*) *method* (Fig. 32.6). As described in the next section, it is very easy to implement this method with the computer.

32.4 COMPUTER PROGRAMS FOR EULER'S METHOD

As seen in Fig. 32.7, it is a relatively simple matter to develop programs for Euler's method. Notice that the heart of both programs is two nested loops. The inner loop performs the actual computation. The outer loop is used so that answers do not have

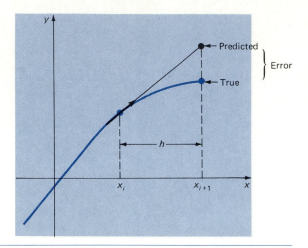

FIGURE 32.6
Graphical depiction of Euler's method.

to be printed out for each computation step. This is important because for applications with a very small step size, the quantity of output would be enormous. Thus the programs in Fig. 32.7 allow you to control the output so that a reasonable number of values are printed.

The programs in Fig. 32.7 do not have user-friendly input or output. In Prob. 32.1 you will have the task of including such features in your version of the program. Additionally, because the graphical capabilities of personal computers are so varied, we have not included a plotting option in Fig. 32.7. However, such an option is useful for the effective interpretation of the results. If your computer system has plotting capabilities, we recommend that you expand your program to include a plot of the solution. The printer plots described at the end of Chap. 22 can be modified for this purpose. In any event, the inclusion of this capability will greatly enhance the program's utility for problem solving.

Figure 32.8 shows some screens generated with a commercial software package that has an Euler's method algorithm very similar to Fig. 32.7. The package is used to solve Example 32.2. The solution is shown in Fig. 32.8a, and a plot of the results is shown in Fig. 32.8b. These screens illustrate how output can be designed in a clear and accessible fashion. You can employ them as models for your own software.

32.5 ADVANCED METHODS

Euler's method is called a first-order approach because it employs straight-line projections to approximate the trajectory of the rate equation. Higher-order methods are available that are capable of more faithfully capturing the curvature of the solution; hence they are more accurate than Euler's approach. The most common of these are called *Runge-Kutta methods*. In addition, Euler's and the higher-order techniques can be used to determine solutions of simultaneous rate equations. These are very impor-

```
100 REM     ***************************************************
110 REM     *               EULER'S METHOD                    *
120 REM     *              (BASIC version)                    *
130 REM     *               by  S.C. Chapra                   *
140 REM     *                                                 *
150 REM     *   This program uses Euler's method to find      *
160 REM     *   the solution of a differential equation.      *
170 REM     ***************************************************
180 REM
190 REM     .---------------------------------------------------.
200 REM     |            DEFINITION OF VARIABLES                 |
210 REM     |                                                   |
220 REM     | X0    = INITIAL VALUE OF INDEPENDENT VARIABLE      |
230 REM     | X1    = FINAL VALUE OF INDEPENDENT VARIABLE        |
240 REM     | Y0    = INITIAL VALUE OF DEPENDENT VARIABLE        |
250 REM     | H     = STEP SIZE                                  |
260 REM     | PRNT  = PRINT INTERVAL                             |
270 REM     | NP    = NUMBER OF PRINT STEPS                      |
280 REM     | NC    = NUMBER OF COMPUTATION STEPS                |
290 REM     | SLOPE = VALUE OF DIFFERENTIAL EQUATION             |
300 REM     '---------------------------------------------------'
310 REM
320 REM          Function Specifying Differential Equation
330 REM
340 DEF FNF(Y) = -2*X^3 + 12*X^2 - 20*X + 8.5
350 REM
360 REM Input Initial Conditions and Integration Parameters
370 REM
380 CLS
390 INPUT "INITIAL, FINAL INDEPENDENT VALUES? ", X0,X1
400 INPUT "INITIAL DEPENDENT VALUE? ", Y0
410 INPUT "STEP SIZE = ? ",H
420 INPUT "PRINT INTERVAL = ? ",PRNT
430 PRINT:PRINT
440 REM
450 REM          Compute Print and Computation Steps
460 REM
470 NP = (X1 - X0)/PRNT
480 NC = PRNT/H
490 REM
500 REM          Assign and Print out Initial Conditions
510 REM
520 X = X0
530 Y = Y0
540 PRINT X,Y
550 REM
560 REM          Nested Loops to Implement and Print out
570 REM                   Results of Euler's Method
580 REM
590 FOR I = 1 TO NP
600    FOR J = 1 TO NC
610       Y = Y + FNF(Y)*H
620       X = X + H
630    NEXT J
640    PRINT X,Y
650 NEXT I
660 END
```

```
C     ***************************************************
C     *               EULER'S METHOD                    *
C     *              (FORTRAN version)                   *
C     *               by  S.C. Chapra                    *
C     *                                                  *
C     *   This program uses Euler's method to solve      *
C     *   a single ordinary differential equation.       *
C     ***************************************************
C
C     .---------------------------------------------------.
C     |            DEFINITION OF VARIABLES                 |
C     |                                                   |
C     | X0    = INITIAL VALUE OF INDEPENDENT VARIABLE      |
C     | X1    = FINAL VALUE OF INDEPENDENT VARIABLE        |
C     | Y0    = INITIAL VALUE OF DEPENDENT VARIABLE        |
C     | H     = STEP SIZE                                  |
C     | PRNT  = PRINT INTERVAL                             |
C     | NP    = NUMBER OF PRINT STEPS                      |
C     | NC    = NUMBER OF COMPUTATION STEPS                |
C     | SLOPE = VALUE OF DIFFERENTIAL EQUATION             |
C     '---------------------------------------------------'
C
C     ****************** MAIN PROGRAM *******************
C
      Input Initial Conditions and Integration Parameters
C
      READ(5,*) X0,X1
      READ(5,*) Y0
      READ(5,*) H
      READ(5,*) PRNT
C
C          Compute Print and Computation Steps
C
      NP = (X1 - X0)/PRNT
      NC = PRNT/H
C
C          Assign and Print out Initial Conditions
C
      X = X0
      Y = Y0
      WRITE(6,*) X,Y
C
C          Nested Loops to Implement and Print out
C                   Results of Euler's Method
C
      DO 20 I=1, NP
         DO 10 J=1, NC
            Y = Y + FNF(Y,X)*H
            X = X + H
10       CONTINUE
         WRITE(6,*) X,Y
20    CONTINUE
      STOP
      END
C     ---------------------------------------------------
C          Function Specifying Differential Equation
C
      REAL FUNCTION FNF(Y,X)
      REAL X,Y
      FNF = -2*X**3 + 12*X**2 - 20*X + 8.5
      RETURN
      END
```

FIGURE 32.7 BASIC and FORTRAN computer programs for Euler's method.

tant in engineering because they provide a way to predict the dynamics of systems of interconnected elements such as the ones shown in Fig. 29.1. Whereas the techniques in Chap. 29 focused on predicting the steady state of such systems, simultaneous differential equations are used to predict how the individual elements of such systems vary dynamically. Additional information on these and other advanced topics related to rate equations can be found elsewhere (Chapra and Canale, 1985).

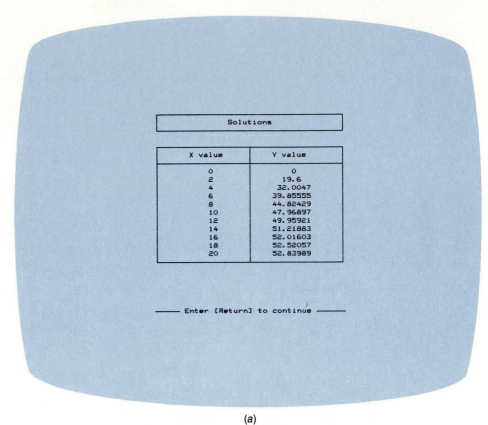

(a)

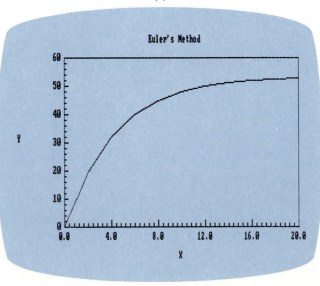

FIGURE 32.8
Computer screens generated by an Euler's method package (NUMERICOMP by Canale and Chapra, 1985). The package is used to solve the problem previously analyzed in Example 32.2.

(b)

PROBLEMS

32.1 Utilize Fig. 32.7 as a starting point for developing your own computer program for Euler's method. Employ structured programming techniques, modular design, and internal and external documentation in your development. Incorporate the capability to plot the solution.

32.2 What is meant by a first-order ordinary differential equation? Give an example. Provide an example of a second-order ordinary differential equation.

32.3 What type of finite-divided difference is employed in Eq. (32.8)?

32.4 Repeat Example 32.2 using 1-s intervals and compare with the exact solution and the 2-s interval numerical solution. Compute the percent relative error for both numerical solutions at $t = 2, 4, 6$, and 8 relative to the analytical solution. What conclusion can be drawn regarding halving the step size?

32.5 Employ Newton's law of cooling (Problem 30.8) to calculate the temperature of a metal ball dropped into water that is held constant at $T_a = 20°C$. Compute the ball's temperature at 1-s intervals from $t = 0$ to 25 min. The ball's temperature at $t = 0$ is 90°C. Use a value of 0.1 min^{-1} for k.

32.6 Repeat Prob. 32.5 but employ a value of 0.2 per minute for k.

32.7 Population-growth dynamics are important in a variety of engineering planning studies. One of the simplest models of such growth incorporates the assumption that the rate of change of the population p is proportional to the existing population at any time t:

$$\frac{dp}{dt} = Gp \qquad \text{(P32.7)}$$

where G is a growth rate (per year). This model makes intuitive sense because the greater the population, the greater the number of potential parents.

At time $t = 0$ an island has a population of 10,000 people. If $G = 0.075$ per year, employ Euler's method to predict the population at $t = 20$ years, using a step size of 0.5 years. Plot p versus t on standard and semilog graph paper. Determine the slope of the line on the semilog plot. Discuss your results. *Note:* Recall Eq. (22.6) and the accompanying discussion.

32.8 Although the model in Prob. 32.7 works adequately when population growth is unlimited, it breaks down when factors such as food shortages, pollution, and lack of space inhibit growth. In such cases, the growth rate itself can be thought of as being inversely proportional to population. One model of this relationship is

$$G = G'(p_{max} - p) \qquad \text{(P32.8)}$$

where G' is a population-dependent growth rate (per people-year) and p_{max} is the maximum sustainable population. Thus when population is small

$(p << p_{max})$, the growth rate will be at a high, constant rate of $G'p_{max}$. For such cases, growth is unlimited and Eq. (P32.8) is essentially identical to Eq. (P32.7). However, as population grows (that is, p approaches p_{max}), G decreases until at $p = p_{max}$ it is zero. Thus the model predicts that when the population reaches the maximum sustainable level, growth is nonexistent, and the system is at a steady state. Substituting Eq. (P32.8) into Eq. (P32.7) yields

$$\frac{dp}{dt} = G'(p_{max} - p)p$$

For the same island studied in Prob. 32.7, employ Euler's method to predict the population at $t = 20$ years, using a step size of 0.5 years. Employ values of $G' = 10^{-5}$ per people-year and $p_{max} = 20{,}000$ people. At time $t = 0$ the island has a population of 10,000 people. Plot p versus t and interpret the shape of the curve.

32.9 For a simple ideal resistor-inductor circuit, the following rate equation holds:

$$L\frac{di}{dt} + Ri = 0$$

where i is current, L is inductance, and R is resistance. Solve for i, if $L = R = 1$ and i at $t = 0$ equals 0.1 A. Employ Euler's method with a step size of 0.1 and solve for i from $t = 0$ to 1. Plot i versus t.

32.10 Real circuits often do not behave in the idealized fashion represented by Prob. 32.9. For example, circuit dynamics might be described by a relationship such as

$$L\frac{di}{dt} + (-i + i^3)R = 0$$

where all parameters are as defined in Prob. 32.9. Solve for i as a function of time under the same conditions specified in Prob. 32.9. Plot the results.

32.11 Suppose that, after falling for 10 s, the parachutist from Example 32.1 and 32.2, pulls the rip cord. At this point, assume that the drag coefficient is instantaneously increased to a constant value of 50 kg/s. Compute the parachutist's velocity from $t = 10$ to 30 s with Euler's method. Employ all the parameters from Examples 32.1 and 32.2 and take the initial condition from the analytical solution. Plot v versus t for $t = 0$ to 30 s. Use the analytical solution for $t = 0$ to 10 s and the numerical solution obtained above for $t = 10$ to 30 s.

32.12 The rate of heat flow (conduction) between two points on a cylinder heated at one end is given by

$$\frac{dQ}{dt} = \lambda A \frac{dT}{dx}$$

where λ is a constant, A is the cylinder's cross-sectional area, Q is heat flow, T is temperature, t is time, and x is the distance from the heated end. Because the equation involves two derivatives, we will simplify this equation by letting

$$\frac{dT}{dx} = \frac{100(L - x)(20 - t)}{100 - xt}$$

where L is the length of the rod. Combining the two equations gives

$$\frac{dQ}{dt} = \lambda A \frac{100(L - x)(20 - t)}{100 - xt}$$

Employ Euler's method to compute the heat flow for $t = 0$ to 15 s, if $\lambda = 0.3$ cal·cm/s·°C, $A = 10$ cm², $L = 20$ cm, and $x = 2.5$ cm. The initial condition is that $Q = 0$ at $t = 0$. Plot your results.

32.13 Recognizing that $dx/dt = v$, solve Prob. 31.11 with Euler's method.

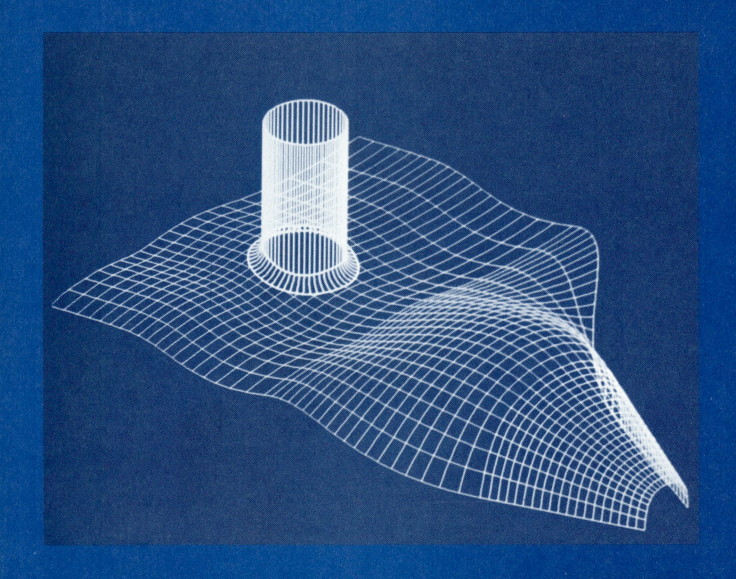

PART EIGHT

ENGINEERING APPLICATIONS

In the previous part of the book, we reviewed numerical methods for solving mathematical problems with the computer. Along with the methods, we included examples and end-of-the chapter problems drawn from engineering practice. The present part of the book is intended to examine such computer-oriented applications in more depth.

All of the following chapters focus on mathematical modeling. Rather than using a general example such as the falling parachutist from Part Seven, each chapter here focuses on specific applications from the four major areas of engineering: chemical, civil, mechanical, and electrical. In each case we present an organizing principle that lies at the heart of computer modeling for the area.

Chapter 33 offers an overview of the major organizing principles of engineering. *Chapter 34* is devoted to mass-balance models for reactors in chemical engineering. In *Chap. 35,* which focuses on civil engineering, force balances are used to analyze a structure. *Chapter 36* also applies a force balance but uses it to predict the dynamics of a machine. *Chapter 37* employs Kirchhoff's laws to perform the electrical engineering problem of network analysis.

CHAPTER33

ORGANIZING PRINCIPLES IN ENGINEERING

Knowledge and understanding lie at the heart of the effective implementation of any tool. No matter how impressive your tool chest, you will be hard-pressed to repair a car if you do not understand how it works.

The same holds true for the tools that are used for engineering problem solving. To this point, you have been introduced to several of these tools: computer hardware and software, graphics, sorting, statistics, and numerical methods. Although each of the tools has great potential utility, they are practically useless without a fundamental understanding of how engineering systems work.

This understanding is gained by empirical means—that is, by observation and experiment. The section of the book on data analysis (Part Six) provides techniques and concepts that play an important role in empirical studies.

Although careful observations and experiments are at the heart of our efforts to understand engineering systems, they are only half the story. Over years and years of observation and experiment, engineers and scientists have noticed that certain aspects of their empirical studies occur repeatedly. Such general behavior can then be expressed as fundamental laws, or organizing principles that essentially embody the cumulative wisdom of past experience. Thus most engineering problem solving employs the two-pronged approach of empiricism and theoretical analysis.

It must be stressed that the two prongs are closely coupled. As new measurements are taken, the generalizations may be modified or new ones developed. Similarly, the generalizations can have a strong influence on the experiments and observations. In particular, generalizations can serve as organizing principles that can be employed to synthesize observations and experimental results into a coherent and comprehensive framework from which conclusions can be drawn. Among the most important of these organizing principles are the conservation laws.

33.1   CONSERVATION LAWS

Although they form the basis for a variety of complicated and powerful mathematical models, the great conservation laws of science and engineering are very simple to understand. They all boil down to

$$\text{Change} = \text{increases} - \text{decreases} \tag{33.1}$$

This equation can be understood using an example from everyday life—your bank account.

Suppose that you apply Eq. (33.1) to your checking account over the course of a month. If you deposit more money than you withdraw, your balance at the end of the month will be greater than at the beginning. According to Eq. (33.1), your account will have a positive change because the increases (deposits) are greater than decreases (withdrawals). Conversely, if you withdraw more than you deposit, the change will be negative, and your account will be less at the month's end. In either case, the change, as calculated by Eq. (33.1), can be used to determine the outcome.

EXAMPLE 33.1   Conservation of Cash

PROBLEM STATEMENT: Suppose you have $150 in your account at the beginning of the month. Use Eq. (33.1) to compute your balance at the end of the month. During this time interval, you make deposits of $40, $30, and $150 and withdraw $75 and $15.

SOLUTION: First we can reformulate Eq. (33.1) in terms of the present problem context:

Change in dollars = total deposits − total withdrawals

This equation is now an expression of the conservation of cash. If the bank stays in operation over the course of the month, the equation provides an organizing principle for computing the change in dollars in your account:

Change in dollars = (40 + 30 + 150) − (75 + 15) = 130

This result can then be used to compute the dollars in the account at the end of the month using the formula

$$\begin{array}{c} \text{Dollars at end} \\ \text{of month} \end{array} = \begin{array}{c} \text{dollars at start} \\ \text{of month} \end{array} + \begin{array}{c} \text{change} \\ \text{in dollars} \end{array}$$

or

$$\begin{array}{c} \text{Dollars at end} \\ \text{of month} \end{array} = 150 + 130 = 280$$

Therefore we have calculated that there is $280 in the account at the end of the month.

Although simple, the foregoing example embodies one of the most fundamental ways in which conservation laws are used in engineering—that is, to predict changes. We will give it a special name—the *transient* or *time-variable computation*.

Aside from predicting changes, another way in which conservation laws are applied is for cases where change is nonexistent. If change is zero, Eq. (33.1) becomes

$$\text{Change} = 0 = \text{increases} - \text{decreases}$$

or

$$\text{Increases} = \text{decreases}$$

Thus, if no change occurs, the increases and decreases must be in balance. This case, which is also given a special name—the *steady-state computation*—has many applications in engineering. For example, for incompressible fluid flow in pipes, the flow into a junction must be balanced by flow going out, as in

$$\text{Flow in} = \text{flow out} \tag{33.2}$$

For the junction in Fig. 33.1, Eq. (33.2) can be used to compute that the flow out of the fourth pipe must be 60.

Although Fig. 33.1 and Example 33.1 might appear trivially simple, they embody the two fundamental ways in which conservation laws are applied in engineering. A major objective of the present part of the book is to develop some examples of how this is done.

33.2 BALANCES IN ENGINEERING

In the previous section we showed how conservation laws can be used to develop balances. These, in turn, can be applied to compute both transient and steady-state solutions for engineering problems. In the following chapters, we will develop four such balances for the major areas of engineering: chemical, civil, mechanical, and

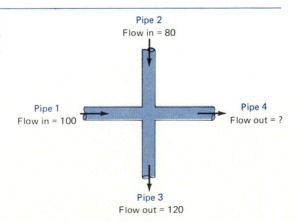

FIGURE 33.1
A flow balance for incompressible fluid flow at the junction of pipes.

Pipe 2
Flow in = 80

Pipe 1
Flow in = 100

Pipe 4
Flow out = ?

Pipe 3
Flow out = 120

electrical. For each case we will focus our discussion on a particular device that is fundamental to that area of engineering. The devices and types of balances are summarized in Table 33.1.

TABLE 33.1 Devices and Types of Balances that Are Commonly Used in the Four Major Areas of Engineering. For each case, the conservation law upon which the balance is based is specified.

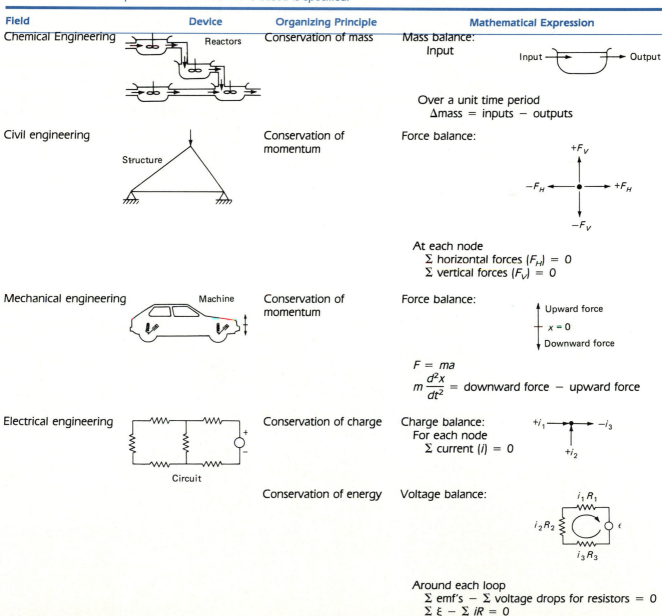

Field	Device	Organizing Principle	Mathematical Expression
Chemical Engineering	Reactors	Conservation of mass	Mass balance: Input Over a unit time period Δmass = inputs − outputs
Civil engineering	Structure	Conservation of momentum	Force balance: At each node Σ horizontal forces $(F_H) = 0$ Σ vertical forces $(F_V) = 0$
Mechanical engineering	Machine	Conservation of momentum	Force balance: $F = ma$ $m\dfrac{d^2x}{dt^2}$ = downward force − upward force
Electrical engineering	Circuit	Conservation of charge	Charge balance: For each node Σ current $(i) = 0$
		Conservation of energy	Voltage balance: Around each loop Σ emf's − Σ voltage drops for resistors = 0 $\Sigma \xi - \Sigma iR = 0$

Chapter 34, which relates to chemical engineering, focuses on mass balances of reactors. The mass balance derives from conservation of mass. It specifies that the change of mass in the reactor depends on the amount of mass flowing in minus the mass flowing out. We will develop equations from this principle that can be employed for both transient and steady-state computations.

Chapters 35 and 36 both employ force balances that are essentially derived from the conservation of momentum. In Chap. 35, a force balance is utilized for the analysis of a truss. For this case, we assume that the truss is not in motion, and therefore the summation of vertical and horizontal forces at each node of the truss must equal zero. Thus this example focuses on steady-state behavior of the structure. In contrast, Chap. 36 applies the same basic principle but uses it to analyze the up-and-down motion or vibrations of an automobile. For this case, the rate of change of the vertical motion is equal to the difference between downward and upward forces. Thus this example focuses on the transient behavior of the vehicle.

Finally, in Chap. 37 both charge and energy balances are used to model an electric circuit. The charge balance, which derives from conservation of charge, is similar in spirit to the flow balance depicted in Fig. 33.1. Just as flow must balance at the junction of pipes, electric current must balance at the junction of electric wires. The energy balance specifies that the change of energy around any loop of the circuit must

FIGURE 33.2
The engineering problem-solving process.

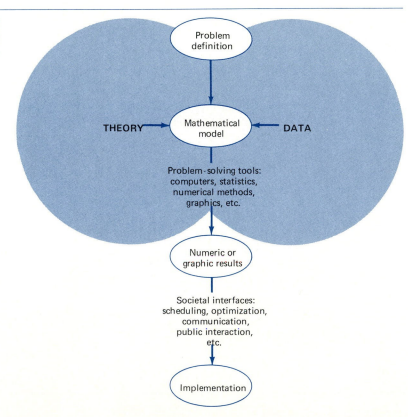

add up to zero. These balances will be used to compute both transient and steady-state solutions for electric circuits.

Although there are many other examples of the application of balances, the four outlined in Table 33.1 are fundamental enough to introduce you to the way in which conservation laws serve as organizing principles in engineering problem solving.

33.3 THE ENGINEERING PROBLEM-SOLVING PROCESS

While laws and idealizations are important organizing concepts, they are but one facet of the total scheme of engineering problem solving (Fig. 33.2). Therefore, although the balances serve as the focus of the upcoming chapters, we will also attempt to demonstrate how other tools such as data analysis, numerical methods, and graphics figure in problem solving. Of course, our primary intent will be to illustrate how computers enhance the entire process.

PROBLEMS

33.1 What is the two-pronged approach to engineering problem solving? Into what category should the conservation laws be placed?

33.2 What is the form of the transient conservation law? What is it for steady state?

33.3 The following information is available for a bank account:

Date	Deposits	Withdrawals	Balance
5/1			512.33
	200.13	427.26	
6/1			
	206.80	328.61	
7/1			
	450.25	206.80	
8/1			
	127.31	380.61	
9/1			

Use the conservation of cash to compute the balance on 6/1, 7/1, 8/1, and 9/1. Show each step in the computation. Is this a steady-state on a transient computation?

33.4 Give examples of conservation laws in engineering and in everyday life.

CHAPTER 34

CHEMICAL ENGINEERING: MASS BALANCE

Chemical engineers use reactors to concoct a variety of products, ranging from synthetic detergents to drugs. In the present chapter we will use numerical methods to analyze the behavior of systems of reactors.

34.1 USING LINEAR ALGEBRAIC EQUATIONS IN THE STEADY-STATE ANALYSIS OF A SERIES OF REACTORS

One of the most important organizing principles in chemical engineering is the *conservation of mass*. In quantitative terms, the principle is expressed as a mass balance that accounts for all sources and sinks of a material that pass in and out of a unit volume (Fig. 34.1). Over a finite period of time, this can be expressed as

$$\text{Accumulation} = \text{inputs} - \text{outputs} \tag{34.1}$$

The mass balance represents a bookkeeping exercise for the particular substance being modeled. For the period of the computation, if the inputs are greater than the outputs, the mass of the substance within the volume increases. If the outputs are greater than the inputs, the mass decreases. If inputs are equal to the outputs, accumulation is zero and mass remains constant. For this stable condition, or *steady-state,* Eq. (34.1) can be expressed as

$$\text{Inputs} = \text{outputs} \tag{34.2}$$

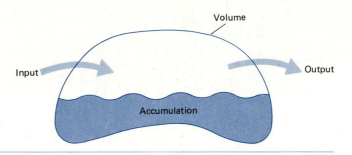

FIGURE 34.1
A schematic representation of mass balance.

The mass balance can be used for engineering problem solving by expressing the inputs and outputs in terms of measurable variables and parameters. For example, if we were performing a mass balance for a conservative substance (that is, one that does not increase or decrease due to chemical transformations) in a reactor (Fig. 34.2), we would have to quantify the rate at which mass flows into the reactor through the two inflow pipes and out of the reactor through the outflow pipe. This can be done by taking the product of the flow rate, Q (in cubic meters per minute), and the concentration c (in milligrams per cubic meter) for each pipe. For example, for pipe 1 in Fig. 34.2, $Q_1 = 2$ m³/min and $c_1 = 25$ mg/m³; therefore the rate at which mass flows into the reactor through pipe 1 is $Q_1c_1 = (2 \text{ m}^3/\text{min})(25 \text{ mg/m}^3) = 50$ mg/min. Thus, 50 mg of chemical flows into the reactor through this pipe each minute. Similarly, for pipe 2 the mass inflow rate can be calculated as $Q_2c_2 = (1.5 \text{ m}^3/\text{min})(10 \text{ mg/m}^3) = 15$ mg/min.

Notice that the concentration out of the reactor through pipe 3 is not specified by Fig. 34.2. This is because we already have sufficient information to calculate it on the basis of the conservation of mass. Because the reactor is at steady state, Eq. (34.2) holds and the inputs should be in balance with the outputs, as in

$$Q_1c_1 + Q_2c_2 = Q_3c_3$$

Substituting known values into this equation yields

$$50 + 15 = 3.5c_3$$

FIGURE 34.2
A steady-state, completely mixed reactor with two inflow pipes and one outflow pipe. The flows Q are in cubic meters per minute, and the concentrations c are in milligrams per cubic meter.

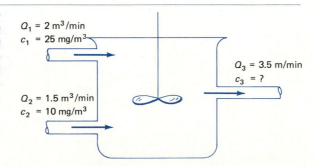

which can be solved for $c_3 = 18.6$ mg/m³. Thus we have determined the concentration in the third pipe. However, the computation yields an additional bonus. Because the reactor is well mixed (as represented by the propeller in Fig. 34.2) the concentration will be uniform, or homogeneous, throughout the tank. Therefore the concentration in pipe 3 should be identical to the concentration throughout the reactor. Consequently, the mass balance has allowed us to compute both the concentration in the reactor and in the outflow pipe. Such information is of great utility to chemical engineers who must design reactors to yield mixtures of a specified concentration.

Because simple algebra was used to determine the concentration for the single reactor in Fig. 34.2, it might not be obvious how computers figure in mass-balance calculations. Figure 34.3 shows a problem setting where computers are not only useful but are a practical necessity. Because there are five interconnected, or coupled, reactors, five simultaneous mass-balance equations are needed to characterize the system. For reactor 1, the rate of mass flow in is

$$5(10) + Q_{31}c_3$$

and the rate of mass flow out is

$$Q_{12}c_1 + Q_{15}c_1$$

Because the system is at steady state, the inflows and outflows must be equal:

$$5(10) + Q_{31}c_3 = Q_{12}c_1 + Q_{15}c_1$$

or, substituting the values for flow from Fig. 34.3,

$$6c_1 - c_3 = 50$$

FIGURE 34.3
Five reactors linked by pipes.

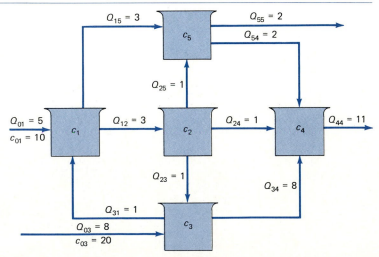

Similar equations can be developed for the other reactors:

$$- 3c_1 + 3c_2 = 0$$

$$- c_2 + 9c_3 = 160$$

$$- c_2 - 8c_3 + 11c_4 - 2c_5 = 0$$

$$- 3c_1 - c_2 + 4c_5 = 0$$

A numerical method such as Gauss elimination and a computer can be used to solve these five equations for the five unknown concentrations:

$$c_1 = 11.51$$
$$c_2 = 11.51$$
$$c_3 = 19.06$$
$$c_4 = 17.00$$
$$c_5 = 11.51$$

34.2 USING RATE EQUATIONS IN THE TRANSIENT ANALYSIS OF A REACTOR

In addition to steady-state computations, we might also be interested in the transient response of a completely mixed reactor. To do this, we have to develop a mathematical expression for the accumulation term in Eq. (34.1). Because accumulation represents the change in mass in the reactor per change in time, accumulation can be simply formulated as

$$\text{Accumulation} = \frac{\Delta M}{\Delta t} \qquad (34.3)$$

where M is the mass of chemical in the reactor. By definition, concentration is defined as mass per unit volume, or

$$c = \frac{M}{V}$$

This equation can be solved for mass and the result ($M = cV$) can be substituted into Eq. (34.3) to reexpress accumulation in terms of concentration:

$$\text{Accumulation} = \frac{\Delta cV}{\Delta t}$$

If we assume that the volume of liquid in the reactor is constant, V can be moved outside the difference:

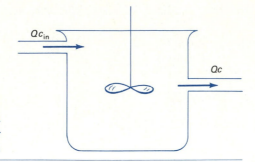

$$\text{Accumulation} = V\frac{\Delta c}{\Delta t}$$

Finally, as Δt approaches zero, the difference can be reexpressed as a derivative:

$$\text{Accumulation} = V\frac{dc}{dt} \tag{34.4}$$

Thus a mathematical formulation for accumulation is volume times the derivative of c with respect to t.

Now Eqs. (34.1) and (34.4) can be used to represent the mass balance for a single reactor such as the one shown in Fig. 34.4:

$$V\frac{dc}{dt} = Qc_{in} - Qc \tag{34.5}$$

This rate equation can be used to determine transient or time-variable solutions for the reactor. For example, if $c = c_0$ at $t = 0$, calculus can be employed to analytically solve Eq. (34.5) for

$$c = c_{in}(1 - e^{-(Q/V)t}) + c_0 e^{-(Q/V)t}$$

If $c_{in} = 50$ mg/m^3, $Q = 5$ m^3/min, $V = 100$ m^3, and $c_0 = 10$ mg/m^3, the equation is

$$c = 50(1 - e^{-0.05t}) + 10e^{-0.05t}$$

Figure 34.5 shows this exact, analytical solution.

Euler's method provides an alternative approach for solving Eq. (34.5). Figure 34.5 includes two solutions with different step sizes. As the step size is decreased, the numerical solution converges on the analytical solution. Thus, for this case, the numerical method can be used to check the analytical result. Although the analytical solution for this case is fairly simple and straightforward, other solutions to rate

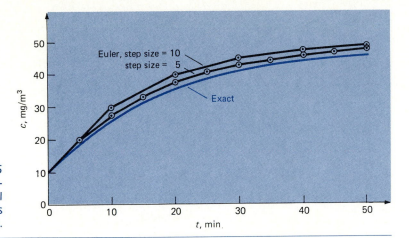

FIGURE 34.5
Plot of analytical and numerical solutions of Eq. (34.5). The numerical solutions are obtained with Euler's method using different step sizes.

equations may be quite complicated or impossible to obtain. In these cases, a numerical method that can be implemented quickly can be quite valuable in verifying the analytical solution or in providing the only viable solution.

Besides checking the results of an analytical solution, numerical solutions have added value in those situations where analytical solutions are impossible or so difficult that they are impractical. Suppose, for example, that the concentration of the inflow to the reactor is not constant but rather varies sinusoidally with time, as in (see Fig. 34.6a)

$$c_{in} = \bar{c}_{in} + c_a \sin \frac{2\pi t}{T} \tag{34.6}$$

where $\bar{c}_{in}$ is the base-inflow concentration, c_a is the amplitude of the oscillation, and T is its period (that is, the time required for a complete cycle). For this case, the mass balance equation is

$$V \frac{dc}{dt} = Q\left(\bar{c}_{in} + c_a \sin \frac{2\pi t}{T}\right) - Qc \tag{34.7}$$

It is possible to solve this equation by calculus, but the solution is time-consuming and complicated:

$$
\begin{aligned}
c = & \, c_0 \, e^{-(Q/V)t} + \bar{c}_{in}(1 - e^{-(Q/V)t}) \\
& + \frac{Qc_a}{V\sqrt{(Q/V)^2 + (2\pi/T)^2}} \left[\sin\left(\frac{2\pi t}{T} - \tan^{-1}\frac{2\pi V}{TQ}\right) \right. \\
& + \left. e^{-(Q/V)t} \sin\left(-\tan^{-1}\frac{2\pi V}{TQ}\right) \right]
\end{aligned}
\tag{34.8}
$$

TABLE 34.1 Values of Concentration Calculated Using Euler's Method to Solve Eq. (34.7) with an Initial Condition of $c_0 = 10$ at $t = 0$. A step size of 0.1 min was used for the computation, and results were output every 5 min.

t, min	c, mg/m^3
0	10.00
5	21.62
10	35.00
15	47.05
20	54.73
25	56.31
30	52.16
35	44.59
40	37.07
45	32.91
50	34.06

For such cases, Euler's approach offers a much easier means of obtaining the solution. Table 34.1 and Fig. 34.6b show the result corresponding to the situation in Fig. 34.6a for $c_0 = 10$ mg/m^3. The solution starts at the initial condition and then oscillates upward.

34.3 USING ROOTS OF EQUATIONS IN THE DETERMINATION OF A REACTOR'S RESPONSE TIME

Aside from the direct solution of the rate equation, other numerical methods can be used to investigate additional aspects of transient behavior. For example, suppose that you were interested in determining the time required for the concentration to reach 50 mg/m^3. This type of information might be needed to establish the approximate start-up time for the reactor. Computation of reactor response time can be formulated as a roots-of-equation problem. From the plot (Fig. 34.6b) we can make an approximate estimate of about 17 min. To obtain a more accurate estimate, bisection can be used to find the root of [Eq. (34.8) with numerical values substituted for the parameters]

$$f(t) = 10\, e^{-0.05t} + 50(1 - e^{-0.05t})$$
$$+ 14.7879[\sin(0.1257t - 1.1921) + 0.9292 e^{-0.05t}] - 50 = 0$$

Using bisection with initial guesses of $x_l = 10$, $x_u = 20$, and $\epsilon_s = 0.01$ percent yields a root of $t = 16.55$ min after 13 iterations, with an estimated error of $\epsilon_a = 7.4 \times 10^{-3}$ percent.

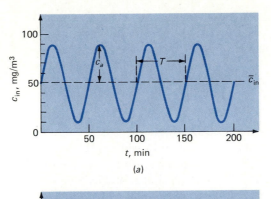

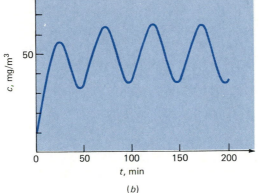

FIGURE 34.6
(a) Sinusoidal inflow concentration and (b) the resulting concentration in the reactor, as computed with Euler's method.

34.4 USING INTEGRATION IN THE DETERMINATION OF TOTAL MASS INPUT OR OUTPUT

A final application of a numerical method to the present problem context would be to compute how much mass entered or left the reactor over a specified time period. Integration provides a means to make this computation, as in

$$M = \int_{t_1}^{t_2} Qc\, dt$$

where t_1 and t_2 are the initial and final times. This formula makes intuitive sense if you recall the analogy between integration and summation. Thus the integral represents the summation of the product of flow times concentration to give the total mass entering or leaving from t_1 to t_2.

Because the flow rate for our example is constant, Q can be moved outside the integral:

$$M = Q \int_{t_1}^{t_2} c\, dt \tag{34.9}$$

Suppose you want to compute the total mass flowing in from $t_1 = 0$ to $t_2 = 50$ min. To do this, Eq. (34.6) is substituted into Eq. (34.9) to yield (with proper parameter values)

$$M_{in} = 5 \int_0^{50} (50 + 40 \sin 0.1257t) \, dt$$

Using calculus, this integral is easy to evaluate, with the result that $M_{in} = 12,500$ mg. This result can be visually verified by examining Fig. 34.6a. Because the integral is equivalent to the area between the curve and the t axis, it is clear that the integral equals 50 mg/m^3 × 50 min, or 2500 (mg/m^3) min. Multiplying this result by the flow of 5 m^3/min yields the result of 12,500 mg.

The evaluation of the outflow of mass is not quite as easy as for the inflow. This involves substituting Eq. (34.8) into Eq. (34.9). The result is difficult to evaluate analytically. However, using the values of c computed previously with Euler's method [Table 34.1, which is the numerical solution of Eq. (34.7)], and a method such as the multiple-segment trapezoidal rule yields an estimate for the integral of 2017.4 (mg/m^3) min. This can be multiplied by the flow of 5 m^3/min to give 10,087 mg. Therefore, over the time from $t = 0$ to 50, 12,500 mg was input to the reactor, while 10,087 mg was output. Thus 2413 mg accumulated in the reactor during this time. Dividing this amount by the reactor's volume gives $c = 2413$ mg/100 m^3 = 24.13 mg/m^3. Adding this to the 10 mg/m^3 that was originally in the reactor gives a total of 34.13 mg/m^3. This value agrees closely with the value of 34.06 computed with Euler's method (see Table 34.1 at $t = 50$). Thus not only does the integration provide us with information regarding how much mass enters and leaves the reactor, it also provides us with an independent check of the validity of our numerical solution to the rate equation.

PROBLEMS

34.1   Write the conservation-of-mass equation. What is its form at steady state?

34.2   Because the system shown in Fig. 34.3 is at steady state, what can be said regarding the four flows: Q_{01}, Q_{03}, Q_{44}, and Q_{55}?

34.3   Recompute the concentrations for the five reactors shown in Fig. 34.3, if the flows are changed to:

$$
\begin{array}{llll}
Q_{01} = 5 & Q_{31} = 2 & Q_{25} = 3 & Q_{23} = 1 \\
Q_{15} = 3 & Q_{55} = 4 & Q_{54} = 2 & Q_{34} = 5 \\
Q_{12} = 4 & Q_{03} = 6 & Q_{24} = 0 & Q_{44} = 7
\end{array}
$$

34.4   Solve the same system as specified in Prob. 34.3 but set Q_{12} and Q_{54} equal to zero. Use conservation of flow to recompute the values for the other flows. What does the answer indicate to you regarding the physical system?

34.5 If $c_{in} = \bar{c}(1 - e^{-t})$, calculate the outflow concentration of a single, completely mixed reactor as a function of time. Use Euler's method to perform the computation. Employ values of $\bar{c} = 50$ mg/m^3, $Q = 5$ m^3/min, $V = 100$ m^3, and $c_0 = 10$ mg/m^3. Perform the computation from $t = 0$ to 50 min.

34.6 The following equation pertains for a concentration of a chemical in a completely mixed reactor:

$$c = c_{in}(1 - e^{-0.05t}) + c_0 e^{-0.05t}$$

If $c_0 = 5$ and $c_{in} = 20$, compute the time required for c to be 95 percent of c_{in}. Employ bisection.

34.7 The outflow chemical concentration from a completely mixed reactor is measured as:

t, min	0	5	10	15	20	30	40	50	60
c, mg/m^3	10	20	30	40	60	80	70	50	60

For an outflow of $Q = 10$ m^3/min, estimate the mass of chemical that exits the reactor from $t = 0$ to 60 min.

34.8 Aside from inflow and outflow, another way by which mass can enter or leave a reactor is by a chemical reaction. For example, if the chemical decays, the reaction can sometimes be characterized as a first-order reaction:

Reaction $= -kVc$

where k is a reaction rate (in minutes^{-1}), which can generally be interpreted as the fraction of the chemical that goes away per unit time. For example, $k = 0.1$ min^{-1} can be thought of as meaning that roughly 10 percent of the chemical in the reactor decays in a minute. The reaction can be substituted into the mass-balance equation [Eq. (34.5)] to give

$$V\frac{dc}{dt} = Qc_{in} - Qc - kVc$$

If $k = 0.1$ min^{-1}, $c_{in} = 50$ mg/m^3, $Q = 5$ m^3/min, and $V = 100$ m^3, what is the steady-state concentration of the reactor?

34.9 Repeat Prob. 34.8 but compute the transient concentration response if $c_0 = 10$ mg/m^3. Compute the response with Euler's method from $t = 0$ to 20 min.

34.10 Duplicate the computation in Secs. 34.2 and 34.4 but change c_0 to 100 mg/m^3, Q to 10 m^3/min, and V to 50 m^3.

CHAPTER35

CIVIL ENGINEERING: STRUCTURAL ANALYSIS

Civil engineering encompasses a broad variety of specialties, including structural, water resources, transportation, geotechnical, and environmental engineering. Any one of these specialties could be employed to illustrate the application of physical laws, numerical methods, and computers for engineering problem solving. However, because of its fundamental place in the education of all civil engineers, we have chosen to focus this chapter on structural analysis.

35.1 USING LINEAR ALGEBRAIC EQUATIONS IN THE ANALYSIS OF A STATICALLY DETERMINANT TRUSS

An important problem in structural engineering is that of finding the forces and reactions associated with a statically determinant truss. Figure 35.1 is an example of such a truss.

The forces (F) represent either tension or compression on the internal members of the truss. External reactions (H_2, V_2, and V_3) are forces that characterize how the truss interacts with the supporting surface. The hinge at node 2 can transmit both horizontal and vertical forces to the surface, whereas the roller at node 3 transmits only vertical forces. It is observed that an external loading of 1000 lb is applied downward at node 1. The object of this problem is to determine how the structure's members and its supports are influenced by this force. To do this, we can describe the structure as a system of coupled linear algebraic equations. Free-body force diagrams are shown for each node in Fig. 35.2. The sum of the forces in both horizontal and vertical directions must be zero at each node because the system is at rest.

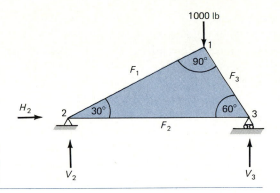

FIGURE 35.1
Forces on a statically determinant truss.

Therefore for node 1,

$$\Sigma F_H = 0 = -F_1 \cos 30° + F_3 \cos 60° + F_{1,h}$$

$$\Sigma F_V = 0 = -F_1 \sin 30° - F_3 \sin 60° + F_{1,v}$$

For node 2,

$$\Sigma F_H = 0 = F_2 + F_1 \cos 30° + F_{2,h} + H_2$$

$$\Sigma F_V = 0 = F_1 \sin 30° + F_{2,v} + V_2$$

For node 3,

$$\Sigma F_H = 0 = -F_2 - F_3 \cos 60° + F_{3,h}$$

$$\Sigma F_V = 0 = F_3 \sin 60° + F_{3,v} + V_3$$

FIGURE 35.2
Free-body force diagram for the nodes.

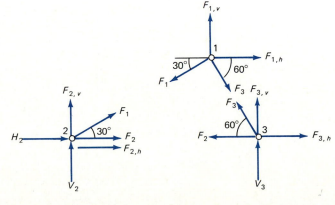

where $F_{i,h}$ is the external horizontal force applied to node i (where a positive force is from left to right) and $F_{i,v}$ is the external vertical force applied to node i (where a positive force is upward). Thus in this problem, the 1000-lb downward force on node 1 corresponds to $F_{1,v} = -1000$. For this case, all other $F_{i,v}$'s and $F_{i,h}$'s are zero. Note that the directions of the forces are unknown, but proper application of Newton's laws requires only that consistent assumptions regarding direction be made. Solutions are negative if the directions are assumed incorrectly. Also note that in this problem, the forces in all members are assumed to be in tension and acting to pull adjoining nodes together. This problem can be written as the following system of six equations and six unknowns by expressing the force balances as:

$$
\begin{aligned}
0.866F_1 \quad - 0.5F_3 \quad\quad\quad\quad &= 0 \\
0.5F_1 \quad + 0.866F_3 \quad\quad\quad\quad &= -1000 \\
-0.866F_1 - F_2 \quad\quad -H_2 \quad\quad &= 0 \\
-0.5F_1 \quad\quad\quad\quad -V_2 \quad\quad &= 0 \\
F_2 + 0.5F_3 \quad\quad\quad\quad &= 0 \\
- 0.866F_3 \quad\quad -V_3 &= 0
\end{aligned}
\tag{35.1}
$$

Notice in Eq. (35.1) that partial pivoting is required to avoid division by zero diagonal elements. If we employ a pivot strategy, the system can be solved using Gauss elimination. The result is

$$
\begin{aligned}
F_1 &= -500 \\
F_2 &= 433 \\
F_3 &= -866 \\
H_2 &= 0 \\
V_2 &= 250 \\
V_3 &= 750
\end{aligned}
$$

The negative signs indicate that members 1 and 3 are in compression. The results are shown in Fig. 35.3. This approach can be used to study the effect of different external forces on the truss. For example, we might want to study the effect of horizontal forces induced by a wind blowing from left to right. If the wind can be idealized as two horizontal point forces of 1000 lb on nodes 1 and 2, the system of equations can be solved for

$$
\begin{aligned}
F_1 &= 866 \\
F_2 &= 250
\end{aligned}
$$

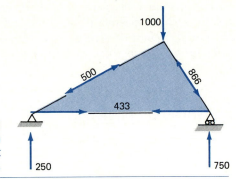

FIGURE 35.3
Forces and reactions due to a
1000-lb force acting downward at
node 1.

$$F_3 = -500$$

$$H_2 = -2000$$

$$V_2 = -433$$

$$V_3 = 433$$

For a wind from the right, $F_{1,h} = -1000$, $F_{3,h} = -1000$, and all other external forces are zero, with the result that

$$F_1 = -866$$

$$F_2 = -1250$$

$$F_3 = 500$$

$$H_2 = 2000$$

$$V_2 = 433$$

$$V_3 = -433$$

FIGURE 35.4
Two test cases showing (a) winds
from the left and (b) winds from
the right.

The results indicate that the winds have markedly different effects on the structure. Both cases are depicted in Fig. 35.4.

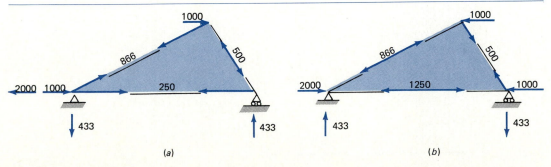

(a)   (b)

The foregoing method becomes particularly useful when applied to large, complicated structures. In engineering practice it may be necessary to solve trusses with hundreds or even thousands of structural members. Linear equations provide a powerful approach for gaining insight into the behavior of these structures.

35.2 USING LINEAR REGRESSION IN THE CALCULATION OF ELONGATION OF MEMBERS

Once the forces in the members of a structure are determined, a related question involves calculating the resulting elongation of an individual beam. This is based on the relationship of stress and strain. Stress s is defined as force per cross-sectional area:

$$s = \frac{F}{A} \tag{35.2}$$

whereas strain e is defined as change in length per total length:

$$e = \frac{\Delta L}{L} \tag{35.3}$$

For small deformations, Hooke's law can be used to relate stress and strain:

$$s = Ee \tag{35.4}$$

where E is a proportionality constant known as the *modulus of elasticity,* or *Young's modulus.* The region where this equation holds is called the *elastic range.* Table 35.1

TABLE 35.1 Experimentally Derived Strain and Stress Data for Beams of the Type Used to Construct the Structure in Fig. 35.1

Strain e, in/in	Stress s, ksi
0.0002	5
0.0002	7
0.0002	6
0.0004	11
0.0004	13
0.0004	12.5
0.0006	19
0.0006	17.5
0.0006	17
0.0008	24.6
0.0008	23.5

and Fig. 35.5 show some strain and stress data that applies to the members in the structure in Fig. 35.1. Linear regression can be used to fit this data with the straight line:

$$s = 29866e + 0.072$$

Thus an estimate of the modulus of elasticity is $E = 29,866$ ksi. Note that the intercept is approximately zero as would be expected from Eq. (35.4).

In order to compute a value for the change in length of a member, Eqs. (35.2) and (35.3) are substituted into Eq. (35.4):

$$\frac{F}{A} = E \frac{\Delta L}{L}$$

which can be solved for

$$\Delta L = \frac{LF}{AE} \tag{35.5}$$

This equation can be used to determine elongation for any of the members in the structure, provided the elongation remains in the elastic range. For example, for the

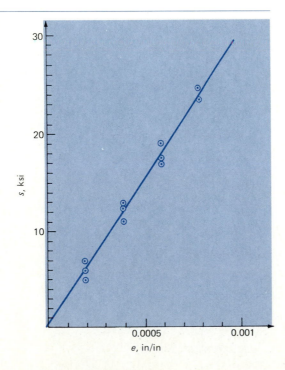

FIGURE 35.5
Plot of stress versus strain along with the best-fit straight line derived with linear regression.

case depicted in Fig. 35.4*b*, member 3 is in tension with a force of 500 lb. Thus, because $L = 60$ ft, $E = 29.9 \times 10^6$ lb/in², and $A = 0.5$ in², Eq. (35.5) can be used to determine that the resulting change in length is

$$\Delta L = \frac{(60 \times 12 \text{ in})(500 \text{ lb})}{(0.5 \text{ in}^2)(29.9 \times 10^6 \text{ lb/in}^2)} = 0.024 \text{ in}$$

35.3   USING INTEGRATION IN THE DETERMINATION OF TOTAL FORCE AND LINE OF ACTION FOR DISTRIBUTED LOADS

Another aspect of structural analysis is that forces are sometimes distributed rather than isolated at a joint. Figure 35.6 and Table 35.2 show the distribution of wind force along the side of a skyscraper.

Analyses of such cases can be greatly simplified if the whole distributed force can be represented as a force acting at a single point on the structure. The location of this single point is called the *line of action*. Integration can be used to determine both the total point force (f_{total}) and the line of action (d) by the following formulas:

$$f_{\text{total}} = \int_0^L F(l) \, dl \tag{35.6}$$

FIGURE 35.6
A distributed wind blowing against the side of a skyscraper.

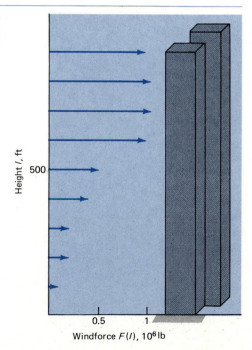

TABLE 35.2 Wind Force $F(l)$ and Height Times Wind Force $lF(l)$ as a Function of Height above Street Level l for the Skyscraper in Fig. 35.6

l, ft	$F(l)$, 10^6 lb/ft	$lF(l)$, 10^6 lb
0	0.00	0
100	0.10	10
200	0.20	40
300	0.20	60
400	0.40	160
500	0.50	250
600	1.00	600
700	1.05	735
800	1.05	840
900	1.00	900

and

$$d = \frac{\int_0^L lF(l)\ dl}{\int_0^L F(l)\ dl} \tag{35.7}$$

where L is the total height. With the trapezoidal rule and the data from Table 35.2, Eq. (35.6) can be used to compute $f_{total} = 500 \times 10^6$ lb. Similarly, $\int l\ F(l)\ dl = 314{,}500$ ft·lb; and Eq. (35.7) can be used to compute $d = 314{,}500/500 = 629$ ft. Therefore the distributed force can be represented as a point force of 500×10^6 lb acting at a point 629 ft above street level.

PROBLEMS

35.1 Calculate the forces and reactions for the truss in Fig. 35.1 if a downward force of 2000 lb and a horizontal force to the right of 1500 lb are applied at node 1.

35.2 What is Hooke's law? Give the equation and state its meaning in your own words.

35.3 In the example for Fig. 35.1, where a 1000 lb downward force is applied at node 1, the external reactions, V_2 and V_3, were calculated. But if the lengths of the truss members had been given, we could have calculated V_2 and V_3 by utilizing the fact that $V_2 + V_3$ must equal 1000 and by summing moments around node 2. However, because we do know V_2 and V_3, we can work backwards to solve for the lengths of the truss members. Note that because there are three unknown lengths and only two equations, we can only solve for the relationship between lengths. Solve for this relationship.

35.4 In the free-body diagrams of Fig. 35.2, all member forces are assumed to act away from the nodes. Is this an assumption of tension or of compression of the members? If the solution to a member's force is negative, what does it indicate about the original assumption?

35.5 Why must the summation of vertical or horizontal forces at a node for a structure such as Fig. 35.1 be equal to zero?

35.6 Employing the same methods as used to analyze Fig. 35.1, determine the forces and reactions for the truss shown in Fig. P35.6.

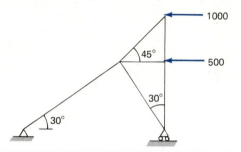

FIGURE P35.6

35.7 Solve for the forces and reaction for the truss in Fig. P35.7. Does the vertical-member force in the middle member seem reasonable? Why?

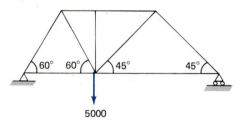

FIGURE P35.7

35.8 If the wind loads in Table 35.2 are changed to

l	0	100	200	300	400	500	600	700	800	900
$F(l)$	0	0.3	0.2	0.4	0.5	0.6	0.9	1.0	1.0	1.2

compute the total force and the line of action.

35.9 In addition to the forces given in Prob. 35.8, a swirling wind produces a small force acting on the opposite side of the structure:

l	0	100	200	300	400
$F(l)$	0	−0.1	−0.2	−0.1	0

where the minus signs indicate that the forces act in opposition to the forces given in Prob. 35.8. Compute the net total force and the line of action due to the combined effect of both forces.

35.10 Verify the results of the linear regression for the data in Table 35.1.

35.11 Perform a linear regression for the following data obtained by subjecting a 2-in-diameter concrete cylinder to various loads and measuring the deflection. The original length of the cylinder is 18 in.

Force	0	100	200	300	400	500	600
ΔL, 10^{-2} in	0	0.011	0.020	0.029	0.035	0.040	0.042

Plot this data and the regression line and suggest improvements that can be made in this analysis.

35.12 A wind force distributed against the side of a skyscraper is measured as

Height l, ft	Force $F(l)$, lb/ft
0	0
100	50
200	155
300	200
400	220
500	400
600	450
700	475
750	490

Compute the net force and the line of action due to this distributed wind.

35.13 Water exerts pressure on the upstream face of a dam as shown in Fig. P35.13. The pressure can be characterized by

$$p(z) = \rho g(D - z) \tag{P35.13}$$

where $p(z)$ is pressure in pascals (or newtons per square meter) exerted at an elevation z meters above the reservoir bottom; ρ is the density of water, which

FIGURE P35.13

Water exerting pressure on the upstream face of a dam: (*a*) side view showing force increasing linearly with depth; (*b*) front view showing width of dam in meters.

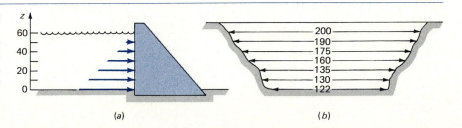

for the present problem is assumed to be a constant value of 10^3 kg/m^3; g is the acceleration due to gravity (9.8 m/s^2); and D is the elevation (in meters) of the water surface above the reservoir bottom. According to Eq. (P35.13), pressure increases linearly with depth, as depicted in Fig. P35.13a. Omitting atmospheric pressure (because it works against both sides of the dam face and essentially cancels out), the total force f_t can be determined by multiplying pressure times the area of the dam face (as shown in Fig. P35.13b). Because both pressure and area vary with elevation, the total force is obtained by evaluating.

$$f_t = \int_0^D \rho \, gw(z)(D - z) \, dz$$

where $w(z)$ is the width of the dam face in meters at elevation z (Fig. P35.13b). The line of action can also be obtained by evaluating

$$d = \frac{\int_0^D \rho \, gzw(z)(D - z) \, dz}{\int_0^D \rho \, gw(z)(D - z) \, dz}$$

Use Simpson's rule to compute f_t and d. Check the results with your computer program for the trapezoidal rule.

35.14 Using Gauss elimination, evaluate the forces and reactions for the structure shown in Fig. P35.14.

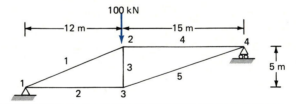

FIGURE P35.14

CHAPTER 36

MECHANICAL ENGINEERING: VIBRATION ANALYSIS

Differential equations are often used to model the behavior of engineering systems. Harmonic oscillators form one class of such models that are broadly applicable to most fields of engineering. Some basic examples of harmonic oscillators are a simple pendulum, a mass on a spring, and an inductance-capacitance electrical circuit (Fig. 36.1). Although these are very different physical systems, their oscillations can all be described by similar mathematical models. Thus, although the present problem deals with the mechanical engineering design of an automobile shock absorber, the general approach is applicable to a variety of other problem contexts in all fields of engineering.

(a)

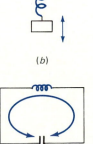

(b)

(c)

36.1 USING ROOTS OF EQUATIONS IN THE DESIGN OF AN AUTOMOBILE SHOCK ABSORBER

As depicted in Fig. 36.2, a sports car of mass M is supported by springs. Shock absorbers offer resistance to motion of the car that is proportional to the vertical speed (up and down motion) of the car. Disturbance of the car from equilibrium causes the system to move with an oscillating motion $x(t)$. At any instant, the net forces acting on M are the resistance of the spring and the damping force of the shock absorber. The resistance of the spring is proportional to a spring constant k and the distance x from equilibrium:

$$\text{Spring force} = -kx \tag{36.1}$$

where the negative sign indicates that the restoring force acts to return the car toward the position of equilibrium. Equation (36.1) is another form of *Hooke's law*, which was introduced in the previous chapter.

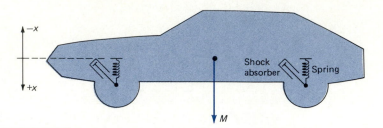

FIGURE 36.2
A sports car of mass M.

The damping force of the shock absorber is given by

$$\text{Damping force} = -c\frac{dx}{dt} \tag{36.2}$$

where c is a damping coefficient and dx/dt is the vertical velocity. The negative sign indicates that the damping force acts in the opposite direction against the velocity.

The equation of motion for the system is given by Newton's second law ($F = ma$), which for the present problem is expressed as the following force balance:

$$
\begin{array}{ccccc}
M & \times & d^2x/dt^2 & = & -c(dx/dt) & - & kx \\
\text{Mass} & \times & \text{acceleration} & = & \text{damping force} & + & \text{spring force}
\end{array}
\tag{36.3}
$$

or

$$\frac{d^2x}{dt^2} + \frac{c}{M}\frac{dx}{dt} + \frac{k}{M}x = 0 \tag{36.4}$$

This is a second-order linear differential equation that can be solved using the methods of calculus. For example, if the car hits a hole in the road at $t = 0$, such that it is displaced from equilibrium a distance $x = x_0$ and $dx/dt = 0$, then the solution is

$$x(t) = e^{-nt}\left(x_0 \cos pt + x_0 \frac{n}{p} \sin pt\right) \tag{36.5}$$

where $n = c/(2M)$, $p = \sqrt{(k/M) - (c^2/4M^2)}$, and $k/M > c^2/4M^2$. Equation (36.5) gives the vertical position of the car as a function of time. The parameter values are $c = 1.4 \times 10^7$ g/s, $M = 1.2 \times 10^6$ g, $k = 1.25 \times 10^9$ g/s², and $x_0 = 0.3$. Substitution of these values into Eq. (36.5) gives

$$x(t) = e^{-5.8333t}[0.3 \cos (31.7433t) + 0.05513 \sin (31.7433t)]$$

Figure 36.3 is a plot of $x(t)$ versus t, as calculated from this equation. Notice that after hitting the bump, the spring's vibrations are damped out. Mechanical-engineering design considerations require that estimates be provided for the first three times the car passes through the equilibrium point.

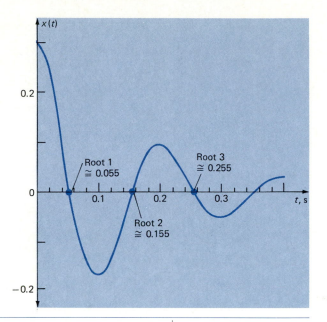

FIGURE 36.3
Plot of the position versus time for a sports car after the wheel hits a hole in the road.

This design problem must be solved using the numerical method of bisection. Estimates of the initial guesses are easily obtained by reference to Fig. 36.3. This example illustrates how graphical methods often provide information that is essential for the successful application of the numerical techniques. The plot indicates that this problem is complicated by the existence of several roots. Thus, in this case, rather narrow bracketing intervals must be used to avoid overlap. Table 36.1 lists the results of using bisection, given a stopping criterion of 0.1 percent.

36.2 USING REGRESSION IN THE DETERMINATION OF THE SPRING CONSTANT

In Eq. (36.3), the spring force is represented by Hooke's law,

$$F = -kx \qquad (36.6)$$

TABLE 36.1 Results of Using Bisection to Locate the First Three Roots for Vibrations of a Shock Absorber. A stopping criterion of 0.1 percent was used to obtain these results. Note that the exact values of the roots are 0.0552894503, 0.15439332, and 0.253497189.

Lower Guess	Upper Guess	Root Estimate	Number of Iterations	Relative Error, Approximate %	Relative Error, True %
0.0	0.1	0.0553222656	10	0.088	0.059
0.1	0.2	0.154394531	9	0.063	0.0008
0.2	0.3	0.253320313	8	0.077	0.070

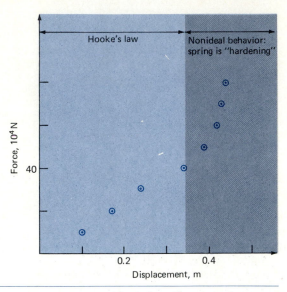

FIGURE 36.4 Plot of force (in 10^4 newtons) versus displacement (in meters) for the spring from an automobile suspension system.

This relationship, which holds when the spring is not stretched too far, signifies that the extension of the spring and the applied force are linearly related. The proportionality is parameterized by the spring constant k. A value for this parameter can be established experimentally by hanging known weights from the spring and measuring the resulting extension. Such data is contained in Table 36.2 and plotted in Fig. 36.4. Notice that above a weight of 40×10^4 N, the linear relationship between the force and displacement breaks down. This sort of behavior is typical of what is termed a "hardening spring."

TABLE 36.2 Experimental Values for Elongation (x) and Force (F) for the Spring on an Automobile Suspension System

Elongation, m	Force, 10^4 N
0.10	10
0.17	20
0.24	30
0.34	40
0.39	50
0.42	60
0.43	70
0.44	80

For the linear region (that is, the first four points), linear regression can be used to fit a straight line*:

$$F = -1.67 + 125.5x$$

Therefore, the value of k according to the regression is 125.5×10^4 N/m, which can be converted to 1.26×10^9 g/s^2. This is similar to the value of 1.25×10^9 g/s^2 used in Sec. 36.1. Note that the intercept for the regression result is not exactly zero as in Eq. (36.6). However, the value of -1.67 is small enough to be disregarded. Be aware of the fact that in the field of statistics, there are rigorous methods available to test the hypothesis that the intercept is zero.

36.3 USING INTEGRATION IN THE CALCULATION OF WORK WITH A VARIABLE FORCE

Many mechanical engineering problems involve the calculation of work, for which the general formula is

Work = force × distance

When you were introduced to this concept in high school physics, simple examples were presented using forces that remained constant throughout the displacement. For example, if a force of 10 lb was used to pull a block a distance of 15 ft, the work was calculated as 150 ft·lb.

Although such a simple computation is useful for introducing the concept, realistic problem settings are usually more complex. For example, suppose that the force varies during the course of the calculation. In such cases, the work equation is expressed as

$$W = \int_{x_0}^{x_n} F(x)\, dx \tag{36.7}$$

where W is work in foot-pounds, x_0 and x_n are the initial and final positions, respectively, and $F(x)$ is a force that varies as a function of position. If $F(x)$ is easy to integrate, Eq. (36.7) can be evaluated with calculus. However, in a realistic problem setting the force might not be expressed in such a manner. In fact, when we are analyzing measured data, the force might only be available in tabular form. For such cases, numerical integration is the only viable option for the evaluation.

*Strictly speaking, we should have regressed displacement versus force because the latter is actually the independent variable (that is, the one that is controlled and precise) in this experiment. However, the end result is not measurably affected by our choosing to regress force versus displacement. For cases with significant scatter, it might make a difference (recall Sec. 26.1).

This can be illustrated by calculating the work done in stretching the spring from the experiment in the last section. For the region for which Hooke's law applies (that is, from $x_0 = 0$ to $x_n = 0.34$ m), Hooke's law can be substituted into Eq. (36.7) to give

$$W = \int_{x_0}^{x_n} kx \, dx$$

The sign is positive because in order to stretch the spring we must exert a force that is equal but opposite to the spring force. Substituting $k = 125.5 \times 10^4$ N/m, $x_0 = 0$, and $x_n = 0.34$ yields

$$W = \int_0^{0.34} 125.5 \times 10^4 x \, dx$$

which can be determined easily with calculus to give $W = 7.25 \times 10^4$ J. We can verify this by independently integrating the data from Table 36.2 with the trapezoidal rule. Because the points are spaced unequally, an individual trapezoidal rule has to be applied to find the area between each pair, and these can be summed to develop the total integral:

$$W = (0.10 - 0)\frac{0 + 10}{2} + (0.17 - 0.10)\frac{10 + 20}{2}$$

$$+ (0.24 - 0.17)\frac{20 + 30}{2} + (0.34 - 0.24)\frac{30 + 40}{2}$$

$$= 6.8 \times 10^4 \text{ J}$$

The slight discrepancy between this and the analytical solution is due to the scatter in the data.

Now suppose that we desire to compute the work to elongate the spring from $x_0 = 0.34$ to $x_n = 0.44$. A simple option is to again apply the trapezoidal rule:

$$W = (0.39 - 0.34)\frac{40 + 50}{2} + (0.42 - 0.39)\frac{50 + 60}{2}$$

$$+ (0.44 - 0.42)\frac{60 + 2(70) + 80}{4} = 5.3 \times 10^4 \text{ J}$$

Note that because the last three points are equally spaced, we have used the multiple-segment trapezoidal rule to compute their integral. Therefore the total work to stretch the spring the entire distance from 0 to 0.44 is

$$W = 6.8 \times 10^4 + 5.3 \times 10^4 = 12.1 \times 10^4 \text{ J}$$

PROBLEMS

36.1 What is Hooke's law for a spring? The spring constant in Eq. (36.1) corresponds to what constant in Eq. (35.4)?

36.2 If all the parameters given in the text for Eq. (36.5) remain the same but $c = 0.3 \times 10^7$ g/s, calculate the first three zeros of the equation to an error level of 0.1 percent. *Note*: Graph the equation first before calculating the roots.

36.3 If c decreases to zero in Eq. (36.5), what is the resulting equation? How would this fact affect the car's motion?

36.4 Because the data points in Fig. 36.4 in the nonideal region resemble a power law relationship, suggest what might be done to perform a linear regression for the data in that region.

36.5 In Eq. (36.4), what happens if the mass of the car M is very much greater than c or k?

36.6 If the mass were zero in Eq. (36.4), the following equation would hold:

$$c\frac{dx}{dt} + kx = 0$$

This equation corresponds to the hypothetical situation where the shock absorber and spring are connected to a body with zero mass. If the spring and shock absorber were displaced $x = 0.5$ initially, solve, using Euler's method, for the resulting motion (x versus t) for $c = 5 \times 10^7$ g/s and $k = 2 \times 10^7$ g/s^2.

36.7 Using the following data, calculate the work done by stretching a spring that has a spring constant of $k \cong 3 \times 10^2$ N/m to $x = 0.4$ m.

F, 10^3 N	0	0.01	0.028	0.046	0.063	0.082	0.11	0.13	0.18
x, m	0	0.05	0.1	0.15	0.2	0.25	0.3	0.35	0.4

Compute the work using the trapezoidal rule and then repeat the analysis using Simpson's 1/3 rule.

36.8 At what point (x distance) does the spring in Prob. 36.7 deviate from Hooke's law?

36.9 For the data points given in Prob. 36.7 that do follow Hooke's law, determine the spring constant by linear regression and compare that result with the approximation in Prob. 36.7.

CHAPTER 37

ELECTRICAL ENGINEERING: CIRCUIT ANALYSIS

A common device or problem setting in electrical engineering is the electric circuit (Fig. 37.1), which is composed of wires and elements interconnected to form a closed path. Electricity is conducted through the circuit in order to convey information or, when the circuit is linked to electromechanical devices, to perform work.

In the present chapter we will study simple passive circuits of the type shown in Fig. 37.1. Such circuits are called "passive" because they consist of elements such as resistors, capacitors, and inductors that store or dissipate electric energy. This is in contrast to active circuits containing elements such as transducers, amplifiers, and transistors that can serve as energy sources.

As is true with systems in all fields of engineering, organizing principles are required to analyze electrical systems. Kirchhoff's laws, along with mechanisms of the sort outlined in Table 37.1, provide the framework for predicting the behavior of passive electric circuits.

FIGURE 37.1
A passive circuit.

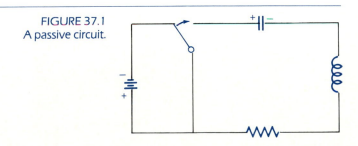

TABLE 37.1 Some Relationships or Mechanisms Defining the Ideal Behavior of the Elements of a Passive Circuit

	Resistor	**Inductor**	**Capacitor**
Function:	Dissipates energy	Stores energy in a magnetic field	Stores energy in an electric field by storing charges on plates separated by a nonconducting medium
Schematic representation:	⏦	⎍⎍⎍	⊣⊢
Ideal relationship or mechanism:	Ohm's law	Henry's law	Faraday's law
Mathematical relationship:	$V = iR$	$V = L\dfrac{di}{dt}$	$V = \dfrac{q}{C}$

37.1 USING LINEAR ALGEBRAIC EQUATIONS IN THE STEADY-STATE ANALYSIS OF A RESISTOR NETWORK

A common problem in electrical engineering involves determining the currents and voltages at various locations in resistor circuits. These problems are solved using Kirchhoff's current and voltage rules. The current (or point) rule states that the algebraic sum of all currents entering a node must be zero (see Fig. 37.2), or

$$\Sigma i = 0$$

where all current entering the node is considered positive in sign. The current rule is an application of the principle of conservation of charge.

The voltage (or loop) rule specifies that the algebraic sum of the potential differences (that is, voltage changes) in any loop must equal zero. For a resistor circuit, this is expressed as

$$\Sigma \xi - \Sigma iR = 0$$

where ξ is the emf (electromotive force) of the voltage sources and R is the resistance of any resistors on the loop. Note that the second term derives from Ohm's law (Table 37.1), which states that the voltage drop across an ideal resistor is equal to the product of the current and the resistance. Kirchhoff's voltage rule is an expression of the conservation of energy.

Application of these rules results in systems of simultaneous linear algebraic equations because the various loops within a circuit are coupled. For example, consider

FIGURE 37.2
Kirchhoff's current rule.

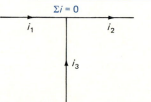

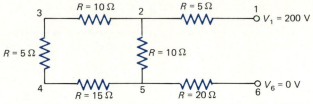

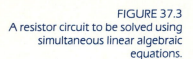

FIGURE 37.3
A resistor circuit to be solved using
simultaneous linear algebraic
equations.

the circuit shown in Fig. 37.3. The currents associated with this circuit are unknown both in magnitude and direction. This presents no great difficulty because one simply assumes a direction for each current. If the resultant solution from Kirchhoff's laws is negative, then the assumed direction was incorrect. For example, Fig. 37.4 shows some assumed currents.

Given these assumptions, Kirchhoff's current rule is applied at each node to yield

$$i_{12} + i_{52} + i_{32} = 0$$

$$i_{65} - i_{52} - i_{54} = 0$$

$$i_{43} - i_{32} = 0$$

$$i_{54} - i_{43} = 0$$

Application of the voltage rule to each of the two loops gives

$$- i_{54}R_{54} - i_{43}R_{43} - i_{32}R_{32} + i_{52}R_{52} = 0$$

$$- i_{65}R_{65} - i_{52}R_{52} + i_{12}R_{12} - 200 = 0$$

or, substituting the resistances from Fig. 37.3 and bringing constants to the right-hand side,

$$- 15i_{54} - 5i_{43} - 10i_{32} + 10i_{52} = 0$$

$$- 20i_{65} - 10i_{52} + 5i_{12} = 200$$

FIGURE 37.4 Assumed currents.

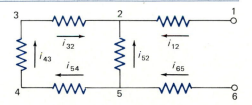

Therefore the problem amounts to solving the following set of six equations with six unknown currents:

$$i_{12} + i_{52} + i_{32} = 0$$

$$- i_{52} + i_{65} - i_{54} = 0$$

$$- i_{32} + i_{43} = 0$$

$$i_{54} - i_{43} = 0$$

$$10i_{52} - 10i_{32} - 15i_{54} - 5i_{43} = 0$$

$$5i_{12} - 10i_{52} - 20i_{65} = 200$$

Although impractical to solve by hand, this system is easily handled using the Gauss elimination method discussed in Chap. 29. Proceeding in this manner, the solution is

$$i_{12} = 6.1538$$

$$i_{52} = -4.6154$$

$$i_{32} = -1.5385$$

$$i_{65} = -6.1538$$

$$i_{54} = -1.5385$$

$$i_{43} = -1.5385$$

Thus, with proper interpretation of the signs of the result, the circuit currents and voltages are as shown in Fig. 37.5. The advantages of using numerical algorithms and personal computers for problems of this type should be evident.

37.2 USING ROOTS OF EQUATIONS IN THE DESIGN OF AN *RLC* CIRCUIT

Aside from steady-state computations, electrical engineers regularly analyze the transient, or time-variable, behavior of circuits. Such a situation occurs when the switch in the circuit shown in Fig. 37.1 is closed. In this case there will be a period of

FIGURE 37.5
The solution of currents and voltages obtained using Gauss elimination.

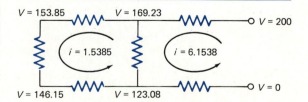

adjustment following the closing of the switch as a new steady state is approached. The length of this adjustment period is closely related to the energy-storing properties of the capacitor and the inductor. Energy storage may oscillate between these two elements during a transient period. However, resistance in the circuit will dissipate or dampen the magnitude of the oscillations.

According to Kirchhoff's voltage law, the sum of the voltage drops around a closed circuit is zero, or

$$V_L + V_R + V_C = 0 \tag{37.1}$$

where V_L, V_R, and V_C are the voltage drops across the inductor, resistor, and capacitor, respectively. Mathematical formulations for each of these voltage drops are summarized in Table 37.1. For the inductor, the voltage drop V_L is specified by *Henry's law*:

$$V_L = L\frac{di}{dt} \tag{37.2}$$

where L is a proportionality constant called the *inductance*, which has units of henrys (1 H = 1 Vs/A). Equation (37.2) is a mathematical expression of Henry's observation that a current change through a coil induces an opposing voltage that is directly proportional to the rate of change of current (that is, di/dt).

For the resistor, the voltage drop V_R is specified by *Ohm's law:*

$$V_R = iR \tag{37.3}$$

where R is the *resistance* of the resistor, measured in ohms (1 Ω = 1 V/A).

Finally, the voltage drop for the capacitor V_C is specified by *Faraday's law*:

$$V_C = \frac{q}{C} \tag{37.4}$$

where C is a proportionality constant called the *capacitance*, which has units of farads (1 F = 1 As/V). Equation (37.4) is a mathematical expression of the fact that the voltage drop across a capacitor is directly proportional to its charge.

Equations (37.2) through (37.4) can be substituted into Eq. (37.1) to yield

$$L\frac{di}{dt} + Ri + \frac{q}{C} = 0$$

However, the current is equivalent to the rate of change of charge per time:

$$i = \frac{dq}{dt}$$

Therefore

$$L\frac{d^2q}{dt^2} + R\frac{dq}{dt} + \frac{q}{C} = 0$$

This is a second-order linear ordinary differential equation that can be solved using the methods of calculus. If $q = q_0 = V_0C$ at $t = 0$ (where V_0 is the voltage from a battery), the solution is

$$q(t) = q_0e^{-Rt/2L}\cos\left[\sqrt{\frac{1}{LC} - \left(\frac{R}{2L}\right)^2}\,t\right] \tag{37.5}$$

Equation (37.5) describes the time rate of change of charge on the capacitor as a function of time. The solution is plotted on Fig. 37.6.

A typical electrical engineering design problem might involve determining the proper resistor to dissipate energy at a specified rate, given known values of L and C. For the present case, assume that charge must be dissipated to 1 percent of its original value ($q/q_0 = 0.01$) in $t = 0.05$ s, with $L = 5$ H and $C = 10^{-4}$ F.

This problem involves determining the root of Eq. (37.5). To do this, move q to the right-hand side and divide by q_0 to yield

$$f(R) = 0 = e^{-Rt/2L}\cos\left[\sqrt{\frac{1}{LC} - \left(\frac{R}{2L}\right)^2}\,t\right] - \frac{q}{q_0}$$

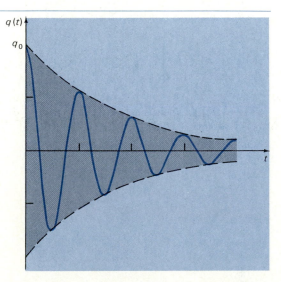

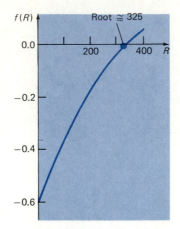

FIGURE 37.7
Plot of Eq. (37.6) used to obtain initial guesses for R that bracket the root.

or, using the numerical values given, the problem reduces to determining the root of

$$f(R) = e^{-0.005R} \cos \left[\sqrt{2000 - 0.01R^2}\,(0.05)\right] - 0.01 \qquad (37.6)$$

Examination of this equation suggests that a reasonable range for R is 0 to 400 Ω (because $2000 - 0.01R^2$ must be greater than zero to avoid a complex number). Figure 37.7—a plot of Eq. (37.6)—confirms this. Twenty-one iterations of bisection give $R = 328.1515\ \Omega$, with an error of less than 0.0001 percent.

Thus you can specify a resistor with this rating for the circuit shown in Fig. 37.1 and expect to achieve a dissipation performance that is consistent with the requirements of the problem. This design problem could not be solved efficiently without a numerical method such as bisection.

37.3   USING INTERPOLATION AND RATE EQUATIONS IN THE TRANSIENT ANALYSIS OF A NONIDEAL *RL* CIRCUIT

Aside from using calculus as in the previous section, we can use numerical techniques such as Euler's method to calculate the transient response of a circuit. Consider the simple *RL* circuit seen in Fig. 37.8. Kirchhoff's voltage law can be applied in conjunction with Eqs. (37.2) and (37.3) to give

$$L \frac{di}{dt} + iR = 0$$

or

$$\frac{di}{dt} = -\frac{R}{L}\,i \qquad (37.7)$$

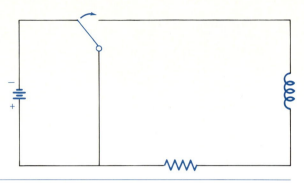

FIGURE 37.8 An *RL* circuit.

This simple linear rate equation can be solved easily by calculus or Euler's method for current as a function of time. However, real resistors may not always obey Ohm's law and the R in Eq. (37.7) is not necessarily constant. Suppose that you performed some very precise experiments to measure the voltage drop and corresponding current for a resistor. The results, as listed in Table 37.2 and Fig. 37.9, suggest a curvilinear relationship rather than the straight line represented by Ohm's law. In order to quantify this relationship, a curve must be fit to the data. Because of measurement error, regression would typically be the preferred method of curve fitting for analyzing such experimental data. However, the smoothness of the relationship suggested by the points as well as the precision of the experimental methods leads you to fit an interpolating polynomial as a first try. When you do this you discover that a third-order polynomial yields almost perfect results. The polynomial, which is developed from the Lagrange form by grouping terms, is

$$V = 40i + 160i^3$$

Kirchhoff's voltage law can be applied in conjunction with Eq. (37.2) to give

$$L\frac{di}{dt} + 40i + 160i^3 = 0$$

or

$$\frac{di}{dt} = -\frac{40i + 160i^3}{L} \tag{37.8}$$

Contrast this equation with the rate equation for the ideal *RL* circuit [Eq. (37.7)]. Euler's method can be used to solve both; the results are displayed in Fig. 37.10. Notice how the nonideal case differs significantly from the solution for Eq. (37.7).

TABLE 37.2 Experimental Data for Voltage Drop across a Resistor Subjected to Various Levels of Current

i	V
−1.00	−200.0
−0.50	−40.0
−0.25	−12.5
0.25	12.5
0.50	40.0
1.00	200.0

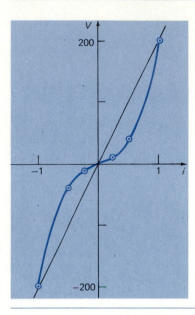

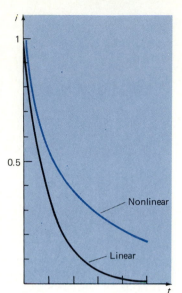

FIGURE 37.10
Numerical solutions for both the
ideal [Eq. (37.7) with $R = 200\ \Omega$]
and the nonideal [Eq. (37.8)]
resistors.

FIGURE 37.9
Plot of voltage versus current for a
resistor. The curve is the actual relationship whereas the straight
line is an idealized version following Ohm's law.

PROBLEMS

37.1 State Kirchhoffs' current and voltage rules for electric circuits.

37.2 Solve the resistor circuit in Fig. 37.3, using Gauss elimination, if $V_1 = 110$ V and $V_6 = -110$ V. Notice that the voltages do not cancel each other out.

37.3 Solve the circuit in Fig. P37.3 for the currents in each wire. Use Gauss elimination with pivoting.

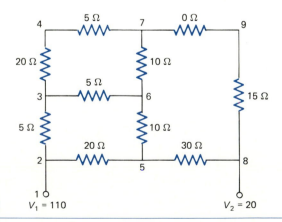

FIGURE P37.3

37.4 Form the mathematical relationship with Kirchhoff's rules for the circuit shown in Fig. P37.4.

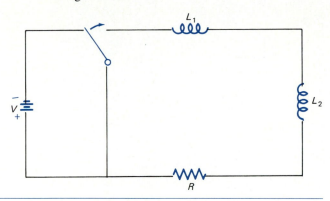

FIGURE P37.4

37.5 Solve for the current variation with time in the circuit shown in Fig. P37.4 if $L_1 = 5$ H, $L_2 = 20$ H, $R = 5$ Ω, and $V = 20$ V. Use Euler's method and determine the time when the current is 50 percent of its maximum value.

37.6 Repeat Prob. 37.5 if R is given by

$R = 20 + 100\ i^2$

37.7 Solve Eq. (37.6) to determine when the charge is 10 percent of its original value. Employ bisection.

37.8 Graph Eq. (37.5) for $R = 200$ Ω and let t vary. (That is, let t be the independent variable.) How many seconds elapse before the ratio of q/q_0 drops permanently below 0.10? Use a graphical approach for a coarse estimate, and use bisection to refine it.

PART NINE

A1:   HELP

Type X A B C D E F Reset View Save Options Name Quit
Set graph type
═══ 224,236 ═══

Graph Type -- Select type of graph to draw

→ Choose a graph type. (default Graph Type: Line)

Bar   Stacked-Bar   Line   XY   Pie Chart

imagine a
circle...

"Format" choices include: Lines, Symbols, Both

Requirements

Bar, Stacked-Bar, Line: One or more data ranges (A-,B-,..F-)
                    XY: X-range, one or more data ranges (A-,B-,..F-)
                   Pie: A-range (B-,C-,..F-ranges are ignored)

Graph Commands   Graph Ranges   Line and XY Formats   Help Index

GENERIC, OR "CANNED," PROGRAMS

To this point you have been introduced to a number of ways in which personal computers can serve your engineering career. We have presented material on BASIC and FORTRAN and have shown you how these languages can be used to develop your own "homemade" computer programs to solve engineering problems. Although the capability to write your own programs will serve you well, there will be many times throughout your career when you will use software created by others.

Because of the sheer number of choices and their expense, the selection of the proper software packages might at first seem overwhelming. Unfortunately, a myth has developed that suggests that a separate piece of software is required for each individual task. As we mentioned in Chap. 1, there are a number of multipurpose or generic programs that can be applied in numerous problem contexts. To this point we have already illustrated two types of such "canned" software: statistics and numerical methods packages. In the present part of the book, we will present short discussions of three other major types of generic software: electronic spreadsheets, computer graphics, and word processors. In addition, we include brief discussions of some of the other types of generic software: database managers, telecommunication systems, and integrated software.

CHAPTER 38
ELECTRONIC SPREADSHEETS

As an engineer, you will have many occasions to perform long, interconnected computations. Often such calculations will be performed on a large sheet of paper called a *spreadsheet*, or *worksheet* (Fig. 38.1). Whether or not they are formally called spreadsheets, large tabulated computations are used in all fields that deal in numbers. The areas where they were first recognized are those related to business and economics. For this reason, the original electronic spreadsheets were created in a business context. However, they also have great utility in engineering. As described in the next section, they are one of the most popular software packages ever invented.

38.1 THE ORIGIN OF SPREADSHEETS

While a student at the Harvard Business School in 1978, Daniel Bricklin was suffering through the task of preparing spreadsheets for some of his courses. Not only were these spreadsheets long and involved, most of the numbers were interdependent. Therefore, if he made a mistake or wanted to change one of the numbers, numerous other computations had to be performed and many results modified to correct the entire sheet.

Bricklin realized that a way out of his dilemma was to write software so that the spreadsheet could be implemented on a microcomputer. Then, if a number was changed, the computer could modify the entire sheet automatically. In this way, the electronic spreadsheet was born. Together with Robert Frankston, a programmer, and Daniel Flystra, a businessman, Bricklin formed a company called Software Arts and in 1979 began selling the program under the trademark VisiCalc. Soon other companies began marketing spreadsheets, the most successful being 1-2-3, offered by the Lotus De-

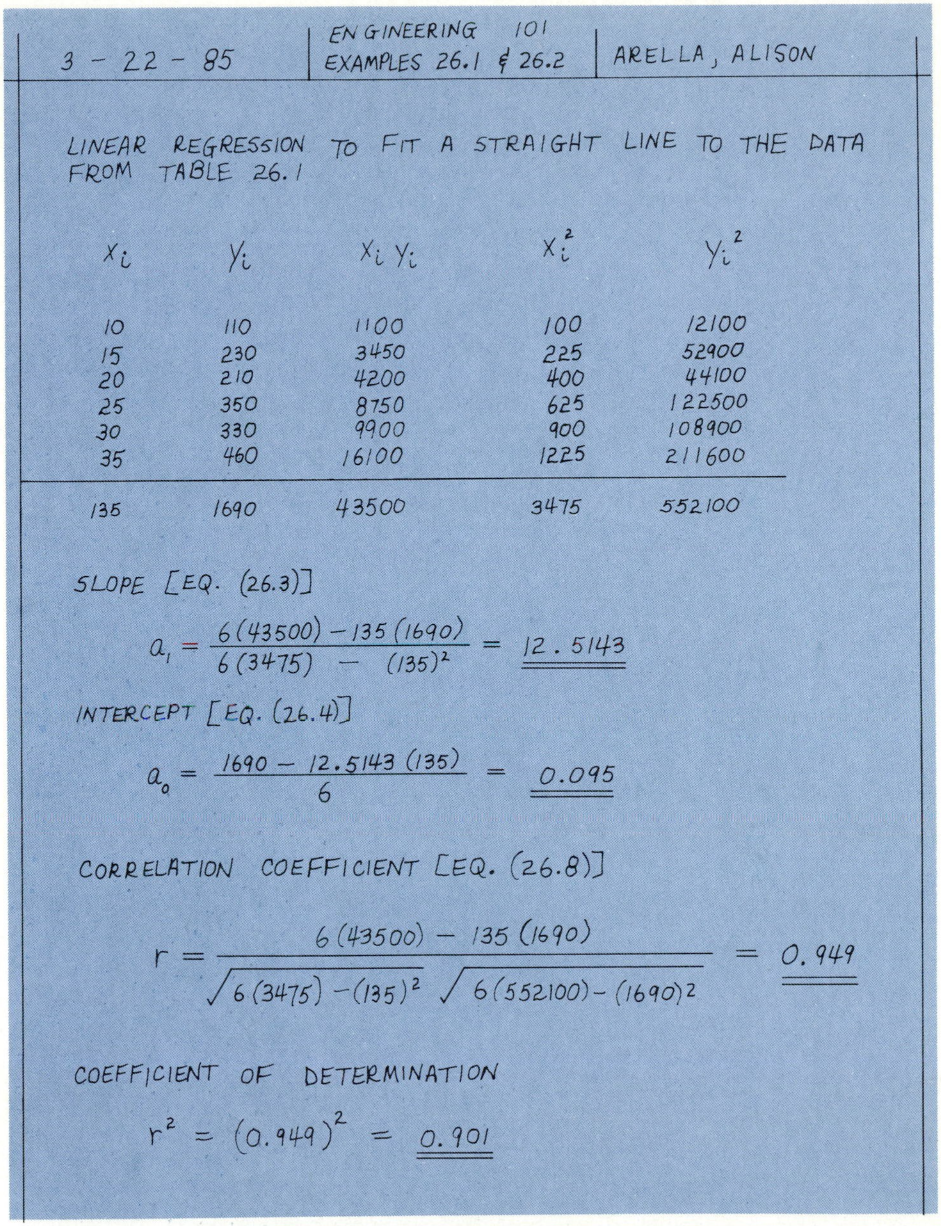

| 3 – 22 – 85 | ENGINEERING 101 EXAMPLES 26.1 & 26.2 | ARELLA, ALISON |

LINEAR REGRESSION TO FIT A STRAIGHT LINE TO THE DATA FROM TABLE 26.1

x_i	y_i	$x_i y_i$	x_i^2	y_i^2
10	110	1100	100	12100
15	230	3450	225	52900
20	210	4200	400	44100
25	350	8750	625	122500
30	330	9900	900	108900
35	460	16100	1225	211600
135	1690	43500	3475	552100

SLOPE [EQ. (26.3)]

$$a_1 = \frac{6(43500) - 135(1690)}{6(3475) - (135)^2} = 12.5143$$

INTERCEPT [EQ. (26.4)]

$$a_0 = \frac{1690 - 12.5143(135)}{6} = 0.095$$

CORRELATION COEFFICIENT [EQ. (26.8)]

$$r = \frac{6(43500) - 135(1690)}{\sqrt{6(3475) - (135)^2}\ \sqrt{6(552100) - (1690)^2}} = 0.949$$

COEFFICIENT OF DETERMINATION

$$r^2 = (0.949)^2 = 0.901$$

FIGURE 38.1
An engineering student's homework problem is an example of a handwritten, preelectronic spreadsheet.

velopment Corporation. (1-2-3 is often simply referred to as Lotus or Lotus 1-2-3.) To date, millions of spreadsheet programs have been sold. Along with word-processing programs, they represent the best-selling software packages of all time.

There are two reasons why Lotus and other spreadsheets have been so popular. First they are extremely flexible. That is, although they are simple in concept, they can be applied in a wide number of problem contexts. In this sense they represent the quintessential example of generic software. Second, they are easy to apply. Thus a businessperson or engineer can obtain practical benefits from the software with reasonable investment of effort. This ease of application will be apparent in the next section, where we describe what electronic spreadsheets are and how they are used.

38.2   WHAT IS AN ELECTRONIC SPREADSHEET?

Because of their popularity, there are now many versions of electronic spreadsheets. Just as was the case with the dialects of BASIC and FORTRAN, each version has its own special characteristics and idiosyncrasies. Therefore the following description is limited to the general features that are typical of most electronic spreadsheets. Additionally, the description is consistent with the SPREADSHEET program on the ENGINCOMP package that accompanies this book. It should be noted that this program has many similarities to Lotus 1-2-3. In a subsequent section we will elaborate on how ENGINCOMP differs from other spreadsheets.

Figure 38.2 shows the main menu for the SPREADSHEET program on ENGIN-COMP. Note that you are allowed to create a new spreadsheet, modify an old spreadsheet, save spreadsheets on a floppy disk, and retrieve spreadsheets from storage.

Selection 1 from the menu causes the display of a new, blank spreadsheet, as shown in Figure 38.3. Electronic spreadsheets divide the screen into rows and columns. Such two-dimensional characterizations should be very familiar to you by now.

FIGURE 38.2
The main menu of the spreadsheet
program on ENGINCOMP.

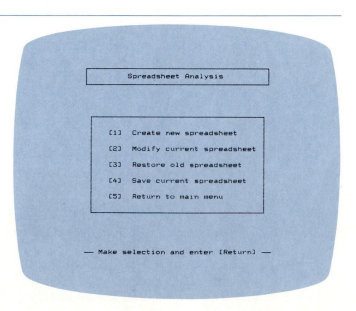

```
            Spreadsheet Analysis


        [1]  Create new spreadsheet

        [2]  Modify current spreadsheet

        [3]  Restore old spreadsheet

        [4]  Save current spreadsheet

        [5]  Return to main menu


     — Make selection and enter [Return] —
```

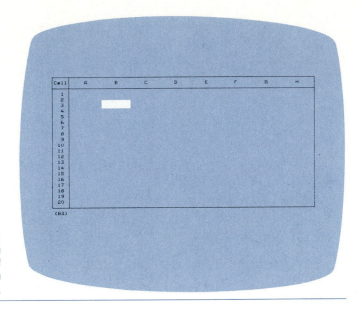

FIGURE 38.3
The layout of a spreadsheet screen showing how a column letter and a row number are used to locate a cell.

In our discussion of computer languages, we employed dimensioned variables where two subscripts could be used to locate a number in a two-dimensional array. In graphing, cartesian (or x, y) coordinates are employed to locate a point on a two-dimensional space. For spreadsheets, the same thinking applies. As shown in Fig. 38.3, the vertical columns are identified by letter and the horizontal rows by numbers. As with two-dimensional arrays or cartesian coordinates, the combination of two values—a letter and a number—can be used to define each location. In spreadsheet jargon, each location is called a *cell*. Thus the highlighted cell in Fig. 38.3 is designated as B3 because it is located in column B and row 3. Notice how your location on the spreadsheet is also indicated by the label [B3] that shows up at the lower-left-hand corner of the screen. If you move to another cell, this label changes automatically to reflect the new position. For the ENGINCOMP spreadsheet, the four arrow cursor keys on the number pad at the right-hand side of your keyboard can be used to move to any other cell.

Now that you know how to move around the spreadsheet, the next step is to enter information into a cell. On most spreadsheets, each cell may contain one of three types of information—a label, a value, or an equation. For the SPREADSHEET program on ENGINCOMP, these can be assigned to each cell, using the horizontal space that runs along the bottom of the screen. This horizontal space is called the *work area*. As seen in Fig. 38.3, the first quantity in the user work area is the cell label that designates your location on the spreadsheet. This label indicates that what you type next will be entered into that particular cell. For example, if you desire to enter a label such as ''homework'' into cell [B3] you move the cursor to that location and type

[B3] Homework

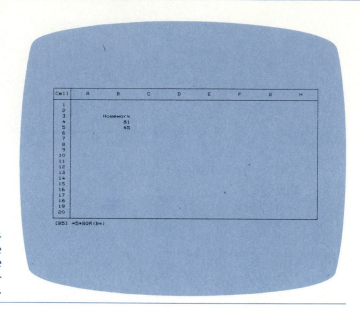

FIGURE 38.4
An example illustrating the three types of information that can be entered into a spreadsheet cell: labels, numbers, and equations.

When you strike RETURN, the label automatically appears in the proper location on the spreadsheet, as shown in Fig. 38.4. Note that the display of labels and values are limited to a field of nine characters.

Next you might want to enter a number into a cell. If you want to enter an 81 into [B4], you merely use the cursor arrows to move to that location and type in the number:

[B4] 81

When RETURN is struck, the quantity is entered into the cell.

Finally an equation can be entered into a cell. This is done by typing an equal sign followed by the equation.* For example, suppose that you want to multiply 5 times the square root of the contents of cell [B4] and put the result in cell [B5]. You first move the cursor to cell [B5] and type

[B5] = 5 * SQR(B4)

When this is entered, the computer will multiply 5 times the square root of 81 and print the result, 45, in cell [B5]. After all the above commands have been entered, the spreadsheet looks like Fig. 38.4. Now that a computational scheme has been established, the power and convenience of the spreadsheet can be exploited. This is accomplished by returning the cursor to cell [B4] and changing the value to 16. As a result, the value of cell [B5] will be automatically updated to 20. If the value of other

*For Lotus 1-2-3, the comparable command would be: +5*SQR(B4).

cells were defined in terms of [B4] and [B5] their values would also be updated. These automatic updating characteristics can be easily extended into many different kinds of complex engineering computations. As seen in the following example, the combination of labels, values, and equations is all that is needed to perform a meaningful analysis with the spreadsheet.

EXAMPLE 38.1   Computing Grades with a Spreadsheet

PROBLEM STATEMENT: Use the spreadsheet from ENGINCOMP to compute the average grade in a course. Have the spreadsheet determine the average quiz and homework grades. Then compute the final grade FG by

$$FG = \frac{WQ * AQ + WH * AH + WF * FE}{100}$$

where WQ, WH, and WF are the weighting factors for quizzes, homework, and the final exam, respectively; AQ and AH are the average quiz and homework grades, respectively; and FE is the final-exam grade. Use the following data: WQ = 35; WH = 25; WF = 40; quizzes = 96, 94, and 85; and homework = 85, 88, 82, 100, 100, 95, and 96. Use the spreadsheet to determine the final-exam grade needed to obtain an A in the course—that is, to receive a final grade of 90 or better.

SOLUTION: The commands and data to set up the spreadsheet are shown in Figs. 38.5 and 38.6. After you have entered all these commands and data, the only missing variable is the final-exam grade. As a first try, you assign a grade of 80 for the final exam. When this value is entered into cell [B15], the computer instantaneously cal-

FIGURE 38.5 The commands to set up a spreadsheet to determine the grade in a course. The dashes are used to make the spreadsheet more coherent and visually appealing.

```
[B1] quizzes           [A13] average
[C1] homeworks         [A15] fin exam=
[B2] ---------         [G1] weight
[C2] ---------         [F2] ---------
[A3] number            [G2] ---------
[B4] ---------         [F3] quizzes=
[C4] ---------         [F4] homework=
[A5] grades            [F5] fin exam=
[B12] ---------        [A18] fin grade
[C12] ---------
[B13] = (B5 + B6 + B7 + B8 + B9 + B10+ B11)/B3
[C13] = (C5 + C6 + C7 + C8 + C9 + C10 + C11)/C3
[B18] = (G3 * B13 + G4 * C13 + G5 * B15)/100
```

[B3] 3	[C3] 7	[C8] 100	[G3] 35
[B5] 96	[C5] 85	[C9] 100	[G4] 25
[B6] 94	[C6] 88	[C10] 95	[G5] 40
[B7] 85	[C7] 82	[C11] 96	

FIGURE 38.6
The data for a spreadsheet to determine a final grade in a course.

culates a ''bottom line'' or final grade of 87.1547 (Fig. 38.7). Therefore, an 80 on the final exam would not earn an A.

Now at this point, the real power of the spreadsheet becomes evident. If you were performing this computation by hand, you would have to recalculate the weighted average to determine the effect of changing the final-exam grade. In contrast, because the computation is already set up on the spreadsheet, all that must be done is to enter another final grade to cell [B15], and the bottom line is instantaneously modified. For example, if you enter an 85, the final grade is automatically upgraded to 89.1547. You could then try another value, say 90, and again the outcome of 91.1547 would be immediate. Thus, within seconds, you have obtained the following results:

Final Exam	Final Grade
80	87.1547
85	89.1547
90	91.1547

FIGURE 38.7
The spreadsheet to determine the final grade for a course.

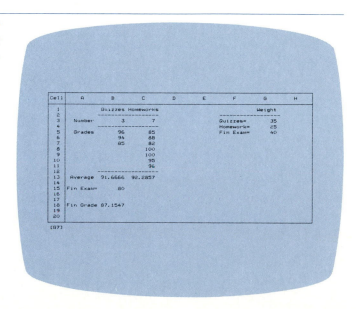

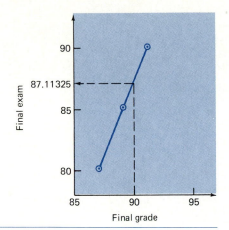

FIGURE 38.8
Plot to determine the final-exam
grade needed to attain a final
grade of 90.

You could plot the final exam versus the final grade (Fig. 38.8) and use an "eyeball" estimate or linear interpolation to determine that a final exam of 87.11325 is needed to obtain a final grade of 90. This can be verified by entering 87.11325 to cell [B15], with the result that cell [B18] shows 90.0000.

It should be noted that the plot is not even necessary. Because the results are almost instantaneous, you can simply try a number of different values in a trial-and-error fashion until you come up with the desired result.

Although the foregoing example is simple, it demonstrates three great strengths of spreadsheets. First, they are extremely easy to use and understand. Second, they provide an organized record of your computation. Third, they are ideal for "what if" computations. As seen in the next section, this latter aspect adds a powerful new dimension to engineering problem solving.

38.3 ENGINEERING APPLICATIONS OF SPREADSHEETS

Engineering problem solving usually involves considerable computational effort. In the precomputer era, this computational burden had a limiting effect on the problem-solving process. As shown in Fig. 38.9, significant amounts of energy were expended on obtaining solutions. This had a number of effects. First, solution techniques were often simplified to make them feasible within the time constraints of a particular problem setting. Second, the number and extent of the computations were limited. Finally, as seen in Fig. 38.9, because the techniques were so time-consuming, other aspects of the problem-solving process suffered.

As we have already seen in Parts Six through Eight, the computer's speed and memory greatly extend our analytical capabilities. By removing some of our computational burden, the computer allows us to place more emphasis on the creative

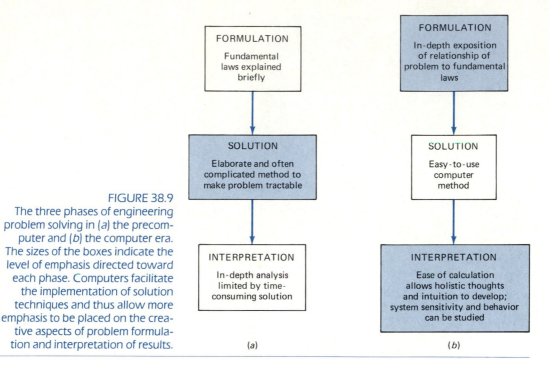

FIGURE 38.9
The three phases of engineering problem solving in (a) the precomputer and (b) the computer era. The sizes of the boxes indicate the level of emphasis directed toward each phase. Computers facilitate the implementation of solution techniques and thus allow more emphasis to be placed on the creative aspects of problem formulation and interpretation of results.

aspects of problem formulation and interpretation of results (see Fig. 38.9b). Nowhere is this benefit more obvious than with the spreadsheet. Although the following example is taken from an academic setting, it illustrates how the spreadsheet can enhance and broaden engineering computations.

EXAMPLE 38.2   Computing Projectile Motion with a Spreadsheet

PROBLEM STATEMENT: Figure 38.10 is a sample problem taken from an engineering textbook. Solve this problem with a spreadsheet.

SOLUTION: Before turning to the spreadsheet, we can first solve the problem by hand in a conventional way. To begin, the given variables and constants are

Height of cliff $= y_0 = -150$ m

Initial velocity $= v_0 = 180$ m/s

Angle $= \theta = 30°$

Acceleration of gravity $= a = -9.8$ m/s^2

The equations used for the solutions are as follows. First the vertical and horizontal components of the initial velocity are   (continued on page 602)

SAMPLE PROBLEM 11.7

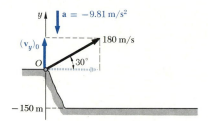

A projectile is fired from the edge of a 150-m cliff with an initial velocity of 180 m/s, at an angle of 30° with the horizontal. Neglecting air resistance, find (a) the horizontal distance from the gun to the point where the projectile strikes the ground, (b) the greatest elevation above the ground reached by the projectile.

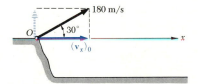

Solution. We shall consider separately the vertical and the horizontal motion.

Vertical Motion. Uniformly accelerated motion. Choosing the positive sense of the y axis upward and placing the origin O at the gun, we have

$$(v_y)_0 = (180 \text{ m/s}) \sin 30° = +90 \text{ m/s}$$
$$a = -9.81 \text{ m/s}^2$$

Substituting into the equations of uniformly accelerated motion, we have

$$
\begin{array}{llr}
v_y = (v_y)_0 + at & v_y = 90 - 9.81t & (1) \\
y = (v_y)_0 t + \tfrac{1}{2}at^2 & y = 90t - 4.90t^2 & (2) \\
v_y^2 = (v_y)_0^2 + 2ay & v_y^2 = 8100 - 19.62y & (3)
\end{array}
$$

Horizontal Motion. Uniform motion. Choosing the positive sense of the x axis to the right, we have

$$(v_x)_0 = (180 \text{ m/s}) \cos 30° = +155.9 \text{ m/s}$$

Substituting into the equation of uniform motion, we obtain

$$x = (v_x)_0 t \qquad x = 155.9t \qquad (4)$$

a. Horizontal Distance. When the projectile strikes the ground, we have

$$y = -150 \text{ m}$$

Carrying this value into Eq. (2) for the vertical motion, we write

$$-150 = 90t - 4.90t^2 \qquad t^2 - 18.37t - 30.6 = 0 \qquad t = 19.91 \text{ s}$$

Carrying $t = 19.91$ s into Eq. (4) for the horizontal motion, we obtain

$$x = 155.9(19.91) \qquad\qquad x = 3100 \text{ m} \quad \blacktriangleleft$$

b. Greatest Elevation. When the projectile reaches its greatest elevation, we have $v_y = 0$; carrying this value into Eq. (3) for the vertical motion, we write

$$0 = 8100 - 19.62y \qquad y = 413 \text{ m}$$
$$\text{Greatest elevation above ground} = 150 \text{ m} + 413 \text{ m}$$
$$= 563 \text{ m} \quad \blacktriangleleft$$

FIGURE 38.10
A problem dealing with projectile motion. (*From Beer and Johnston, Vector Mechanics for Engineers: Dynamics, 1984.*)

$$(v_y)_0 = v_0 \sin \theta$$

$$(v_x)_0 = v_0 \cos \theta$$

The equation for vertical motion is [Eq.(2) from Fig. 38.10]

$$y = (v_y)_0 t + \frac{1}{2} at^2$$

The projectile hits the ground when y equals the cliff height. Substituting y_0 for y and rearranging the equation gives

$$\frac{1}{2} at^2 + (v_y)_0 t - y_0 = 0$$

The root of this quadratic is the value of time that the projectile is in flight:

$$t = \frac{-(v_y)_0 - \sqrt{(v_y)_0^2 + 2ay_0}}{a}$$

Substituting this result into the equation for uniform motion [Eq. (4) from Fig. 38.10] provides the horizontal distance traveled by the projectile:

$$x = (v_x)_0 t$$

The maximum elevation can be computed on the basis of Eq. (3) from Fig. 38.10,

$$v_y^2 = (v_y)_0^2 + 2ay$$

The maximum elevation occurs when $v_y = 0$, or

$$0 = (v_y)_0^2 + 2ay_{max}$$

which can be solved for

$$y_{max} = -\frac{(v_y)_0^2}{2a}$$

This result can be used to compute the maximum elevation above the ground by subtracting the cliff height:

$$y'_{max} = -\frac{(v_y)_0^2}{2a} - y_0$$

where the prime is used to indicate that this is the height above the ground.

The boxed formulas above are the relationships that must be solved in sequence to obtain the answers. For example, using the given variables and constants results in

$$(v_y)_0 = 180 \sin 30 = 90 \text{ m/s}$$

$$(v_x)_0 = 180 \cos 30 = 155.88 \text{ m/s}$$

$$t = \frac{-90 - \sqrt{90^2 + 2(-9.81)(-150)}}{-9.81} = 19.886 \text{ s}$$

$$x = 155.88(19.886) = 3100.0 \text{ m}$$

$$y'_{max} = -\frac{90^2}{2(-9.81)} - (-150) = 562.84 \text{ m}$$

Aside from minor discrepancies due to significant figure differences, these results are consistent with those of Fig. 38.10.

The same computation can be set up as a spreadsheet. The commands for accomplishing this with ENGINCOMP are listed in Fig. 38.11, and the resulting spreadsheet is shown in Fig. 38.12.

To this point, the manual and the spreadsheet computations demand similar levels of effort. A little more work is necessary to set up the spreadsheet, but this effort is compensated for by the fact that a neat report of the computation has been developed.

In addition, once the spreadsheet is set up, it can be easily employed to investigate how changes in the given variables will affect the outcome. For example, suppose that we desired to increase the range of the cannon and had to decide whether to build a platform in order to increase the initial height or to purchase a higher-powered cannon in order to increase the initial velocity.

To gain insight into this question, we determine how much the maximum range is lengthened by a 10 percent increase in either the initial height or the initial velocity. First we determine the maximum range under present conditions. This can be done by trying a number of different angles with the following results:

Angle	Distance
30	3099.98
35	3304.75
40	3422.46
45	3446.49
50	3373.91

The computations suggest a maximum at about 45°. We try other angles in this vicinity, with the final result of 3449.48 m for an angle of 43.8° determined as the maximum.

```
[A1]  Given             [D1]  Interim           [G1]  Answers
[A2]  ---------         [D2]  ---------         [G2]  ---------
[A3]  Height            [D3]  Vyo               [G3]  Distance
[A5]  Velocity          [D5]  Vxo               [G5]  Elevation
[A7]  Angle             [D7]  Time
[A9]  Gravity
[B1]  Values
[B2]  ---------         [E3] = B5 * SIN (B7 * 3.14159/180)

[B3]  -150              [E5] = B5 * COS (B7 * 3.14159/180)
[B5]  180               [E7] = (-E3 - SQR (E3^2 + 2 * B9 * B3))/B9
[B7]  30                [H3] = E5 * E7
[B9]  -9.81             [H5] = (E3^2)/(2 * B9) - B3


[E10] Time           [F10] x              [G10] y
[E11] ---------      [F11]---------       [G11] ---------
[E12] = C15          [F12] = E5 * E12     [G12] = E3 * E12 + 0.5 * B9 * E12 ^ 2
[E13] = E12 + C17    [F13] = E5 * E13     [G13] = E3 * E13 + 0.5 * B9 * E13 ^ 2
[E14] = E13 + C17    [F14] = E5 * E14     [G14] = E3 * E14 + 0.5 * B9 * E14 ^ 2
[E15] = E14 + C17    [F15] = E5 * E15     [G15] = E3 * E15 + 0.5 * B9 * E15 ^ 2
[E16] = E15 + C17    [F16] = E5 * E16     [G16] = E3 * E16 + 0.5 * B9 * E16 ^ 2
[E17] = E16 + C17    [F17] = E5 * E17     [G17] = E3 * E17 + 0.5 * B9 * E17 ^ 2
[E18] = E17 + C17    [F18] = E5 * E18     [G18] = E3 * E18 + 0.5 * B9 * E18 ^ 2
[E19] = E18 + C17    [F19] = E5 * E19     [G19] = E3 * E19 + 0.5 * B9 * E19 ^ 2
[E20] = E19 + C17    [F20] = E5 * E20     [G20] = E3 * E20 + 0.5 * B9 * E20 ^ 2


[B15] INIT TIME      [C15] 0
[B17] TIME INCR      [C17] 5
```

FIGURE 38.11
Commands needed to program the projectile problem on the ENGINCOMP spreadsheet program.

Now we can analyze how increasing the height or the initial velocity will increase the range. For example, increasing the height to 165 m increases the range to 3463.82 m for an angle of 43.6°. This means that a 10 percent increase in height yields about a 0.4 percent increase in distance.

For the 10 percent increase in initial velocity, a much greater range is obtained. For an initial velocity of 198 m/s, the maximum distance is 4143.61 m, with an angle of 44°. This represents a 20.1 percent increase in range. Therefore, all other things (such as cost) being equal, the increase in firepower is a much more effective option.

The above is called a *sensitivity analysis*. Its intent is to build up insight regarding the behavior of the system being studied. Once the spreadsheet is set up and entered into the personal computer, the entire analysis takes less than 5 min. The ease with which such analyses can be implemented suggests the great potential of spreadsheets for enhancing and broadening the range of our analytical work.

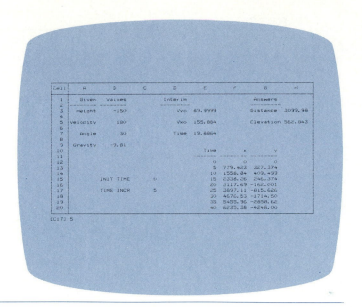

FIGURE 38.12
The spreadsheet for the projectile motion problem.

38.4   ADVANCED CAPABILITIES OF SPREADSHEETS

The SPREADSHEET program on ENGINCOMP is perfectly adequate for many routine classroom engineering problems. It is very easy to learn to use and interpret. Its simplicity, ease of use, and low cost are its primary advantages. In order to achieve this goal, many advanced capabilities available on other spreadsheet programs have been omitted. For example, the ENGINCOMP spreadsheet has a single page of 8 columns and 20 rows of cells. This is an order of magnitude smaller than some other spreadsheet programs. Furthermore, the mathematical functions available on the ENGINCOMP spreadsheet are limited to a few commonly used in engineering. Other spreadsheet programs like 1-2-3 by Lotus have more built-in mathematical, logical, financial, and statistical functions. Finally, more-advanced spreadsheet programs have a variety of special commands to facilitate the input of data.

ENGINCOMP should be thought of as a learning tool to introduce the concept and operation of a spreadsheet program. However, it is also quite adequate for simple engineering applications and classroom problems.

As a professional practicing engineer, you will probably have occasion to purchase or use larger and more-complicated spreadsheet programs. Some of these programs link the spreadsheet to graphics, data-management, and word-processing programs. This is called *integrated software* and will be discussed in another section.

This software is quite powerful, but it will cost several hundred dollars and may be difficult and time-consuming to learn. Thus, some trade-offs may be associated with the advanced capability. Therefore it is to your advantage to familiarize yourself with the available alternatives before making an expensive software purchase.

PROBLEMS

38.1 Use the SPREADSHEET program on ENGINCOMP to solve for the real roots of the parabola

$$ax^2 + bx + c = 0$$

where the roots are given by

$$\text{Roots} = \frac{-b \pm \sqrt{b^2 - 4ac}}{2a}$$

Input various values for a, b, and c and compute the roots. Note the results of the computations when $a = 0$ and when $4ac > b^2$.

38.2 A boom of length L supports a weight W of 5000 lb, as shown in Fig. P38.2. Solve for the stress in the cable and the boom as a function of θ and W. Set the solution up on a spreadsheet program such as the one on ENGINCOMP. Vary θ and the length of the boom in a manner such that the cable maintains an angle of 90° with the wall. Output values for the length of the boom, the length of the cable, and the stresses in the boom and cable as θ varies from 0 to 75°.

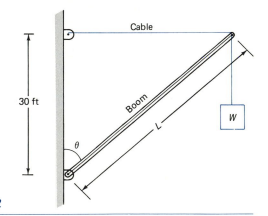

FIGURE P38.2

38.3 A block of mass M is at rest (Fig. P38.3). The coefficient of static friction between the block and the horizontal surface is μ. Investigate the forces between the block and the surface and the motion of the block as the angle between the horizontal surface and the plane varies between 0 and 75°. Try various values of M and μ, using a spreadsheet program such as the one on ENGINCOMP.

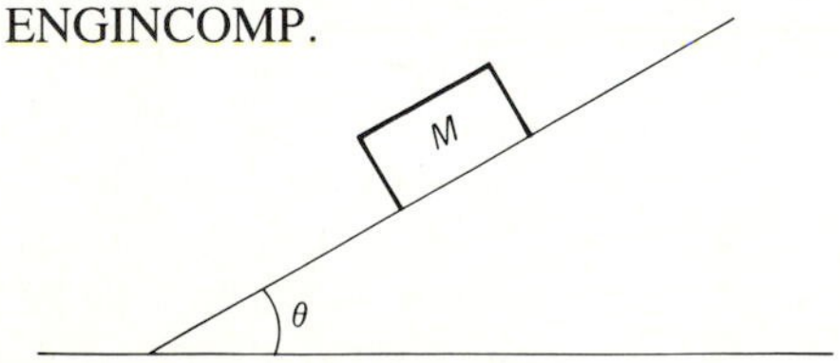

FIGURE P38.3

38.4   A current I divides between two parallel resistances of R_1 and R_2 (Fig. P38.4). Determine the voltage across the parallel combination and the current in each resistor for the following values:

$$I = 2, 5, 10$$

$$R_1 = 2, 4, 6$$

$$R_2 = 4, 7, 9$$

Use a spreadsheet program such as the one on ENGINCOMP to perform your analysis.

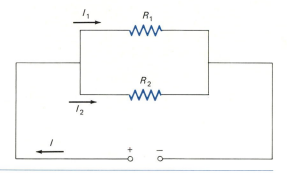

FIGURE P38.4

38.5   Use a spreadsheet program such as the one on ENGINCOMP to monitor your weekly expenses and income. Adjust the data, if necessary, to show a balanced monthly budget.

38.6   A triangular field is defined by three points A, B, and C, with coordinates (ax, ay), (bx, by), and (cx, cy), as seen in Fig. P38.6. Determine how much fence is required to enclose the field for various values of (ax, ay), (bx, by), and (cx, cy). Use a spreadsheet program such as the one on ENGINCOMP to facilitate your computations.

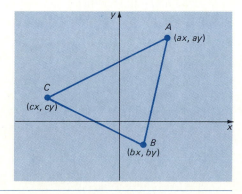

FIGURE P38.6

38.7   A chemical reactor converts a toxic organic compound dissolved in water to carbon dioxide, as shown in Fig. P38.7. The efficiency of the system depends on the flow rate of the water (Q), the volume of the reactor (V), and the decomposition rate of the toxic material (K):

$$Eff = \frac{Q}{Q + VK}$$

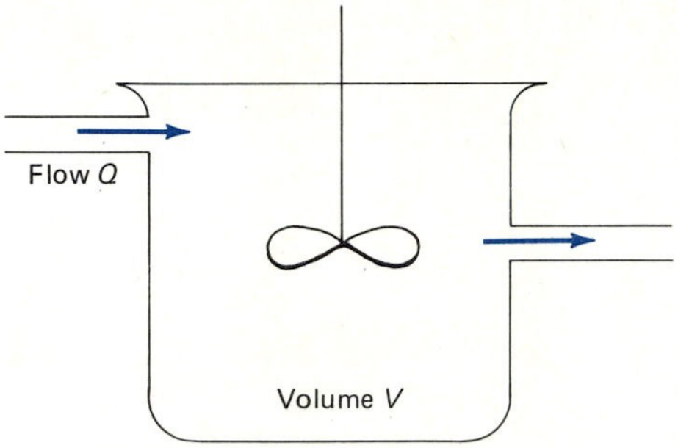

FIGURE P38.7

Use a spreadsheet program such as the one on ENGINCOMP to complete the following table.

Efficiency	Volume, in³	Flow Rate, m³/day	Decomposition Rate, 1/day
0.9	100	?	0.1
0.95	?	90	0.08
0.98	150	50	?
?	200	120	0.08
0.99	150	?	0.08
0.99	150	20	?

38.8   Use a spreadsheet to program Fig. 38.1. Examine how a 10 percent increase of the y value at $x = 35$ affects r. Perform the same sensitivity analysis for the y value at $x = 25$ (remember to return the value at $x = 35$ to its original value prior to doing this). What do the results suggest regarding the sensitivity of a regression analysis to the location of the points?

CHAPTER 39
COMPUTER GRAPHICS

One of the most exciting aspects of the personal computer is its graphics capabilities. In Chap. 22 you were provided an initial idea of this capability in text mode. Now we will delve more deeply into the topic. Here, as in Chap. 22, our focus will be on line graphs. However, we will introduce some new approaches that will allow you to program refined versions of your own graphs. Additionally, we will present material on ENGINCOMP software that is available for this purpose. Finally, in the latter sections of this chapter, we will discuss other types of computer graphics and their implications for engineering.

39.1 GRAPHICS EQUIPMENT AND TERMINOLOGY

As mentioned in Chap. 3, most personal-computer monitors are based on the technology of the *cathode-ray tube*, or *CRT*. A beam of electrons is directed at a coat of phosphors on the interior surface of the screen. These phosphors glow momentarily wherever the beam strikes the surface. It is the composite of all the individual glowing phosphors that produces the image that we view on the other side of the screen.

Because the glow fades when the beam moves elsewhere on the screen, the image must be redrawn or refreshed continuously. There are two basic ways that the beam moves about the screen to refresh the images. The first, called a *vector display*, produces images by moving the beam between two points on the screen to create a straight line or vector. The final image consists of a composite of these individual vectors. Figure 39.1*a* indicates how a set of axes is created on a vector display.

The second mode of operation, called the *raster display*, produces images by moving the beam over a set of horizontal lines. The beam's intensity is heightened to

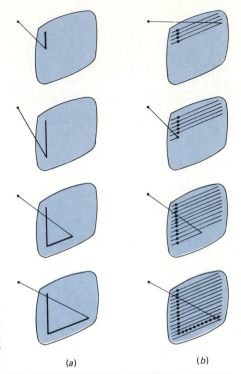

FIGURE 39.1
The fundamental difference between the ways that images are created for (*a*) vector and (*b*) raster displays.

leave a glow wherever an image is desired. The common television set uses this technology. Figure 39.1*b* shows how a set of axes is created by a raster display.

Vector displays were originally the most common mode used for graphics. They have a number of advantages, including the fact that they require little memory, produce very crisp images, and are well suited for dynamic or moving graphics. However, they are also relatively costly. Because they make use of television technology, the raster displays are much less expensive. As a consequence, they are the type of display used in personal computers.

As shown in Fig. 39.1*b*, the image on a raster display is composed of the individual energized points. These are sometimes referred to as *picture elements*, or *pixels*. One deficiency of raster displays is that images composed of discrete pixels can sometimes appear jagged. As seen in Fig. 39.2, this effect, which is also called *staircasing*, is dependent on the density or number of pixels that make up the screen. The degree of density is referred to as the screen's *resolution*. A larger number of pixels, i.e., the higher the resolution, yields smoother and crisper images. However, the better the resolution, the more expensive the monitor. Also, as the number of pixels increases, the required memory increases and the speed of operation decreases. Thus there are trade-offs that must be considered when choosing a monitor. If the primary use of the system is for word processing or simple computations, a low-resolution monitor might be adequate. However, if the main purpose is to obtain graphics displays, a high-resolution screen will be required.

FIGURE 39.2
An exaggerated example of how the use of discrete pixels to create an image can lead to jagged or "staircased" surfaces. In this case, because of low resolution, the circle appears very jagged.

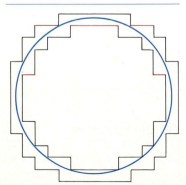

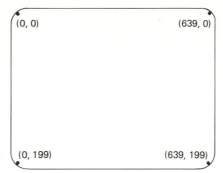

FIGURE 39.3
The dimensions for the IBM PC's
high-resolution graphics mode.

A screen's resolution is specified quantitatively by its dimensions—that is, by the number of columns and rows of pixels it contains. For example, the text mode used for the plots in Chap. 22 had dimensions of 80 columns by 25 rows, or 80 $\times$ 25 cells. Note that, when resolution is characterized, the columns are placed before the rows. This differs from our previous discussions of the LOCATE statement (or of arrays or simultaneous equations), where the row was designated first followed by the column.

As its name implies, the text mode is not intended for intricate graphic displays but is designed for dealing with textual material, as in word processing or simple computer programming. Higher-resolution graphics modes typically range from 280 $\times$ 192 to 1000 $\times$ 1000 pixels. In the present chapter, we will focus on the IBM PC's high-resolution graphics mode, which has 640 $\times$ 200 pixels per screen (Fig. 39.3).

39.2 HIGH-RESOLUTION GRAPHICS

Because of the differences between graphics commands for computers, it is inappropriate to present a comprehensive exposition on the topic here. We have, therefore, chosen to focus the present discussion on a single system, the IBM PC. We have chosen this personal computer because of its widespread availability and, also, because its high-resolution graphics mode has most of the essential features of many computer-graphics systems. We will limit our discussion to the most fundamental of the IBM's graphics commands. These are (1) placing the system in the graphics mode, (2) moving to any location on the screen, (3) creating a point at a location on the screen, (4) drawing a line between two locations on the screen, (5) drawing a box anywhere on the screen, and (6) drawing a circle anywhere on the screen. In addition, we will limit the discussion primarily to black-and-white graphics.

39.2.1 Placing the System in the Graphics Mode

The IBM PC has three possible modes: the text mode, the medium-resolution graphics mode, and the high-resolution graphics mode. The SCREEN statement is used to select the mode, as in

ln SCREEN *mn*

where *ln* is the line number and *mn* is the mode number (*mn* = 0, 1, and 2 for text, medium resolution, and high resolution, respectively). In the present treatment, we will ignore the text mode and the medium-resolution mode (which is primarily used for color graphics) and concentrate on the high-resolution mode.

An additional statement that bears mentioning is the COLOR statement, which is of the general form

ln COLOR *foreground, background, border*

where the *foreground* parameter specifies the color of the characters in the text mode, *background* specifies the color against which they are displayed, and *border* (which is used with the color/graphics adapter) specifies the color of the unused area of the screen. The parameters can take on one of the sixteen values given in Table 39.1. If a 16 is added to any of these numbers it will cause the characters to blink. For example, a foreground color of 20 will cause blinking red characters. If, as is assumed here, you are employing a monochrome adapter, the background is limited to 0 (black) and (7) white. For the foreground, the options are 0 (black), 1 (white with character underlines), 7 (white), and 15 (high-intensity white). The default setting for COLOR is

COLOR 7, 0

which results in normal white-on-black characters. If you wanted black on white you would employ

COLOR 0, 7

39.2.2 Moving to a Location

It is essential that we have the capability of positioning ourselves at any location on the screen. Recall from Chap. 22 that the LOCATE (or HTAB or VTAB) statement is used for this purpose in the text mode. For high-resolution graphics, the comparable statement is

ln PRESET *(nc,nr), color*

where *nc* is the column number (from 0 to 639), *nr* is the row number (from 0 to

TABLE 39.1   Parameters Used in Text Mode to Represent 16
Colors in the COLOR Statement

0 Black	5 Magenta	10 Light green
1 Blue	6 Brown	11 Light cyan
2 Green	7 White	12 Light red
3 Cyan	8 Gray	13 Light magenta
4 Red	9 Light blue	14 Yellow
	15 High-intensity white	

199), and *color* specifies the color to be used.* If no COLOR statement is included, the default value is the same as the default color of the background. This is why no point is displayed by PRESET. Notice that the sequencing of the column and row numbers is the reverse of that for the LOCATE statement, where the row is first. In addition, notice that the numbering starts at 0 rather than at 1, as was the case with the text mode.

39.2.3 Creating a Point

Although the statements in the previous section allow us to move anywhere on the screen, they do not energize any pixels, and so no point will be displayed. The high-resolution statement, PSET, can be used for this purpose:

> *ln* PSET *(nc,nr), color*

This statement will place a small dot about the size of a period at the pixel in column *nc* and row *nr*.

Note that when *color* is specified, there is no difference between PSET and PRESET. However, as shown in the examples in the present chapter, if no *color* is designated, the defaults make PSET into the foreground color and PRESET into the background. Thus the former is a dot, whereas the latter cannot be seen.

39.2.4 Drawing a Line

As was done previously with asterisks and text mode in Chap. 22, a line can be generated by stringing together a series of points. Although they can certainly be constructed in this manner, lines are so commonplace that a special statement is available to draw them directly. This is the LINE statement, which is of the general form

> *ln* LINE *(nc$_1$,nr$_2$)* − *(nc$_2$,nr$_2$)*

This statement permits us to draw a straight line between a pair of coordinates *nc$_1$,nr$_1$* and *nc$_2$,nr$_2$*.

39.2.5 Drawing a Box

The most primitive way to draw a box is to use PSET. This is accomplished by using FOR/NEXT loops to move across the screen. Figures 39.4 and 39.5 show how a box is constructed in this manner. A similar box could be constructed with four LINE commands

```
10 LINE (50,70)-(70,70)
20 LINE (70,70)-(70,90)
30 LINE (50,70)-(50,90)
40 LINE (50,90)-(70,90)
```

---

*Note that the parameters for high-resolution graphics differ from those in Table 39.1. Consult your user's manual for the proper values for your system.

```
100 KEY OFF
110 CLS
120 SCREEN 2
130 FOR J = 50 TO 100 STEP 50
140     FOR I = 100 TO 300
150         PSET(I,J)
160     NEXT I
170 NEXT J
180 FOR I = 100 TO 300 STEP 200
190     FOR J = 50 TO 100
200         PSET(I,J)
210     NEXT J
220 NEXT I
230 END
```

FIGURE 39.4
A BASIC program to draw a box.

or with one special LINE command with the suffix B

$$10\ \text{LINE}\ (50,70)-(70,90),,B$$

where the ,, is an implicit reference to a color hue. This form of the LINE statement can only be used to draw a square or rectangle where sides are parallel to the edges of the screen. These commands can be used to draw a four-sided figure with any orientation.

39.2.6 A High-Resolution Line Graph

The program shown in Fig. 39.4 can be easily expanded to plot a high-resolution graph of an arbitrary function. A program to plot the function

$$y = 10e^{0.03x}$$

FIGURE 39.5
The box drawn by the program from Fig. 39.4.

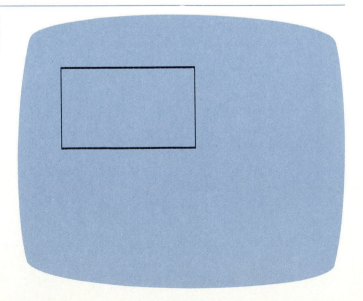

```
100 KEY OFF
110 CLS
120 SCREEN 2
130 FOR J = 50 TO 150 STEP 100
140     FOR I = 100 TO 500
150         PSET(I,J)
160     NEXT I
170 NEXT J
180 FOR I = 100 TO 500 STEP 400
190     FOR J = 50 TO 150
200         PSET(I,J)
210     NEXT J
220 NEXT I
230 FOR X = 5 TO 95
240     Y = 10 * EXP(.03*X)
250     XP = 100 + X*(400/100)
260     YP = 150 - Y*(100/200)
270     PSET(XP,YP)
280 NEXT X
290 END
```

FIGURE 39.6
A BASIC program to create a line graph of a function.

is shown in Fig. 39.6, with corresponding results given in Fig. 39.7. Only six lines are added (lines 230 to 280) to Fig. 39.4 to plot the function. Note carefully the introduction of the variables XP and YP (that is, X plot and Y plot) that are scaled to be consistent with the size of the box.

This program could also be extended to include labels and tick marks for the axes. This plot should be compared to the one in text mode in Chap. 22.

39.2.7   Other Graphics Commands

Many forms of BASIC (such as Microsoft BASIC) have special graphics commands to allow you to conveniently draw figures on the screen. An example is the special LINE command with the B suffix to draw a box, discussed in Sec. 39.2.5. This command can replace either four separate LINE commands or a whole program using FOR/NEXT and PSET such as Fig. 39.4.

FIGURE 39.7
An example of the type of plot generated with the program from Fig. 39.6.

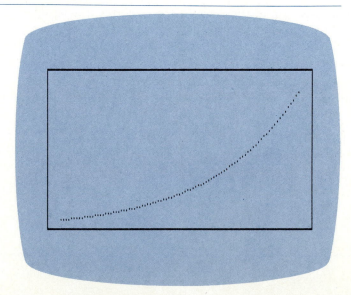

The CIRCLE statement of BASIC is another example. A circle can be drawn on the screen using FOR/NEXT and PSET, provided you employ the properly scaled mathematical function for a circle, that is,

$$x^2 + y^2 = r^2$$

where x and y are the two axes and r is the radius.

However, let's look at the CIRCLE statement of BASIC, a command that can be used to place a circle of any size or color in any location on the screen:

```
CIRCLE (120,90), 50
```

This command draws a circle at 120, 90 with a radius of 50.

Circle commands can also be used to draw arcs and ellipses. In addition, many forms of BASIC have a special DRAW command that can be used to display a complex figure with a single command. These are advanced graphics commands that are covered by more specialized texts or user's manuals.

39.2.8 High-Resolution Graphics Commands for Other Computers

Table 39.2 summarizes the statements needed to perform high-resolution graphics for a number of personal and mainframe computer-graphics systems in both BASIC and FORTRAN. Together with your own user's manual, this table should allow you to translate the programs from this chapter into a form that is compatible with these machines.

39.3 CANNED GRAPHICS SOFTWARE

The material in the foregoing section and in Chap. 22 provides you with the capability of creating images on a monitor screen or printer page. Although our focus has been on line graphs, the fundamental ideas can be applied to concoct a wide variety of other engineering-oriented graphics.

TABLE 39.2 Commands to Perform High-Resolution Graphics

	BASIC/IBM PC	BASIC/Apple II	FORTRAN/Tektronix PLOT10
Put system in high-resolution graphics mode	SCREEN 2	HGR2 COLOR=3	CALL INITT(960) CALL TERM(1,1024)
Move to column nc and row nr	PRESET(nc,nr)	HCOLOR=0 HPLOT nc, nr	CALL MOVABS(nc,nr)
Create point in column nc and row nr	PSET(nc,nr)	HCOLOR=3 HPLOT nc,nr	CALL PNTABS(nc,nr)
Draw a line from nc_1,nr_1 to nc_2,nr_2	LINE$(nc_1,nr_1)-(nc_2,nr_2)$	HPLOT nc_1,nr_1 TO nc_2,nr_2	CALL DRWLIN(nc_1,nr_1,nc_2,nr_2)

In addition to your own ability to create images, prepackaged or canned software is also available for computer graphics. As with other forms of generic software, these programs have been developed to insulate you from the kinds of ''nuts-and-bolts'' programming details discussed in the previous section. Thus the graphics can be generated effortlessly so you can concentrate on the more creative aspects of the process.

The ENGINCOMP software that accompanies this text includes a data- and function-plotter program that is representative of such software. As shown in the next example, it represents a potent tool for gaining visual insight into engineering-related problems.

EXAMPLE 39.1   Locating Roots with Computer Graphics

PROBLEM STATEMENT: In Chap. 28, we studied methods for determining the roots of an equation. A graphical definition of a root is the value of an independent variable x where a function $f(x)$ crosses the x axis. Such a function is

$$f(x) = \sin 10x + \cos 3x$$

This type of function frequently occurs in the analysis of vibrating systems such as harmonic oscillators. Use computer graphics to gain insight into the behavior of this function.

SOLUTION: Figure 39.8a shows the main menu for the GRAPHICS program from ENGINCOMP. Notice that you have five options: (1) enter a function, (2) enter new data, (3) modify old data, (4) plot the function or data, and (5) return to the main menu of ENGINCOMP. Select option 1 to enter the function. Type in the function, as shown in Fig. 39.8b, and strike RETURN to enter it to the computer. Note that X is the only variable permitted. Standard mathematical operators (such as $+$, $-$, $*$, and $/$) and intrinsic functions (such as EXP, SIN, COS, etc.) are allowed. The function can contain up to 256 characters.

The program checks for syntax errors when you enter the function. If syntax errors are detected, a diagnostic message is displayed at the bottom of the screen. In these situations, the program will also display the error-free part of your function at the bottom of the screen. This gives you a clue to the character that prompted the error message. Upon learning of the error, you can make corrections by inputting characters with appropriate modifications. You can also use the BACKSPACE key to erase a character or the DEL key to delete the entire function. This process of entering and editing the function is a convenient feature of the ENGINCOMP software. It is a fine example of the type of time-saving features that are characteristic of high-quality canned software.

When the function has been successfully entered to the program, you are automatically transferred back to the main menu. Now you can choose option 4, where you can select a linear, semilog, or log-log form of the plot. In addition, you can opt to plot a function alone, data alone, or both. Before the plot can be constructed, it is also necessary to supply minimum and maximum values for both the independent

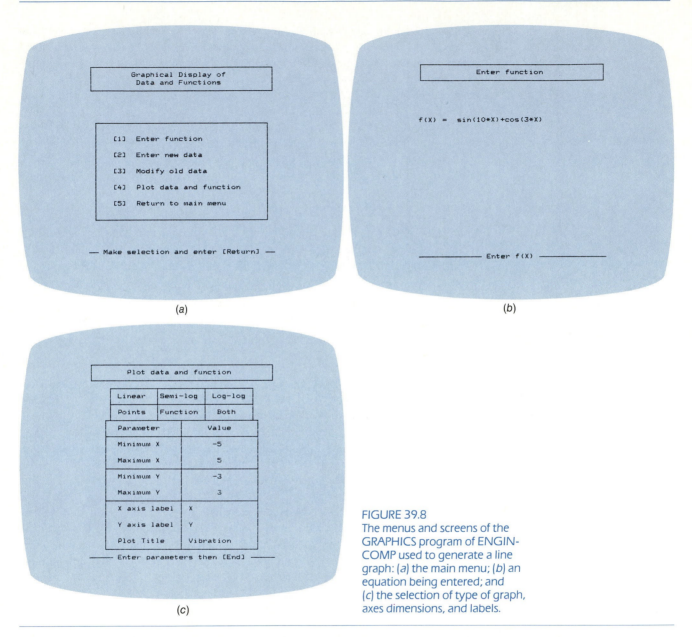

FIGURE 39.8
The menus and screens of the GRAPHICS program of ENGIN-COMP used to generate a line graph: (*a*) the main menu; (*b*) an equation being entered; and (*c*) the selection of type of graph, axes dimensions, and labels.

variable *x* (called X in the program) and the dependent variable *f*(*x*) (called Y in the program). The program prompts you for the dimensions with the wide horizontal bar. Figure 39.8*c* shows the screen after the limits are entered. Any or all of these values can be changed by moving the cursor to them and entering new numbers. If you enter an erroneous limit—for example, if minimum X ≥ maximum X—the program will alert you with an error message at the bottom of the screen. This is another example of how high-quality software facilitates your efforts.

When you are satisfied that you have entered the desired limits, the program gives you the opportunity to label the x and y axes and give the plot a title. The input session is terminated using the END key. When you have struck this key, a plot is automatically constructed, as shown in Fig. 39.9a. Notice how this plot has many of the features of a good line graph discussed in Chap. 22. The axes are numbered and scaled, and the plot is drawn in high-resolution mode. Notice also how secondary dashed axes are included to designate the zero values of the variables. The axes are labeled and the plot has a title.

The utility of the software can be demonstrated by using it to gain insight into the behavior of the function. Close inspection of Fig. 39.9a indicates that the function crosses the x axis at various points, connoting several roots in the interval from $f(x) = -5$ to 5. In addition, the function appears to be tangential to the axis at about $x = 4.2$. Such behavior is referred to as a *double root*.

In order to examine this possibility in more detail, the plot can be "blown up" by entering a new set of axis dimensions. The result, corresponding to a narrowed range of the x axis from $x = 3$ to $x = 5$, is shown in Fig. 39.9b. This version still suggests a multiple root at about $x = 4.2$.

Finally, as seen in Fig. 39.9c, the vertical scale is narrowed further to $f(x) = -0.15$ to $f(x) = 0.15$ and the horizontal scale to $x = 4.2$ to $x = 4.3$. The plot shows clearly that a double root does not exist in this region and that in fact there are two distinct roots at about $x = 4.23$ and $x = 4.26$.

The time and effort required to manually construct the graphs in Fig. 39.9 would be enormous and might even discourage you from performing the analyses. In contrast, the ENGINCOMP graphics program accomplishes the same task easily. The effort to develop your own homemade software of comparable quality would be considerable. Thus the canned program provides a significant resource that might warrant its inclusion in your arsenal of problem-solving tools.

ENGINCOMP's graphics program is but one example of the types of computer graphics software that are available or will be available for engineering problem solving. Other examples range from three-dimensional plotters to computer-aided design, or CAD, systems. Additionally, sophisticated graphics are often an important component of other types of canned software. For example, some spreadsheets have plotting options. As described in the next section, these capabilities represent an immense increase in our ability to produce visual expressions of our work.

39.4 COMPUTER GRAPHICS AND ENGINEERING

In the present chapter, we have stressed how computers can be employed to easily generate visual depictions of data and mathematical functions. Because visualization is often a critical step in gaining insight into otherwise abstract information, we have chosen to emphasize this quality of computer graphics. Beyond line graphs, however, there are many other areas and applications where computer-generated graphics can benefit engineering.

Examples are found in the areas of computer-aided design and computer-aided

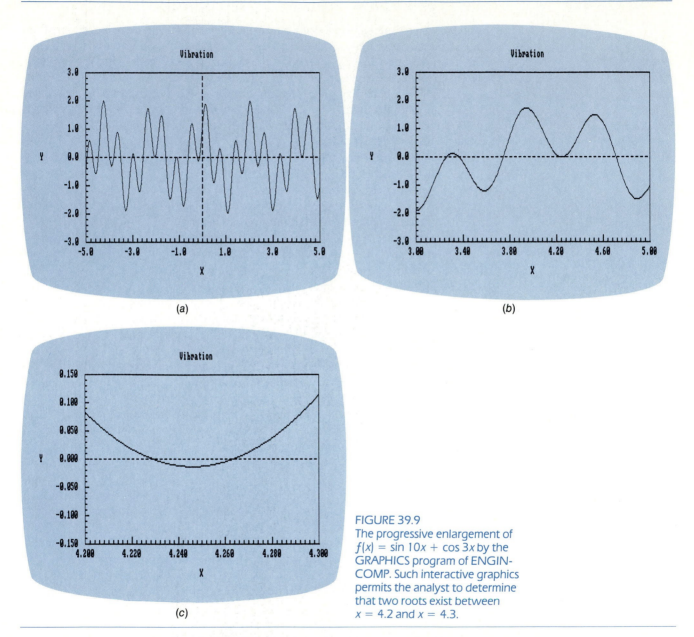

FIGURE 39.9
The progressive enlargement of
$f(x) = \sin 10x + \cos 3x$ by the
GRAPHICS program of ENGIN-
COMP. Such interactive graphics
permits the analyst to determine
that two roots exist between
$x = 4.2$ and $x = 4.3$.

manufacture (CAD/CAM) briefly discussed in Chap. 1. Graphics figure prominently
in several facets of the processes but are particularly relevant to CAD. Computer-
aided drafting facilitates the preparation of blueprints and other drawings that are
needed to bring engineered works and products to fruition.

Aside from two-dimensional graphs, an interesting capability offered by computers is the rendering of drawings in three dimensions. Among other things, the

coordinates of an object can be entered and the computer employed to display it from a number of different perspectives.

Finally, computer graphics can be employed for animation—that is, to add the dimension of time to the spatial dimensions. "Moving pictures" can be of great value in understanding the dynamics of objects and systems. Because computers provide the vehicle for producing such images, graphics will figure prominently in the development and advancement of engineering in the coming years.

PROBLEMS

39.1 Employ the most advanced graphics capabilities available on your computer to develop a graphics program similar to the one on ENGINCOMP. That is, develop a program that can plot a rectilinear line graph of discrete data points, a continuous function, or both. If possible, include the option to plot semilog and log-log graphs. This program can be employed in place of or along with ENGINCOMP to solve many of the following problems.

39.2 Use ENGINCOMP or your own program to obtain linear plots of the following functions:

(a) $y = \sin x + \cos x$ $\qquad$ $x = -10$ to 10
(b) $y = 10e^{-x} \cos 5x$ $\qquad$ $x = 0$ to 3
(c) $y = 10x/(5 + x)$ $\qquad$ $x = 0$ to 50
(d) $y = 10e^{-x} + 15(1 - e^{-0.5x})$ $\qquad$ $x = 0$ to 8
(e) $y = 5 + 4x$ $\qquad$ $x = -5$ to 5
(f) $y = x^2 + 2x - 8$ $\qquad$ $x = -10$ to 10

In each case try different scales for the vertical and horizontal axes. Change the labels on the x and y axes. Use your own title for the plots. Use a printer if available to obtain a hard copy of any of the above.

39.3 Use ENGINCOMP or your own program to obtain both linear and semilog plots of the following functions:

(a) $y = e^{-3x}$ $\qquad$ $x = 0$ to 1
(b) $y = 3e^{-x/2}$ $\qquad$ $x = 0$ to 10
(c) $y = 10^{x/10}$ $\qquad$ $x = 0$ to 10
(d) $y = 2^x$ $\qquad$ $x = 0$ to 10

Discuss the relationship between the linear and semilog plots.

39.4 Use ENGINCOMP or your own program to obtain both linear and log-log plots of the following functions:

(a) $y = 2x^2$ $\qquad$ $x = 0.1$ to 10
(b) $y = x^{-0.5}$ $\qquad$ $x = 0.1$ to 10
(c) $y = 4.1x^{1.3}$ $\qquad$ $x = 0.1$ to 10

Discuss the relationship between the linear and the log-log plots.

39.5 The following is experimentally derived data that describes the relationship between the growth rate of a fermentation yeast (r) and the concentration of an organic compound (c):

Concentration, mg/L	0	1	2	3	4	8	12	16
Growth rate, day^{-1}	0	0.05	0.15	0.15	0.12	0.10	0.07	0.05

The following model has been proposed to represent the data:

$$r = \frac{c}{c^2 + 2c + 5}$$

Use ENGINCOMP or your own program to plot both the data and the model. How well do the data and the plot compare on the basis of the plot?

39.6 The population of a city p_c is decreasing according to the formula

$$p_c = 100,000e^{-t/30} + 40,000$$

for $t = 0$ to 100, where t is in years and $t = 0$ corresponds to 1985. The population of a suburb p_s is increasing according to the formula

$$p_s = 50,000 + 70,000(1 - e^{-t/15})$$

for $t = 0$ to 100. Employ ENGINCOMP on your own program to graphically estimate the year in which the two populations will be equal. (Note that this problem can also be solved with a root-location method such as bisection.)

39.7 The following function characterizes the oscillations of an electric relay:

$$y = \sin 1.1t + \sin 1.4t$$

Use ENGINCOMP or your own program to determine how many times the relay opens or closes as t varies from 0 to 30 s. The relay is closed when y is positive and open when y is negative. Thus the number of times the relay opens or closes is equal to the number of times the function crosses the t axis.

39.8 Use ENGINCOMP or your own program to determine the number of real roots of the following functions in the range of $x = -10$ to $x = 10$:
(a) $y = 4x^3 - 3x^2 + x - 5$
(b) $y = x^4 - 3x^3 + 2x^2 - x - 8$
(c) $y = x^5 + 3x^4 - 5x^3 - 15x^2 + 4x + 10$
(d) $y = x^4 - 16x^3 + 60x^2 - 50x + 4$
(e) $y = x^5 - 8x^4 - 40x^3 + 300x^2 + 100x - 500$

39.9 Graphically approximate the smallest positive root of each of the functions from Prob. 39.8. Employ ENGINCOMP or your own program to estimate the root to within two significant figures.

39.10 Write a computer program using the LINE statement of IBM PC BASIC (or a comparable statement for your computer) that connects the following points with a straight line:
(a) (50, 50), (500, 100), (25, 160)
(b) (60, 60), (450, 90), (30, 180)
(c) (100, 50), (500, 50), (50, 150), (450, 150)
(d) (400, 30), (600, 30), (100, 180), (300, 180)

CHAPTER 40
WORD PROCESSING

Aside from spreadsheets and graphics software, there is a variety of other generic programs that have utility for engineers. Among the most useful for both students and professionals are the word-processing packages that can be employed to prepare written documents such as reports and papers. The present chapter provides a general review of such programs, along with specific examples of the use of a popular word-processing package—WordStar.

Word processing is the use of a computer to create, view, edit, store, retrieve, and print text. The heart of word processing is the "captured keystroke." When you employ a conventional typewriter to prepare a document such as a term paper or a report, you strike each key and that's the end of it. If an error is made, you must physically undo the error with correction tape or fluid or, all too often, retype the document. In contrast, when you are typing with a word processor, the computer generates a pattern of electronic bits to represent each character. The most direct use of these bits is to cause the character's image to form on the computer's monitor screen. However, beyond this immediate use, the electronic "capturing" of the character unlocks an immense potential for facilitating the preparation of documents. In essence, the character exists as an entity that can be manipulated or, in computer jargon, processed. It can be shifted to make room for other characters, or it can be moved to an entirely different location in the document. It can be transmitted electronically to a printer to generate a hard copy. Most importantly, it can be saved on secondary storage media, such as a diskettes, for later use.

The implications of these capabilities for report generation are profound. An error or omission in the early pages of a report no longer spells disaster. Rather than having to "cut and paste" or to retype many pages of manuscript to incorporate the change, you can electronically insert the modification, and a revised hard copy can be generated on the spot. Suppose that when you show your report to your boss or teacher, you are informed that the 1-in margins you have employed are unacceptable and that $1\frac{1}{2}$-in margins are required. If the report was typed on a conventional typewriter, such a modification would obviously require retyping the entire manuscript. In contrast, most word processors allow such format changes to be made with a few simple computer commands.

These, and other capabilities, make word processing a potent tool for communicating information. And the word processor will figure prominently in the careers of professionals such as engineers who routinely present their ideas and findings to others. The following sections provide a brief overview of the types, operation, and capabilities of some current word-processing systems.

40.1. TYPES OF WORD PROCESSORS

Word processors are made up of two primary components: a text editor and a text formatter. A *text editor* allows you to enter or modify text on a screen. The text editor is either line- or screen-oriented. As the name implies, a *line-oriented* text editor handles text on a line-by-line basis. Each line is stored in memory as an entity. Thus if you want to change a particular word, you must determine the line where it is located, make the change, and then resave the line in memory. In contrast, for *screen-oriented* text editors, each character on the screen represents one entity in memory. Consequently, a series of lines can be displayed on the screen and you can directly move a cursor to any location on the screen where editing is to be performed. Thus screen-oriented editors are usually more convenient and efficient than line editors.

In order to produce a hard copy of the document, the text editor must be used in conjunction with a *text formatter*. This is another program that allows you to transmit the document to a peripheral device such as a printer and control the style and format in which it is printed.

Most early word processors had separate text editors and text formatters. Today both functions are usually combined into a single package. These *full-featured word processors* are much easier to use because the fundamental capabilities are simultaneously available without changing program environments. In addition, most full-featured word processors have a variety of other user-friendly features to expedite the creation, editing, and printing of documents.

Today there are hundreds of word processors available on the market. Some are dedicated word processors—that is, computers whose sole purpose is to create documents. In addition, software is available to allow your personal or mainframe computer to act as a word processor. The most widely used software is the personal computer package called WordStar. Introduced in 1977 and available on almost every personal computer system, WordStar is by no means the most efficient or friendly package available. However, because it is so ubiquitous (over a million copies have been sold and probably many more pirated), it has become a de facto industry standard. For this reason, and because it has all the essential features of a typical package, we will use it as the focus for the following discussion of the operation of a word processor.

40.2 THE OPERATION OF A WORD-PROCESSING PACKAGE

As is typical of any effective word-processing package, WordStar offers a wide variety of capabilities for manipulating text. Because of the limited scope of this book, we cannot begin to explore all these capabilities. We will, therefore, limit the following discussion to some of the more fundamental operations involved in word processing. These are

1. Creating and saving
2. Retrieving and editing
3. Printing

Creating and saving a document.   The first step is to get the computer into a word-processing mode. The following example shows how this is done for WordStar and gives an illustration of how a short document can be created and stored.

EXAMPLE 40.1   Creating and Saving a Document with WordStar

PROBLEM STATEMENT: In this example we will describe the procedure for starting WordStar and for creating and for saving a document on a diskette.

SOLUTION: This application was performed on the IBM PC. Other computers may use slightly different implementations of WordStar. However, the essential commands should be very similar.

1. Starting WordStar. After booting the computer, place the disk containing WordStar in drive A and a formatted data disk in drive B. The document will be saved on this data disk.
Type the letters WS following the logged drive prompt (A>), as in

```
A> WS
```

and strike RETURN. WordStar will be loaded and the OPENING MENU, along with the file directory of the disk in drive A, will be displayed.
As seen in Fig. 40.1, the OPENING MENU lists a number of commands that can be implemented by typing the letter associated with the command. WordStar does not require that you hit RETURN after typing these letters. For example, type an F and watch the resulting action on the screen. The file directory should disappear. Type F again and the file directory will reappear. Such commands that turn on and off are called *toggle-switch commands*.
Before beginning to type the document you have to check that the computer is aware of the location of your data disk. Type L (without return) and the screen will prompt you with

```
THE LOGGED DISK DRIVE IS NOW A:
NEW LOGGED DISK DRIVE (letter, colon, RETURN)?
```

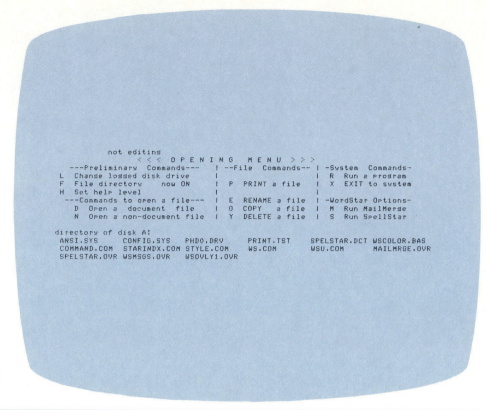

FIGURE 40.1
The OPENING MENU of Word-
Star, along with the file directory
of the disk in drive A.

This prompt is intended to switch the designation of the disk drive on which you want to store your data file. Because the data disk is in drive B, you would type B: and then strike RETURN. Notice that this action causes the file directory to be changed to the directory for drive B.

2. Creating a file. Next you must create a file to serve as the repository for your text. WordStar distinguishes between files it calls ''documents'' and ''nondocuments.'' The former are used for actual word-processed texts, whereas the latter are used for data (such as names and addresses) that you might want to employ in conjunction with documents (such as letters). Because you are interested in creating a word-processed document, type a D. The screen will prompt you with (Fig. 40.2)

NAME OF FILE TO EDIT?_

If you respond with a name that already exists, WordStar will recall the file and display it on the screen. If the file does not already exist, as is the case for the present example, then supplying a new name will cause WordStar to create a new file. Following the standard conventions for naming files (which are actually delineated in Fig. 40.2):

NAME OF FILE TO EDIT? EXAMPLE.TXT

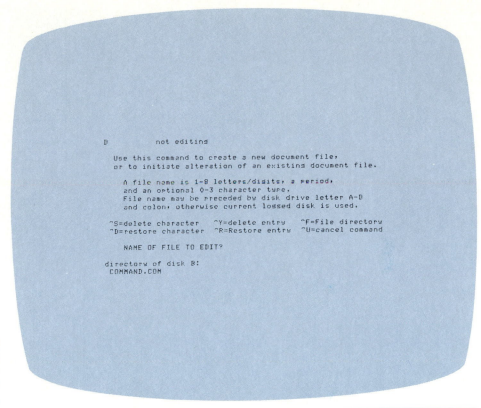

```
D:              not editing

    Use this command to create a new document file,
    or to initiate alteration of an existing document file.

       A file name is 1-8 letters/digits, a period,
       and an optional 0-3 character type.
       File name may be preceded by disk drive letter A-D
       and colon, otherwise current logged disk is used.

    ^S=delete character    ^Y=delete entry     ^F=File directory
    ^D=restore character   ^R=Restore entry    ^U=cancel command

       NAME OF FILE TO EDIT?

  directory of disk B:
    COMMAND.COM
```

FIGURE 40.2
Typing a D in the OPENING
MENU (Fig. 40.1) causes WordStar
to query you regarding the name
of the file you wish to edit.

After you have struck RETURN, WordStar will create a new file named EXAM-PLE.TXT. Notice how this name was chosen to reflect the file's contents. The extension TXT is employed to indicate that the file holds text.

At this point the screen should show the MAIN MENU with a few helpful additions (Fig. 40.3). At the top of the page is the *status line*, which should read

```
B:EXAMPLE.TXT  PAGE 1 LINE 1 COL 01        INSERT ON
```

This tells you the location and name of your data file (B:EXAMPLE.TXT), your location on the page of text you are going to type (because you have not started to type you are obviously at the beginning—PAGE 1 LINE 1 COL 01), and the status of INSERT. As will be described later, INSERT ON means that WordStar is ready to insert text rather than replace it.

Below the opening menu is the *ruler line*

```
L----!---!----!----!---!---!----!----!----!----R
```

Among other things, the ruler line defines the margins of the document. The L marks the left-hand margin setting, and the R marks the right. At this point you are ready to begin typing the document.

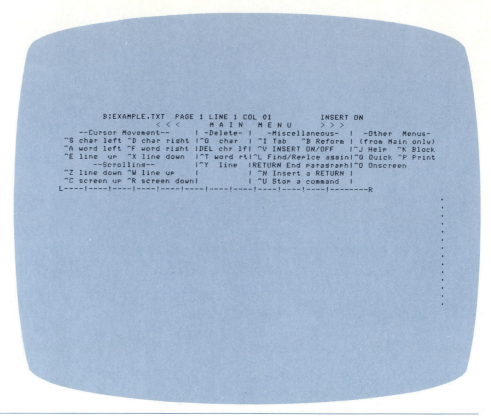

B:EXAMPLE.TXT   PAGE 1 LINE 1 COL 01   INSERT ON
                  < < <   M A I N   M E N U   > > >
     --Cursor Movement--   I -Delete- I   -Miscellaneous- I -Other Menus-
^S char left ^D char right I^G char I ^I Tab   ^B Reform I (from Main only)
^A word left ^F word right IDEL chr lfI ^V INSERT ON/OFF   I^J Help ^K Block
^E line up ^X line down I^T word rtI^L Find/Replce againI^Q Quick ^P Print
     --Scrolling--   I^Y line IRETURN End paragraphI^O Onscreen
^Z line down ^W line up   I   I ^N Insert a RETURN I
^C screen up ^R screen downI   I ^U Stop a command I
L----!----!----!----!----!----!----!----!----!----!--------R

3. Creating a document.   Before you begin to type, there are a few conventions you should know. When entering text, do not hit RETURN at the end of each line. If a word extends beyond the right margin, WordStar automatically places it on the next line. This capability is called *word wrap* and is available on all but the most primitive word processors. The return key is only depressed to signal the end of a paragraph or to skip a line. Also, while you are typing, occasionally glance up at the status line to see how the page, line, and column numbers change in order to provide you with the location of the cursor in the document. Now type the following paragraph:

The practice of engineering is inextricably tied to the evolution of new tools and technologies. Today microelectronic computer technology is providing a host of powerful new tools that will have a profound effect on our profession.

4. Saving a document.   You can save this text on the disk. To do this, press the CTRL key while simultaneously pressing the K key. WordStar uses ^K, where the ^ is used to signify the CTRL key, to designate this operation. Notice that the menu is now changed to the BLOCK MENU (Fig. 40.4). The commands for saving a file are listed on the left side of the BLOCK MENU:

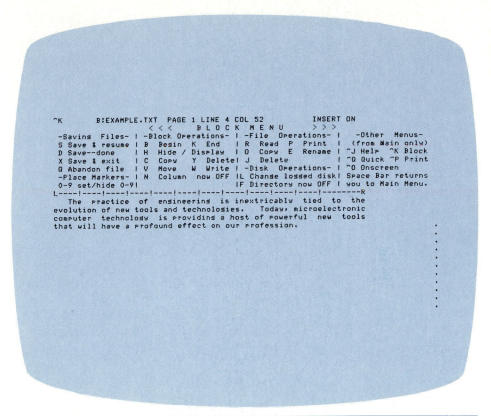

```
^K        B:EXAMPLE.TXT  PAGE 1 LINE 4 COL 52        INSERT ON
               < < <    B L O C K  M E N U    > > >
-Saving Files- I -Block Operations- I -File Operations- I  -Other Menus-
S Save & resume I B  Begin  K  End   I R  Read  P  Print I (from Main only)
D Save--done    I H  Hide / Display  I O  Copy  E  Rename I ^J Help ^K Block
X Save & exit   I C  Copy   Y  Delete I J  Delete        I ^Q Quick ^P Print
Q Abandon file  I V  Move   W  Write I -Disk Operations- I ^O Onscreen
-Place Markers- I N  Column  now OFF IL Change logged disk I Space Bar returns
0-9 set/hide 0-9 I                   IF Directory now OFF I you to Main Menu.
L----!----!----!----!----!----!----!----!----!----!----!--------R
     The  practice  of  engineering  is inextricably  tied  to  the
evolution of new tools and technologies.   Today, microelectronic
computer  technology  is providing a host of powerful  new  tools
that will have a profound effect on our profession.
```

FIGURE 40.4
In order to save a document, the command ^K is entered. This action causes the BLOCK MENU to be displayed. This menu lists a number of options, including the commands for saving a document.

```
S Save & resume
D Save--done
X Save & exit
```

S is used when the file is to be saved in the middle of an editing session. D is used when you are finished working on a particular file but still want to use WordStar. X is employed when you are finished working on a file and want to exit WordStar and return to the system. For the purposes of the present example, type X. At this point, the logged drive prompt (>B) appears at the bottom of the screen, indicating that the file has been saved, you have exited WordStar, and are back in the system. To ensure that the file has been saved, type DIR ® and observe that EXAMPLE.TXT is included on the file directory for the data disk in drive B.

Retrieving and editing a document. Example 40.1 has shown how WordStar can be used to create and store a short document. Suppose that you return to the computer at a later time and desire to retrieve that document and make some modifications. The following example illustrates how this is done.

EXAMPLE 40.2   *Retrieving and Editing a Document with WordStar*

PROBLEM STATEMENT: In this example we will perform some standard word-processing manipulations on the file EXAMPLE.TXT that was created in Example 40.1. This involves retrieving the file and editing and resaving it on diskette.

SOLUTION: First follow the procedure described at the beginning of Example 40.1 for starting WordStar. When you have finished this procedure, the screen should display the OPENING MENU, together with the directory of the data disk in drive B.

1. Retrieving a file.   A file is retrieved by typing a D (without return). The computer will prompt

```
NAME OF FILE TO EDIT?_
```

If you type in the file name, EXAMPLE.TXT, and strike RETURN, the file will be retrieved and displayed in a form that is ready for editing.

2. Editing a file.   We will perform two simple editing operations: inserting a sentence and correcting a mistyped word. First we will insert the following sentence between the first and second sentences of the paragraph created in Example 40.1 (notice that the word "pulleys" is deliberately misspelled):

From the levers, ramps and pullees of ancient times to the machines of the industrial revolution, engineers have always been quick to capitalize on new devices and concepts that extend their capabilities.

To insert this sentence move the cursor so that it is positioned at the letter *T* of the first word of the second sentence: Today. As delineated at the left of the MAIN MENU, the cursor can be moved by a number of commands (Fig. 40.5). For example, ^X moves the cursor to the next line down. Similarly, ^D moves it a character to the

FIGURE 40.5 There are two different ways to move the cursor when using WordStar on the IBM PC. One option is to simultaneously depress the CTRL key along with the letters E (up), X (down), S (left), and D (right). Alternatively, the arrow keys on the typing pad at the right of the keyboard can be depressed to accomplish the identical actions.

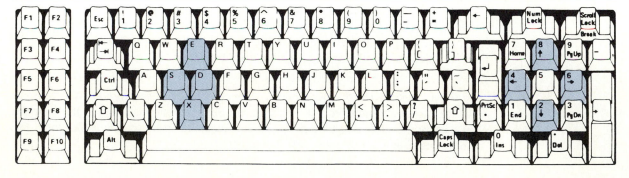

right. The same moves can be accomplished with the arrow keys on the typing pad at the right of the keyboard.

After locating the cursor at the beginning of the second sentence, check that INSERT ON is displayed on the status line at the top of the screen. The command ^V is a toggle switch to turn the insert ON and OFF. Strike ^V a number of times and observe how INSERT ON appears and disappears with each strike.

With INSERT ON, you can now begin to type the sentence. Notice how, as each character is typed, the word ''Today'' and the rest of the final sentence is shifted to the right to make room for the new characters.

When the sentence has been completed, the paragraph may look a bit disorganized, with some lines extending beyond the right margin and other lines not quite reaching it. To clean up the paragraph, enter ^O. This will change the menu to the ONSCREEN MENU. Check to ensure that Hyph-help in the menu is OFF. If not, type an H. Otherwise, WordStar will inquire whether you want to hyphenate long words that fall at the end of lines. When Hyph-help is off, implement the paragraph reform command ^B, and the paragraph should be reformed automatically. (You may have to hit ^B more than once to accomplish this.)

Next we must correct the misspelled word ''pullees.'' Move the cursor to the second e at the end of the word. Strike ^V to switch from the insert to the replace mode. You can tell this has been done correctly if the words INSERT ON are no longer visible at the end of the status line. Now type a lower case y. In constrast to the insert mode, the replace mode causes the y to be printed over, or in place of, the erroneous e.

At this point the paragraph should be typed correctly. To save this corrected version, strike ^K to switch to the BLOCK MENU. Then strike X to save the file and exit from WordStar.

Printing a document.   Examples 40.1 and 40.2 have illustrated how a document is created and edited. Now we will turn to the procedure by which a hard copy of the document can be obtained.

EXAMPLE 40.3   *Printing a Document with WordStar*

PROBLEM STATEMENT: In this example, we will demonstrate how a hard copy of the file generated in Examples 40.1 and 40.2, EXAMPLE.TXT, can be printed. We will also illustrate how WordStar can implement a special printing effect, such as underscoring, on most printers.

SOLUTION: Only documents that are not being edited may be printed. If you desire to print a file that you are editing, save the file with ^K and D to return to the OPENING MENU. If, as in the present example, you are not already in WordStar, follow the procedure described at the beginning of Example 40.1 for starting WordStar. At the end of the procedure, the screen should be displaying the OPENING MENU together with the directory of the data disk in drive B.

After readying the printer—that is, loading it with paper and placing it on line—select P from the OPENING MENU. The computer will prompt

```
NAME OF FILE TO PRINT?_
```

At this point you type EXAMPLE.TXT; you then have two options available for obtaining a printout:

1. After typing the file name, press ESC. With this option, the entire document is output without pauses between pages. This assumes you are using continuous form paper with a tractor feed.

2. After typing the file name, press RETURN. WordStar will then ask a number of questions that will allow you to control the printout. For example, it will query you for a beginning and ending page number in the event that you only desire a portion of the document. If you desire single-sheet printing, answer Y to WordStar's question,

```
PAUSE FOR PAPER CHANGE BETWEEN PAGES?_
```

After you have answered the remaining questions, the printout is generated (Fig. 40.6a) and WordStar returns to the OPENING MENU.

On most printers WordStar allows special printing effects, including boldfacing, underscoring, and subscripts and superscripts. To implement these effects, use the D command to retrieve EXAMPLE.TXT for editing. For the present example, we will underscore the words ''microelectronic computer technology.'' Move the cursor to the letter m at the beginning of this phrase. Type the command ^P. This will change the menu to the PRINT MENU (Fig. 40.7). Next type the letter S. Notice that a ^S has been inserted before the phrase. Now move the cursor just beyond the y at the end of the phrase and strike ^P and S. This will cause a ^S to be inserted at the end of the phrase. The result should look like

```
^Smicroelectronic computer technology^S is providing a host of
```

FIGURE 40.6
(a) A WordStar printout of the example paragraph. (b) The same paragraph but with underscoring.

```
     The  practice  of  engineering is inextricably tied  to  the
evolution  of  new tools and technologies.  From the  levers  and
pulleys  of  ancient  times to  the  machines  of  the  industrial
revolution, engineers have always been quick to capitalize on the
new devices and concepts that extend their capabilities.  Today,
microelectronic  computer  technology  is  providing  a   host  of
powerful  new  tools  that  will have a profound effect  on  our
profession.
```

(a)

```
     The  practice  of  engineering is inextricably tied  to  the
evolution  of  new tools and technologies.  From the  levers  and
pulleys  of  ancient  times to  the  machines  of  the  industrial
revolution, engineers have always been quick to capitalize on the
new devices and concepts that extend their capabilities.  Today,
microelectronic  computer  technology  is  providing  a   host  of
powerful  new  tools  that  will have a profound effect  on  our
profession.
```

(b)

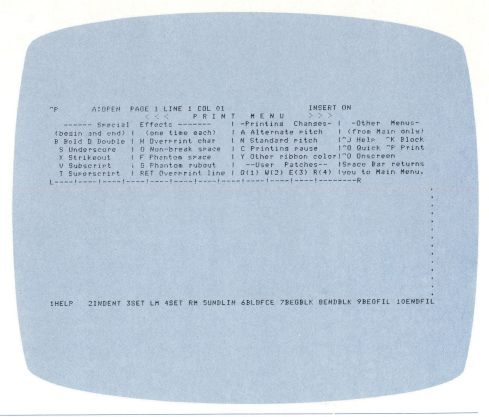

FIGURE 40.7
The PRINT MENU of WordStar.

The addition of ^S tells the computer that this phrase is to be underscored. At this point, the line might appear to extend beyond the right border. This is actually not the case, as WordStar ignores the ^S when the paragraph is printed, leaving a perfectly justified right margin. This can be seen when the printout is generated (Fig. 40.6b).

Although the foregoing examples should serve to introduce you to the operation of WordStar, they only begin to illustrate the powers of this tool. Additional information can be obtained from the WordStar manual and some of the books devoted to its operation (e.g., Brown, 1985). In the next section, we will discuss some of the other capabilities that are available on WordStar and other full-featured word processors.

40.3 CAPABILITIES OF WORD PROCESSORS

In the previous section, you have been introduced to some of the fundamental capabilities of WordStar. These are word wrap, cursor control, insertion, replacement, underscoring, and printing. Beyond these basic features, there are a variety of other capabilities available on WordStar and other full-feature word-processing packages. Some of these are outlined here.

Search and replace. Suppose that you desire to change the spelling of a person's name throughout a lengthy text. Rather than locating each occurrence and making the changes individually, the computer can implement the modifications all at once. Such a search and replace is one example of what is referred to as *global editing*. You can also implement a simple search where all locations of a particular word or phrase can be identified. A search has a number of applications, including the compilation of an index or glossary.

Block moves. Many word processors can move or copy large blocks of text from one location in a document to another. This capability is sometimes referred to as "cutting and pasting" because of its similarity to the physical cutting and pasting that was required to implement block moves prior to the advent of word processors.

Formatting. There are a whole host of capabilities related to controlling the appearance of the text. As with conventional typing, these include the setting of margins, tabs, and line spacing. However, with a word-processed document, the advantage is that these changes can be automatically implemented after the fact. On a normal typewriter you must retype a document to change it from single to double spacing. With a word processor such changes can be made with a single command.

Headers, footers, and autopagination. Many programs have the ability to automatically place titles and page numbers at the top or bottom of every page.

Spelling checks. These are usually separate programs that can be employed in conjunction with a word processor to check the spelling of words in a text. The spelling-checker program includes a dictionary of thousands of words. The words in a text are compared with those in the dictionary. If a word is not found, it will be marked or flagged in some way for review so that you can check whether it is correct.

Mail merge. This feature allows you to produce customized form letters. First a letter is composed with areas such as the address and salutation left blank or marked with a special merging command. Next you create a file of the names and addresses of those you wish to receive the letter. Finally the mail-merge program will automatically place each name and address into a letter and either store the resulting group or print them out.

40.4 WORD PROCESSING AND ENGINEERING

As mentioned at this chapter's start, word processing has direct application to the engineers' traditional task of communicating their ideas and findings to others. A word-processing capability can make report preparation and correspondence economical and efficient. Beyond these direct applications, word processing has other benefits. For example, its superior editing capabilities can be employed to facilitate the development and translation of large computer programs.

Aside from such professional applications, it should be obvious that word processing can represent a great resource to you as a student. Benefits range from the preparation of lab reports and term papers to writing job-seeking letters for the summer or after graduation.

Additionally, we believe that learning to word process may actually improve the quality of your writing. Of course, as with all software, you will still be operating under the GIGO rule—that is, garbage in, garbage out. However, the capabilities to cut and paste, edit, and check spelling available on most processors offer great potential for improving the quality of all the documents you produce. In addition, the fact that you do not have to worry about every little detail when typing a draft (because you can come back and fix it easily) should allow you to concentrate more on the coherent expression of your thoughts.

PROBLEMS

40.1 What is meant by a ''captured keystroke''? Why is it important?

40.2 What is the difference between a text editor and a text formatter? Distinguish between line-oriented and screen-oriented text editors.

40.3 Employ a word-processing package to produce your own version of Fig. 40.6*a*.

40.4 Employ a word-processing package to produce your own version of Fig. 40.6*b*. However, rather than underscoring ''microelectronic computer technology,'' highlight the phrase by printing it in boldface type. *Note:* For WordStar the commands ^P and B are used to specify boldface.

40.5 Employ a word processor to develop a letter requesting summer employment from a firm.

40.6 If you have a mail-merge capability, repeat Prob. 40.5 but generate letters for five different prospective employers.

40.7 Use a word-processing package to generate a resume. You will be graded on how professional looking the document appears and how well it represents your capabilities.

CHAPTER 41

DATABASE MANAGEMENT, TELEPROCESSING, AND INTEGRATED SOFTWARE

Besides spreadsheets, graphics generators, and word processors, the other members of the software "big five" are database managers and teleprocessors. The present chapter provides an introduction to these tools as well as a brief discussion of integrated software packages that combine several capabilities into one framework.

41.1 DATABASE MANAGEMENT

Engineers are continually organizing information and must be able to access it in a systematic fashion for easy retrieval. The most direct example is the compilation of the technical data that is typically amassed during an engineering project. Beyond such technical information, other examples range from the names and addresses of clients and contacts to bibliographies of primary reference materials. Today computers are providing powerful tools to allow engineers to access such information in an effective and efficient manner.

A *database* can be generally defined as a collection of data organized in a fashion that permits retrieval and use of that data. The database is composed of files. As noted in our previous discussion in Sec. 14.2.1, a *file* is a collection of logically related information. Each individual piece of information on a file is called an *item*. Examples might be a name or social security number. A *record* is a group of items that relate to the same object or individual. For example, in your university's mainframe computer there is undoubtedly a record referring to your academic standing. It probably consists of a number of items, including your name, social security or identification number, class, department, grades, etc. This record is in turn part of the database that contains the academic records of all the students at your school.

In Chaps. 14 and 21 you were introduced to the BASIC and FORTRAN statements needed to store and retrieve files on a computer. In Chap. 23 you were introduced to techniques that can be used to order such data. These chapters provided you with a first indication of how a computer might be used to manage a database. Employing methods that are similar in spirit, but much more efficient and powerful, software manufacturers have developed specialized computer programs for the express purpose of effectively managing databases. Called *database management systems*, or *DBMS*, these programs serve as an interface between the user and the database. As such, they are designed to expedite and simplify the process of creating, accessing, and maintaining databases.

Most of the database management systems for personal computers can, in fact, be conceived of as programming languages. Recall from Part Two that computers are only capable of responding directly to machine-language instructions consisting of bits. Because machine language is ill-suited for human use, high-level programming languages such as BASIC and FORTRAN were developed to condense groups of machine-language instructions into single, Englishlike statements. Thus high-level languages serve as the interface between the human and the machine. Similarly, many database management systems consist of a number of human-oriented statements. In contrast to general computer languages such as BASIC and FORTRAN, the DBMS statements are expressly designed to manipulate files. The DBMS program takes each of these statements and translates them into the machine language necessary for the computer to perform the operation. Figure 41.1 shows an example of the appearance of a typical DBMS dialect.

A benefit of the language orientation of the DBMS is that it gives users the flexibility to tailor databases to their specific needs. The disadvantage is that many of the more popular DBMS languages are relatively difficult to master. You will gain an inkling of this when we discuss the operation of a database in a later section. However, first we will briefly review the manner in which databases are logically organized.

FIGURE 41.1
Statements of a typical database management program— dBASE II—to set up a mailing list.

```
. CREATE
ENTER FILENAME: B:MAILIST
ENTER RECORD STRUCTURE AS FOLLOWS:
  FIELD      NAME,TYPE,WIDTH,DECIMAL PLACES
  001        FIRST,C,12
  002        SECOND,C,15
  003        STREET,C,25
  004        CITY,C,12
  005        STATE,C,12
  006        ZIP,C,5
  007
```

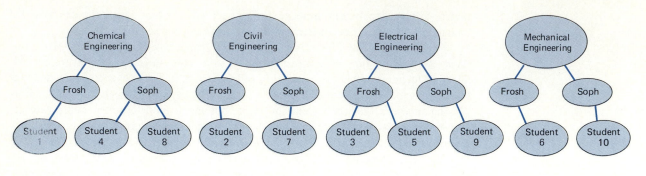

(a)

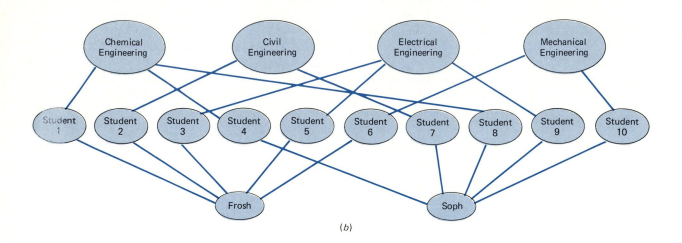

(b)

Student	Class	Department
1	Frosh	Chemical
2	Frosh	Civil
3	Frosh	Electrical
4	Soph	Chemical
5	Frosh	Electrical
6	Frosh	Mechanical
7	Soph	Civil
8	Soph	Chemical
9	Soph	Electrical
10	Soph	Mechanical

(c)

FIGURE 41.2 Three database management models: (a) hierarchical, (b) network, and (c) relational.

41.1.1 Database Organization

There are a number of fundamental approaches for the logical organization of databases. At present there are three major models of database structure. The two oldest, which have been the most common on mainframe computers, are the *hierarchical* and *network models*. The third is the *relational model*, widely used on microcomputers and increasingly employed on minicomputers and mainframes.

The hierarchical model. In this approach, the data items are ordered in a top-down fashion—that is, they always take the form of one to many. In Fig. 41.2a note that each department may have several students, whereas each student is only affiliated with one department. The terms *parent* and *children* are often used to describe these relationships. For example, each department in Fig. 41.2a is a parent and each class is a child. Similarly, the class is a parent relative to the student as a child. For the hierarchical model, a parent can have more than one child but a child can never have more than one parent.

To process a hierarchical database, one normally enters at the highest-level parent item (in Fig. 41.2a, the department) and progresses downward through the structure until the desired data item is located. Because of the logical ordering of the structure, information searches are relatively fast. However, modifications are complicated by the fact that the lines of connection must be considered when items are added or deleted.

The network model. The major difference between the network model (Fig. 41.2b) and the hierarchical model is that the network model allows children to have more than one parent. For example, a student (child) belongs to both a class and a department (parents). However, a child cannot have two or more parents from the same item type. Thus a student cannot belong to two departments or two classes.

The more complex links of the network model make data searches even more efficient than in the hierarchical model. For example, it makes it easy to locate all the students in a particular class or department. However, the increased complexity makes it all the more difficult to add and delete items.

The relational model. In this model, the data is set up in a format that is similar to a table (Fig. 41.2c). Each horizontal row represents a record, and each vertical column represents an item. In a true relational data base

1. Each row is unique—that is, no row can exactly duplicate another.
2. There must be an item in one column or group of columns that is unique. This is called the *key item* (recall Sec. 14.2.1) and serves as an identifier for each row.
3. There must be one and only one entry for each item, although the entry may be zero.
4. The rows are in no particular order.

The last requirement—the arbitrary ordering of the rows—is both the major strength and the major weakness of the model. Adding a new record to the database consists of merely adding a new row to the bottom of the table. Thus there is no need to fit the new data into a well-defined and complex structure, as is the case with the hierarchical or network models. On the other hand, the computer may have to tediously search the entire table to locate a particular item.

The lack of a strong built-in structure means that the relational database is the most flexible of the three approaches. Whereas the hierarchical and network models are expressly designed to expedite certain kinds of searching, the relational structure has no predetermined scheme. Thus the user has the flexibility to request any data, and the relational DBMS can perform the necessary operations to obtain that data.

41.1.2 The Operation of a DBMS

Today there are a variety of database management systems available on the market. The most widely used is dBASE II. Created and produced by Ashton-Tate, this package is available on most personal computer systems.

dBASE II is by no means the easiest DBMS to use. This is primarily because it is closer to a programming language than a typical menu-driven software package. Consequently, it is sometimes difficult to use at first. However, once its vocabulary is mastered, it is actually quite easy to operate and provides a powerful tool for setting up and manipulating customized databases. Additionally, because it is so ubiquitous, it has become a de facto industry standard. For this season, dBASE II will be used as the focus for the following examples.

Because of the scope of our text, the following discussion is limited to a few of the more fundamental operations involved in database management:

1. Creating a database
2. Interrogating the database
3. Generating a report

Creating a database.   The first step is to get the computer into a database-management mode. The following example shows how this is done for dBASE II and serves as an illustration of how a short file can be created and stored.

EXAMPLE 41.1   Creating and Saving a Data file with dBASE II

PROBLEM STATEMENT: In this example we will describe the procedure for using dBASE II to set up a file to store a name, an I.D. number, and a grade for each one of a group of students (Table 41.1).

SOLUTION: This application was performed on the IBM PC. Other computers may use slightly different implementations of dBASE II. However, the essential commands should be very similar.

    1. Starting dBASE II.   After booting the computer, place the disk containing dBASE II in drive A and a formatted disk in drive B. The database will be saved on

TABLE 41.1   Student Records Stored and Manipulated on a
Database Management System such as dBASE
II

Name	ID	Grade
Martin, Pat	7831313	65
Taylor, Nancy	7840019	100
Brown, Wayne	7925444	93
Cox, Mary	7812516	90
Collins, Peggy	7915252	81
Howard, Don	7952746	88
Duncan, John	7833421	62
Lester, James	7840978	69

this data disk. Type dBASE after you are given the logged drive prompt (A>), as in

```
A> DBASE
```

and strike RETURN. dBASE II will be loaded and a dot with a flashing cursor
(.___) will appear to indicate that the computer is ready.

2. Creating a file structure.   Before data can be entered, the structure of the file
must be set up. This involves exactly specifying the characteristics of the items (or
in the parlance of dBASE II, the fields) that will compose each record in the file. To
do this, type the word CREATE and strike RETURN. This tells the computer that
you are ready to begin a file. The computer will respond

```
ENTER FILENAME:
```

This can be done following the standard convention for naming files, as in

```
ENTER FILENAME: B:STUDENT
```

The B: alerts the computer that the file is to be saved on drive B. If the B: is omitted,
the program will store it on drive A. Notice that our choice of a file name, STUDENT,
is descriptive of its contents. The extension .DBF is appended by the computer to
indicate that it is a database file. When working in dBASE II, you do not have to
include the extension when referring to a file.

Next, the computer responds,

```
ENTER RECORD STRUCTURE AS FOLLOWS:
  FIELD  NAME, TYPE, WIDTH, DECIMAL PLACES
  001
```

Thus the dBASE II program requires that you specify four pieces of information—NAME, TYPE, WIDTH, and DECIMAL PLACES—to define each field. The *field name* should be descriptive of the contents of the field. The *type* refers to whether the field will hold character (C), numerical (N), or logical (L) information. Character data can be either alphabetic or numeric. Any number that will not be employed in mathematical operations—for example, zip codes or ID numbers—is usually input as character data. Logical data is that which is either true or false or yes or no. The *width* specifies the maximum number of characters in a field. *Decimal places* refers to the number of places to be reserved. The *decimal* specification is obviously only pertinent to numerical information and is omitted for character and logical types. For our example file, STUDENT.DBF, the specifications for the fields could be set up as

```
001     NAME,C,18
002     ID,C,7
003     GRADE,N,3
```

At this point, the desired structure is complete. If you press RETURN without an entry, the structure is saved on the data disk. Henceforth, when you input or add to STUDENT.DBF, this structure will be employed. Consequently, it should be created carefully. For example, if you fail to designate enough space when specifying the width, it is very difficult to enlarge the structure after the fact. Conversely, if you set aside too much space, the capacity of your disk will be exhausted sooner.

After setting up the structure, dBASE II will print out the query

```
INPUT DATA NOW?_
```

Type Y and the computer responds,

```
RECORD #00001
NAME    :                          :
ID      :                :
GRADE   :      :
```

The items for the first record from Table 41.1 can be entered. If the information takes up the entire field, the computer ''beeps'' and automatically jumps to the next line. Otherwise, you strike RETURN to advance to the next field. The BACKSPACE key can be employed to correct errors prior to pressing RETURN.

The remaining records can be entered in the same fashion. After the last record has been input, press RETURN. The .__ prompt will appear, and everything that you have typed is automatically saved in the file STUDENT.DBF on the disk in drive B.

At this point, or at any other time after the file has been created, you can obtain a summary of the structure by typing DISPLAY STRUCTURE, with the result:

```
DISPLAY STRUCTURE
STRUCTURE FOR FILE:   B:STUDENT.DAT
```

```
NUMBER OF RECORDS:    00008
DATE OF LAST UPDATE: 01/01/80
PRIMARY USE DATABASE
FLD          NAME        TYPE WIDTH     DEC
001          NAME          C     018
002          ID            C     007
003          GRADE         N     003
** TOTAL **                    00029
```

You can now sign off dBASE II by typing QUIT. This will return you to the system.

Interrogating a database.   Example 41.1 has demonstrated how dBASE II can be used to create a data file. Now we can investigate how this file can be manipulated to obtain select information.

EXAMPLE 41.2   *Interrogating a Database with dBASE II*

PROBLEM STATEMENT: In this example we will perform some standard manipulations on the file, STUDENT.DBF, that was created in Example 41.1. These operations involve (1) moving to various locations in the file and displaying records, (2) listing all or part of the file according to the values of the fields, and (3) sorting the file.

SOLUTION: Recall from our previous discussion of READ/DATA and sequential files, that our location in a file can be thought of as a pointer that indicates a particular record (Fig. 41.3). We can now investigate how the pointer is moved to various locations and how the resulting information is displayed.

1. Moving about in the file.   Whenever you desire to learn where the pointer is located, type DISPLAY. After the computer has read in all the data in Example 41.1, the pointer is situated at the last record and the result is

```
DISPLAY
00008 LESTER, JAMES     7840978  69
```

Figure 41.4 shows some other manipulations that might be performed. Typing DISPLAY ALL will cause the entire file to be listed. GO TOP causes the pointer to move to the first record in the file. GO followed by a record number causes the pointer to move to that record. SKIP followed by a number will cause the pointer to skip down that number of records.

2. Listing part or all of the file.   Aside from moving about the file, you might also desire to list the whole file or parts of it according to particular values of the fields. Figure 41.5 shows some examples. Typing LIST results in a listing of the entire file. As seen in Fig. 41.5*a*, the result is identical to DISPLAY ALL. Typing LIST OFF performs the same operation but omits the record numbers. Typing LIST

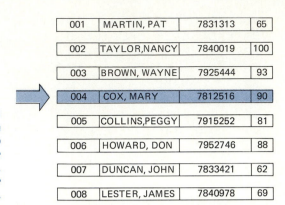

FIGURE 41.3
The pointer defines your location in a file. It indicates or "points" to the record at which the database management system is situated. For example, in this illustration the pointer is at the fourth record.

FOR followed by a logical condition involving items will result in the appropriate records being listed. For example, as shown in Fig. 41.5c, entering LIST FOR GRADE > 70 will list the records for all the students with grades better than 70. Compound logical expressions can also be used for the same purpose, as shown in Fig. 41.5d.

3. Sorting the file. If you desire to print the students in alphabetical order, type INDEX ON NAME TO B:STUDLALPH and strike RETURN (Fig. 41.6a). This instructs dBASE II to sort the students alphabetically and hold this list in another file called STUDLALPH on drive B. The computer will respond 00008 RECORDS INDEXED to let you know that the sort was successful. You can then type LIST to see the result (Fig. 41.6a). Similar manipulations can be accomplished on the basis of the other fields. For example, Fig. 41.6b shows the result of sorting the students according to their ID numbers.

FIGURE 41.4
A part of dBASE II session showing how the pointer is moved to various locations of the file.

```
. DISPLAY ALL
00001    Martin, Pat        7831313      65
00002    Taylor, Nancy      7840019     100
00003    Brown, Wayne       7925444      93
00004    Cox, Mary          7812516      90
00005    Collins, Peggy     7915252      81
00006    Howard, Don        7952746      88
00007    Duncan, John       7833421      62
00008    Lester, James      7840978      69
. GO TOP
. DISPLAY
00001    Martin, Pat        7831313      65
. GO 4
. DISPLAY
00004    Cox, Mary          7812516      90
. GO TOP
. SKIP 2
RECORD: 00003
. DISPLAY
00003    Brown, Wayne       7925444      93
. SKIP 2
RECORD: 00005
. DISPLAY
00005    Collins, Peggy     7915252      81
.
```

```
(a)  . LIST
     00001   Martin, Pat        7831313    65
     00002   Taylor, Nancy      7840019   100
     00003   Brown, Wayne       7925444    93
     00004   Cox, Mary          7812516    90
     00005   Collins, Peggy     7915252    81
     00006   Howard, Don        7952746    88
     00007   Duncan, John       7833421    62
     00008   Lester, James      7840978    69
(b)  . LIST OFF
     Martin, Pat        7831313    65
     Taylor, Nancy      7840019   100
     Brown, Wayne       7925444    93
     Cox, Mary          7812516    90
     Collins, Peggy     7915252    81
     Howard, Don        7952746    88
     Duncan, John       7833421    62
     Lester, James      7840978    69
(c)  . LIST FOR GRADE > 70
     00002   Taylor, Nancy      7840019   100
     00003   Brown, Wayne       7925444    93
     00004   Cox, Mary          7812516    90
     00005   Collins, Peggy     7915252    81
     00006   Howard, Don        7952746    88
     .
(d)  . LIST FOR GRADE < 80 .AND. ID < '7840000'
     00001   Martin, Pat        7831313    65
     00007   Duncan, John       7833421    62
```

FIGURE 41.5
(a) and (b) Parts of a dBASE II session showing how the whole file can be listed. (c) and (d) Parts can also be listed selectively according to the values of the fields.

Generating a report.   Examples 41.1 and 41.2 have illustrated how a database can be created and manipulated. Now we will turn to the procedures for generating printed reports.

EXAMPLE 41.3   Generating Reports with dBASE II

PROBLEM STATEMENT: In this example we will show how dBASE II can be employed to generate reports (in this case, labeled tables).

SOLUTION: Before illustrating how labeled tables can be generated, we can first show how a rough copy of your file can be printed. First, hold down the CTRL key (which we represent by the symbol ^) and press P. This activates the printer. From this point forward, everything you type will be output on the printer. For example, Figs. 41.4 through 41.6 were all output by first striking ^P. Striking ^P again will deactivate the printer.

FIGURE 41.6
Two examples of sorting a database: (a) alphabetical order and (b) ID order.

```
INDEX ON NAME TO B:STUDLAPH             . INDEX ON ID TO B:IDSORT
00008 RECORDS INDEXED                   00008 RECORDS INDEXED
. LIST                                  . LIST
00003   Brown, Wayne     7925444   93   00004   Cox, Mary        7812516    90
00005   Collins, Peggy   7915252   81   00001   Martin, Pat      7831313    65
00004   Cox, Mary        7812516   90   00007   Duncan, John     7833421    62
00007   Duncan, John     7833421   62   00002   Taylor, Nancy    7840019   100
00006   Howard, Don      7952746   88   00008   Lester, James    7840978    69
00008   Lester, James    7840978   69   00005   Collins, Peggy   7915252    81
00001   Martin, Pat      7831313   65   00003   Brown, Wayne     7925444    93
00002   Taylor, Nancy    7840019  100   00006   Howard, Don      7952746    88
              (a)                                     (b)
```

(a)
```
. REPORT FORM B:REP1
  ENTER OPTIONS, M=LEFT MARGIN, L=LINES/PAGE, W=PAGE WIDTH M=15, L=55, W=70
  PAGE HEADING? (Y/N) Y
  ENTER PAGE HEADING: STUDENT LIST
  DOUBLE SPACE REPORT? (Y/N) N
  ARE TOTALS REQUIRED? (Y/N) N
  COL     WIDTH,CONTENTS
  001     18,NAME
  ENTER HEADING: NAME
  002      7,ID
  ENTER HEADING: ID
  003      5,GRADE
  ENTER HEADING: GRADE
  004
```

(b)
```
                         PAGE NO. 00001
                         05/22/85

                                           STUDENT LIST

                         NAME              ID    GRADE

                         Martin, Pat       7831313    65
                         Taylor, Nancy     7840019   100
                         Brown, Wayne      7925444    93
                         Cox, Mary         7812516    90
                         Collins, Peggy    7915252    81
                         Howard, Don       7952746    88
                         Duncan, John      7833421    62
                         Lester, James     7840978    69
```

```
. USE B:STUDENT
. REPORT FORM B:ASTUDENT TO PRINT FOR GRADE >= 90
ENTER OPTIONS, M=LEFT MARGIN, L=LINES/PAGE, W=PAGE WIDTH M=15, L=55, W=70
PAGE HEADING? (Y/N) Y
ENTER PAGE HEADING: STUDENTS WITH A'S
DOUBLE SPACE REPORT? (Y/N) N
ARE TOTALS REQUIRED? (Y/N) N
COL     WIDTH,CONTENTS
001     18,NAME
ENTER HEADING: NAME
002      7,ID
ENTER HEADING: ID
003      5,GRADE
ENTER HEADING: GRADE
004
```

(c)
```
                         PAGE NO. 00001
                         01/01/80

                                           STUDENTS WITH A'S

                         NAME              ID    GRADE

                         Taylor, Nancy     7840019   100
                         Brown, Wayne      7925444    93
                         Cox, Mary         7812516    90
```

FIGURE 41.7
Report generation with dBASE II: (a) setting the headings; (b) the resulting report for the whole file; (c) printing part of the file. In this case, the students with grades higher than 90 are output.

Although rough printouts of the sort shown in Figs. 41.4 through 41.6 are certainly useful, they are not in a format suitable for professional purposes. dBASE II is capable of generating such professional reports as illustrated in Fig. 41.7. First the format for the table must be specified. This is accomplished by typing REPORT FORM B: REP1 and striking RETURN. As shown in Fig. 41.7a, the computer responds with a number of queries that are used to specify the headings, spacings, etc. Next, to obtain the printout, type REPORT FORM B:REP1 TO PRINT and strike RETURN. As shown in Fig. 41.7*b*, the result is a professional quality report. Figure 41.7*c* shows how parts of the file can be selectively printed in report form.

Although the foregoing examples should serve to introduce you to the operation of dBASE II, they only begin to illustrate the powers of this tool. Additional information can be obtained from the dBASE II user's manual and some of the books devoted to its operation (for example, Byers, 1982, or Krum, 1984). The capabilities of dBASE II as well as other DBMS can be found in a variety of references. Two invaluable sources of information are Glossbrenner (1984) and the *Whole Earth Software Catalog* (Brand, 1984). An advanced version—dBASE III—is also available that has capabilities beyond dBASE II.

41.2 TELEPROCESSING

The development of communications technology over the last century has had an awesome impact on the rate at which information is transmitted around our world. In the nineteenth century, it took days to weeks for news from North America to reach Europe. Today, because of the rapid evolution of telecommunications, information moves almost instantaneously between most parts of the globe.

Despite these advances, delays in the transfer of information still occur. Most of these relate to the communication of detailed, printed data. Suppose that a local sports team plays an away game and you desire detailed statistics over and above the score. You will probably have to wait for the next day's newspaper to obtain these statistics.

Or suppose that you are collaborating on a project with an engineering firm in a distant city. Report preparation and other information-related tasks could be hindered by mail delays of one to several days.

Today the computer is being employed more and more as a communications tool to expedite the transmission of detailed information. The general name for this application of computers is *teleprocessing*. This label is a composite of the words "telecommunications" (communicating, usually over long distances, via electronically transmitted waves) and "data processing." As shown in the next section, teleprocessing consists of using telecommunications devices, such as telephones, to allow two computers to converse with each other.

41.2.1 How Computers Converse

Although electronic networks can be expressly created to link computers, any large-scale system would involve great expense. Rather than develop a new network, a more economical strategy would be to utilize already existing communications systems. In the 1960s, this fact was recognized by the pioneers of timesharing computers (recall Sec. 2.4), who used the telephone system as their primary linkage mechanism. Today hardware and software exist to allow personal computers to communicate via the telephone.

Although the extent and economy of the phone system make it a logical choice as the communications medium for teleprocessing, it is not without disadvantages.

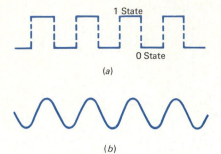

FIGURE 41.8
The fundamental difference between (*a*) digital and (*b*) analog, or continuous, signals.

The primary shortcoming relates to the fact that it was developed for voice rather than data transmission. As a consequence, computers and telecommunications systems handle and transmit information in two fundamentally different modes.

As mentioned in Chap. 2, computers are *digital* devices that employ discrete bits to characterize information (Fig. 41.8*a*). In contrast, the telephone represents signals in an *analog* fashion (Fig. 41.8*b*) by a continuously varying sine wave. This mode is well-suited to transmit the variations in the pitch of a human voice.

Because of this fundamental difference, some means is required to translate signals from one mode to the other. Translation of a signal from digital to continuous waves is called *modulation* and the reverse process *demodulation*. A single device called a *modem* (a composite of *mo*dulation and *dem*odulation) is used to perform both operations. Figure 41.9 provides a visual depiction of how modems are used to link two personal computers via the telephone system.

Notice that there are two types of modems in Fig. 41.9. As the name implies, the *direct-connect modem* is attached directly to the phone line using a modular phone jack. A wire is then employed to connect the modem to a port in the back of the microcomputer. A *port* is a socket, also called a *serial interface*, into which a plug is inserted to transmit a stream of bits in and out of a computer.

In contrast, an *acoustic coupler* is a device with two suction cups into which a telephone handset is fit. Because of the increased possibility of interference from outside noise, these are usually inferior to direct-connect modems.

The speed of a modem is represented by the number of binary digits or *bits per second* (*bps*) it can transmit. The number of bits per second is sometimes referred to as the modem's *baud rate*.* Today the most commonly used modems are 300 and 1200 bps. Three hundred bits per second translates roughly to about 300 words per minute, which is a lot faster than the average typist but just slow enough to read if the text is displayed on your monitor. Twelve hundred bits per second is four times faster but still slow enough to skim. Because of their speed, 1200-bps modems are usually preferred. However, in areas where phone transmission is poor, a slower rate may be needed to ensure correct reception.

---

*Strictly speaking there is a distinction between bits per second and baud rate. The latter actually signifies the number of signal events per second. In the early days of data communication, the two terms were equivalent because each signal event corresponded to a bit. Today signal events may sometimes transmit two or more bits. Therefore bps is a more precise measure of the actual rate at which information is passed.

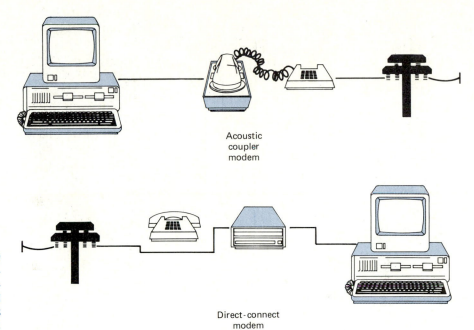

Acoustic
coupler
modem

Direct-connect
modem

FIGURE 41.9
The use of modems and the telephone system to forge a teleprocessing link between two personal
computers. Two types of modems
are employed.

In addition to the modem, a teleprocessing system also needs software. This software, called a *terminal program*, controls the modem, properly managing the transfer of information between it and your screen, disks, and printer.

A final aspect of data transmission that bears mention are the directional capabilities of the system. *Simplex transmission* (Fig. 41.10a), where data is sent in a single, prespecified direction, is rarely used for common teleprocessing. Because most applications involve two-way communication, *duplex transmission* is predominant.

FIGURE 41.10
Graphical depiction of (a) simplex,
(b) half-duplex, and (c) full-duplex
transmission.

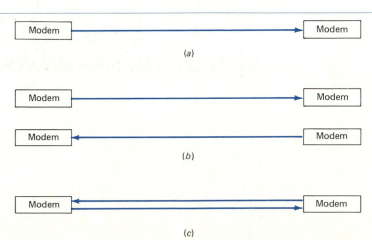

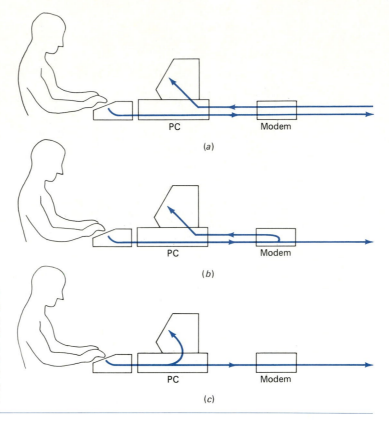

FIGURE 41.11
Some ways in which your typing is displayed on your monitor when operating in (*a*) full-duplex and (*b*) and (*c*) half-duplex modes. (*b*) and (*c*) are distinguished by whether the terminal software or the modem governs the display on the monitor.

In *half-duplex transmission* (Fig. 41.10*b*), messages are sent in either direction but only one way at a time. For *full-duplex transmission* (Fig. 41.10*c*), information can be sent in both directions simultaneously. Most commercial databases and information services operate in full-duplex mode, but many mainframe computers do not.

The distinction between half- and full-duplex is important because either one mode or the other may be required in order to link a particular computer. Therefore for greatest flexibility your modem should have both options. In addition, the type of transmission has a bearing on what you see on your monitor screen. When you type a message in full-duplex mode, the words that appear on the screen do not come directly from your keyboard. Rather, they leave your computer, travel to the computer you are communicating with, then "echo" back to your system, and are displayed on your screen. Consequently, if you want to see what you are typing, either your modem or the software must make certain that a *local copy* is displayed on your screen (Fig. 41.11).

41.2.2 Teleprocessing and Engineering

Teleprocessing is an umbrella term that encompasses a broad range of capabilities. In this section we will mention some of these capabilities and suggest their relevance to engineering.

Mainframe data exchange.   Earlier in this book we mentioned that the modern engineer has access to a wide variety of computing tools. Although the emphasis of the present part of the book has been slanted toward personal computers, large mainframe machines can also figure in your work. Teleprocessing provides a means to link your personal computer system with a mainframe and in the process greatly enhance your capabilities. For example, you might assemble data and make preliminary calculations on your personal computer as a prelude to a large number-crunching computation. When the calculation is adequately prepared you can access a mainframe for speedy implementation.

Electronic mail.   There are a variety of systems available to allow you to transmit messages via the computer. As an engineer you will often find yourself in the position of having to communicate with colleagues and clients over large distances. Electronic mail is one way to facilitate this process.

Information retrieval.   Today there are a variety of databases that can be accessed by teleprocessing. Although many of them have general appeal, such as those related to the stock market and news, others have direct relevance to engineering. One example is the Environmental Protection Agency's STORET system, which civil and chemical engineers can access for information related to water- and waste-treatment plant design.

Conferencing.   Computer and video screens can be used to set up discussions between parties. Although teleconferencing has yet to be greatly exploited, it provides a means to coordinate activities and make decisions involving parties located at great distances. Among other things, it is extremely cost-effective, considering the expense of travel. As the technology evolves and becomes more common, it could figure prominently in such engineering activities as project management and proposal submission.

Aside from the above areas, there are a variety of other aspects of teleprocessing that could affect both the professional and personal areas of your life. These include electronic bulletin boards, home banking and shopping, investment services, and electronic publishing. As communicators, we engineers will undoubtedly capitalize on the opportunities offered by this new technology. In particular, for both professionals and students, teleprocessing provides a means to network with colleagues and clients.

41.3   INTEGRATED SOFTWARE

You have now been introduced to several types of software that can prove useful in your professional and personal endeavors. These are spreadsheets, graphic packages, word processors, database managers, and teleprocessors. Once you become familiar with their operation, it will not be uncommon for you to employ several of them during the course of a day. Every time you switch from one to the other, you will have to spend time loading the new program into the computer. If you were to add it all up, you would probably be amazed at the total time spent every week searching for disks and waiting for programs to load.

Beyond the time element, you may also be frustrated by the fact that a file generated by one of these programs is incompatible with one of the other software packages. For example, you might calculate some results on a spreadsheet that you would like to see in graphical form. Unfortunately, in many cases the data generated will not be in a form that can be directly transferred to a graphics program. Thus you would probably have to enter the data manually into a file that the graphics program can read. Further, when you finally produce the graphs you might not be able to integrate them directly into a report you wish to prepare on your word processor.

Integrated software overcomes such time and compatibility problems by combining several programs into one coherent package. Moving from one program to another is as simple as changing the channels on your television. In addition, all the output is completely compatible. Thus you can generate numbers on the spreadsheet and immediately see how they look in graphical form. In the same sense, the graph can be easily integrated into a report produced by the package's word processor.

41.3.1   Types and Capabilities of Integrated Software

Lotus Development Corporation's 1-2-3 was the first widely successful integrated package. Aside from its spreadsheet capability, 1-2-3 also includes graphics and a limited database capability.

In order to provide a full-blown integrated package, Lotus has developed an enhanced version of 1-2-3. This package, called Symphony, includes spreadsheet, word-processor, graph-generator, database, and teleprocessing capabilities (Fig. 41.12). You might say that Lotus 1-2-3 has become Symphony 1-2-3-4-5. Symphony can be entirely contained in the computer's RAM, and the data it can manipulate is limited

FIGURE 41.12
A screen showing an integrated software package—Symphony—in use. (*Courtesy of Lotus Corporation.*)

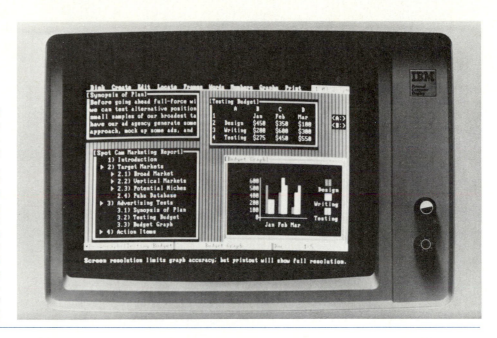

FIGURE 41.13
Framework, Ashton-Tate's new software program, allows an unlimited number of "frames" to be displayed on a "desk" work area. The photo above shows four frames displaying (clockwise from the upper left): a word-processing document; a spreadsheet; a bar graph; and an outline frame, which in turn contains a number of other frames. All screens are generated on a standard IBM monochrome display using the IBM monochrome display adapter. (*Courtesy of Ashton-Tate.*)

to the amount of available RAM. Thus one of its strengths is that it does not require a hard-disk system. In addition, it has the advantage that its data files are downward-compatible with 1-2-3. This means that files developed using 1-2-3 can be employed by Symphony.

Ashton-Tate, the developers of dBASE II, have created an integrated package called Framework (Fig. 41.13). It is also a RAM-resident program that does not require a hard disk. The first versions of Framework have all the capabilities of Symphony, with the exception of teleprocessing. In addition, Framework has the advantage that its data files are downward-compatible with its predecessor, dBASE II.

41.3.2 Integrated Software and Engineering

By tying together a group of fundamental capabilities, integrated software packages elevate information processing to a higher plane. Whether the task is report generation or data analysis, the enhanced power and flexibility offered by these packages will both facilitate and inspire your efforts. For professionals, the benefits should be obvious. As a student, you should use these packages both to prepare yourself for your future career and to enhance your productivity. One area where integrated packages have direct application is in producing lab reports. These documents typically involve several capabilities—word processing, analysis (spreadsheets), data (DBMS), and graphics—that are included on most integrated packages. By utilizing these tools, you will profit from both the time savings and from the improved quality of the finished product. In addition, they will help you to realize the rationale underlying projects such as lab reports—that is, to foster your ability to communicate ideas and data in a clear and coherent fashion.

PROBLEMS

41.1   Identify the three major models of database structure. What are their relative advantages and disadvantages?

41.2   The following relational database summarizes information regarding some major software packages:

Package	Type	Company	1984 list price
InfoStar	DBMS	MicroPro	$595
WordStar	Wordprocessing	MicroPro	$495
1-2-3	Spreadsheet	Lotus	$495
Symphony	Integrated	Lotus	$695
dBASE II	DBMS	Ashton-Tate	$495
Framework	Integrated	Ashton-Tate	$695
Multiplan	Spreadsheet	Microsoft	$195
Microsoft Word	Word processing	Microsoft	$375
CalcStar	Spreadsheet	MicroPro	$195

Reexpress this database in terms of the other two major models of database structure in a fashion similar to that seen in Fig. 41.2.

41.3   Employ a DBMS package such as dBASE II to create a database to store the information listed in Table 41.1. Generate printed reports of (*a*) all students listed in ID order; (*b*) all students with grades of 80 through 89; and (*c*) all students with both grades greater than 70 and ID numbers less than 7900000.

41.4   Employ a DBMS package such as dBASE II to create a database to store the information from Prob. 41.2. Generate printed reports of (*a*) the package in alphabetical order (interpret your results if anything seems odd): (*b*) all packages costing more than $250; and (*c*) all spreadsheets costing less than $250.

41.5   List the differences between digital and analog signals; describe how they are translated in teleprocessing.

41.6   Distinguish between a direct-connect and an acoustic-coupler modem. Which is preferable and why?

41.7   Distinguish between bps and baud rate. Which might tend to be higher?

41.8   Distinguish between simplex, half-duplex, and full-duplex transmission.

41.9   Discuss the benefits of integrated software packages.

CHAPTER 42

FOR STUDENTS ONLY

If you are a professor, we would appreciate it if you would stop reading at this point. Thank you and have a nice day.

Now that we're alone, can we talk? Although we are sure that you have enjoyed this course immensely, you must now acknowledge the fact that it is time to move on. Although you will undoubtedly have the option of taking other computer-related courses, from this time forward, self-education will play a more prominent role in your acquisition of computer skills. We would like to discuss a few issues that we feel are important in this process.

42.1 ACQUIRING HARDWARE

We have tried to design this book so that it can be used in a number of different computing environments. Thus some of you may have employed it in conjunction with your university's mainframe machine, whereas others may have used a personal computer as your primary tool. As we have stated a number of times throughout this book, we believe that as a modern engineer, you must be comfortable in a range of computing environments. Only in this way can you capitalize on all the capabilities that are at your disposal. However, regardless of what environment you have employed for this course, we believe that sooner or later you will purchase your own personal computer.

From the perspective of your development as an engineering student, we believe that the sooner you obtain one the better. Obtaining a personal computer early in your college career will have a number of benefits. First and foremost, the gain in efficiency will allow you to devote more of your time and energy to developing intellectual skills rather than to such arduous chores as manual computations and hand-typed term papers and lab reports. Second, the process of understanding all the ins and outs of your computer will be an education in itself. Although this learning will be machine-specific, there are enough similarities among different types of computers that the knowledge you gain will have broad application. In addition, the fact that you become comfortable with a machine will give you more confidence regarding your ability to function in other computing environments. Finally, we believe that the very act of interacting with computers fosters and develops logical thought and mental discipline that will carry over to the other areas of your development as an engineer. With these facts in mind, we can now present some ideas that you should consider when purchasing hardware and software.

42.1.1 Securing the Funds

After you have decided that owning a computer is in your best interest, your next step is to secure the funds to make the purchase. First, you might consider Ma and Pa. A sizable number of graduating high school students each year are fortunate (and affluent) enough to receive objects as costly as automobiles for graduation gifts. Although it might have less immediate value on Friday and Saturday nights, there is no question that a gift of a personal computer would be immeasurably more beneficial in the long run.

If your folks cannot come up with the cash, it's up to you to secure the capital. Quite a few universities have set up programs to allow their students to purchase hardware at a reduced price. Some of these have permitted the purchase price to be spread over the four years of the student's matriculation. Thus, rather than paying, say $2000, up front, you pay on the order of $500 per year. This is a particularly appealing arrangement for those students who have difficulty securing loans from commercial lending institutions.

Short of affluent parents or a special purchase program, it will be up to you to decide whether you want to spend your hard-earned cash on a personal computer. As with any economic decision, you will have to weigh the costs and benefits against other purchases such as food, shelter, clothing, transportation, tuition, and luxuries like spring break at the beach. Just do not forget that the cost is an investment that will reap you benefits throughout the remainder of your education and professional career.

42.1.2 Selecting the Right Computer for Your Needs

Not too many years ago, it was not so difficult to buy a personal computer. There were three or four brands to choose from and only a few stores from which to buy. With so few choices, you could probably go through the entire process in a few hours and be confident that you had made a sound decision.

Today you could also go through the same process in a few hours. However, with hundreds of brands and stores to choose from, odds are that you would be lucky to find just the right system for your needs.

Although at first glance, the situation might seem hopeless, it's not much different from buying an automobile. As with buying a car, a little research and thought on

TABLE 42.1   Eight Points to Consider before Buying a Personal Computer

1. **Determine your needs.**   The student and professional engineer have different computing needs. As a student, your primary requirements are for relatively small-scale computations and word processing. On the other hand, professional engineers usually need increased speed, memory, and precision for large-scale computations. In addition, many professional engineers require enhanced capabilities in order to effectively manage their businesses.

2. **Identify your software requirements.**   As with hardware, engineers have their own special software needs. In particular, our computational and language requirements differ from most other professionals. We will elaborate more on these special needs in Sec. 42.2. For the time being, it is important to note the minimum memory and processing-related requirements of the software. These specifications could dictate or at least limit your hardware choices. In addition, some very popular software packages are not available for certain computers.

3. **Prepare an outline of your system.**   Based on points 1 and 2, make a sketch of the hardware system that will fulfill your needs. This will involve specifying the power of your central processing unit (that is, its speed, memory, and precision) as well as all the peripheral devices, such as disk drives (floppy or hard), monitors (monochrome or color), printers (dot-matrix or letter-quality), modems (300- or 1200-baud), etc.

4. **Is the system expandable?**   Try to anticipate your future needs. Some computers can be expanded easily as your computer needs and capabilities grow.

5. **How compatible is the system?**   This point has both local and global aspects. In a local sense, it pays to have a system that is compatible with your colleagues and associates. For example, if your school has committed to a particular type of personal computer, it is advisable that you employ a compatible system. On a global scale, there are benefits in obtaining a popular brand of computer. In particular, machines that have been widely accepted tend to accrue large and varied software inventories.

6. **Comparison-shop.**   Aside from a home and an automobile, a computer system might represent the largest purchase you will ever make. Just as with any investment, it is advisable to visit as many dealers as possible before making a purchase. Ask questions and take notes until a clear picture of the options and the trade-offs emerge. To supplement your comparison-shopping, background information can be obtained from the many fine computer magazines that are available at most newsstands or bookstores. These often publish critical evaluations and comparisons of machines. Remember, do not jump at the first offer or go strictly for bargains. At all costs, do not sacrifice service and performance for price.

7. **"Test-drive" some systems.**   If at all possible, operate some of the systems that seem to fulfill your needs. Sometimes, factors such as the feel of the keyboard or the look of the monitor can figure prominently in your enjoyment of a personal computer. Only by "test-driving" the system will you be able to evaluate these factors and determine whether the machine "fits" you.

8. **How convenient is it to repair and service?**   Just as with an automobile, repairs and service are an essential component of the effective operation of your system. Make sure that your service options are reliable, quick, and cost-effective.

your part can make it a successful, if not a pleasurable, experience. It is hoped that the study of this book has made you a more informed consumer with respect to computers. In addition, the book should have helped you to clarify your needs.

Beyond this, Table 42.1 lists a number of steps you can follow in buying a personal computer. In particular, check whether your school or department is committed to using one specific brand over the next four years. If so, this is one very strong reason for considering that brand. In most cases, you would want to be compatible with your professors and other students. There are a number of practical reasons for having this compatibility. One of the most important is that you will be able to share information with others regarding the effective use of your machine.

42.2 ACQUIRING SOFTWARE

As outlined in Table 42.1, software needs are an important factor influencing your purchase of a personal computer. For an engineer, these needs differ somewhat from those of other professionals. Table 42.2 lists the seven major areas that cover the fundamental needs of most engineers. The first five—word processor, spreadsheet, database manager, graphics generator, and teleprocessor—are the "big five" of any software library. They can be acquired separately or as part of the integrated software packages discussed in Sec. 41.3. As noted in Chap. 1, your special needs as an engineer could influence your choice of these items. For example, you might prefer a particular word processor because of the ease with which it handles mathematical equations.

After the big five, you will probably require one or more language translators so that you can compose your own programs. The acquisition of a language is one area where engineers diverge from other fields. Whereas a businessman or doctor may never write a computer program, engineers require this capability in order to develop customized software to solve the many diverse computations that they routinely face.

As for the choice of a particular language, this is partially a matter of personal preference and partially a function of your area of engineering. The most common languages among engineers today are BASIC, FORTRAN, and Pascal. If you choose BASIC, you will also want to obtain a compiler. Compiling a BASIC program can usually make it run on the order of 5 to 10 times faster than the original interpreted version.

Aside from languages, the other area where our needs diverge from most other disciplines relates to computations. Although spreadsheets are capable of handling a wide variety of relevant computations, there are many other computational needs that

TABLE 42.2   *Seven Types of Software that Meet the Fundamental Computing Needs of Engineers*

- Word processor
- Spreadsheet
- Database manager
- Graphics generator
- Teleprocessor
- Programming languages
- Computational tools

require special software. Examples are the numerical and statistical methods discussed in earlier parts of this book. As set forth in Tables 13.1 and 20.1, we hope you have used this book as a vehicle for beginning your computational library. In addition, commercially available software is being developed to facilitate computations. The NUMERICOMP and ENGINCOMP packages discussed in this book are examples. In addition, powerful equation solvers as well as statistics and optimization software are available.

Before buying software, a little research can help clarify your needs and inform your decision. Glossbrenner (1984) has written an excellent book on the subject. Another reference of immense value is the *Whole Earth Software Catalog* (Brand, 1984), which offers knowledgeable and candid evaluations of personal computer software as well as of hardware and accessories.

42.2.1 Software Piracy

Just as you have to spend money to acquire a personal computer, there is also a cost associated with obtaining software. Today a significant fraction of the programs are being pirated. That is, they have been copied without compensating the developers for their investment of time, effort, and creativity. Because of the high price of software and the ease with which piracy can usually be accomplished, the theft of software has reached epidemic proportions. As revealed in the following story, widespread piracy reduces the incentive for people to create software.

About a year ago, we met a fellow professor at a conference. He boasted about how he had never purchased a single piece of software since buying his own personal computer 2 years before. We mentioned something about ethics, but he was so enthused about his exploits that he did not seem to hear us. Recently, we met this same fellow and half-jokingly asked him what he had stolen lately. Unexpectedly, his eyes blazed with righteous indignation as he proceeded to tell us that software pirates were the scum of the earth and that their proper punishment should include thumbscrews, the rack, and a bed of hot coals. We were somewhat at a loss to explain his turnabout until we learned that shortly after our previous conversation he had formed his own software company. After investing a substantial amount of time, money, and creative energy in the endeavor, he was dismayed to realize that he was not receiving just compensation for his efforts because users were pirating his work.

The above anecdote serves to illustrate the primary objection raised against software piracy. That is, that if it continues unchecked, the quality and diversity of commercial software will suffer because the best and the brightest minds will turn their energies to more profitable endeavors. It is an immense problem, which, along with other information-related crimes, is just beginning to be addressed by our social, legal, and economic systems.

We are not going to mount the podium at this point and preach to you regarding the evils of stealing software. If you have owned a personal computer for more than 2 months and have not stolen a single piece of software you are probably ready for canonization. However, be aware that today's ill-gotten gain could become tomorrow's loss.

42.3   THE FUTURE: ARTIFICIAL INTELLIGENCE

Now that we have come to the end of this book, we hope that we have accomplished our objective of demonstrating how computers can serve your professional and personal development. We hope that we have given you some sense of the history of computing and, in particular, of the exciting developments of the past decade.

In our description of the four generations of computing in Chap. 2, we deliberately omitted future trends. We felt that it would be preferable to defer such speculation to the end of the book, when you would be in a better position to appreciate the new developments that are just emerging. Now we are ready for this discussion.

As described in Chap. 2, the development of computational tools can be perceived as a progression of generations (Fig. 42.1). This progression was linear up to the third generation, where, with the introduction of timesharing, a divergence began. Whereas earlier computers had been developed expressly for large institutions and organizations, the third and fourth generations brought machines tailored to the needs of the individual. Today almost all of the computing needs of an engineer or a small-businessperson can be satisfied by a personal computer system costing less than $10,000.

At the same time that the microelectronic revolution was placing more emphasis on personal computing, the big, mainframe computers were also gaining power and strength. Today, in fact there exists a new class, called *supercomputers*, that transcends the mainframes and represents awesome computing capability. For example, a powerful supercomputer such as the Cray-2 can perform on the order of a billion

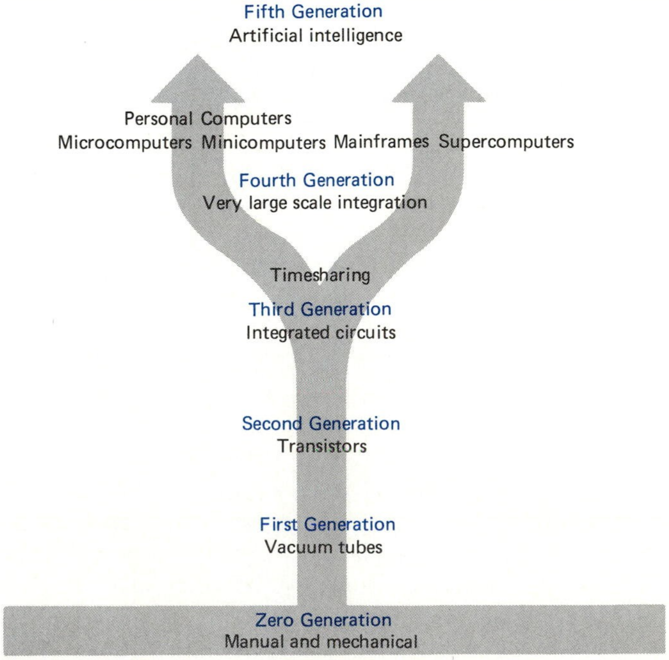

FIGURE 42.1
Schematic diagram of the evolution of modern computers. The progression seems to be moving toward a fifth generation in which artificial intelligence will figure prominently.

Fifth Generation
Artificial intelligence

Personal Computers
Microcomputers Minicomputers Mainframes Supercomputers

Fourth Generation
Very large scale integration

Timesharing

Third Generation
Integrated circuits

Second Generation
Transistors

First Generation
Vacuum tubes

Zero Generation
Manual and mechanical

arithmetic operations per second, whereas an Apple II or IBM PC performs thousands. The supercomputers are beginning to allow scientists and engineers to perform extremely large computations that were previously impossible or impractical to implement.

Because of continuing advances in microelectronics, the improvements in computing tools will undoubtedly be sustained in the foreseeable future. It is conceivable that in the coming years, personal computers will approach the capabilities of today's mainframes. At the same time, supercomputers will have advanced to the point that they can readily handle most of today's imaginable research calculations.

With the bulk of our computing needs more than satisfied, you might wonder what will be done with all this excess computing power. One significant possibility that is already being investigated is *artificial intelligence*—that is, the application of computers to the simulation of thought processes.

As you have learned in this book, computers are not smart; they merely have great memories and can perform data manipulations very quickly. Therefore the movement to make computers intelligent represents a profound step in their evolution.

Although it is unclear how far the concept can be advanced, research on artificial intelligence, or AI, is being funded at a significant level. In particular, the Japanese government has and will invest considerable resources into what is called the *Japanese fifth-generation project*. Among its goals is a computer, hundreds of times faster than today's most advanced supercomputer, that contains AI software. Research is also being directed at areas such as computer vision and voice recognition, which will facilitate the linkage of AI to human activities. Comparable efforts are being launched in the West.

The quest for AI is already having profound effects on computer science itself. Since the late 1940s, the principles of computer design have been unchanged. These principles are called *Von Neumann architecture* after John Von Neumann. As mentioned in Chap. 2, Von Neumann supplied the last piece in the computer design puzzle when he suggested that programs could be entered into a computer rather than physically rewiring the computer for each application. There are a few other basic characteristics that constitute von Neumann architecture. Most prominent of these is that there is sequential, centralized control of processing by a single computer that is a processor, communications device, and memory. Although computers have changed very rapidly over the past 40 years, they still conform to this and other principles of von Neumann's design. The situation is analogous to today's automobiles, which are similar in many fundamental respects to the Model T.

Today's interest in AI is forcing computer scientists to question the von Neumann architecture, and some truly revolutionary ideas are being put forward. Chief among these is the idea of parallel processing, whereby computers work on problems in parallel. This type of design is thought to be closer to the way in which the human mind operates. The introduction of parallel processing will lead to radical changes in the architecture of future computers and could have an immense effect on their powers and capabilities.

Although artificial intelligence is in its formative stages, preliminary applications are already beginning to emerge. One area with great potential for application in engineering is the so-called *expert system*. This is a software package that consists of (1) a stored *knowledge base* in a specialized area. This knowledge base is usually a set of rules of the IF/THEN type that represent expertise in a particular field. (2) An *inference engine* then provides a means of applying the rules to imitate reasoning. Thus, the software has the capability to arrange the rules and work through them in a logical order. In operation, the user supplies input facts to which the expert system responds as an intelligent consultant by giving advice and suggesting possible decisions.

At least as important as these emerging expert systems are the research efforts to develop intelligent computer interfaces in areas such as speech, pattern, and language recognition. Although it is unclear where these innovations will lead, the capital and the intellectual curiosity that have already been invested will ultimately have a great impact on our world.

Those of you reading this book are in the enviable position of entering the engineering profession at the beginning of this era. Engineers will be at the forefront of efforts to apply the new tools to society's needs. We hope our book will serve you well in contributing to those efforts.

PROBLEMS

42.1 Write a brief eassay on the pros and cons of software piracy. At the end, state and defend your position on the matter as it relates to your own behavior regarding the issue. You will not be graded on the basis of your opinion but rather on the scope of your knowledge and your insight. If possible, type the essay on a word processor.

42.2 Assume that you have unlimited financial resources. List the hardware and software that you would obtain to configure your own personal computer system. Plan from the perspective of your professional as well as your academic needs. Be as specific as you can and, if possible, estimate a cost for the entire system.

42.3 Identify the five generations of the electronic computer era.

42.4 What is artificial intelligence? Speculate on how it might be applied in engineering.

APPENDIX A

ASCII CHARACTER SET FOR THE IBM PC

Numeric Code	Character	Numeric Code	Character	Numeric Code	Character
000	(null)	026	→	052	4
001	☺	027	←	053	5
002	●	028	(cursor right)	054	6
003	♥	029	(cursor left)	055	7
004	♦	030	(cursor up)	056	8
005	♣	031	(cursor down)	057	9
006	♠	032	(space)	058	:
007	(beep)	033	!	059	;
008	■	034	''	060	<
009	(tab)	035	#	061	=
010	(line feed)	036	$	062	>
011	(home)	037	%	063	?
012	(form feed)	038	&	064	@
013	(carriage return)	039	'	065	A
014	♫	040	(	066	B
015	☼	041	)	067	C
016	►	042	*	068	D
017	◄	043	+	069	E
018	↕	044	,	070	F
019	‼	045	-	071	G
020	¶	046	.	072	H
021	§	047	/	073	I
022	▬	048	0	074	J
023	↨	049	1	075	K
024	↑	050	2	076	L
025	↓	051	3	077	M

Numeric Code	Character	Numeric Code	Character	Numeric Code	Character
078	N	126	~	173	¡
079	O	127	⌂	174	«
080	P	128	Ç	175	»
081	Q	129	ü	176	▒
082	R	130	é	177	▒
083	S	131	â	178	▓
084	T	132	ä	179	│
085	U	133	à	180	┤
086	V	134	å	181	╡
087	W	135	ç	182	╢
088	X	136	ê	183	╖
089	Y	137	ë	184	╕
090	Z	138	è	185	╣
091	[	139	ï	186	║
092	\	140	î	187	╗
093	]	141	ì	188	╝
094	∧	142	Ä	189	╜
095	—	143	Å	190	╛
096	`	144	É	191	┐
097	a	145	æ	192	└
098	b	146	Æ	193	┴
099	c	147	ô	194	┬
100	d	148	ö	195	├
101	e	149	ò	196	—
102	f	150	û	197	┼
103	g	151	ù	198	╞
104	h	152	ÿ	199	╟
105	i	153	Ö	200	╚
106	j	154	Ü	201	╔
107	k	155	¢	202	╩
108	l	156	£	203	╦
109	m	157	¥	204	╠
110	n	158	Pt	205	═
111	o	159	ƒ	206	╬
112	p	160	á	207	╧
113	q	161	í	208	╨
114	r	162	ó	209	╤
115	s	163	ú	210	╥
116	t	164	ñ	211	╙
117	u	165	Ñ	212	╘
118	v	166	ª	213	╒
119	w	167	º	214	╓
120	x	168	¿	215	╫
121	y	169	⌐	216	╪
122	z	170	¬	217	┘
123	{	171	½	218	┌
124	¦	172	¼	219	█
125	}			220	▄

Numeric Code	Character	Numeric Code	Character	Numeric Code	Character
221	▌	233	⊖	245	⌡
222	▐	234	Ω	246	÷
223	▬	235	δ	247	≈
224	α	236	∞	248	°
225	β	237	∅	249	•
226	Γ	238	∈	250	·
227	π	239	∩	251	√
228	Σ	240	≡	252	ⁿ
229	σ	241	±	253	²
230	μ	242	≥	254	■
231	τ	243	≤	255	(blank 'FF')
232	Φ	244	⌠		

APPENDIX B

ADDITIONAL BASICA FUNCTIONS

Special arithmetic functions of BASICA. The argument X can be a numeric constant, a variable, or an expression of any numeric data type—integer, single-precision, or double-precision.

Function	Result Returned
CDBL(X)	Double: the double-precision equivalent of X.
CINT(X)	Integer: X rounded to a whole number; X must be from −32768 to 32767.
CSNG(X)	Real: X is returned in single-precision.

String functions of BASICA. The arguments X$, X1$, and X2$ can be string constants or variables. The arguments N and M can be numeric constants, variables, or expressions. Unless otherwise stated, N and M should be from 0 to 255.

Function	Result Returned
ASC(X$)	Integer: the ASCII code for the first character of X$.
CHR$(N)	String: the character corresponding to the ASCII code of N.
DATE$	String: the date in the form, mm-dd-yyyy.
INSTR(N,X1$,X2$)	Integer: the beginning position of X2$ in X1$. N is the position in X1$ at which the comparison begins. N must be from 1 to 255. 0 is returned if X1$ is not found.
LEFT$(X$,N)	String: the left N characters of the string X$.
LEN(X$)	Integer: the number of characters in X$.
MID$(X$,M,N)	String: the portion of X$ is returned from position M of X$ for the next N characters. N is optional; if omitted, then the portion of X$ from M to the end is returned. M must be an integer from 1 to 255.
RIGHT$(X$,N)	String: the right N characters of the string X$.
SPACE$(N)	String: N blank spaces.
STR$(X)	String: the string equivalent of the numeric argument X.
STRING$(N,M)	String: a string of length N whose characters all have the ASCII code M.
STRING$(N,X$)	String: the first character of X$ repeated N times.
TIME$	String: the time of day in the form, hh:mm:ss, where hh is from 00 to 23, mm is from 00 to 59, and ss is 00 to 59.
VAL(X$)	Real: converts X$ to a number.

Miscellaneous functions of BASICA. The argument N can be a numeric constant, variable, or expression.

Function	Action
INKEY$	Allows the program to accept a single character from the keyboard without the ENTER key being pressed.
INPUT$(N)	Allows the program to accept N characters from the keyboard without the ENTER key being pressed.

APPENDIX C

USING MS-DOS AND BASIC ON THE IBM PC

This appendix consists of a series of exercises to orient you to the IBM PC. It is primarily written in terms of the two-drive system shown in Fig. C.1. However, information on utilizing one-drive models is also provided.

C.1 Turning on the Microcomputer

The disk operating system (DOS) must be loaded into the computer's memory at the beginning of any session on an IBM PC. In other words, the computer must read DOS from a diskette. Once this is done, the IBM PC will be capable of manipulating files that exist on a diskette.

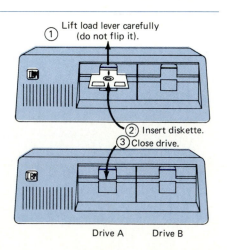

FIGURE C.1
The system unit of a two-drive IBM PC showing the proper way to insert a diskette into drive A. Note that some IBM PC compatibles have the disk drives stacked vertically. For these units the top drive is usually drive A.

The process of loading DOS is called *booting* the system. To do this, insert a disk containing DOS into drive A and close the disk-drive door (Fig. C.1). If this is the first time you are using the IBM PC, employ the master DOS disk that comes with the machine. Once the diskette is inserted, the system can be booted in one of two ways.

1. If the computer is off, turn it on and the system will automatically boot.
2. If the computer is already on, press the Ctrl-Alt-Del keys simultaneously (see Fig. C.2). This is referred to as a *warm boot*.

Note that, in both cases, the diskette in drive A must be formatted and must contain DOS. Otherwise the system will not boot. We will describe how to format and put DOS on a diskette shortly.

When the system is booted, the computer will respond with something like

```
Current date is Tue 1-01-1980
Enter new date: 1-22-85                          ←You enter today's date.
Current time is 0:00:15.10
Enter new time: 14:05                            ←You enter correct time.


The IBM Personal Computer DOS
Version 2.00 (C) Copyright IBM Corp 1981, 1982, 1983

A>
```

FIGURE C.2 The keyboard of the IBM PC showing the Ctrl, Alt, and Del keys that are depressed simultaneously to implement a "warm boot."

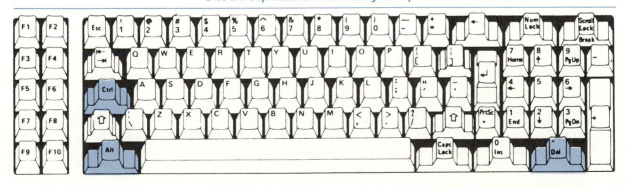

Function keys

Alphanumeric and control keys

Number pad

The A> is called a *system prompt*. It tells you that the computer is ready to receive your instructions and that it is operating out of disk drive A. If you desire to operate out of the second disk drive you would type

```
A> B: ®
```

(The symbol ® is used to denote the striking of the carriage return.) The computer responds,

```
B>
```

indicating that it is now operating on disk drive B. To switch back, type

```
B> A: ®
```

and the computer responds,

```
A>
```

You can go back and forth between drives A and B as often as you wish. For simplicity, you will probably be operating out of drive A most of the time.

Before proceeding, we want to make two suggestions:

1. If you are going to employ a computer that is already on and has been used by someone else, reboot the system with *your* system diskette. The previous user may have loaded a DOS version different from the one that is compatible with your software.

2. It is very important to input the proper date and time whenever you boot the system. Whenever you save a program, the computer records the date and time. This information can be invaluable at a later date when you may need to distinguish between several versions of your programs.

C.2 Formatting a Diskette

Whenever you use a new diskette or one that has been employed on another machine, it must be formatted for the IBM PC. *Note:* Whenever you format a diskette, any programs or files on the diskette will be lost forever, *and we do mean forever*!

To format and place DOS on your diskette, make sure the master DOS disk (that is, the one that comes with your machine) is in drive A. Then type

```
A> FORMAT B:/S ®
```

and the computer responds,

```
Insert new diskette for drive B:
and strike any key when ready
```

Insert the diskette to be formatted in drive B and strike any key. After about a minute the computer will respond with something like

```
Formatting...Format complete
System transferred

    362496 bytes total disk space
     40960 bytes used by system
    321536 bytes available on disk


Format another (Y/N)?_
```

The number of bytes will vary depending on your system and diskettes. If you want to format another diskette, type Y and repeat the process. Otherwise, type N and the computer will respond with A>, indicating that it is ready to receive instructions.

Remove the master DOS diskette from drive A and put it back into its protective cover. Now take the diskette out of drive B, insert it into drive A, and close the drive door. Type

```
A> DIR ®
```

This command will give you a listing of all the files that currently exist on the disk. At this point, the computer should respond with something like

```
Volume in drive A has no label
Directory of A:\

COMMAND   COM   17664   3-08-83   12:00p
          1 File(s)      321536 bytes free
```

The file COMMAND.COM is part of DOS.

For a computer with a single disk drive the procedure is similar. Place the DOS disk in drive A and type

```
A> FORMAT A:/S ®
```

and the system will respond with

```
Insert new diskette for drive A:
and strike any key when ready
```

You must then remove the DOS disk and insert the diskette to be formatted into drive A. After you have struck any key, the diskette will be formatted.

There are times when you will not want to store DOS but will merely want to format a diskette. Such would be the case when you want to employ the diskette to store data files such as those generated by a word processor or a database management system. In these instances, DOS would take up space unnecessarily. To format a disk simply to hold data, omit the /S, as in

```
A> FORMAT B: ®
```

or for a one-drive computer,

```
A> FORMAT A: ®
```

C.3 Copying Files from One Diskette to Another

Copying one file.   The following procedure can be employed to copy a file from a diskette in drive A to one in drive B.

Insert the disk holding the files to be copied into drive A. For the present example, this can be your DOS disk. Insert the disk to which you want to copy into drive B. The general command to copy a file is

```
A> COPY filename.ext B: ®
```

where *filename* is an eight-character name that identifies the file and *ext* is a three-character extension that serves as a secondary identifier. As an example, suppose we copy the file containing BASICA, as in

```
A> COPY BASICA.COM B: ®
```

The computer will respond with a message when the copy is completed successfully. At this point, the file containing BASICA would reside on the diskette in drive B. To verify this, type

```
A> DIR B:®
```

and the system will respond with something like

```
Volume in drive B has no label
Directory of B:\

COMMAND   COM     17664    3-08-83   12:00P
BASICA    COM     16256    3-08-83   12:00P
          2 file(s)     305280 bytes free
```

Thus we can see that BASICA.COM has been added to your file directory.

Copying an entire diskette.   Aside from copying individual files, there will also be

times when you may desire to copy the entire contents of one diskette to another.*
To do this, insert the master DOS disk in drive A and type

```
A> DISKCOPY A: B: ®
```

The computer will respond,

```
Insert source diskette in drive A:
Strike any key when ready
```

Before striking any key, remove the master disk from drive A (unless you are copying
it) and replace it with the disk to be copied (called the *source* diskette). Strike any
key and the computer will respond,

```
Insert target diskette in drive B:
Strike any key when ready
```

Insert the disk which is to receive the copy (called the *target* diskette) into drive B
and strike any key. When the copy is complete, the computer responds,

```
Copy complete
Copy another (Y/N)?
```

Select N if you are finished or Y if you want to repeat the process.
For a single-drive system, the procedure is similar, but you initially type

```
A> DISKCOPY ®
```

Thereafter the computer will repeatedly instruct you to insert either the source or the
target disk into the drive and to strike any key until the copy is completed.
Note that diskettes should be copied for personal use only. Do not violate copyright laws.

C.4   Using BASIC and BASIC Files

As mentioned previously, when you see the A> or B> prompt, the computer is
telling you that it is currently operating in the system. In this state, the computer can
be employed to copy programs, list directories, format, delete files, etc. However, at
this point, the computer cannot run a BASIC program. To do this, you must first
instruct the computer to read the BASIC language interpreter off the disk and into the
computer's memory. Type

```
A> BASICA ®
```

*Note that once this is accomplished, the information on the diskette in drive A will be intact and unmodified. However,
any files that were originally on the drive B diskette will be destroyed and replaced by those on the drive A diskette.

This command only works if BASICA is on the disk you are using. Recall that you just copied it onto your disk, so the computer should respond with something like

```
The IBM Personal Computer Basic
Version D2.00 Copyright IBM Corp. 1981, 1982, 1983
61330 Bytes free

ok
```

```
1LIST 2RUN 3LOAD" 4SAVE" 5CONT 6,"LPT1 7TRON 8TROFF 9KEY ØSCREEN
```

The OK is called the BASIC *prompt*. It is similar to the system prompt A> in the sense that it indicates that the computer is waiting for instructions from you.

Entering and saving a program.   As described in more detail in the beginning of Chap. 8, a BASIC program can now be entered into the computer. For present purposes, type in the simple addition program:

```
10 A = 28
20 B = 15
30 C = A + B
40 PRINT C
50 END
```

After typing each line, strike RETURN. When you are finished entering the program type RUN ®; the computer should execute the program and print out the answer: 43. Note that pressing the special function key, F2, on the left-hand side of the keyboard is equivalent to typing RUN ®. Try it.

Now we are ready to save the simple addition program on diskette. Type

```
SAVE "SIMPADD1" ®
```

The special function key F4 can be employed to type out the SAVE'' part of the above command automatically. Try it. In order to verify that the program has been successfully saved, you would like to check the file directory. However, because you are in BASIC you cannot accomplish this in the same way you would if you were in the system—that is, by typing A>DIR. Instead, you must type FILES ®. The computer will respond with

```
A:\
COMMAND .COM       BASICA   .COM       SIMPADD1.BAS
  305185 bytes free
OK
```

Notice how the computer has automatically affixed the extension BAS to SIMPADD1 in order to let you know that it contains a BASIC program.

An alternative way to check the directory would be to return to the operating system. To do this, type SYSTEM ® and the screen should show the system prompt A>. Now type DIR ® and the file directory will again be displayed.

Loading and deleting a program.   Now that you are back in the system, you cannot run BASIC programs. Therefore you have to get back into BASIC, as in

        A> BASICA ®

Once you are back, there are no BASIC programs in the computer's primary memory. To load a BASIC program, such as SIMPADD1 from your disk back to the computer, type

        LOAD "SIMPADD1" ®

and the program will be loaded. Note that the F3 key will automatically type LOAD" to facilitate loading a program. To verify that the program is in primary memory, type LIST (or, alternatively, press the F1 key) and strike RETURN.

The final step in this exercise is to delete a program from your disk. When you are in BASIC, this is done by typing

        KILL "SIMPADD1.BAS" ®

Notice that for this particular command, the extension BAS must be included.

Alternatively, if you return to the system, the file can be deleted by

        A> DEL SIMPADD1.BAS ®

APPENDIX D

USING A BASIC COMPILER

Compiling a BASIC program will allow it to execute much faster than the original interpreted version. In addition, compiling will help protect your source program from unauthorized alteration or disclosure. The present appendix outlines how this is done with the Microsoft BASIC compiler for the MS-DOS operation system on the IBM PC. Note that different compilers and even different implementations of the same type of compiler can utilize different instructions. Check your compiler's manual to learn the specifics of its procedure.

Table D.1 summarizes the files needed to compile a BASIC program.

D.1 Creating the Source Module

The BASIC program can be created with any available text editor. However, one of the best ways is to employ the editing capabilities of your BASIC interpreter. For the present exercise, use the simple addition program from Fig. 8.1 and store the program on a diskette with a name such as SIMPADD1.BAS. The program must be saved in

TABLE D.1 Files for Compiling a BASIC Program on the IBM PC with the Microsoft Compiler

File	Purpose
BASCOM.COM	The compiler
LINK.EXE	The linker program
BASRUN.LIB	The function library used in conjunction with the run-time module
BASCOM.LIB	The function library used without the run-time module
BASRUN.EXE	The run-time module

ASCII format in order to be used by the compiler. This is done by typing SAVE ''SIMPADD1'', A and striking return.

D.2 Compiling the Program

To compile the program, insert the diskette holding the file BASCOM.COM into drive A and the disk holding SIMPADD1.BAS into drive B. Make drive B the default drive by typing

```
A> B:
```

Then type

```
B> A:BASCOM SIMPADD1,SIMPADD1,SIMPADD1;
```

The A:BASCOM tells the computer to run the file BASCOM.COM off of drive A; the first SIMPADD1 is the source filename, the second is the name for the object file to be created (the compiler will automatically attach the extension .OBJ), and the third is the name for the source listing file (the compiler will automatically attach the extension .LST). If, as in this case, the successive names are identical, the command can be abbreviated as

```
B> A:BASCOM SIMPADD1,,;
```

If your program has no errors, the computer will respond with something like

```
Microsoft BASIC77 V3.13 8/05/83
Pass one    No errors detected
            6 Source lines
```

If errors occur, you must correct them and repeat the compiling step.

D.3 Linking the Program

To implement the link phase, insert the disk holding both LINK.EXE and BAS-RUN.LIB into drive A and then type

```
B> A:LINK SIMPADD1,SIMPADD1,SIMPADD1,A:BASRUN.LIB;
```

where the three SIMPADD1s are the names we are giving to the relocatable object file, the executable program, and the linker listing, respectively, and the A:BASRUN.LIB tells the program that the library file is on drive A. Again, because the names are duplicated we could also have written this line as

```
B> A:LINK SIMPADD1,,,A:BASRUN.LIB;
```

If this phase is completed successfully, the computer responds

```
Microsoft Object Linker V2.01 (large)
(C) Copyright 1982, 1983 by Microsoft Inc.
```

D.4   Executing the Program

At this point, the program is fully compiled. Before you run the program, you must load the file BASRUN.EXE onto the diskette holding the file SIMPADD1.EXE. Once this is done, the program can be run by merely typing

```
B> SIMPADD1
```

with the result

```
43
```

D.5   Compiling Without the Run-time Module

An alternative to the foregoing is to compile the program with the command

```
B> A:BASCOM SIMPADD1,SIMPADD1,SIMPADD1/O;
```

Adding the /O tells the computer to compile the program in a manner so that it does not require the run-time module. The remainder of the procedure is identical, with the exception that the link is implemented using BASCOM.LIB:

```
B> A:LINK SIMPADD1,SIMPADD1,SIMPADD1,A:BASCOM.LIB;
```

There are trade-offs between the two methods, and there are other approaches for compiling programs. You should consult the user's manual for your compiler to acquaint yourself with these options and capabilities.

APPENDIX E

USING APPLE-DOS AND BASIC ON THE APPLE II

This appendix consists of a series of exercises to orient you to the Apple II computer. It is primarily written in terms of a two-drive system. However, information on utilizing one-drive models is also provided.

E.1. Turning on the Microcomputer

The disk operating system (DOS) must be loaded into the computer's memory at the beginning of any session on an Apple II. In other words, the computer must read DOS from a diskette. Once this is done, the APPLE II will be capable of manipulating files that exist on a diskette.

The process of loading DOS is called *booting* the system. To do this, insert a disk containing DOS into drive 1 and close the disk drive door. If this is the first time that you are using the APPLE II, employ the master DOS disk that comes with the machine. Once the diskette is inserted, it can be booted in one of two ways.

1. If the computer is off, turn it on and the system will automatically boot.
2. If the computer is already on, type IN#6 ®. This is referred to as a *warm boot*.

Note that in both cases, the diskette in drive 1 must be formatted and must contain DOS. Otherwise the system will not boot. We will describe how to format and put DOS on a diskette shortly.

When the system is booted, the computer will respond with something like

```
APPLE ][

DOS VERSION 3.3 08/25/80

APPLE II PLUS OR ROMCARD SYSTEM MASTER

(LOADING INTEGER INTO LANGUAGE CARD)

]
```

The] is called a *prompt*. It tells you that the computer is ready to receive your instructions.

Before proceeding, we want to make a suggestion. If you are going to employ a computer that is already on and has been used by someone else, reboot the system with *your* system diskette. The previous user may have loaded a DOS version different from the one that is compatible with your software.

E.2   Formatting a Diskette

Whenever you are using a new diskette or one that has been employed on another machine, you must format it for the APPLE II. *Note:* Whenever you format a diskette, any programs or files on the diskette will be lost forever, *and we do mean forever*!

To format a diskette, insert the diskette to be initialized in drive 1. Type the NEW command to clear any previous program in memory. A "greeting" program must be present on the diskette in order for it to be booted. This program will run automatically each time the disk is booted. An example greeting program might be

```
10 REM JOHN DOE, INITIALIZED 5/10/85
20 PRINT "INITIALIZED BY JOHN DOE"
30 PRINT "ON 5/10/85"
40 END
```

where you should substitute your name for John Doe. Next type

```
]INIT HELLO ®.
```

(The symbol ® is used to denote the striking of the carriage return.) After about a minute, the computer will respond with the prompt,].

The next time you boot this diskette, the program HELLO will run and the message

```
INITIALIZED BY JOHN DOE
ON 5/10/85


]
```

will appear on the screen. Try a warm boot to demonstrate this fact.

Next, to see what is contained on the disk, type

```
]CATALOG ®
```

and the computer should respond with

```
DISK VOLUME 254
A 002 HELLO
```

This indicates that the program HELLO is stored on this diskette. The A indicates that the program is in Applesoft BASIC and the 002 indicates that the program takes up two sectors. A diskette can store 496 sectors of information. The disk volume number, 254, is a default value that the Apple II supplies in the absence of any direction by you. To specify a different number, say 200, you would employ

```
]INIT HELLO V200 ®
```

when you initialize the diskette. Volume numbers of from 1 to 254 are permitted.

E.3   Copying a Diskette

To copy one diskette to another, insert the master DOS disk in drive 1 and type

```
]RUN COPYA ®
```

The computer will respond with messages. Keep striking RETURN until the computer prints

```
ORIGINAL SLOT:    6
        DRIVE:    1

DUPLICATE SLOT:   6
         DRIVE:   2

--PRESS "RETURN" KEY TO BEGIN COPY--
```

Before striking RETURN, you must insert the disk to be copied into drive 1 and the new disk in drive 2. Now strike RETURN. When the copy is complete, the computer responds,

```
DO YOU WISH TO MAKE ANOTHER COPY?
```

Type N ® if you are finished and Y ® if you want to repeat the process.

For a single-drive system, the procedure is similar but before striking RETURN to begin the copy, you must insert the source disk into drive 1. After you strike RETURN, the computer will read a portion of the source disk and then will instruct you to insert the destination disk into drive 1. The process is repeated until all the material is copied.

E.4   Using BASIC and BASIC Files

During a session on the Apple II you should be able to execute, save, retrieve, and delete programs. In the present section we will review the commands employed for this purpose: RUN, SAVE, LOAD, and DELETE.

Entering and saving a program.   Type NEW to erase the program in primary memory and then enter the following simple addition program to the computer:

```
10 A = 28
20 B = 15
30 C = A+B
40 PRINT C
50 END
```

After typing each line, strike RETURN. When you are finished entering the program, type RUN ®; the computer should execute the program and print out the answer: 43.

Now we are ready to save the simple addition program on diskette. Type

```
]SAVE SIMPADD1 ®
```

To verify that the program has been saved, type

```
]CATALOG
```

The computer will respond with

```
DISK VOLUME 254

A 002 HELLO
A 002 SIMPADD1
```

indicating that the program is successfully saved.

To load a program from the diskette back into primary memory, type

Loading and deleting a program.

```
]LOAD SIMPADD1 ®
```

and the program will be loaded. To verify that retrieval was successful, type LIST ®.

The final step in this exercise is to delete a program from your disk. This can be done by typing

```
]DELETE SIMPADD1 ®
```

APPENDIX F

SEQUENTIAL FILE STATEMENTS FOR THE APPLE II

The following statements are consistent with the examples in Sec. 14.2.

To open a file:

```
PRINT CHR$(4); "OPEN filename"*
```

To transfer data from a program to a file:

```
PRINT CHR$(4); "WRITE filename"
PRINT YEAR;",";FLOW
PRINT CHR$(4)
```

To transfer data to a program from a file:

```
PRINT CHR$(4); "READ filename"
INPUT YEAR, FLOW
PRINT CHR$(4)
```

To append data to a file:

```
PRINT CHR$(4); "APPEND filename"
PRINT YEAR; ","; FLOW
PRINT CHR$(4)
```

*Filename refers to the name you have given the file; the name may consist of up to 30 characters and must start with a letter.

To close a file:

```
PRINT CHR$(4); "CLOSE filename"
```

To delete a file:

```
PRINT CHR$(4); "DELETE filename"
```

To rename a file:

```
RENAME CHR$(4); "RENAME oldfilename,newfilename"
```

Figure F.1 shows several Apple II subroutines to perform particular file manipulations. These subroutines are directly analogous to Figs. 14.6 through 14.9 for the IBM PC. The discussion in Sec. 14.2 related to the figures is also relevant to Fig. F.1.

FIGURE F.1
Program fragments in Applesoft BASIC to (*a*) create a file, (*b*) print the contents of a file, (*c*) append a new record to the end of a file, and (*d*) modify a record on a file.

```
1000   REM        CREATE A SEQUENTIA
       L FILE
1010   INPUT "CREATE A FILE NAMED"
       ;FILNAM$
1020   PRINT CHR$ (4);"OPEN";FILN
       AM$
1025 SENTINEL = 0
1030   REM        BEGIN WHILE LOOP
1032   IF SENTINEL < > - 999 THEN
       1040
1034   GOTO 1080
1040   INPUT "YEAR = ";YEAR
1050   INPUT "FLOW = ";FLOW
1060   PRINT CHR$ (4);"WRITE";FIL
       NAM$
1062   PRINT YEAR;",";FLOW
1064   PRINT CHR$ (4)
1070   INPUT "LAST RECORD (ENTER -
       999)";SENTINEL
1075   GOTO 1030
1080   REM        END LOOP
1085 SENTINEL = 0
1090   PRINT CHR$ (4);"CLOSE";FIL
       NAM$
1100   END
```
                          (*a*)

```
2000   REM        PRINT OUT FILE CON
       TENTS
2002   ONERR GOTO 2080
2010   INPUT "LIST A FILE NAMED";F
       ILNAM$
2020   HOME
2030   PRINT "YEAR"; TAB( 20);"FLO
       W"
2040   PRINT CHR$ (4);"OPEN";FILN
       AM$
2050   REM        BEGIN WHILE LOOP
2060   PRINT CHR$ (4);"READ";FILN
       AM$
2062   INPUT YEAR,FLOW
2064   PRINT CHR$ (4)
2070   PRINT YEAR; TAB( 20);FLOW
2075   GOTO 2050
2080   REM        END LOOP
2090   PRINT CHR$ (4);"CLOSE";FIL
       NAM$
2100   END
```
                          (*b*)

```
3000   REM          APPEND A SEQUENTIA
       L FILE
3010   INPUT "APPEND DATA TO A FIL
       E NAMED";FILNAM$
3020   PRINT CHR$ (4);"APPEND";FI
       LNAM$
3025 SENTINEL = 0
3030   REM          BEGIN WHILE LOOP
3032   IF SENTINEL < > - 999 THEN
       3040
3034   GOTO 3080
3040   INPUT "YEAR = ";YEAR
3050   INPUT "FLOW = ";FLOW
3060   PRINT CHR$ (4);"WRITE";FIL
       NAM$
3062   PRINT YEAR;",";FLOW
3064   PRINT CHR$ (4)
3070   INPUT "LAST RECORD (ENTER -
       999)";SENTINEL
3075   GOTO 3030
3080   REM          END LOOP
3090   PRINT CHR$ (4);"CLOSE";FIL
       NAM$
3100   END
```

(c)

```
4000   REM          CORRECT A VALUE
4002   ONERR GOTO 4150
4010   PRINT CHR$ (4);"OPEN FLOW.
       DAT"
4020   PRINT CHR$ (4);"OPEN FLOW.
       DUM"
4030   REM          INPUT YEAR TO BE C
       ORRECTED
4040   INPUT "YEAR TO BE CORRECTED
       = ";YR
4050   REM          BEGIN WHILE LOOP
4060   PRINT CHR$ (4);"READ FLOW.
       DAT"
4062   INPUT YEAR,FLOW
4064   PRINT CHR$ (4)
4070   IF YR < > YEAR THEN 4080
4072   GOTO 4100
4080   PRINT CHR$ (4);"WRITE FLOW
       .DUM"
4082   PRINT YEAR;",";FLOW
4084   PRINT CHR$ (4)
4090   GOTO 4140
4100   REM          ELSE
4110   PRINT "CURRENT VALUE = ";FL
       OW
4120   INPUT "CHANGE TO ";FLOW
4130   PRINT CHR$ (4);"WRITE FLOW
       .DUM"
4132   PRINT YEAR;",";FLOW
4134   PRINT CHR$ (4)
4140   REM          ENDIF
4145   GOTO 4050
4150   REM          END LOOP
4160   PRINT CHR$ (4);"CLOSE FLOW
       .DAT"
4170   PRINT CHR$ (4);"CLOSE FLOW
       .DUM"
4180   PRINT CHR$ (4);"DELETE FLO
       W.DAT"
4190   PRINT CHR$ (4);"RENAME FLO
       W.DUM,FLOW.DAT"
4200   END
```

(d)

FIGURE F.1 (continued)

APPENDIX G

ADDITIONAL FORTRAN 77 FUNCTIONS

Intrinsic functions of FORTRAN for type conversion. The argument X can be a real, integer, double-precision, or complex constant, variable, or expression, and Y can be a real or double-precision constant, variable, or expression.

Function	Result Returned
CMPLX(X)	Complex: converts X to complex
INT(X)	Integer: converts X to integer
NINT(Y)	Integer: rounds Y to integer
REAL(X)	Real: converts X to real

Intrinsic functions of FORTRAN to manipulate characters. The arguments L, L1, and L2 can be string constants or variables. N can be an integer constant, variable, or expression.

Function	Result Returned
CHAR(N)	Character: the character corresponding to the ASCII code of N.
ICHAR(L)	Integer: the EBCDIC code for the first character of L.
INDEX(L1,L2)	Integer: the position of L2 in L1; 0 is returned if L2 is not found.
LEN(L)	Integer: the number of characters in L.
LGT(L1,L2)	Logical: true if L1 is greater than L2 using ASCII codes.
LLT(L1,L2)	Logical: true if L1 is less than L2 using ASCII codes.

Intrinsic functions to manipulate complex variables.   The argument CX is a complex variable and A and B are real.

Function	Result returned
CABS (CX)	Complex; absolute value.
CSQRT (CX)	Complex; square root.
CCOS (CX)	Complex; cosine.
CSIN (CX)	Complex; sine.
CLOG (CX)	Complex; natural logarithm.
CEXP (CX)	Complex; exponential function.
REAL (CX)	Real; returns real part of CX.
AIMAG (CX)	Real; returns imaginary part of CX.
CONJ (CX)	Complex; converts CX to its complex conjugate.
CMPLX (A,B)	Complex; converts two real variables, A and B, into a complex value.

APPENDIX H

USING FORTRAN 77
ON THE IBM PC

Just as depicted in Figure 15.6, there are a number of steps to prepare and execute a FORTRAN program. The present appendix outlines how this is done with the Microsoft FORTRAN compiler for the MS-DOS operation system on the IBM PC. Note that different compilers and even different implementations of the same type of compiler can utilize different instructions. Check your compiler's manual to learn the specifics of your procedure.

H.1 Creating the Source Module

The FORTRAN program can be created with any available text editor. For example, the EDLIN program on your system disk or WordStar could be used to create a file containing the simple addition program from Fig. 15.5. Store the program on a diskette with a name such as SIMPADD1.FOR. Note that the extension FOR is required by the MS-FORTRAN compiler.

H.2 Compiling the Program

The compiling phase consists of two passes. (See Table H.1.) The first pass translates the source program into two intermediate files, PASIBF.SYM and PASIBF.BIN, which are then employed on the second pass. It also produces a program-listing file, which for the present example would be SIMPADD1.LST. This file is important because it contains information on any syntax errors in the program.

TABLE H.1 *Files for Compiling a FORTRAN Program on the IBM PC with the Microsoft FORTRAN Compiler.*

File	Purpose
FOR1.EXE	First pass of compiler.
PAS2.EXE	Second pass of compiler.
FORTRAN.LEM	Function libraries; the version .LEM is used unless your computer
FORTRAN.L87	has an 8087 coprocessor, in which case .L87 is employed.
LINK.V1	Linker programs; the version .V1 is used with the early versions of
LINK.V2	DOS, whereas .V2 is employed with DOS 2.0. Whichever one is used, it must be renamed LINK.EXE.

To implement the first pass, insert the diskette holding the MS–FORTRAN file FOR1.EXE into drive A and the disk holding SIMPADD1.FOR into drive B. Make drive B the default drive by typing

```
A> B:
```

Then type

```
B> A:FOR1 SIMPADD1,SIMPADD1,SIMPADD1;
```

If, as in this case, the successive names are identical, the command can be abbreviated as

```
B> A:FOR1 SIMPADD1,,;
```

If your program has no errors, the computer will respond with something like

```
Microsoft FORTRAN77 V3.13 8/05/83
Pass one    No errors detected
             8 Source lines
```

If errors occur, diagnostics will be printed. For this situation, you must edit the source code and then rerun the first pass.

The second pass takes the intermediate files created in the first pass and optimizes the code. The output of the second pass is the object module, which for the present case would be SIMPADD1.OBJ. It also destroys the intermediate files from the first pass and creates two new ones that can be used in a third pass to create an object listing file. Because such a file is rarely used by most programmers (it is a symbolic, assemblerlike listing of the object code), we will skip the third pass and move directly to the fourth and final pass—the linker.

However, before doing that, we must first implement the second pass. Insert the disk holding PAS2.EXE into drive A and type

```
B> A:PAS2
```

If the pass is successful, the computer will respond with something like

```
Code Area Size = #0110 (    272)
Code Area Size = #0018 (     24)
Code Area Size = #0012 (     18)
Pass Two  No errors detected.
```

If errors occur, you must correct them and repeat the first and second passes.

H.3   Linking the Program

This takes the object module created in the second pass and transforms it into an executable program by linking it with the necessary library functions and reassigning memory addresses. The output from this phase is the load module, which for the present case would be named SIMPADD1.EXE.

To implement the link phase, insert the disk holding both LINK.EXE and FORTRAN.LEM (see Table H.1) into drive A and then type

```
B> A:LINK SIMPADD1,SIMPADD1,SIMPADD1,A:FORTRAN.LEM;
```

where the three SIMPADD1s are the names we are giving to the relocatable object file, the executable program, and the linker listing, respectively, and the A:FORTRAN.LEM tells the program that the library file is on drive A. Again, because the names are duplicated we could also have written this line as

```
B> A:LINK SIMPADD1,,,A:FORTRAN.LEM;
```

If this phase is completed successfully, the computer responds

```
Microsoft Object Linker V2.01 (large)
(C) Copyright 1982, 1983 by Microsoft Inc.
```

H.4   Executing the Program

At this point, the program is fully compiled. The program can be run by merely typing

```
B> SIMPADD1
```

with the result

```
43.0000000
```

H.5   Using Batch Files to Make Compilation More Convenient

Batch files can be employed to make compilation easier by consolidating several commands into one or two batch files. If a hard disk is used, all the compiler files can be easily loaded and a simple batch file developed to transform the compilation into a one-step process. For a system using conventional floppy disks, however, all the compiler files usually will not fit on a single diskette, and therefore the compilation must be reduced to a two-step process.

To do this, place FOR1.EXE and PAS2.EXE on a first diskette and LINK.EXE and FORTRAN.LEM on a second. Then add the following batch file program, COMPIL.BAT, to the first:

```
A:FOR1 %1, %1, %1;
IF ERRORLEVEL 1 GOTO FAIL
A:PAS2
:FAIL
```

and add the following batch file program LINKER.BAT, to the second:

```
A:LINK %1, %1, %1,A:FORTRAN.LEM
```

Next, place the first diskette on drive A and the diskette holding the source program in drive B and type

```
B> A:COMPIL SIMPADD1
```

This will cause the first two passes to be implemented. If errors occur, the process will stop and corrections must be made.

When the program runs without errors, remove the first diskette and replace it with the second. Then type

```
A:LINKER SIMPADD1
```

If no errors are present, the remainder of the compilation will be completed.

APPENDIX I

A RANDOM-NUMBER GENERATOR FOR FORTRAN

Because all FORTRAN compilers do not have intrinsic functions to compute random numbers, the following material is included so that everyone will have the capability to perform the examples and problems in this text. Efforts to generate random numbers date back to the first electronic computers. Although there are a variety of different schemes for accomplishing this, the most popular are of the general form (Knuth, 1981)

$$x_{n+1} = (ax_n) \bmod m$$

for $n \geq 0$. The term "mod" is short for *modulo*, and the terms a and m are parameters which are called the *multiplier* and the *modulus*, respectively. The formula means that a number x_n is multiplied by a and then divided by m. The remainder of this division is then assigned to x_{n+1}. On the next pass, this result is assigned to x_n, and the process is repeated to generate another value, x_{n+1}. By implementing the procedure iteratively (that is, over and over again), a series of random numbers is generated. These numbers can, in turn, be divided by m so that the random numbers range between 0 and 1.

The numbers that result from this procedure, which is commonly referred to as the *multiplicative congruential method*, are called *pseudorandom numbers*. They are called "pseudorandom" because eventually they will start to repeat. Thus they are not truly random. This is because they are generated with a formula. However, by wise choices of a and m, a sequence of numbers can be generated which for all practical purposes, are random. The following user-defined function subprogram should be adequate for the types of examples employed in the present text:

```
FUNCTION RNDNUM(SEED)
INTEGER A,X,SEED
DATA A,M,I/1027,1048576,1/
IF (I.EQ.1) THEN
    X = SEED
    I = 0
    XM = M
ENDIF
X = MOD(A*X,M)
XX = X
RNDNUM = XX/XM
RETURN
END
```

where MOD (I,J) is an intrinsic function that produces the integer remainder when I is divided by J. The value, SEED, is a starting value.

APPENDIX J
DERIVED UNITS FOR THE SI SYSTEM

Table J.1 presents some of the derived units employed in this book. These units are composites of the fundamental units of the SI system: length (in meters), time (in seconds), mass (in kilograms), electric current (in amperes), temperature (in kelvin), molecular substance (in moles), and luminous intensity (in candela). Note that two sets of units are tabulated: one in terms of the base units and another in terms of base and derived units. Note also that the unit sr is a steradian, which is the solid angle with its vertex at the center of a sphere. The solid angle is subtended by an area on the sphere's surface that is equal to a square, with sides equal in length to the sphere's radius.

TABLE J.1

Quantity	Name	SI Symbol	Conventional Units	Base Units
General and Mechanical				
Acceleration			m/s^2	
Area			m^2	
Density			kg/m^3	
Energy	joule	J	$N \cdot m$	$kg \cdot m^2/s^2$
Force	newton	N	$kg \cdot m/s^2$	
Frequency	hertz	Hz	$cycle/s^2$	
Mass*			kg	
Momentum			$kg \cdot m/s$	
Power	watt	W	J/s	$N \cdot m/s$
Specific volume			m^3/kg	
Torque			$N \cdot m$	$kg \cdot m^2/s^2$
Velocity			m/s	
Volume			m^3	
Work	joule	J	$N \cdot m$	$kg \cdot m^2/s^2$

TABLE J.1
continued

Quantity	Name	SI Symbol	Conventional Units	Base Units
Chemical				
Concentration (mass)			kg/m^3	
Concentration (molar)			n/m^3	
Mass flux rate			$kg/m^2/s$	
Electrical				
Capacitance	farad	F	$A·s/V$	$A·s^4/kg/m^2$
Charge	coulomb	C	$A·s$	
Current*	ampere	A	A	
Inductance	henry	H	$V·s/A$	$kg·m^2/s^2A^2$
Magnetic flux	weber	Wb	$kg·m^2/s^2/A$	
Magnetic flux density	tesla	T	$kg/s/A$	
Potential difference (or voltage)	volt	V	W/A or J/C	$kg·m^2/s^3/A$
Resistance	ohm	Ω	V/A	$kg·m^2/s^3/A^2$
Optics				
Luminous flux	lumen	lm	$cd·sr$	
Illuminance	lux	lx	$cd·sr/m^2$	
Fluids				
Dynamic viscosity			$Pa·s$	$kg/s/m$
Flow rate			m^3/s	
Kinematic viscosity			m^2/s	
Pressure	pascal	Pa	N/m^2	
Thermodynamics				
Heat capacity			J	$kg·m^2/s^2$
Thermal conductivity			W/m/K	$kg·m/s^3/K$

*Fundamental units.

REFERENCES

Ang, A. H-S., and W. H. Tang, *Probability Concepts in Engineering Planning and Design, Vol. 1, Basic Principles,* Wiley, New York, 1975.

Barron, J. C., *Basic Programming with Structured Modules,* Holt, New York, 1983.

Beer, Ferdinand P., and E. Russell Johnston, Jr., *Vector Mechanics for Engineers: Statics and Dynamics,* 4th ed., McGraw-Hill, New York, 1984.

Box, G. E. P., W. G. Hunter, and J. S. Hunter, *Statistics for Experimenters: An Introduction to Design, Data Analysis and Model Building,* Wiley, New York, 1978.

Bradbeer, R., P. DeBono, and P. Laurie, *The Beginner's Guide to Computers,* Addison-Wesley, Reading, Mass., 1982.

Brand, S., *Whole Earth Software Catalog,* Quantum/Doubleday, Garden City, N.Y., 1984.

Branscomb, L. M., "Electronics and Computers: An Overview," *Science,* **215**:755 (1982).

Brown, C., *Essential WordStar,* Brooks/Cole, Monterery, Calif., 1985.

Byers, R., *Everyman's Database Primer,* Ashton-Tate, Culver City, Calif., 1982.

Canale, R. P., and S. C. Chapra, *ENGINCOMP: Student Version User's Manual for IBM PC Computers,* McGraw-Hill, New York, 1986.

——— and ———, *NUMERICOMP: Student Version User's Manual for IBM PC Computers,* McGraw-Hill, New York, 1985.

Chapra, S. C., and R. P. Canale, *Numerical Methods for Engineers with Personal Computer Applications,* McGraw-Hill, New York, 1985.

Cougar, J. D., and F. R. McFadden, *First Course in Data Processing,* Wiley, New York 1981.

Draper, N. R., and H. Smith, *Applied Regression Analyses,* 2d ed., Wiley, New York, 1981.

Gerald, C. F., and P. O. Wheatley, *Applied Numerical Analysis,* 3d ed., Addison-Wesley, Reading, Mass., 1984.

Glossbrenner, A., *How to Buy Software,* St. Martin's Press, New York, 1984.

Kay, A., "A Personal Computer for Children of All Ages," *Proceedings of the ACM National Conference,* 1972.

Kelly, J. E., Jr., *The IBM PC and 1-2-3,* Banbury, Wayne, Pa., 1983.

Kelly-Bootle, S., *The Devil's DP Dictionary*, McGraw-Hill, New York, 1981.

Kemeny, J. G. *True BASIC,* Addison-Wesley, Reading, Mass., 1985.

Kemeny, J. G., and T. E. Kurtz, *Back to BASIC,* Addison-Wesley, Reading, Mass., 1985.

Knuth, D. E., *The Art of Computer Programming, Vol. 2, Seminumerical Algorithms*, 2d ed.,
    Addison-Wesley, Reading, Mass., 1981.

———, *The Art of Computer Programming, Vol. 3, Sorting and Searching*, Addison-Wesley,
    Reading, Mass., 1973.

Krum, R., *Understanding and Using dBASE II,* Brady Communications Co., Inc., Bowie,
    Md., 1984.

Ralston, A., and P. Rabinowitz, *A First Course in Numerical Analysis*, 2d ed., McGraw-Hill,
    New York, 1978.

Snedecor, G. W., and W. G. Cochran, *Statistical Methods*, Iowa State University Press, Ames,
    Iowa, 1967.

Toong, H. D., and A. Gupta, ''Personal Computers,'' *Scientific American*, **247**:86 (1982).

Wonnacott, T. H., and R. J. Wonnacott, *Introductory Statistics*, Wiley, New York, 1972.

INDEX